Electronic and Optoelectronic Properties of Semiconductor Structures presents the underlying physics behind devices that drive today's technologies. The book covers important details of structural properties, bandstructure, transport, optical and magnetic properties of semiconductor structures. Effects of low-dimensional physics and strain – two important driving forces in modern device technology – are also discussed. In addition to conventional semiconductor physics the book discusses self-assembled structures, mesoscopic structures and the developing field of spintronics.

The book utilizes carefully chosen solved examples to convey important concepts and has over 250 figures and 200 homework exercises. Real-world applications are highlighted throughout the book, stressing the links between physical principles and actual devices.

Electronic and Optoelectronic Properties of Semiconductor Structures provides engineering and physics students and practitioners with complete and coherent coverage of key modern semiconductor concepts. A solutions manual and set of viewgraphs for use in lectures is available for instructors.

JASPRIT SINGH received his Ph.D. from the University of Chicago and is Professor of Electrical Engineering and Computer Science at the University of Michigan, Ann Arbor. He has held visiting positions at the University of California, Santa Barbara and the University of Tokyo. He is the author of over 250 technical papers and of seven previous textbooks on semiconductor technology and applied physics.

Electronic and Optoelectronic Properties of Semiconductor Structures

Jasprit Singh

University of Michigan, Ann Arbor

CAMBRIDGE
UNIVERSITY PRESS

CAMBRIDGE UNIVERSITY PRESS
Cambridge, New York, Melbourne, Madrid, Cape Town, Singapore, São Paulo

Cambridge University Press
The Edinburgh Building, Cambridge CB2 2RU, UK

Published in the United States of America by Cambridge University Press, New York

www.cambridge.org
Information on this title: www.cambridge.org/9780521823791

First published 2003
This digitally printed first paperback version 2007

A catalogue record for this publication is available from the British Library

ISBN-13 978-0-521-82379-1 hardback
ISBN-10 0-521-82379-X hardback

ISBN-13 978-0-521-03574-3 paperback
ISBN-10 0-521-03574-0 paperback

CONTENTS

3 BANDSTRUCTURE MODIFICATIONS 109

4 TRANSPORT: GENERAL FORMALISM 152

5 DEFECT AND CARRIER–CARRIER SCATTERING

6 LATTICE VIBRATIONS: PHONON SCATTERING

9 OPTICAL PROPERTIES OF SEMICONDUCTORS 345

PREFACE

Semiconductor-based technologies continue to evolve and astound us. New materials, new structures, and new manufacturing tools have allowed novel high performance electronic and optoelectronic devices. To understand modern semiconductor devices and to design future devices, it is important that one know the underlying physical phenomena that are exploited for devices. This includes the properties of electrons in semiconductors and their heterostructures and how these electrons respond to the outside world. This book is written for a reader who is interested in not only the physics of semiconductors, but also in how this physics can be exploited for devices.

The text addresses the following areas of semiconductor physics: i) electronic properties of semiconductors including bandstructures, effective mass concept, donors, acceptors, excitons, etc.; ii) techniques that allow modifications of electronic properties; use of alloys, quantum wells, strain and polar charge are discussed; iii) electron (hole) transport and optical properties of semiconductors and their heterostructures; and iv) behavior of electrons in small and disordered structures. As much as possible I have attempted to relate semiconductor physics to modern device developments.

There are a number of books on solid state and semiconductor physics that can be used as textbooks. There are also a number of good monographs that discuss special topics, such as mesoscopic transport, Coulomb blockade, resonant tunneling effects, etc. However, there are few single-source texts containing "old" and "new" semiconductor physics topics. In this book well-established "old" topics such as crystal structure, band theory, etc., are covered, along with "new" topics, such as lower dimensional systems, strained heterostructures, self-assembled structures, etc. All of these topics are presented in a textbook format, not a special topics format. The book contains solved examples, end-of-chapter problems, and a discussion of how physics relates to devices. With this approach I hope this book fulfills an important need.

I would like to thank my wife, Teresa M. Singh, who is responsible for the artwork and design of this book. I also want to thank my editor, Phil Meyler, who provided me excellent and timely feedback from a number of reviewers.

Jasprit Singh

INTRODUCTION

Semiconductors and devices based on them are ubiquitous in every aspect of modern life. From "gameboys" to personal computers, from the brains behind "nintendo" to world wide satellite phones—semiconductors contribute to life perhaps like no other manmade material. Silicon and semiconductor have entered the vocabulary of newscasters and stockbrokers. Parents driving their kids cross-country are grudgingly grateful to the "baby-sitting service" provided by ever more complex "gameboys." Cell phones and pagers have suddenly brought modernity to remote villages. "How exciting," some say. "When will it all end?" say others.

The ever expanding world of semiconductors brings new challenges and opportunities to the student of semiconductor physics and devices. Every year brings new materials and structures into the fold of what we call semiconductors. New physical phenomena need to be grasped as structures become ever smaller.

I.1 SURVEY OF ADVANCES IN SEMICONDUCTOR PHYSICS

In Fig. I.1 we show an overview of progress in semiconductor physics and devices, since the initial understanding of the band theory in the 1930s. In this text we explore the physics behind all of the features listed in this figure. Let us take a brief look at the topics illustrated.

- **Band theory:** The discovery of quantum mechanics and its application to understand the properties of electrons in crystalline solids has been one of the most important scientific theories. This is especially so when one considers the impact of band theory on technologies such as microelectronics and optoelectronics. Band theory and its outcome—effective mass theory—has allowed us to understand the difference between metals, insulators, and semiconductors and how electrons respond to external forces in solids. An understanding of electrons, holes, and carrier transport eventually led to semiconductor devices such as the transistor and the demonstration of lasing in semiconductors.

- **Semiconductor Heterostructures:** Initial work on semiconductors was carried out in single material systems based on Si, Ge, GaAs, etc. It was then realized that if semiconductors could be combined, the resulting structure would yield very interesting properties. Semiconductors heterostructures are now widely used in electronics and optoelectronics. Heterostructures are primarily used to confine electrons and holes and to produce low dimensional electronic systems. These low dimensional systems, including quantum wells, quantum wires and quantum dots have density of states and other electronic properties that make them attractive for many applications.

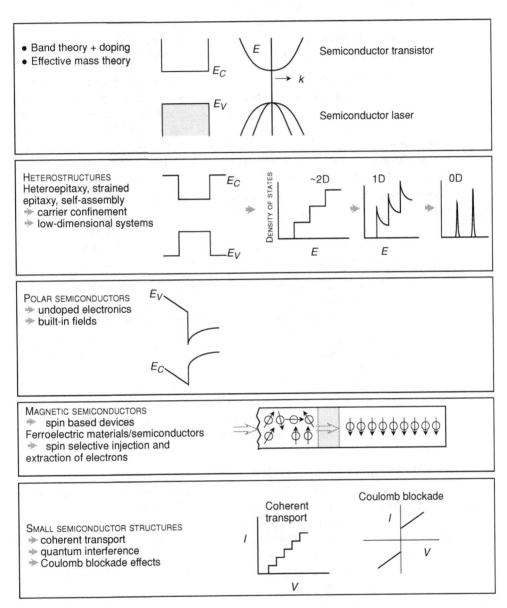

Figure I.1: Evolution of semiconductor physics and phenomena. These topics are discussed in this book.

Advances in heterostructures include strain epitaxy and self-assembled structures. In strained epitaxy it is possible to incorporate a high degree of strain in a thin layer. This can be exploited to alter the electronic structure of heterostructures. In self-assembled structures lateral structures are produced by using the island growth mode or other features in growth processes. This can produce low-dimensional systems without the need of etching and lithography.

- **Polar and Magnetic Heterostructures:** Since the late 1990s there has been a strong push to fabricate heterostructures using the nitride semiconductors (InN, GaN, and AlN). These materials have large bandgaps that can be used for blue light emission and high power electronics. It is now known that these materials have spontaneous polarization and a very strong piezoelectric effect. These features can be exploited to design transistors that have high free charge densities without doping and quantum wells with large built-in electric fields.

  In addition to materials with fixed polar charge there is now an increased interest in materials like ferroelectrics where polarization can be controlled. Some of these materials have a large dielectric constant, a property that can be exploited for design of gate dielectrics for very small MOSFETs. There is also interest in semiconductors with ferromagnetic effects for applications in spin selective devices.

- **Small Structures:** When semiconductor structures become very small two interesting effects occur: electron waves can propagate without losing phase coherence due to scattering and charging effects become significant. When electron waves travel coherently a number of interesting characteristics are observed in the current-voltage relations of devices. These characteristics are qualitatively different from what is observed during incoherent transport.

  An interesting effect that occurs in very small capacitors is the Coulomb blockade effect in which the charging energy of a single electron is comparable or larger than $k_B T$. This effect can lead to highly nonlinear current-voltage characteristics which can, in principle, be exploited for electronic devices.

I.2   PHYSICS BEHIND SEMICONDUCTORS

Semiconductors are mostly used for information processing applications. To understand the physical properties of semiconductors we need to understand how electrons behave inside semiconductors and how they respond to external stimuli. Considering the complexity of the problem—up to 10^{22} electrons cm^{-3} in a complex lattice of ions —it is remarkable that semiconductors are so well understood. Semiconductor physics is based on a remarkably intuitive set of simplifying assumptions which often seem hard to justify rigorously. Nevertheless, they work quite well.

The key to semiconductor physics is the band theory and its outcome—the effective mass theory. As illustrated in Fig. I.2, one starts with a perfectly periodic structure as an ideal representation of a semiconductor. It is assumed that the material can be represented by a perfectly periodic arrangement of atoms. This assumption although not correct, allows one to develop a band theory description according to which electrons act as if they are in free space except their effective energy momentum

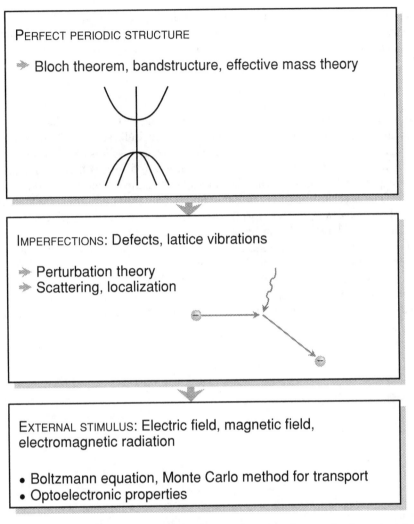

Figure I.2: A schematic of how our understanding of semiconductor physics proceeds.

relation is modified. This picture allows one to represent electrons near the bandedges of semiconductors by an "effective mass."

In real semiconductors atoms are not arranged in perfect periodic structures. The effects of imperfections are treated perturbatively—as a correction to band theory. Defects can localize electronic states and cause scattering between states. A semiclassical picture is then developed where an electron travels in the material, every now and then suffering a scattering which alters its momentum and/or energy. The scattering rate is calculated using the Fermi golden rule (or Born approximation) if the perturbation is small.

The final step in semiconductor physics is an understanding of how electrons respond to external stimuli such as electric field, magnetic field, electromagnetic field, etc. A variety of techniques, such as Boltzmann transport equations and Monte Carlo computer simulations are developed to understand the response of electrons to external stimulus.

I.3   ROLE OF THIS BOOK

This book provides the underlying physics for the topics listed in Fig. I.1. It covers "old" topics such as crystal structure and band theory in bulk semiconductors and "new" topics such as bandstructure of stained heterostructures, self-assembled quantum dots, and spin transistors. All these topics have been covered in a coherent manner so that the reader gets a good sense of the current state of semiconductor physics.

In order to provide the reader a better feel for the theoretical derivations a number of solved examples are sprinkled in the text. Additionally, there are end-of-chapter problems. This book can be used to teach a course on semiconductors physics. A rough course outline for a two semester course is shown in Table I.1. In a one semester course some section of this text can be skipped (e.g., magnetic field effects from Chapter 11) and others can be covered in less detail (e.g., Chapter 8). If a two semester course is taught, all of the material in the book can be used. It is important to note that this book can also be used for special topic courses on heterostructures or optoelectronics.

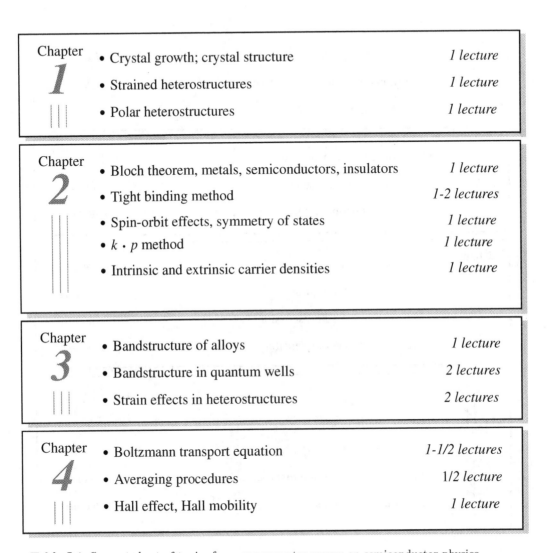

Chapter 1	• Crystal growth; crystal structure	*1 lecture*
	• Strained heterostructures	*1 lecture*
	• Polar heterostructures	*1 lecture*
Chapter 2	• Bloch theorem, metals, semiconductors, insulators	*1 lecture*
	• Tight binding method	*1-2 lectures*
	• Spin-orbit effects, symmetry of states	*1 lecture*
	• $k \cdot p$ method	*1 lecture*
	• Intrinsic and extrinsic carrier densities	*1 lecture*
Chapter 3	• Bandstructure of alloys	*1 lecture*
	• Bandstructure in quantum wells	*2 lectures*
	• Strain effects in heterostructures	*2 lectures*
Chapter 4	• Boltzmann transport equation	*1-1/2 lectures*
	• Averaging procedures	*1/2 lecture*
	• Hall effect, Hall mobility	*1 lecture*

Table I.1: Suggested set of topics for a one semester course on semiconductor physics.

Chapter 5 |||	• Ionized impurity scattering	*1 lecture*
	• Alloy, neutral impurity scattering	*1 lecture*
	• Carrier-carrier scattering	*1 lecture*

Chapter 6 |||	• Phonon dispersion and statistics	*2 lectures*
	• Phonon scatteringã general	*1 lecture*
	• Acoustic phonon scattering, optical phonon scattering	*2 lectures*

Chapter 7 |||	• Low field mobility	*1 lecture*
	• Monte Carlo techniques	*2 lectures*
	• Velocity-field result discussion	*1 lecture*
	• Transport in lower dimensions	*1 lecture*

Chapter 8 |||	Optional Chapter	
	• Bloch oscillations	*1 lecture*
	• Resonant tunneling	*1 lecture*
	• Localization issues and disorder	*1 lecture*
	• Mesoscopic systems	*2 lectures*

Table I.2: Suggested set of topics for a one semester course on semiconductor physics (con't.).

Chapter **9**	• Interband transitions: Bulk and 2D	*2 lectures*
	• Intraband transitions in quantum wells	*1 lecture*
	• Charge injection and light emission	*1 lecture*
	• Nonradiative processes	*1 lecture*

Chapter **10**	• Excitonic states in 3D and lower dimensions	*2 lectures*
	• Modulation of optical properties	*2 lectures*

Chapter **11**	Optional Chapter	
	• Semiclassical theory of magnetotransport	*1 lecture*
	• Landau levels	*1 lecture*
	• Aharonov Bohm effect	*1/2 lecture*
	• Magnetooptic effect	*1/2 lecture*
	• "Spintronics"	

Appendix **B**: Reading assignments

Table I.3: Suggested set of topics for a one semester course on semiconductor physics (con't.).

I.4   Properties of Key Materials

Material	Crystal Structure	Bandgap (eV)	Static Dielectric Constant	Lattice Constant (Å)	Density (gm-cm^{-3})
C	DI	5.50, I	5.570	3.5668	3.5153
Si	DI	1.1242, I	11.9	5.431	2.3290
SiC	ZB	2.416, I	9.72	4.3596	3.166
Ge	DI	0.664, I	16.2	5.658	5.323
AlN	W	6.2, D	$\bar{\varepsilon} = 9.14$	$a = 3.111$ $c = 4.981$	3.255
AlP	ZB	2.45, I	9.8	5.4635	2.401
AlAs	ZB	2.153, I	10.06	5.660	3.760
GaN	W	3.44, D	$\varepsilon_{\parallel} = 10.4$ $\varepsilon_{\perp} = 9.5$	$a = 3.175$ $c = 5.158$	6.095
GaP	ZB	2.272, I	11.11	5.4505	4.138
GaAs	ZB	1.424, D	13.18	5.653	5.318
GaSb	ZB	0.75, D	15.69	6.0959	5.6137
InN	W	1.89, D	—	$a = 3.5446$ $c = 8.7034$	6.81
InP	ZB	1.344, D	12.56	5.8687	4.81
InAs	ZB	0.354, D	15.15	6.058	5.667
InSb	ZB	0.230, D	16.8	6.479	5.775
ZnS	ZB	3.68, D	8.9	5.4102	4.079
ZnS	W	3.9107, D	$\bar{\varepsilon} = 9.6$	$a = 3.8226$ $c = 6.6205$	4.084
ZnSe	ZB	2.822, D	9.1	5.668	5.266
ZnTe	ZB	2.394, D	8.7	6.104	5.636
CdS	W	2.501, D	$\bar{\varepsilon} = 9.38$	$a = 4.1362$ $c = 6.714$	4.82
CdS	ZB	2.50, D	—	5.818	—
CdSe	W	1.751, D	$\varepsilon_{\parallel} = 10.16$ $\varepsilon_{\perp} = 9.29$	$a = 4.2999$ $c = 7.0109$	5.81
CdTe	ZB	1.475, D	10.2	6.482	5.87
PbS	R	0.41, D*	169.	5.936	7.597
PbSe	R	0.278, D*	210.	6.117	8.26
PbTe	R	0.310, D*	414.	6.462	8.219

Data given are room temperature values (300 K).
KEY: DI: diamond; R: rocksalt; W: wurtzite; ZB: zinc-blende
*: gap at L point; D: direct; I: indirect; $\varepsilon_{\parallel}$: parallel to c-axis; $\varepsilon_{\perp}$: perpendicular to c-axis

FREQUENTLY USED QUANTITIES

QUANTITY	SYMBOL	VALUE
Planck's constant	h $\hbar = h/2\pi$	6.626×10^{-34} J-s 1.055×10^{-34} J-s
Velocity of light	c	2.998×10^{8} m/s
Electron charge	e	1.602×10^{-19} C
Electron volt	eV	1.602×10^{-19} J
Mass of an electron	m_0	9.109×10^{-31} kg
Permittivity of vacuum	$\varepsilon_0 = \dfrac{10^7}{4\pi c^2}$	8.85×10^{-14} F cm^{-1} $= 8.85 \times 10^{-12}$ F m^{-1}
Boltzmann constant	k_B	8.617×10^{-5} eVK^{-1}
Thermal voltage at 300 K	$k_B T/e$	0.026 V

Wavelength – energy relation: $\lambda(\mu m) = \dfrac{1.24}{E(eV)}$

Linewidth $\delta E = 1$ meV $\implies$ $\delta \nu = 0.243$ THz

Linewidth $\delta E = 1$ meV $\implies$ $\begin{cases} \delta\lambda = 19.4 \text{ Å } @ \lambda = 1.55 \ \mu m \\ \delta\lambda = 6.2 \text{ Å } @ \lambda = 0.88 \ \mu m \end{cases}$

MATERIAL	CONDUCTION BAND EFFECTIVE DENSITY (N_c)	VALENCE BAND EFFECTIVE DENSITY (N_v)	INTRINSIC CARRIER CONCENTRATION $(n_i = p_i)$
Si (300 K)	2.78×10^{19} cm^{-3}	9.84×10^{18} cm^{-3}	1.5×10^{10} cm^{-3}
Ge (300 K)	1.04×10^{19} cm^{-3}	6.0×10^{18} cm^{-3}	2.33×10^{13} cm^{-3}
GaAs (300 K)	4.45×10^{17} cm^{-3}	7.72×10^{18} cm^{-3}	1.84×10^{6} cm^{-3}

Chapter

1

STRUCTURAL PROPERTIES OF SEMICONDUCTORS

1.1 INTRODUCTION

Semiconductors form the basis of most modern information processing devices. Electronic devices such as diodes, bipolar junction transistors, and field effect transistors drive modern electronic technology. Optoelectronic devices such as laser diodes, modulators, and detectors drive the optical networks. In addition to devices, semiconductor structures have provided the stages for exploring questions of fundamental physics. Quantum Hall effect and other phenomena associated with many-body effects and low dimensions have been studied in semiconductor structures.

It is important to recognize that the ability to examine fundamental physics issues and to use semiconductors in state of the art device technologies depends critically on the purity and perfection of the semiconductor crystal. Semiconductors are often associated with clean rooms and workers clad in "bunny suits" lest the tiniest stray particle get loose and latch onto the wafer being processed. Indeed, semiconductor structures can operate at their potential only if they can be grown with a high degree of crystallinity and if impurities and defects can be controlled. For high structural quality it is essential that a high quality substrate be available. This requires growth of bulk crystals which are then sliced and polished to allow epitaxial growth of thin semiconductor regions including heterostructures.

In this chapter we start with a brief discussion of the important bulk and epitaxial crystal growth techniques. We then discuss the important semiconductor crystal structures. We also discuss strained lattice structures and the strain tensor for such crystals. Strained epitaxy and its resultant consequences are now widely exploited in semiconductor physics and it is important to examine how epitaxial growth causes distortions in the crystal lattice.

1.2   CRYSTAL GROWTH

1.2.1   Bulk Crystal Growth

Semiconductor technology depends critically upon the availability of high quality substrates with as large a diameter as possible. Bulk crystal growth techniques are used mainly to produce substrates on which devices are eventually fabricated. While for some semiconductors like Si and GaAs (to some extent for InP) the bulk crystal growth techniques are highly matured; for most other semiconductors it is difficult to obtain high quality, large area substrates. Several semiconductor technologies are dependent on substrates that are not ideal. For example, the nitrides GaN, AlN, InN are grown on SiC or sapphire substrates, since there is no reliable GaN substrate. The aim of the bulk crystal growth techniques is to produce single crystal boules with as large a diameter as possible and with as few defects as possible. In Si the boule diameters have reached 30 cm with boule lengths approaching 100 cm. Large size substrates ensure low cost device production.

For the growth of boules from which substrates are obtained, one starts out with a purified form of the elements that are to make up the crystal. One important technique that is used is the Czochralski (CZ) technique. In the Czochralski technique shown in Fig. 1.1, the melt of the charge (i.e., the high quality polycrystalline material) is held in a vertical crucible. The top surface of the melt is just barely above the melting temperature. A seed crystal is then lowered into the melt and slowly withdrawn. As the heat from the melt flows up the seed, the melt surface cools and the crystal begins to grow. The seed is rotated about its axis to produce a roughly circular cross-section crystal. The rotation inhibits the natural tendency of the crystal to grow along certain orientations to produce a faceted crystal.

The CZ technique is widely employed for Si, GaAs, and InP and produces long ingots (boules) with very good circular cross-section. For Si up to 100 kg ingots can be obtained. In the case of GaAs and InP the CZ technique has to face problems arising from the very high pressures of As and P at the melting temperature of the compounds. Not only does the chamber have to withstand such pressures, also the As and P leave the melt and condense on the sidewalls. To avoid the second problem one seals the melt by covering it with a molten layer of a second material (e.g., boron oxide) which floats on the surface. The technique is then referred to as liquid encapsulated Czochralski, or the LEC technique.

A second bulk crystal growth technique involves a charge of material loaded in a quartz container. The charge may be composed of either high quality polycrystalline material or carefully measured quantities of elements which make up a compound crystal. The container called a "boat" is heated till the charge melts and wets the seed crystal. The seed is then used to crystallize the melt by slowly lowering the boat temperature starting from the seed end. In the gradient-freeze approach the boat is pushed into a furnace (to melt the charge) and slowly pulled out. In the Bridgeman approach, the boat is kept stationary while the furnace temperature is temporally varied to form

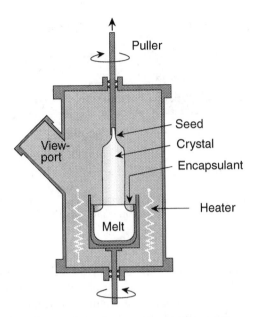

Figure 1.1: Schematic of Czochralski-style crystal grower used to produce substrate ingots. The approach is widely used for Si, GaAs and InP.

the crystal. The approaches are schematically shown in Fig. 1.2.

The easiest approach for the boat technique is to use a horizontal boat. However, the shape of the boule that is produced has a D-shaped form. To produce circular cross-sections vertical configurations have now been developed for GaAs and InP.

In addition to producing high purity bulk crystals, the techniques discussed above are also responsible for producing crystals with specified electrical properties. This may involve high resistivity materials along with n- or p-type materials. In Si it is difficult to produce high resistivity substrated by bulk crystal growth and resistivities are usually $<10^4$ Ω-cm. However, in compound semiconductors carrier trapping impurities such as chromium and iron can be used to produce material with resistivities of $\sim 10^8$ Ω cm. The high resistivity or semi-insulating (SI) substrates are extremely useful in device isolation and for high speed devices. For n- or p-type doping carefully measured dopants are added in the melt.

1.2.2   Epitaxial Crystal Growth

Once bulk crystals are grown, they are sliced into substrates or wafers about 250 μm thick. These are polished and used for growth of epitaxial layers a few micrometers thick. All active devices are produced on these epitaxial layers. As a result the epi-taxial growth techniques are very important. The epitaxial growth techniques have a very slow growth rate (as low as a monolayer per second for some techniques) which allow one to control very accurately the dimensions in the growth direction. In fact, in techniques like molecular beam epitaxy (MBE) and metal organic chemical vapor

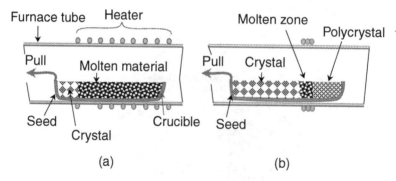

Figure 1.2: Crystal growing from the melt in a crucible: (a) solidification from one end of the melt (horizontal Bridgeman method); (b) melting and solidification in a moving zone.

deposition (MOCVD), one can achieve monolayer ($\sim$ 3 Å) control in the growth direction. This level of control is essential for the variety of heterostructure devices that are being used in optoelectronics. The epitaxial techniques are also very useful for precise doping profiles that can be achieved. In fact, it may be argued that without the advances in epitaxial techniques that have occurred over the last two decades, most of the developments in semiconductor physics would not have occurred. Table 1.1 gives a brief view of the various epitaxial techniques used along with some of the advantages and disadvantages.

Liquid Phase Epitaxy (LPE)

LPE is a relatively simple epitaxial growth technique which was widely used until 1970s when it gradually gave way to approaches such as MBE and MOCVD. It is a less expensive technique (compared to MBE or MOCVD), but it offers less control in interface abruptness when growing heterostructures. LPE is still used for growth of crystals such as HgCdTe for long wavelength detectors and AlGaAs for double heterostructure lasers. As shown in Table 1.1, LPE is a close to equilibrium technique in which the substrate is placed in a quartz or a graphite boat and covered by a liquid of the crystal to be grown (see Fig. 1.3). The liquid may also contain dopants that are to be introduced into the crystal. LPE is often used for alloy growth where the growth follows the equilibrium solid-liquid phase diagram. By precise control of the liquid composition and temperature, the alloy composition can be controlled. Because LPE is a very close to equilibrium growth technique, it is difficult to grow alloy systems which are not miscible or even grow heterostructures with atomically abrupt interfaces. Nevertheless heterostructures where interface is graded over 10-20 Å can be grown by LPE by sliding the boat over successive "puddles" of different semiconductors. For many applications such interfaces are adequate and since LPE is a relatively inexpensive growth technique, it is used in many commercial applications.

Vapor Phase Epitaxy (VPE)

A large class of epitaxial techniques rely on delivering the components that form the crystal from a gaseous environment. If one has molecular species in a gaseous form with

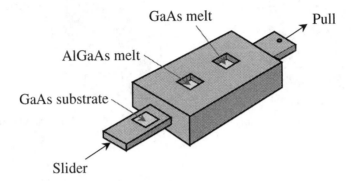

Figure 1.3: A schematic of the LPE growth of AlGaAs and GaAs. The slider moves the substrate, thus positioning itself to achieve contact with the different melts to grow heterostructures.

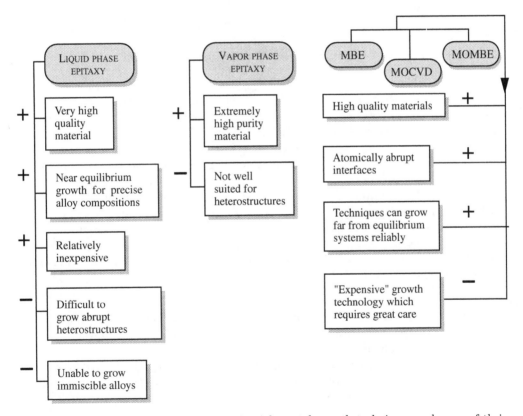

Table 1.1: A schematic of the various epitaxial crystal growth techniques and some of their positive and negative aspects.

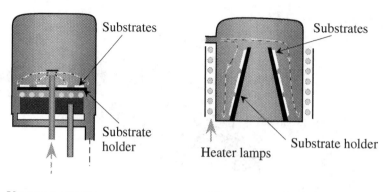

VERTICAL REACTOR   HORIZONTAL REACTOR

Figure 1.4: Reactors for VPE growth. The substrate temperature must be maintained uniformly over the area. This is achieved better by lamp heating. A pyrometer is used for temperature measurement.

partial pressure P, the rate at which molecules impinge upon a substrate is given by

$$F = \frac{P}{\sqrt{2\pi m k_B T}} \sim \frac{3.5 \times 10^{22} P(\text{torr})}{\sqrt{m(g)T(K)}} \text{mol./cm}^2\text{s} \qquad (1.1)$$

where m is the molecular weight and T the cell temperature. For most crystals the surface density of atoms is $\sim 7 \times 10^{14}$ cm^{-2}. If the atoms or molecules impinging from the vapor can be deposited on the substrate in an ordered manner, epitaxial crystal growth can take place.

The VPE technique is used mainly for homoepitaxy and does not have the additional apparatus present in techniques such as MOCVD for precise heteroepitaxy. As an example of the technique, consider the VPE of Si. The Si containing reactant silane (SiH$_4$) or dichlorosilane (SiH$_2$Cl$_2$) or trichlorosilane (SiHCl$_3$) or silicon tetrachloride (SiCl$_4$) is diluted in hydrogen and introduced into a reactor in which heated substrates are placed as shown in Fig. 1.4. The silane pyrolysis to yield silicon while the chlorine containing gases react to give SiCl$_2$, HCl and various other silicon-hydrogen-chlorine compounds. The reaction

$$2\text{SiCl}_2 \rightleftharpoons \text{Si} + \text{SiCl}_4 \qquad (1.2)$$

then yields Si. Since HCl is also produced in the reaction, conditions must be tailored so that no etching of Si occurs by the HCl. Doping can be carried out by adding appropriate hydrides (phosphine, arsine, etc.,) to the reactants.

VPE can be used for other semiconductors as well by choosing different appropriate reactant gases. The reactants used are quite similar to those employed in the MOCVD technique discussed later.

Molecular Beam Epitaxy (MBE)

MBE is capable of controlling deposition of submonolayer coverage on a substrate and has become one of the most important epitaxial techniques. Almost every semiconductor

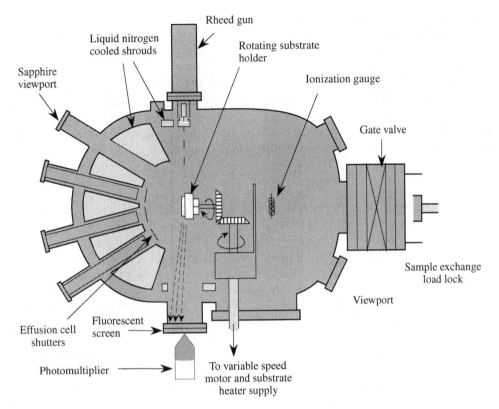

Sapphire viewport

Liquid nitrogen cooled shrouds

Rheed gun

Rotating substrate holder

Ionization gauge

Gate valve

Sample exchange load lock

Viewport

Effusion cell shutters

Fluorescent screen

Photomultiplier

To variable speed motor and substrate heater supply

Figure 1.5: A schematic of the MBE growth system.

has been grown by this technique. MBE is a high vacuum technique ($\sim 10^{-11}$ torr vacuum when fully pumped down) in which crucibles containing a variety of elemental charges are placed in the growth chamber (Fig. 1.5). The elements contained in the crucibles make up the components of the crystal to be grown as well as the dopants that may be used. When a crucible is heated, atoms or molecules of the charge are evaporated and these travel in straight lines to impinge on a heated substrate.

The growth rate in MBE is ~ 1.0 monolayer per second and this slow rate coupled with shutters placed in front of the crucibles allow one to switch the composition of the growing crystal with monolayer control. Since no chemical reactions occur in MBE, the growth is the simplest of all epitaxial techniques and is quite controllable. However, since the growth involves high vacuum, leaks can be a major problem. The growth chamber walls are usually cooled by liquid N_2 to ensure high vacuum and to prevent atoms/molecules to come off from the chamber walls.

The low background pressure in MBE allows one to use electron beams to monitor the growing crystal. The reflection high energy electron diffraction (RHEED) techniques relies on electron diffraction to monitor both the quality of the growing substrate and the layer by layer growth mode.

Metal Organic Chemical Vapor Deposition (MOCVD)

Metal organic chemical vapor deposition (MOCVD) is another important growth technique widely used for heteroepitaxy. Like MBE, it is also capable of producing monolayer abrupt interfaces between semiconductors. A typical MOCVD system is shown in Fig. 1.6. Unlike in MBE, the gases that are used in MOCVD are not made of single elements, but are complex molecules which contain elements like Ga or As to form the crystal. Thus the growth depends upon the chemical reactions occurring at the heated substrate surface. For example, in the growth of GaAs one often uses triethyl gallium and arsine and the crystal growth depends upon the following reaction:

$$Ga(CH_3)_3 + AsH_3 \rightleftharpoons GaAs + 3CH_4 \tag{1.3}$$

One advantage of the growth occurring via a chemical reaction is that one can use lateral temperature control to carry out local area growth. Laser assisted local area growth is also possible for some materials and can be used to produce new kinds of device structures. Such local area growth is difficult in MBE.

There are several varieties of MOCVD reactors. In the atmospheric MOCVD the growth chamber is essentially at atmospheric pressure. One needs a large amount of gases for growth in this case, although one does not have the problems associated with vacuum generation. In the low pressure MOCVD the growth chamber pressure is kept low. The growth rate is then slower as in the MBE case.

The use of the MOCVD equipment requires very serious safety precautions. The gases used are highly toxic and a great many safety features have to be incorporated to avoid any deadly accidents. Safety and environmental concerns are important issues in almost all semiconductor manufacturing since quite often one has to deal with toxic and hazardous materials.

In addition to MBE and MOCVD one has hybrid epitaxial techniques often called MOMBE (metal organic MBE) which try to combine the best of MBE and MOCVD. In MBE one has to open the chamber to load the charge for the materials to be grown while this is avoided in MOCVD where gas bottles can be easily replaced from outside. Additionally, in MBE one has occasional spitting of material in which small clumps of atoms are evaporated off on to the substrate. This is avoided in MOCVD and MOMBE.

EXAMPLE 1.1 Consider the growth of GaAs by MBE. The Ga partial pressure in the growth chamber is 10^{-5} Torr, and the Ga cell temperature is 900 K. Calculate the flux of Ga atoms on the substrate. The surface density of Ga atoms on GaAs grown along (001) direction is 6.3×10^{14} cm^{-2}. Calculate the growth rate if all of the impinging atoms stick to the substrate.

The mass of Ga atoms is 70 g/mole. The flux is (from Eqn. 1.1)

$$F = \frac{3.5 \times 10^{22} \times 10^{-5}}{\sqrt{70 \times 900}} = 5.27 \times 10^{14} \text{atoms/cm}^2$$

Note that the surface density of Ga atoms on GaAs is $\sim 6.3 \times 10^{14}$ cm^{-2}. Thus, if all of the Ga atoms were to stick, the growth rate would be $\sim$0.8 monolayer per second. This assumes that there is sufficient arsenic to provide As in the crystal. This is a typical growth rate for epitaxial films. It would take nearly 10 hours to grow a 10 μm film.

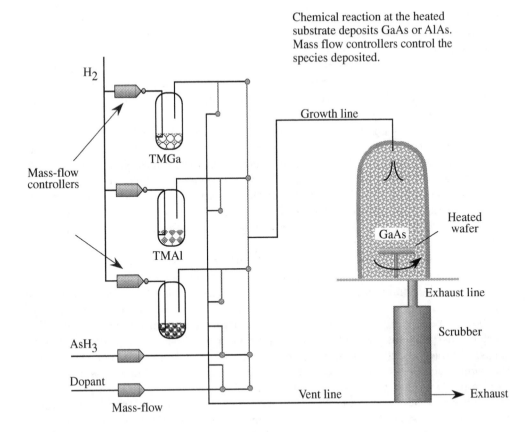

Chemical reaction at the heated substrate deposits GaAs or AlAs. Mass flow controllers control the species deposited.

TMGa : Gallium containing organic compound
TMAl : Aluminum containing organic compound
AsH$_3$: Arsenic containing compound

Figure 1.6: Schematic diagram of an MOCVD system employing alkyds (trimethyl gallium (TMGa) and trimethyl aluminum (TMAl) and metal hydride (arsine) material sources, with hydrogen as a carrier gas.

1.2.3 Epitaxial Regrowth

The spectacular growth of semiconductor microelectronics owes a great deal to the concept of the integrated circuit. The ability to fabricate transistors, resistors, inductors and capacitors on the same wafer is critical to the low cost and high reliability we have come to expect from microelectronics. It is natural to expect similar dividents from the concept of the optoelectronic integrated circuit (OEIC). In the OEIC, the optoelectronic device (the laser or detector or modulator) would be integrated on the same wafer with an amplifier or logic gates.

One of the key issues in OEICs involves etching and regrowth. As we will see

later, the optoelectronic devices have a structure that is usually not compatible with
the structure of an electronic device. The optimum layout then involves growing one
of the device structures epitaxially and then masking the region to be used as, say,
the optoelectronic device and etching away the epitaxial region. Next a regrowth is
done to grow the electronic device with a different structure. The process is shown
schematically in Fig. 1.7. While this process looks simple conceptually, there are serious
problems associated with etching and regrowth.

A critical issue in the epitaxial growth of a semiconductor layer is the quality of
the semiconductor-vacuum interface. This semiconductor surface must be "clean," i.e.,
there should be no impurity layers (e.g., an oxide layer) on the surface. Even if a fraction
of a monolayer of the surface atoms have impurities bonded to them, the quality of the
epitaxial layer suffers drastically. The growth may occur to produce microcrystalline
regions separated by grain boundaries or may be amorphous in nature. In either case,
the special properties arising from the crystalline nature of the material (to be discussed
in the next chapter) are then lost.

The issue of surface cleanliness and surface reconstruction can be addressed
when one is doing a single epitaxial growth. For example, a clean wafer can be loaded
into the growth chamber and the remaining impurities on the surface can be removed by
heating the substrate. The proper reconstruction (which can be monitored by RHEED)
can be ensured by adjusting the substrate temperature and specy overpressure. Now
consider the problems associated with etching after the first epitaxial growth has occurred. As the etching starts, foreign atoms or molecules are introduced on the wafer as
the semiconductor is etched. The etching process is quite damaging and as it ends, the
surface of the etched wafer is quite rough and damaged. In addition, in most growth
techniques the wafer has to be physically moved from the high purity growth chamber
to the etching system. During this transportation, the surface of the wafer may collect
some "dirt." During the etching process this "dirt" may not be etched off and may
remain on the wafer. As a result of impurities and surface damage, when the second
epitaxial layer is grown after etching, the quality of the layer suffers.

A great deal of processing research in OEICs focusses on improving the etch-
ing/regrowth process. So far the OEICs fabricated in various laboratories have performances barely approaching the performance of hybrid circuits. Clearly the problem of
etching/regrowth is hampering the progress in OEIC technology.

It may be noted that the etching regrowth technology is also important in
creating quantum wires and quantum dots which require lateral patterning of epitaxial
layers.

1.3   CRYSTAL STRUCTURE

Essentially all high performance semiconductor devices are based on crystalline materials. there are some devices that use low cast amorphous or polycrystalline semiconductors, but their performance is quite poor. Crystals are made up of identical building
blocks, the block being an atom or a group of atoms. While in "natural" crystals the
crystalline symmetry is fixed by nature, new advances in crystal growth techniques
are allowing scientists to produce artificial crystals with modified crystalline structure.

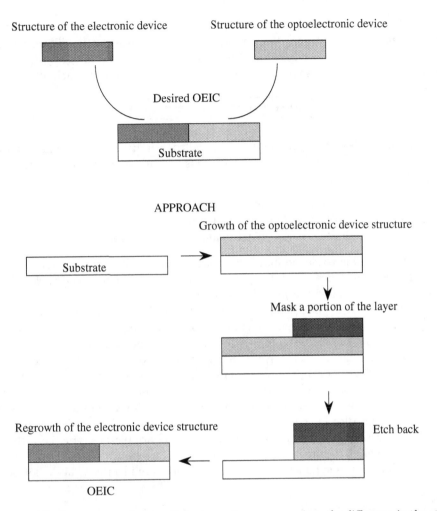

Figure 1.7: The importance of regrowth is clear when one examines the difference in the structure of electronic and optoelectronic devices. Etching and regrowth is essential for fabrication of optoelectronic integrated circuits (OEIC).

These advances depend upon being able to place atomic layers with exact precision and control during growth, leading to "superlattices". To define the crystal structure, two important concepts are introduced. The *lattice* represents a set of points in space which form a periodic structure. Each point sees an exact similar environment. The lattice is by itself a mathematical abstraction. A building block of atoms called the *basis* is then attached to each lattice point yielding the crystal structure.

An important property of a lattice is the ability to define three vectors $\mathbf{a}_1$, $\mathbf{a}_2$, $\mathbf{a}_3$, such that any lattice point $\mathbf{R}'$ can be obtained from any other lattice point $\mathbf{R}$ by a translation

$$\mathbf{R}' = \mathbf{R} + m_1\mathbf{a}_1 + m_2\mathbf{a}_2 + m_3\mathbf{a}_3 \tag{1.4}$$

where m_1, m_2, m_3 are integers. Such a lattice is called Bravais lattice. The entire lattice can be generated by choosing all possible combinations of the integers m_1, m_2, m_3 . The crystalline structure is now produced by attaching the basis to each of these lattice points.

$$\boxed{\text{lattice} + \text{basis} = \text{crystal structure}} \qquad (1.5)$$

The translation vectors $\mathbf{a}_1$, $\mathbf{a}_2$, and $\mathbf{a}_3$ are called primitive if the volume of the cell formed by them is the smallest possible. There is no unique way to choose the primitive vectors. One choice is to pick

$\mathbf{a}_1$ to be the shortest period of the lattice
$\mathbf{a}_2$ to be the shortest period not parallel to $\mathbf{a}_1$
$\mathbf{a}_3$ to be the shortest period not coplanar with $\mathbf{a}_1$ and $\mathbf{a}_2$

It is possible to define more than one set of primitive vectors for a given lattice, and often the choice depends upon convenience. The volume cell enclosed by the primitive vectors is called the *primitive unit cell.*

Because of the periodicity of a lattice, it is useful to define the symmetry of the structure. The symmetry is defined via a set of point group operations which involve a set of operations applied around a point. The operations involve rotation, reflection and inversion. The symmetry plays a very important role in the electronic properties of the crystals. For example, the inversion symmetry is extremely important and many physical properties of semiconductors are tied to the absence of this symmetry. As will be clear later, in the diamond structure (Si, Ge, C, etc.), inversion symmetry is present, while in the Zinc Blende structure (GaAs, AlAs, InAs, etc.), it is absent. Because of this lack of inversion symmetry, these semiconductors are piezoelectric, i.e., when they are strained an electric potential is developed across the opposite faces of the crystal. In crystals with inversion symmetry, where the two faces are identical, this is not possible.

1.3.1   Basic Lattice Types

The various kinds of lattice structures possible in nature are described by the symmetry group that describes their properties. Rotation is one of the important symmetry groups. Lattices can be found which have a rotation symmetry of 2π, $\frac{2\pi}{2}$, $\frac{2\pi}{3}$, $\frac{2\pi}{4}$, $\frac{2\pi}{6}$. The rotation symmetries are denoted by 1, 2, 3, 4, and 6. No other rotation axes exist; e.g., $\frac{2\pi}{5}$ or $\frac{2\pi}{7}$ are not allowed because such a structure could not fill up an infinite space.

There are 14 types of lattices in 3D. These lattice classes are defined by the relationships between the primitive vectors a_1, a_2, and a_3, and the angles α, β, and γ between them. The general lattice is triclinic ($\alpha \neq \beta \neq \gamma, a_1 \neq a_2 \neq a_3$) and there are 13 special lattices. Table 1.2 provides the basic properties of these three dimensional lattices. We will focus on the cubic lattice which is the structure taken by all semiconductors.

There are 3 kinds of cubic lattices: simple cubic, body centered cubic, and face centered cubic.

System	Number of lattices	Restrictions on conventional cell axes and singles
Triclinic	1	$a_1 \neq a_2 \neq a_3$ $\alpha \neq \beta \neq \gamma$
Monoclinic	2	$a_1 \neq a_2 \neq a_3$ $\alpha = \gamma = 90^\circ \neq \beta$
Orthorhombic	4	$a_1 \neq a_2 \neq a_3$ $\alpha = \beta = \gamma = 90^\circ$
Tetragonal	2	$a_1 = a_2 \neq a_3$ $\alpha = \beta = \gamma = 90^\circ$
Cubic	3	$a_1 = a_2 = a_3$ $\alpha = \beta = \gamma = 90^\circ$
Trigonal	1	$a_1 = a_2 = a_3$ $\alpha = \beta = \gamma < 120^\circ, \neq 90^\circ$
Hexagonal	1	$a_1 = a_2 \neq a_3$ $\alpha = \beta = 90^\circ$ $\gamma = 120^\circ$

Table 1.2: The 14 Bravais lattices in 3-dimensional systems and their properties.

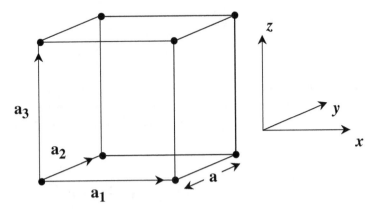

Figure 1.8: A simple cubic lattice showing the primitive vectors. The crystal is produced by repeating the cubic cell through space.

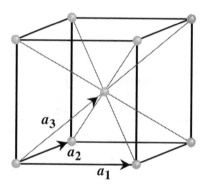

Figure 1.9: The body centered cubic lattice along with a choice of primitive vectors.

Simple cubic: The simple cubic lattice shown in Fig. 1.8 is generated by the primitive vectors

$$a\mathbf{x}, a\mathbf{y}, a\mathbf{z} \tag{1.6}$$

where the **x**, **y**, **z** are unit vectors.

Body-centered cubic: The bcc lattice shown in Fig. 1.9 can be generated from the simple cubic structure by placing a lattice point at the center of the cube. If $\hat{\mathbf{x}}, \hat{\mathbf{y}}$, and $\hat{\mathbf{z}}$ are three orthogonal unit vectors, then a set of primitive vectors for the body-centered cubic lattice could be

$$a_1 = a\hat{\mathbf{x}}, a_2 = a\hat{\mathbf{y}}, a_3 = \frac{a}{2}(\hat{\mathbf{x}} + \hat{\mathbf{y}} + \hat{\mathbf{z}}) \tag{1.7}$$

A more symmetric set for the bcc lattice is

$$a_1 = \frac{a}{2}(\hat{\mathbf{y}} + \hat{\mathbf{z}} - \hat{\mathbf{x}}), a_2 = \frac{a}{2}(\hat{\mathbf{z}} + \hat{\mathbf{x}} - \hat{\mathbf{y}}), a_3 = \frac{a}{2}(\hat{\mathbf{x}} + \hat{\mathbf{y}} - \hat{\mathbf{z}}) \tag{1.8}$$

Face Centered Cubic: Another equally important lattice for semiconductors is the *face-centered cubic* (fcc) Bravais lattice. To construct the face-centered cubic Bravais lattice add to the simple cubic lattice an additional point in the center of each square face (Fig. 1.10).

A symmetric set of primitive vectors for the face-centered cubic lattice (see Fig. 1.10) is

$$a_1 = \frac{a}{2}(\hat{y} + \hat{z}), a_2 = \frac{a}{2}(\hat{z} + \hat{x}), a_3 = \frac{a}{2}(\hat{x} + \hat{y}) \tag{1.9}$$

The face-centered cubic and body-centered cubic Bravais lattices are of great importance, since an enormous variety of solids crystallize in these forms with an atom (or ion) at each lattice site. Essentially all semiconductors of interest for electronics and optoelectronics have fcc structure.

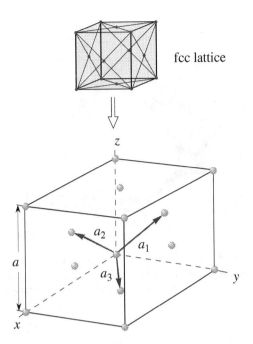

Figure 1.10: Primitive basis vectors for the face centered cubic lattice.

1.3.2   Basic Crystal Structures

Diamond and Zinc Blende Structures
Most semiconductors of interest for electronics and optoelectronics have an underlying fcc lattice. However, they have two atoms per basis. The coordinates of the two basis atoms are

$$(000) \text{ and } (\frac{a}{4}, \frac{a}{4}, \frac{a}{4}) \tag{1.10}$$

Since each atom lies on its own fcc lattice, such a two atom basis structure may be thought of as two inter-penetrating fcc lattices, one displaced from the other by a translation along a body diagonal direction $(\frac{a}{4} \frac{a}{4} \frac{a}{4})$.

Figure 1.11 gives details of this important structure. If the two atoms of the basis are identical, the structure is called diamond. Semiconductors such as Si, Ge, C, etc., fall in this category. If the two atoms are different, the structure is called the Zinc Blende structure. Semiconductors such as GaAs, AlAs, CdS, etc., fall in this category. Semiconductors with diamond structure are often called elemental semiconductors, while the Zinc Blende semiconductors are called compound semiconductors. The compound semiconductors are also denoted by the position of the atoms in the periodic chart, e.g., GaAs, AlAs, InP are called III-V (three-five) semiconductors while CdS, HgTe, CdTe, etc., are called II-VI (two-six) semiconductors.

Hexagonal Close Pack Structure   The hexagonal close pack (hcp) structure is an important lattice structure and many metals have this underlying lattice. Some

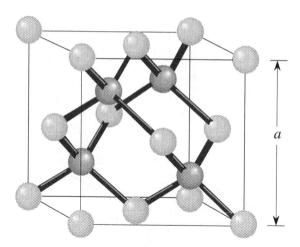

Figure 1.11: The zinc blende crystal structure. The structure consists of the interpenetrating fcc lattices, one displaced from the other by a distance $(\frac{a}{4} \frac{a}{4} \frac{a}{4})$ along the body diagonal. The underlying Bravais lattice is fcc with a two atom basis. The positions of the two atoms is (000) and $(\frac{a}{4} \frac{a}{4} \frac{a}{4})$.

semiconductors such as BN, AlN, GaN, SiC, etc., also have this underlying lattice (with a two-atom basis). The hcp structure is formed as shown in Fig. 1.12a. Imagine that a close-packed layer of spheres is formed. Each sphere touches six other spheres, leaving cavities, as shown. A second close-packed layer of spheres is placed on top of the first one so that the second layer sphere centers are in the cavities formed by the first layer. The third layer of close-packed spheres can now be placed so that center of the spheres do not fall on the center of the starting spheres (left side of Fig. 1.12a) or coincide with the centers of the starting spheres (right side of Fig. 1.12b). These two sequences, when repeated, produce the fcc and hcp lattices.

In Fig. 1.12b we show the detailed positions of the lattice points in the hcp lattice. The three lattice vectors are a_1, a_2 a_3, as shown. The vector a_3 is denoted by c and the term c-axis refers to the orientation of a_3. In an ideal structure, if $\mid a \mid = \mid a_1 \mid = \mid a_2 \mid$,

$$\frac{c}{a} = \sqrt{\frac{8}{3}} \tag{1.11}$$

In Table 1.3 we show the structural properties of some important materials. If two or more semiconductors are randomly mixed to produce an alloy, the lattice constant of the alloy is given by Vegard's law according to which the alloy lattice constant is the weighted mean of the lattice constants of the individual components.

1.3.3   Notation to Denote Planes and Points in a Lattice: Miller Indices

A simple scheme is used to describe lattice planes, directions and points. For a plane, we use the following procedure:

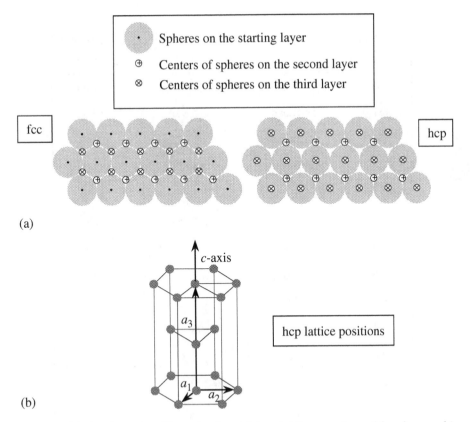

Figure 1.12: (a) A schematic of how the fcc and hcp lattices are formed by close packing of spheres. (b) Arrangement of lattice points on an hcp lattice.

(1) Define the x, y, z axes (primitive vectors).
(2) Take the intercepts of the plane along the axes in units of lattice constants.
(3) Take the reciprocal of the intercepts and reduce them to the smallest integers.

The notation (hkl) denotes a family of parallel planes.

The notation (hkl) denotes a family of equivalent planes.

To denote directions, we use the smallest set of integers having the same ratio as the direction cosines of the direction.

In a cubic system the Miller indices of a plane are the same as the direction perpendicular to the plane. The notation [ ] is for a set of parallel directions; < > is for a set of equivalent direction. Fig. 1.13 shows some examples of the use of the Miller indices to define planes.

EXAMPLE 1.2 The lattice constant of silicon is 5.43 Å. Calculate the number of silicon atoms in a cubic centimeter. Also calculate the number density of Ga atoms in GaAs which

Material	Structure	Lattice Constant (Å)	Density (gm/cm^3)
C	Diamond	3.5668	3.5153
Si	Diamond	5.431	2.329
Ge	Diamond	5.658	5.323
GaAs	Zinc Blende	5.653	5.318
AlAs	Zinc Blende	5.660	3.760
InAs	Zinc Blende	6.058	5.667
GaN	Wurtzite	$a = 3.175; c = 5.158$	6.095
AlN	Wurtzite	$a = 3.111; c = 4.981$	3.255
SiC	Zinc Blende	4.360	3.166
Cd	hcp	$a = 2.98; c = 5.620$	8.65
Cr	bcc	2.88	7.19
Co	hcp	$a = 2.51; c = 4.07$	8.9
Au	fcc	4.08	19.3
Fe	bcc	2.87	7.86
Ag	fcc	4.09	10.5
Al	fcc	4.05	2.7
Cu	fcc	3.61	8.96

Table 1.3: Structure, lattice constant, and density of some materials at room temperature.

has a lattice constant of 5.65 Å.

Silicon has a diamond structure which is made up of the fcc lattice with two atoms on each lattice point. The fcc unit cube has a volume a^3. The cube has eight lattice sites at the cube edges. However, each of these points is shared with eight other cubes. In addition, there are six lattice points on the cube face centers. Each of these points is shared by two adjacent cubes. Thus the number of lattice points per cube of volume a^3 are

$$N(a^3) = \frac{8}{8} + \frac{6}{2} = 4$$

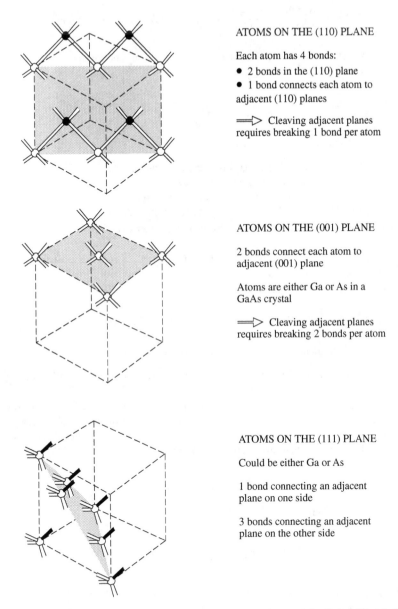

ATOMS ON THE (110) PLANE

Each atom has 4 bonds:
- 2 bonds in the (110) plane
- 1 bond connects each atom to adjacent (110) planes

⟹ Cleaving adjacent planes requires breaking 1 bond per atom

ATOMS ON THE (001) PLANE

2 bonds connect each atom to adjacent (001) plane

Atoms are either Ga or As in a GaAs crystal

⟹ Cleaving adjacent planes requires breaking 2 bonds per atom

ATOMS ON THE (111) PLANE

Could be either Ga or As

1 bond connecting an adjacent plane on one side

3 bonds connecting an adjacent plane on the other side

Figure 1.13: Some important planes in the cubic system along with their Miller indices. This figure also shows how many bonds connect adjacent planes. This number determines how easy or difficult it is to cleave the crystal along these planes.

In silicon there are two silicon atoms per lattice point. The number density is, therefore,

$$N_{Si} = \frac{4 \times 2}{a^3} = \frac{4 \times 2}{(5.43 \times 10^{-8})^3} = 4.997 \times 10^{22} \text{ atoms/cm}^3$$

In GaAs, there is one Ga atom and one As atom per lattice point. The Ga atom density is, therefore,

$$N_{Ga} = \frac{4}{a^3} = \frac{4}{(5.65 \times 10^{-8})^3} = 2.22 \times 10^{22} \text{ atoms/cm}^3$$

There are an equal number of As atoms.

EXAMPLE 1.3 In semiconductor technology, a Si device on a VLSI chip represents one of the smallest devices while a GaAs laser represents one of the larger devices. Consider a Si device with dimensions $(5 \times 2 \times 1)$ μm^3 and a GaAs semiconductor laser with dimensions $(200 \times 10 \times 5)$ μm^3. Calculate the number of atoms in each device.

From Example 1.1 the number of Si atoms in the Si transistor are

$$N_{Si} = (5 \times 10^{22} \text{ atoms/cm}^3)(10 \times 10^{-12} \text{ cm}^3) = 5 \times 10^{11} \text{ atoms}$$

The number of Ga atoms in the GaAs laser are

$$N_{Ga} = (2.22 \times 10^{22})(10^4 \times 10^{-12}) = 2.22 \times 10^{14} \text{ atoms}$$

An equal number of As atoms are also present in the laser.

EXAMPLE 1.4 Calculate the surface density of Ga atoms on a Ga terminated (001) GaAs surface.

In the (001) surfaces, the top atoms are either Ga or As leading to the terminology Ga terminated (or Ga stabilized) and As terminated (or As stabilized), respectively. A square of area a^2 has four atoms on the edges of the square and one atom at the center of the square. The atoms on the square edges are shared by a total of four squares. The total number of atoms per square is

$$N(a^2) = \frac{4}{4} + 1 = 2$$

The surface density is then

$$N_{Ga} = \frac{2}{a^2} = \frac{2}{(5.65 \times 10^{-8})^2} = 6.26 \times 10^{14} \text{ cm}^{-2}$$

EXAMPLE 1.5 Calculate the height of a GaAs monolayer in the (001) direction.

In the case of GaAs, a monolayer is defined as the combination of a Ga and As atomic layer. The monolayer distance in the (001) direction is simply

$$A_{m\ell} = \frac{a}{2} = \frac{5.65}{2} = 2.825 \text{ Å}$$

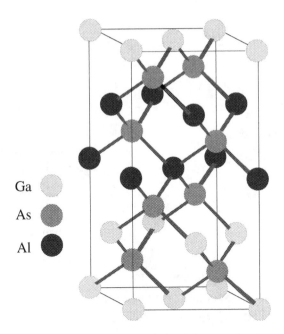

Figure 1.14: Arrangement of atoms in a $(GaAs)_2(AlAs)_2$ superlattice grown along (001) direction.

1.3.4 Artificial Structures: Superlattices and Quantum Wells

It is known that electrons and optical properties can be altered by using heterostructures, i.e., combinations of more that one semiconductor. MBE or MOCVD are techniques which allow monolayer ($\sim$3 Å) control in the chemical composition of the growing crystal. Nearly every semiconductor extending from zero bandgap (α-Sn,HgCdTe) to large bandgap materials such as ZnSe,CdS, etc., has been grown by epitaxial techniques such as MBE and MOCVD. Heteroepitaxial techniques allow one to grow heterostructures with atomic control, one can change the periodicity of the crystal in the growth direction. This leads to the concept of superlattices where two (or more) semiconductors A and B are grown alternately with thicknesses d_A and d_B respectively. The periodicity of the lattice in the growth direction is then $d_A + d_B$. A $(GaAs)_2$ $(AlAs)_2$ superlattice is illustrated in Fig. 1.14. It is a great testimony to the precision of the new growth techniques that values of d_A and d_B as low as monolayer have been grown.

It is important to point out that the most widely used heterostructures are not superlattices but quantum wells, in which a single layer of one semiconductor is sandwiched between two layers of a larger bandgap material. Such structures allow one to exploit special quantum effects that have become very useful in electronic and optoelectronic devices.

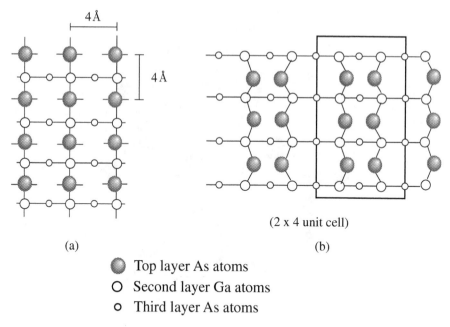

Figure (a) with "4 Å" dimensions labeled; Figure (b) labeled "(2 x 4 unit cell)"; "(a)" and "(b)"

🔘 Top layer As atoms
○ Second layer Ga atoms
∘ Third layer As atoms

Figure 1.15: The structure (a) of the unreconstructed GaAs (001) arsenic-rich surface. The missing dimer model (b) for the GaAs (001) (2×4) surface. The As dimers are missing to create a 4 unit periodicity along one direction and a two unit periodicity along the perpendicular direction.

1.3.5   Surfaces: Ideal Versus Real

The crystalline and electronic properties are quite different from the properties of the bulk material. The bulk crystal structure is decided by the internal chemical energy of the atoms forming the crystal with a certain number of nearest neighbors, second nearest neighbors, etc. At the surface, the number of neighbors is suddenly altered. Thus the spatial geometries which were providing the lowest energy configuration in the bulk may not provide the lowest energy configuration at the surface. Thus, there is a readjustment or "reconstruction" of the surface bonds towards an energy minimizing configuration.

An example of such a reconstruction is shown for the GaAs surface in Fig. 1.15. The figure (a) shows an ideal (001) surface where the topmost atoms form a square lattice. The surface atoms have two nearest neighbor bonds (Ga-As) with the layer below, four second neighbor bonds (e.g., Ga-Ga or As-As) with the next lower layer, and four second neighbor bonds within the same layer. In a "real" surface, the arrangement of atoms is far more complex. We could denote the ideal surface by the symbol C(1×1), representing the fact that the surface periodicity is one unit by one unit along the square lattice along [110] and [$\bar{1}$10]. The reconstructed surfaces that occur in nature are generally classified as C(2×8) or C(2×4) etc., representing the increased periodicity along the [$\bar{1}$10] and [110] respectively. The C(2×4) case is shown schematically in Fig.

1.15b, for an arsenic stabilized surface (i.e., the top monolayer is As). The As atoms on the surface form dimers (along [$\bar{1}$10] on the surface to strengthen their bonds. In addition, rows of missing dimers cause a longer range ordering as shown to increase the periodicity along the [110] direction to cause a C(2×4) unit cell. The surface periodicity is directly reflected in the x-ray diffraction pattern.

A similar effect occurs for the (110) surface of GaAs. This surface has both Ga and As atoms (the cations and anions) on the surface. A strong driving force exists to move the surface atoms and minimize the surface energy. Reconstruction effects also occur in silicon surfaces, where depending upon surface conditions a variety of reconstructions are observed. Surface reconstructions are very important since often the quality of the epitaxial crystal growth depends critically on the surface reconstruction.

EXAMPLE 1.6 Calculate the planar density of atoms on the (111) surface of Ge.

As can be seen from Fig. 1.13, we can form a triangle on the (111) surface. There are three atoms on the tips of the triangle. These atoms are shared by six other similar triangles. There are also 3 atoms along the edges of the triangle which are shared by two adjacent triangles. Thus the number of atoms in the triangle are

$$\frac{3}{6} + \frac{3}{2} = 2$$

The area of the triangle is $\sqrt{3}a^2/2$. The density of Ge atoms on the surface is then 7.29×10^{14} cm^{-2}.

1.3.6   Interfaces

Like surfaces, interfaces are an integral part of semiconductor devices. We have already discussed the concept of heterostructures and superlattices which involve interfaces between two semiconductors. These interfaces are usually of high quality with essentially no broken bonds, except for dislocations in strained structures (to be discussed later). There is, nevertheless, an *interface roughness* of one or two monolayers which is produced because of either non-ideal growth conditions or imprecise shutter control in the switching of the semiconductor species. The general picture of such a rough interface is as shown in Fig. 1.16 for epitaxially grown interfaces. The crystallinity and periodicity in the underlying lattice is maintained, but the chemical species have some disorder on interfacial planes. Such a disorder is quite important in many electronic and opto-electronic devices.

One of the most important interfaces in electronics is the Si/SiO$_2$ interface. This interface and its quality is responsible for essentially all of the modern consumer electronic revolution. This interface represents a situation where two materials with very different lattice constants and crystal structures are brought together. However, in spite of these large differences the interface quality is quite good. In Fig. 1.17 we show a TEM cross-section of a Si/SiO$_2$ interface. It appears that the interface has a region of a few monolayers of amorphous or disordered Si/SiO$_2$ region creating fluctuations in the chemical species (and consequently in potential energy) across the interface. This interface roughness is responsible for reducing mobility of electrons and holes in MOS devices. It can also lead to "trap" states, which can seriously deteriorate device performance if the interface quality is poor.

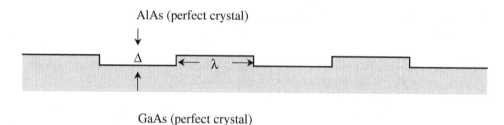

Figure 1.16: A schematic picture of the interfaces between materials with similar lattice constants such as GaAs/AlAs. No loss of crystalline lattice and long range order is suffered in such interfaces. The interface is characterized by islands of height Δ and lateral extent λ.

Finally, we have the interfaces formed between metals and semiconductors. Structurally, these important interfaces are hardest to characterize. These interfaces are usually produced in presence of high temperatures and involve diffusion of metal elements along with complex chemical reactions. The "interfacial region" usually extends over several hundred Angstroms and is a complex non-crystalline region.

1.3.7   Defects in Semiconductors

In the previous section we have discussed the properties of the perfect crystalline structure. In real semiconductors, one invariably has some defects that are introduced due to either thermodynamic considerations or the presence of impurities during the crystal growth process. In general, defects in crystalline semiconductors can be characterized as i) point defects; ii) line defects; iii) planar defects and iv) volume defects. These defects are detrimental to the performance of electronic and optoelectronic devices and are to be avoided as much as possible. We will give a brief overview of the important defects.

Point Defects

A point defect is a highly localized defect that affects the periodicity of the crystal only in one or a few unit cells. There are a variety of point defects, as shown in Fig. 1.18. Defects are present in any crystal and their concentration is given roughly by the thermodynamics relation

$$\frac{N_d}{N_{Tot}} = k_d \exp\left(-\frac{E_d}{k_B T}\right) \tag{1.12}$$

where N_d is the vacancy density, N_{Tot} the total site density in the crystal, E_d the defect formation energy, k_d is a dimensionless parameter with values ranging from 1 to 10 in semiconductors, and T, the crystal growth temperature. The vacancy formation energy is in the range of an eV for most semiconductors.

An important point defect in compound semiconductors such as GaAs is the anti-site defect in which one of the atoms, say Ga, sits on the arsenic sublattice instead of the Ga sublattice. Such defects (denoted by Ga_{As}) can be a source of reduced device performance.

Other point defects are interstitials in which an atom is sitting in a site that is in between the lattice points as shown in Fig. 1.18, and impurity atoms which involve a

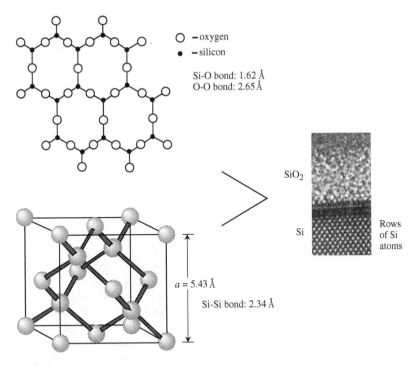

O – oxygen
• – silicon

Si-O bond: 1.62 Å
O-O bond: 2.65 Å

SiO₂

Si

Rows of Si atoms

$a = 5.43$ Å

Si-Si bond: 2.34 Å

Figure 1.17: The tremendous success of Si technology is due to the Si/SiO₂ interface. In spite of the very different crystal structure of Si and SiO₂, the interface is extremely sharp, as shown in the TEM picture in this figure.

wrong chemical species in the lattice. In some cases the defect may involve several sites forming a defect complex.

Line Defects or Dislocations

In contrast to point defects, line defects (called dislocations) involve a large number of atomic sites that can be connected by a line. Dislocations are produced if, for example, an extra half plane of atoms are inserted (or taken out) of the crystal as shown in Fig. 1.19. Such dislocations are called edge dislocations. Dislocations can also be created if there is a slip in the crystal so that part of the crystal bonds are broken and reconnected with atoms after the slip.

Dislocations can be a serious problem, especially in the growth of strained heterostructures (to be discussed later). In optoelectronic devices, dislocations can ruin the device performance and render the device useless. Thus the control of dislocations is of great importance.

Planar Defects and Volume Defects

Planar defects and volume defects are not important in single crystalline materials, but can be of importance in polycrystalline materials. If, for example, silicon is grown on a glass substrate, it is likely that polycrystalline silicon will be produced. In the polycrystalline material, small regions of Si (∼ a few microns in diameter) are perfectly

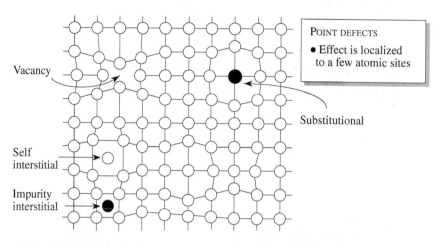

Figure 1.18: A schematic showing some important point defects in a crystal.

crystalline, but are next to microcrystallites with different orientations. The interface between these microcrystallites are called grain boundaries. Grain boundaries may be viewed as an array of dislocations.

Volume defects can be produced if the crystal growth process is poor. The crystal may contain regions that are amorphous or may contain voids. In most epitaxial techniques used in modern optoelectronics, these defects are not a problem. However, the developments of new material systems such as diamond (C) or SiC are hampered by such defects.

EXAMPLE 1.7 Consider an equilibrium growth of a semiconductor at a temperature of 1000 K. The vacancy formation energy is 2.0 eV. Calculate the vacancy density produced if the site density for the semiconductor is 2.5×10^{22} cm^{-3}. Assume that $k_d = 1$.

The vacancy density is

$$
\begin{aligned}
N_{vac} &= N_{Tot} \exp\left(-\frac{E_{vac}}{k_B T}\right) \\
&= (2.5 \times 10^{22} \text{ cm}^{-3}) \exp\left(-\frac{2.0 \text{ eV}}{0.0867 \text{ eV}}\right) \\
&= 2.37 \times 10^{12} \text{ cm}^{-3}
\end{aligned}
$$

This is an extremely low density and will have little effect on the properties of the semiconductor. The defect density would be in mid 10^{15} cm^{-3} range if the growth temperature was 1500 K. At such values, the defects can significantly affect device performance.

1.4 STRAINED HETEROSTRUCTURES

In an epitaxial process, the overlayer that is grown on the substrate could have a lattice constant that may differ from that of the substrate. Such epitaxy is called strained epitaxy and is one of the important emerging areas of crystal growth studies. The motivation for strained epitaxy is two fold:

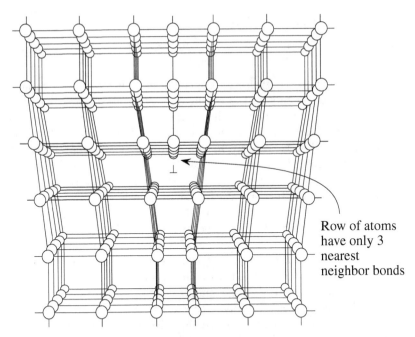

Row of atoms
have only 3
nearest
neighbor bonds

Figure 1.19: A schematic showing the presence of a dislocation. This line defect is produced by adding an extra half plane of atoms. At the edge of the extra plane, the atoms have a missing bond.

i) Incorporation of built-in strain: When a lattice mismatched semiconductor is grown on a substrate and the thickness of the overlayer is very thin (this will be discussed in detail later), the overlayer has a built-in strain. This built-in strain has important effects on the electronic and optoelectronic properties of the material and can be exploited for high performance devices.

ii) Generation of a new effective substrate: We have noted that in semiconductor technology, high quality substrates are only available for Si, GaAs and InP (sapphire and quartz substrates are also available and used for some applications). Most semiconductors are not lattice-matched to these substrates. How can one grow these semiconductors epitaxially? One solution that has emerged is to grow the overlayer on a mismatched substrate. If the conditions are right, a lot of dislocations are generated and eventually the overlayer forms its own substrate. This process allows a tremendous flexibility in semiconductor technology. Not only can it, in principle, resolve the substrate availability problem, it also allows the possibility of growing GaAs on Si, CdTe on GaAs, etc. Thus different semiconductor technologies can be integrated on the same wafer.

Coherent and Incoherent Structures

Consider a case where an overlayer with lattice constant a_L is grown on a substrate with lattice constant a_S. This situation is shown schematically in Fig. 1.20. The strain between the two materials is defined as

$$\epsilon = \frac{a_S - a_L}{a_L} \tag{1.13}$$

Consider a conceptual exercise where we deposit a monolayer of the overlayer on the substrate. If the lattice constant of the overlayer is maintained to be a_L, it is easy to see that after every $1/\epsilon$ bonds between the overlayer and the substrate, either a bond is missing or an extra bond appears as shown in Fig. 1.20b. In fact, there would be a row of missing or extra bonds since we have a 2-dimensional plane. These defects are the dislocations. The presence of these dislocations costs energy to the system since a number of atoms do not have proper chemical bonding at the interface.

An alternative to the incoherent case is shown in Fig. 1.20c. Here all the atoms at the interface of the substrate and the overlayer are properly bonded by adjusting the in-plane lattice constant of the overlayer to that of the substrate. This causes the overlayer to be under strain and the system has a certain amount of strain energy. *This strain energy grows as the overlayer thickness increases.* In the strained epitaxy, the choice between the state of the structure shown in Fig. 1.20b and the state shown in Fig. 1.20c is decided by free energy minimization considerations. Theoretical and experimental studies have focussed on these considerations for over six decades, and the importance of these studies has grown since the advent of heteroepitaxy. The general observations can be summarized as follows:

For small lattice mismatch ($\epsilon < 0.1$), the overlayer initially grows in perfect registry with the substrate, as shown in Fig. 1.20c. However, as noted before, the strain energy will grow as the overlayer thickness increases. As a result, it will eventually be favorable for the overlayer to generate dislocations. In simplistic theories this occurs at an overlayer thickness called the critical thickness, d_c, which is approximately given by

$$d_c \cong \frac{a_S}{2|\epsilon|} \tag{1.14}$$

In reality, the point in growth where dislocations are generated is not so clear cut and depends upon growth conditions, surface conditions, dislocation kinetics, etc. However, one may use the criteria given by Eqn. 1.14 for loosely characterizing two regions of overlayer thickness for a given lattice mismatch. Below critical thickness, the overlayer grows without dislocations and the film is under strain. Under ideal conditions above critical thickness, the film has a dislocation array, and after the dislocation arrays are generated, the overlayer grows without strain with its free lattice constant.

While strained epitaxy below critical thickness is an extremely powerful tool for tailoring the optoelectronic properties of semiconductors, epitaxy beyond the critical thickness is important to provide new effective substrates for new material growth. For these applications the key issues center around ensuring that the dislocations generated stay near the overlayer-substrate interface and do not propagate into the overlayer as shown in Fig. 1.21. A great deal of work has been done to study this problem. Often thin superlattices in which the individual layers have alternate signs of strain are grown to "trap" or "bend" the dislocations. It is also useful to build the strain up gradually.

EXAMPLE 1.8 Estimate the critical thickness for $In_{0.3}Ga_{0.7}As$ grown on a GaAs substrate.

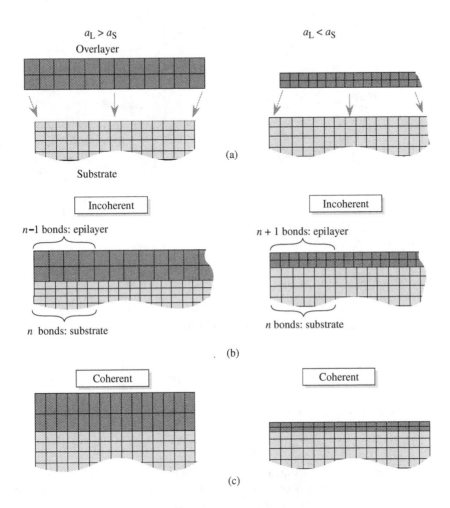

$a_L > a_S$

Overlayer

Substrate

(a)

Incoherent

$n{-}1$ bonds: epilayer

n bonds: substrate

$a_L < a_S$

Incoherent

$n + 1$ bonds: epilayer

n bonds: substrate

(b)

Coherent

Coherent

(c)

Figure 1.20: (a) The conceptual exercise in which an overlayer with one lattice constant is placed without distortion on a substrate with a different lattice constant. (b) Dislocations are generated at positions where the interface bonding is lost. (c) The case is shown where the overlayer is distorted so that no dislocation is generated.

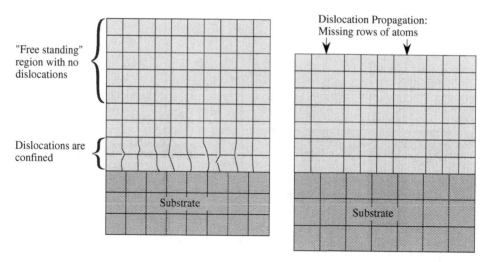

Figure 1.21: Strained epitaxy above critical thickness. On the left hand side is shown a structure in which the dislocations are confined near the overlayer-substrate interface. This is a desirable mode of epitaxy. On the right hand side, the dislocations are penetrating the overlayer, rendering it useless for most optoelectronic applications.

The lattice constant of an alloy is given by the Vegard's law:

$$
\begin{aligned}
a(In_{0.3}Ga_{0.7}As) &= 0.3a_{InAs} + 0.7a_{GaAs} \\
&= 5.775 \ \mathring{A}
\end{aligned}
$$

The strain is

$$
\epsilon = \frac{5.653 - 5.775}{5.653} = -0.022
$$

The critical thickness is approximately

$$
d_c = \frac{5.653 \ \mathring{A}}{2(0.022)} = 128 \ \mathring{A}
$$

This thickness is quite adequate for most devices and can be used to make useful quantum well devices. If, on the other hand, the strain is, say, 5%, the critical thickness is ~ 50 Å, which is too thin for most useful device applications.

Self-Assembed Structures

When a lattice mismatched structure is grown on a substrate (which for most cases can be regarded as semi-infinite) a number of energetic and kinetic demands come into play. There is strain energy that is created in the system if the overlayer is under strain. This has to compete against the chemical bonding energy created by bond formation. Additionally, in real growth surface effects and the ability of the system to reach the free energy minimum state play an important role.

In Fig. 1.22 we show three kinds of growth mechanisms that occur when a strained overlayer is grown on a substrate under near equilibrium conditions. In Fig. 1.22

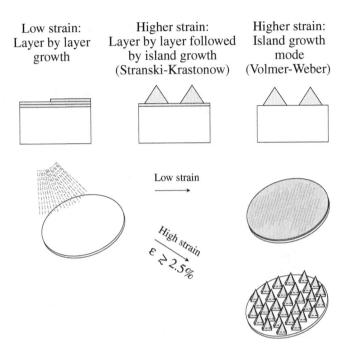

Low strain: Layer by layer growth

Higher strain: Layer by layer followed by island growth (Stranski-Krastonow)

Higher strain: Island growth mode (Volmer-Weber)

Low strain

High strain $\varepsilon \gtrsim 2.5\%$

Lateral feature sizes can be controlled from 100Å -1000Å
$\Longrightarrow 10^{12}$ features per wafer can be produced without lithography

Figure 1.22: Growth modes in strained epitaxy. The island mode growth can be exploited to make "self-assembled" quantum dot structures.

we have a case where the lattice mismatch is very small ($\epsilon \leq 2\%$). The overlayer grows in the monolayer by monolayer mode since this allows the maximum chemical bonding to occur. If the lattice mismatch is increased, the growth occurs by a mode known as the Stranski-Krastanow mode where the initial growth starts out in the monolayer by monolayer growth, but then the overlayer grows in an island mode. The island growth provides fewer chemical per atom for the growing layer (since the surface area is larger), but the strain energy is minimized, since the bond lengths do not have to adjust as much to fit the substrate. Finally, at higher lattice mismatch the growth initiates directly in the island mode (the Volmer-Weber mode).

If heterostructures are to be grown with atomically abrupt interfaces between two semiconductors, one should be in the layer-by-layer growth mode described schematically in Fig. 1.22. There are, however, some advantages of growing in the island mode. By growing islands and then imbedding them by another material it is possible to grow quasi-zero dimensional systems in which electron (holes) are confined in all threee directions. Since such quantum dots are self-organized and regime no lithography/etching/regrowth they are very attractive for many applications. Such self-organized quantum dots have been grown with InGaAs/GaAs, SiGe/Si, etc.

1.5   STRAIN TENSOR IN LATTICE MISMATCHED EPITAXY

In order to study the effect of strain on electronic properties of semiconductors, it is first essential to establish the strain tensor produced by epitaxy. In Appendix A we discuss important issues in strain and stress in materials. The reader who is unfamiliar with these issues should go over this appendix which is the basis for the results given in this section. As noted above, careful growth of an epitaxial layer whose lattice constant is close, but not equal, to the lattice constant of the substrate can result in a coherent strain. If the strain is small one can have layer-by-layer growth as shown in Fig. 1.22. In this case the lattice constant of the epitaxial layer in the directions parallel to the interface is forced to be equal to the lattice constant of the substrate. The lattice constant of the epitaxial perpendicular to the substrate will be changed by the Poisson effect. If the parallel lattice constant is forced to shrink, or a compressive strain is applied, the perpendicular lattice constant will grow. Conversely, if the parallel lattice constant of the epitaxial layer is forced to expand under tensile strain, the perpendicular lattice constant will shrink. These two cases are depicted in Fig. 1.20c. This type of coherently strained crystal is called pseudomorphic.

For layer-by-layer growth, the epitaxial semiconductor layer is biaxially strained in the plane of the substrate, by an amount $\epsilon_{\parallel}$, and uniaxially strained in the perpendicular direction, by an amount $\epsilon_{\perp}$. For a thick substrate, the in-plane strain of the layer is determined from the bulk lattice constants of the substrate material, a_S, and the layer material, a_L:

$$
\begin{aligned}
e_{\parallel} &= \frac{a_S}{a_L} - 1 \\
&= \epsilon
\end{aligned}
\tag{1.15}
$$

Since the layer is subjected to no stress in the perpendicular direction, the perpendicular strain, $\epsilon_{\perp}$, is simply proportional to $\epsilon_{\parallel}$:

$$
\epsilon_{\perp} = \frac{-\epsilon_{\parallel}}{\sigma}
\tag{1.16}
$$

where the constant σ is known as Poisson's ratio.

Noting that there is *no stress* in the direction of growth it can be simply shown that for the strained layer grown on a (001) substrate (for an *fcc* lattice)

$$
\sigma = \frac{c_{11}}{2c_{12}}
\tag{1.17}
$$

$$
\begin{aligned}
\epsilon_{xx} &= \epsilon_{\parallel} \\
\epsilon_{yy} &= \epsilon_{xx} \\
\epsilon_{zz} &= \frac{-2c_{12}}{c_{11}}\epsilon_{\parallel} \\
\epsilon_{xy} &= 0 \\
\epsilon_{yz} &= 0 \\
\epsilon_{zx} &= 0
\end{aligned}
$$

while in the case of strained layer grown on a (111) substrate

$$\sigma = \frac{c_{11} + 2c_{12} + 4c_{44}}{2c_{11} + 4c_{12} - 4c_{44}}$$

$$\epsilon_{xx} = \left[\frac{2}{3} - \frac{1}{3}\left(\frac{2c_{11} + 4c_{12} - 4c_{44}}{c_{11} + 2c_{12} + 4c_{44}}\right)\right]\epsilon_{\parallel}$$

$$\epsilon_{yy} = \epsilon_{xx}$$

$$\epsilon_{zz} = \epsilon_{xx}$$

$$\epsilon_{xy} = \left[\frac{-1}{3} - \frac{1}{3}\left(\frac{2c_{11} + 4c_{12} - 4c_{44}}{c_{11} + 2c_{12} + 4c_{44}}\right)\right]\epsilon_{\parallel}$$

$$\epsilon_{yz} = \epsilon_{xy}$$

$$\epsilon_{zx} = \epsilon_{yz} \tag{1.18}$$

The general strain tensor for arbitrary orientation is shown in Fig. 1.23.

In general, the strained epitaxy causes a distortion of the cubic lattice and, depending upon the growth orientation, the distortions produce a new reduced crystal symmetry. It is important to note that for (001) growth, the strain tensor is diagonal while for (111), and several other directions, the strain tensor has nondiagonal terms. The nondiagonal terms can be exploited to produce built-in electric fields in certain heterostructures as will be discussed in the next section.

An important heterostructure system involves growth of *hcp* lattice-based Al-GaN or InGaN on a GaN substrate along the c-axis. In this case the strain tensor is given by (a_L is the substrate lattice constant, a_S is overlayer lattice constant)

$$\epsilon_{xx} = \epsilon_{yy} = \frac{a_S}{a_L} - 1$$

$$\epsilon_{zz} = -2\frac{c_{13}}{c_{33}}\epsilon_{xx} \tag{1.19}$$

This strain is exploited to generate piezoelectric effect based interface charge as discussed in the next section. In Table 1.4 we provide values of elastic constant of several important semiconductors.

Strained Tensor for Self-Organized Dots
We have noted earlier that under high strain conditions we can have growth occur in the Stranski-Krastonow mode where the initial epilayer (one or two monolayers) grows in the layer by layer mode and then the growth occurs by the island mode. The islands that are produced often have a pyramidal shape as shown in Fig. 1.22. In some systems the islands have the shape of a truncated pyramid or even of a "lens." Such "self-assembled" dots can be used to form quantum dots where a small bandgap material is enclosed completely by a large bandgap material. Such quantum dots have been exploited for zero dimensional physics and devices.

The strain tensor of the self-assembled dots is different from that in a layer that is atomically flat. In Fig. 1.24 we show the strain tensor calculated for a self-assembled dot. The wetting layer (region A in Fig. 1.24) has the usual biaxial strain, but the pyramidal dot has a strong hydrostatic and biaxial component. There is also a large shear component at the edges of the pyramid.

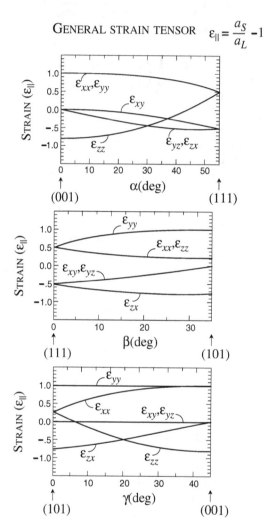

Figure 1.23: The general strain tensor produced when an overlayer is grown on different substrate orientations. The strain between the two materials is $\epsilon_\parallel$.

Material	$C_{11}(N/m^2)$	$C_{12}(N/m^2)$	$C_{41}(N/m^2)$
Si	1.66×10^{11}	0.64×10^{11}	0.8×10^{11}
Ge	1.29×10^{11}	0.48×10^{11}	0.67×10^{11}
GaAs	1.2×10^{11}	0.54×10^{11}	0.59×10^{11}
C	10.76×10^{11}	1.25×10^{11}	5.76×10^{11}

Material	$C_{13}(N/m^2)$	$C_{33}(N/m^2)$
GaN	10.9×10^{11}	35.5×10^{11}
AlN	12×10^{11}	39.5×10^{11}

Table 1.4: Elastic constant for some fcc and hcp based semiconductors. (For Si, Ge, GaAs see H. J. McSkimin and P. Andreatch, *J. Appl. Phys.*, **35**, 2161 (1964) and D. I. Bolef and M. Meres, *J. Appl. Phys.*, **31**, 1010 (1960). For nitrides see J. H. Edgar, *Properties of III-V Nitrides*, INSPEC, London (1994) and R. B. Schwarz, K. Khachaturyan, and E. R. Weber, *Appl. Phys. Lett.*, **74**, 1122 (1997).)

1.6 POLAR MATERIALS AND POLARIZATION CHARGE

In compound semiconductors there is a shift of charge from one atom to another in the basis atoms producing a negatively charged cation and a positively charged anion. In unstrained zinc blende structures the cation and anion sublattices are arranged in such a way that there is no net polarization in the material. On the other hand in the wurtzite crystal (like InN, GaN, AlN) the arrangement of the cation and anion sublattices can be such that there is a relative movement from the ideal wurtzite position to produce a "spontaneous polarization" in the crystal which becomes very important for heterostructures. This effect is illustrated in Fig. 1.25. Also given in Fig. 1.25 are the values of the spontaneous polarization which is aligned along the c-axis of the crystal.

In addition to spontaneous polarization there are two other phenomena which can lead to polarization in the material. Strain can cause a relative shift between the cation and anion sublattices and create net polarization in the material. This is the piezoelectric effect. Finally, in materials known as ferroelectrics, an external applied field can cause net polarization. In Fig. 1.26 we show how the movement of rows can cause polarization effect by looking at the structural arrangements of atoms in barium titanate.

Polar Charge at Heterointerfaces

If there is a net movement of one sublattice against each other, a polarization field is set up. This results in a positive and negative polar charge. Under most conditions the polar charge on the free surfaces is neutralized by charges present in the atmosphere. This causes depolarization of the material. If, however, a heterostructure is synthesized and the two materials forming the structure have different values for the polarization, there is a net polar charge (and polarization) at the interface as shown in Fig. 1.27. In

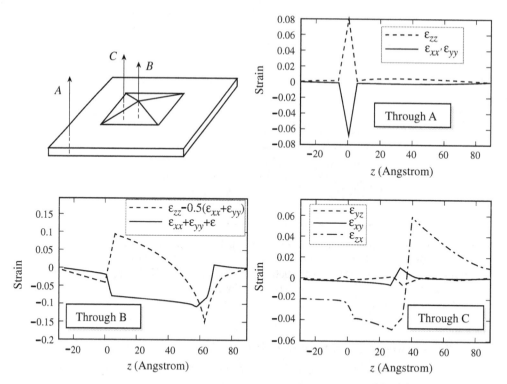

Figure 1.24: Strain tensor in a pyramidal InAs on GaAs self-assembled quantum dot.

semiconductors this polar charge can cause a built-in electric field

$$F = \frac{P}{\epsilon} \tag{1.20}$$

The interface charge $P_A - P_B$ and the built-in interface field (see Fig. 1.27) can be exploited in device design since for most applications this fixed polar charge can act as dopant.

Piezoelectric Effect

As noted above, when a structure is under strain a net polarization can arise—a phenomenon called *piezoelectric effect*. The value of the polar charge induced by strain depends upon the strain tensor. In Section 1.5 we have discussed the nature of the strain tensor in strained epitaxy (i.e., in the coherent growth regime).

Nitride heterostructures have polarization charges at interfaces because of strain related piezoelectric effect as well as from spontaneous polarization. For growth along (0001) orientation the strain tensor for coherently strained wurtzite crystals is given by Eqn. 1.19 in the previous section. The piezoelectric polarization is related to the strain tensor by the following relation

$$P_{pz} = e_{33}\epsilon_{zz} + e_{31}(\epsilon_{xx} + \epsilon_{yy}) \tag{1.21}$$

NITRIDES IN WURTZITE STRUCTURES:

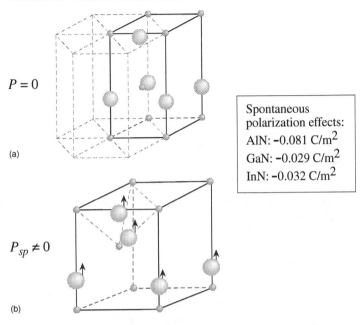

$P = 0$

(a)

Spontaneous
polarization effects:
AlN: −0.081 C/m^2
GaN: −0.029 C/m^2
InN: −0.032 C/m^2

$P_{sp} \neq 0$

(b)

Figure 1.25: (a) The wurtzite crystal in the absence of any cation-anion sublattice shift. (b) A relative shift of the cation and anion sublattices leads to a net polarization in the material. Spontaneous polarization values for InN, GaN, and AlN are given.

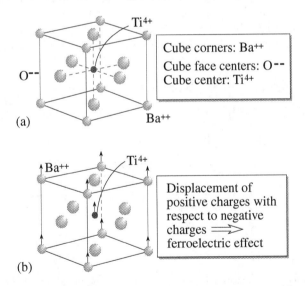

Ti^{4+}

O^{--}

(a)

Ba^{++}

Cube corners: Ba^{++}
Cube face centers: O^{--}
Cube center: Ti^{4+}

Ba^{++}

Ti^{4+}

(b)

Displacement of
positive charges with
respect to negative
charges $\Longrightarrow$
ferroelectric effect

Figure 1.26: (a) The structure of a typical perovskite crystal illustrated by examining barium titanate. (b) The ferroelectric effect is produced by a net displacement of the positive ions with respect to the negative ions.

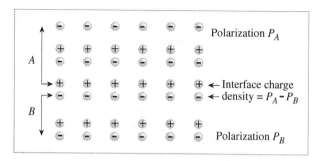

Figure 1.27: A schematic showing how interface charge density can be produced at heterointerfaces of two polar materials.

Piezoelectric effect is also present in zinc blende structures. However, the piezoelectric effect only occurs when the strain tensor has off-diagonal components. The polarization values are given by

$$\begin{aligned}
P_x &= e_{14}\epsilon_{yz} \\
P_y &= e_{14}\epsilon_{xz} \\
P_z &= e_{14}\epsilon_{xy}
\end{aligned}$$
(1.22)

As can be seen from the discussion of the previous section the strain tensor is diagonal for growth along (001) direction. As a result there is no piezoelectric effect. However for other orientations, notably for (111) growth there is a strong piezoelectric effect.

Piezoelectric effect can be exploited to create interface charge densities as high as 10^{13} cm^{-2} in materials. In Table 1.5 we provide the values of piezoelectric constants for some semiconductors. In addition to the polarization induced by strain, the cation

ZINC BLENDE		WURTZITE (*c*-axis growth)		
Material	e_{14} (C/m^2)	Material	e_{31} (C/m^2) $\quad$ e_{33} (C/m^2)	P_{sp} (C/m^2)
AlAs	−0.23	AlN	−0.6 $\qquad\qquad$ 1.46	−0.081
GaAs	−0.16	GaN	−0.49 $\qquad\qquad$ 0.73	−0.029
GaSb	−0.13	InN	−0.57 $\qquad\qquad$ 0.97	−0.032
GaP	−0.10			
InAs	−0.05			
InP	−0.04			

Table 1.5: Piezoelectric constants in some important semiconductors. For the nitrides the spontaneous polarization values are also given. (Data for zinc-blende material from S. Adachi, *J. Appl. Phys.* vol. 58, **R1** (1985). For nitrides see E. Bernardini, V. Fiorentini, and D. Vanderbilt, *Phys. Rev. B* vol. 56, **R10024** (1997).)

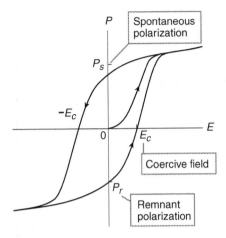

Figure 1.28: A schematic of the ferroelectric hysterisis loop.

and anion sublattices are spontaneously displaced with respect to each other producing an additional polarization. For heterostructures the difference of the spontaneous polarization appears at the interfaces, as noted earlier. In Table 1.5 we also provide the values of spontaneous polarization for AlN, GaN, and InN.

Ferroelectric Materials

Ferroelectric materials have the property that they can have nonzero electric dipole moment even in the absence of an applied electric field. Moreover an external electric field can alter the polarization of the material. In materials like the nitrides discussed above the polarization cannot be altered once the crystal is grown, but in ferroelectrics it is possible for the external field to physically alter the crystal and alter the polarization.

Ferroelectric crystals can be classified into two main categories: order-disorder and displacive. In the order-disorder type, the ferroelectric effect is associated with ordering of the dipoles in the crystal. In displacive type, there is a net displacement of one sublattice against the other. In Fig. 1.26 we have shown how the lattice of BaTiO$_3$ distorts to create net polarization.

Polarization in a ferroelectric material shows hysterisis as shown in Fig. 1.28. The polarization value depends not only on the applied field, but on the history of the material. If the field is increasing from a large negative value to a large positive value (the forward cycle) the polarization may be described by

$$P^+(E) = P_s \tanh\left(\frac{E - E_0}{2\delta}\right) \tag{1.23}$$

where

$$\delta = E_c \left[\ell n \left(\frac{1 + P_r/P_s}{1 - P_r/P_s}\right)\right]^{-1} \tag{1.24}$$

Here E_c is called the coercive field and is the field at which the polarization switches sign. The quantities P_r and P_s are called the remnant and spontaneous polarization,

MATERIAL		T_c (K)	P_s (μCcm^{-2})	POLAR CHARGE (cm^{-2})
KDP type	KH$_2$PO$_4$	123	5.33	3.3 x 10^{13}
	KH$_2$AsO$_4$	96	5.0	3.1 x 10^{13}
Perovskites	BaTiO$_3$	393	26.0	1.62 x 10^{14}
	SrTiO$_3$	32(?)	3.0	1.87 x 10^{13}
	PbTiO$_3$	763	>50.0	3.1 x 10^{14}
	KNbO$_3$	712	30.0	1.87 x 10^{14}
	LiNbiO$_3$	1470	300.0	1.87 x 10^{15}
	LiTaO$_3$		23.3	1.45 x 10^{14}

Table 1.6: A list of several ferroelectric materials and their properties (from *Introduction to Solid State Physics*, C. Kittel, John Wiley and Sons, New York (1971).).

respectively. In the negative field loop the polarization is given by

$$P^-(E) = -P^+(-E) \qquad (1.25)$$

Ferroelectricity disappears above a temperature called the transition temperature (or Curie temperature). In Table 1.6 we show the Curie temperature and the spontaneous polarization values (and the charge per unit area) for several different ferroelectric materials. It can be seen that in materials like BaTiO$_3$ and LiNbO$_3$ very large polarization values are reached. If the coercive fields are very large, fields found in many device applications may not alter the polarization of the ferroelectric material. The polarization may then be fixed at P_s or $-P_s$. The actual sign of the polarization can be chosen by a process called "poling" in which a large electric field is used to fix the polarization direction.

EXAMPLE 1.9 A thin film of Al$_{0.3}$Ga$_{0.7}$N is grown coherently on a GaN substrate. Calculate the polar charge density and electric field at the interface.

The lattice constant of Al$_{0.3}$Ga$_{0.7}$N is given by Vegard's law

$$a_{all} = 0.3a_{AlN} + 0.7a_{GaN} = 3.111 \ \mathring{A}$$

The strain tensor is

$$\epsilon_{xx} = 0.006$$

Using the elastic constant values from Table 1.4

$$\epsilon_{zz} = -0.6 \times 0.006 = 0.0036$$

The piezoelectric effect induced polar charge then becomes

$$P_{pz} = 0.0097 \ \text{C/m}^2$$

This corresponds to a density of 6.06×10^{12} cm^{-2} electronic charges.

In addition to the piezoelectric charge the spontaneous polarization charge is

$$P_{sp} = 0.3(0.089) + 0.7(0.029) - 0.029 = 0.018 \text{ C/m}^2$$

which corresponds to a density of 1.125×10^{13} cm^{-2} charges. The total charge (fixed) arising at the interface is the sum of the two charges.

1.7 TECHNOLOGY CHALLENGES

Compared to metal technology and insulator technology (swords and glasses were made thousands of years ago), semiconductor technology is relatively new. The reason is very simple from a technology point of view—semiconductors need to be extremely "pure" if they are to be useful. Defect densities of a percent may have minimal effect on metals and insulators, but will ruin a semiconductor device. For most high performance devices, defect densities of less than 10^{15} cm^{-3} are needed (i.e., less than one part in 100 million). Apart from demands on purity of crystal growth chambers and starting materials this places an enormous demand on substrates.

Currently semiconductor substrate technology is available (i.e., bulk crystals can be grown in sufficient size/purity) for a handful of materials. These include Si, GaAs, InP, and Ge, which are widely available and SiC, Al$_2$O$_3$, and GaSb, etc., which are available only in small pieces (a few square centimeters) and are very expensive. Since most semiconductors do not have a substrate available from either bulk crystal growth or another lattice matched substrate, this severely restricts the use of a wide range of semiconductors. In Fig. 1.29 we show an overview of some important substrates.

The key challenge facing the development of new technologies (e.g., InAs, GaN, InSb...) is the availability of substrates that can be used without a large dislocation density propagating through the structure. An approach that is widely being studied is based on starting with substrate A, then growing a thick buffer of material B to create a new effective substrate. As shown in Fig. 1.29 the challenge is to contain the dislocations within the buffer region. The buffer region can be material B for which the substrate is needed and may also contain an intermediate material that helps in "trapping" dislocations.

1.8 PROBLEMS

Sections 1.2–1.3

1.1 Consider the (001) MBE growth of GaAs by MBE. Assuming that the sticking coefficient of Ga is unity, calculate the Ga partial pressure needed if the growth rate has to be 1 μm/hr. The temperature of the Ga cell is 1000 K.

1.2 In the growth of GaAs/AlAs structures in a particular MBE system, the background pressure of Ga when the Ga shutter is off is 10^{-7} Torr. If the growth rate of AlAs is 1 μm/hr, what fraction of Ga atoms are incorporated in the AlAs region? The Ga cell is at 1000 K.

1.3 A 5.0 μm Si epitaxial layer is to be grown. The Si flux is 10^{14} cm^{-2} s^{-1}. How long will it take to grow the film if the sticking coefficient is 0.95?

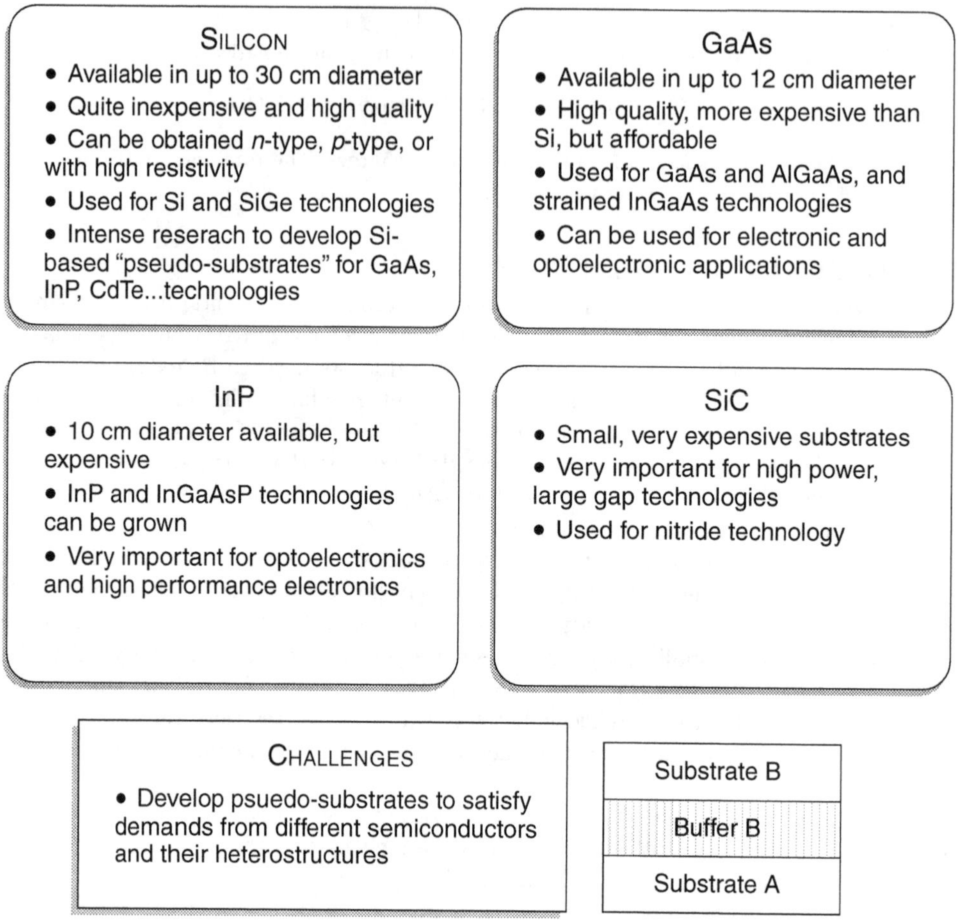

SILICON	GaAs
• Available in up to 30 cm diameter • Quite inexpensive and high quality • Can be obtained *n*-type, *p*-type, or with high resistivity • Used for Si and SiGe technologies • Intense reserach to develop Si-based "pseudo-substrates" for GaAs, InP, CdTe...technologies	• Available in up to 12 cm diameter • High quality, more expensive than Si, but affordable • Used for GaAs and AlGaAs, and strained InGaAs technologies • Can be used for electronic and optoelectronic applications
InP	**SiC**
• 10 cm diameter available, but expensive • InP and InGaAsP technologies can be grown • Very important for optoelectronics and high performance electronics	• Small, very expensive substrates • Very important for high power, large gap technologies • Used for nitride technology

CHALLENGES	Substrate B
• Develop psuedo-substrates to satisfy demands from different semiconductors and their heterostructures	Buffer B
	Substrate A

Figure 1.29: A brief overview of important substrates available in semiconductor technology.

1.4 a) Find the angles between the tetrahedral bonds of a diamond lattice.
b) What are the direction cosines of the (111) oriented nearest neighbor bond along the x,y,z axes.

1.5 Consider a semiconductor with the zinc blende structure (such as GaAs).
a) Show that the (100) plane is made up of either cation or anion type atoms.
b) Draw the positions of the atoms on a (110) plane assuming no surface reconstruction.
c) Show that there are two types of (111) surfaces: one where the surface atoms are bonded to three other atoms in the crystal, and another where the surface atoms are bonded to only one. These two types of surfaces area called the *A* and *B* surfaces, respectively.

1.6 Suppose that identical solid spheres are placed in space so that their centers lie on the atomic points of a crystal and the spheres on the neighboring sites touch each other. Assuming that the spheres have unit density, show that density of such spheres is the following for the various crystal structures:

$$
\begin{aligned}
\text{fcc} &: \quad \sqrt{2}\pi/6 = 0.74 \\
\text{bcc} &: \quad \sqrt{3}\pi/8 = 0.68 \\
\text{sc} &: \quad \pi/6 = 0.52 \\
\text{diamond} &: \quad \sqrt{3}\pi/16 = 0.34
\end{aligned}
$$

1.7 Calculate the number of cells per unit volume in GaAs (a = 5.65 Å). Si has a 4% larger lattice constant. What is the unit cell density for Si? What is the number of atoms per unit volume in each case?

1.8 A Si wafer is nominally oriented along the (001) direction, but is found to be cut 2° off, towards the (110) axis. This off axis cut produces "steps" on the surface which are 2 monolayers high. What is the lateral spacing between the steps of the 2° off-axis wafer?

1.9 Conduct a literature search to find out what the lattice mismatch is between GaAs and AlAs at 300 K and 800 K. Calculate the mismatch between GaAs and Si at the same temperatures.

1.10 In high purity Si crystals, defect densities can be reduced to levels of 10^{13} cm^{-3}. On an average, what is the spacing between defects in such crystals? In heavily doped Si, the dopant density can approach 10^{19} cm^{-3}. What is the spacing between defects for such heavily doped semiconductors?

1.11 A GaAs crystal which is nominally along (001) direction is cut at an angle θ off towards (110) axis. This produces one monolayer high steps. If the step size is to be no more than 100 Å, calculate θ.

1.12 Assume that a Ga-As bond in GaAs has a bond energy of 1.0 eV. Calculate the energy needed to cleave GaAs in the (001) and (110) planes.

1.13 Consider a hcp structure shown in Fig. 1.12. Prove the relation given by $c/a = \sqrt{8/3} = 1.633$.

1.14 A HgCdTe alloy is to be grown with 10% Hg. The sticking coefficient of Hg is 10^{-2} and that of Cd is 1.0. Te is present in excess and does not limit the growth of the crystal. Calculate the Hg and Cd fluxes and partial pressures needed to grow the film at a rate of 1.0 μm/hour.

1.15 A serious problem in the growth of a heterostructure made from two semiconductors is due to the difficulty in finding a temperature at which both semiconductors can grow with high quality. Consider the growth of HgTe and CdTe which is usually grown at $\sim$ 600 K. Assume that the defect formation energy in HgTe is 1.0 eV and in CdTe is 2.0 eV. Calculate the density of defects in the heterostructure with equal HgTe and CdTe. Assume that the constant K_d in Eqn. 1.12 is unity.

1.16 Calculate the defect density in GaAs grown by LPE at 1000 K. The defect formation energy is 2.0 eV. Assume that K_d in Eqn. 1.12 is unity.

1.17 Why are entropy considerations unimportant in dislocation generation?

Sections 1.4–1.5

1.18 A coherently strained quantum well laser has to made from In$_x$Ga$_{1-x}$As on a

GaAs substrate. If the minimum thickness of the region is 50 Å, calculate the maximum composition of In that can be tolerated. Assume that the lattice constant of the alloy can be linearly interpolated from its components.

1.19   Assume that in a semiconductor alloy, the lattice constant scales as a linear weighted average. Find the composition of the $In_x Ga_{1-x}As$ alloy that lattice matches with an InP substrate.

1.20   Calculate the critical thickness for the growth of AlAs on a GaAs substrate.

1.21   A diamond (C) crystal is compressed by hydrostatic pressure so that its lattice constant decreases by 0.1%. Calculate the strain energy per cm^3 in the distorted crystal. See Appendix A for the relevant strain energy expression.

1.22   A 100 Å $In_{0.2}Ga_{0.8}As$ film is grown on a GaAs substrate. The film is coherent. Calculate the strain energy per cm^2 in the film.

1.23   Consider a coherently grown film of $Si_{0.8}Ge_{0.2}$ grown on a Si substrate. Calculate the thickness of the film at which the strain energy density (eV cm^{-2}) becomes equal to the energy density arising from a square array of dislocations in the film.

Assume that the dislocations are on a planar square grid with one broken bond per spacing of a/ϵ where a is the film lattice constant and ϵ is the strain. The energy per broken bond is 1.0 eV.

Section 1.6

1.24   A coherent film of $In_{0.25}Ga_{0.75}As$ is grown on a (111) GaAs substrate. Calculate the fixed charge at the $GaAs/In_{0.25}Ga_{0.75}As$ interface. Also calculate the electric field produced across the interface by this charge. Assume that the relative dielectric constant is 12.0.

1.25   Consider a $In_{0.1}Ga_{0.9}N/GaN$ interface matched to a GaN substrate. The structure is produced by c-axis growth. Calculate the polarization charge produced by spontaneous polarization and by the piezoelectric effect.

1.26   For a particular application it is required that an interface charge of 3×10^{13} cm^{-2} (in units of electron charge) be produced at a $Al_x Ga_{1-x}N/GaN$ structure. The structure is grown on x-axis GaN and is coherent. Calculate the Al composition needed.

1.27   In barium titanate the observed saturation polarization is 2.6×10^{-5} Ccm^{-2}. Estimate the relative displacement of positive and negative ions in the crystal that would lead to such a polarization. The volume of a unit cell is 6.4×10^{-23} cm^3.

1.9   REFERENCES

- **Crystal Growth**

    - V. Markov, *Crystal Growth for Beginners: Fundamentals of Nucleation, Crystal Growth, and Epitaxy*, World Scientific Publication (1995).

    - A. W. Vere, *Crystal Growth: Principles and Progress*, Plenum Publishing Company (1987).

- **Crystal Structure**

  - B. Hyde and S. Andersson, *Inorganic Crystal Structures*, John-Wiley and Sons (1989).

  - M. M. Woolfson, *An Introduction to Crystallography*, Cambridge University Press (1997).

  - *McGraw-Hill Encyclopedia of Science and Technology*, Volume 4, McGraw-Hill (1997).

  - A. C. Gossard (ed.), *Epitaxial Microstructures in Semiconductors and Semi metals*, Volume 40, Academic Press (1994).

  - G. Benedek (ed.), *Point and Extended Defects in Semiconductors*, Plenum Publishing Press (1989).

  - C. Kittel, *Introduction to Solid State Physics*, John-Wiley (1986).

  - Landolt-Bornstein, *Numerical Data and Functional Relationships in Science and Technology*, (O. Madelung, M. Schultz, and H. Weiss, eds.), Springer (1985).

- **Strained Structures**

  - J. F. Nye, *Physical Properties of Crystals: Their Representation by Tensors and Matrices*, Oxford University Press (1987).

  - D. Bimberg, M. Grundman, and N. Ledentsov, *Quantum Dot Heterostructures*, John-Wiley and Sons (1999).

  - P. Harrison, *Quantum Wells, Wires, and Dots: Theoretical and Computational Physics*, John-Wiley and Sons (2000).

  - S. F. Borg, *Fundamentals of Engineering: Elasticity*, Polar Structures (1990).

  - T. Ikeda, *Fundamentals of Piezoelectricity*, Oxford University Press (1990).

  - M. E. Lines and A. M. Glass, *Principles and Applications of Ferroelectrics and Related Materials*, Oxford University Press (2001).

  - E. Bernardini, V. Fiorentini, and D. Vanderbilt, *Spontaneous Polarization, and Piezoelectric Constant of III-V Nitrides*, *Physical Review B*, vol. 56, p. R10024 (1997).

  - J. H. Edgar, *Properties of Group III Nitrides*, INSPEC, London (1994).

Chapter

2

SEMICONDUCTOR BANDSTRUCTURE

2.1 INTRODUCTION

The properties of electrons inside semiconductors are described by the solution of the Schrödinger equation appropriate for the crystal. The solutions provide us the bandstructure or the electronic spectrum for electrons. The problem of finding the electronic spectrum is an enormously complicated one. Solids have a large number of closely spaced atoms providing the electrons a very complex potential energy profile. Additionally electrons interact with each other and in a real solid atoms are vibrating causing time dependent variations in the potential energy. To simplify the problem the potential fluctuations created by atomic vibrations (lattice vibrations) and scattering of electrons from other electrons are removed from the problem and treated later on via perturbation theory. These perturbations cause scattering of electrons from one state to another.

   The problem of bandstructure becomes greatly simplified if we are dealing with crystalline materials. An electron in a rigid crystal structure sees a periodic background potential. As a result the wavefunctions for the electron satisfy Bloch's theorem as discussed in the next section.

   There are two main categories of realistic bandstructure calculation for semiconductors:

1. Methods which describe the entire valence and conduction bands.

2. Methods which describe near bandedge bandstructures.

The techniques in the second category are simpler and considerably more accurate if one is interested only in phenomena near the bandedges. Techniques such as the tight binding method, the pseudopotential method, and the orthogonalized plane wave methods fall in the first category. On the other hand, perturbative techniques (the so called $\mathbf{k} \cdot \mathbf{p}$

methods) fall in the second category. We will develop the tight binding method and the $\mathbf{k}\cdot\mathbf{p}$ method in some detail, since both of these techniques are widely used for describing real semiconductors and their heterostructures.

Before starting our discussion of electronic bandstructure we will summarize some important features of electrons in crystalline materials. We will discuss Bloch theorem which tells us about electron wavefunctions in a crystal and basic differences between metals, semiconductors, and insulators.

2.2 BLOCH THEOREM AND CRYSTAL MOMENTUM

To understand the electronic properties of a material we need to know what the electron wavefunctions and energies are inside a solid. We are only interested in crystalline materials here. The description of electrons in a periodic material has to be via the Schrödinger equation

$$\left[\frac{-\hbar^2}{2m_0}\nabla^2 + U(\mathbf{r})\right]\psi(\mathbf{r}) = \mathbf{E}\psi(\mathbf{r}) \tag{2.1}$$

where $U(\mathbf{r})$ is the background potential seen by the electrons. Due to the crystalline nature of the material, the potential $U(\mathbf{r})$ has the same periodicity, R, as the lattice

$$U(\mathbf{r}) = U(\mathbf{r} + \mathbf{R}) \tag{2.2}$$

If the background potential is zero, the electronic function in a volume V is

$$\psi(\mathbf{r}) = \frac{e^{i\mathbf{k}\cdot\mathbf{r}}}{\sqrt{V}}$$

and the electron momentum and energy are

$$\mathbf{p} = \hbar\mathbf{k}$$
$$E = \frac{\hbar^2 k^2}{2m_0}$$

The wavefunction is spread in the entire sample and has equal probability ($\psi^*\psi$) at every point in space.

Let us examine the periodic crystal. We expect the electron probability to be same in all unit cells of the crystal because each cell is identical. If the potential was random, this would not be the case, as shown schematically in Fig. 2.1a. If R is a periodic vector of the lattice we expect

$$|\psi(\mathbf{r})|^2 = |\psi(\mathbf{r} + R)|^2$$

If this equality is not solid we would be able to distinguish one unit cell from another. Note that the wavefunction itself is not periodic, it is the probability that is periodic. The wavefunction has to be of a special form described by Bloch's theorem. Bloch's theorem states that the eigenfunctions of the Schrödinger equation for a periodic potential are

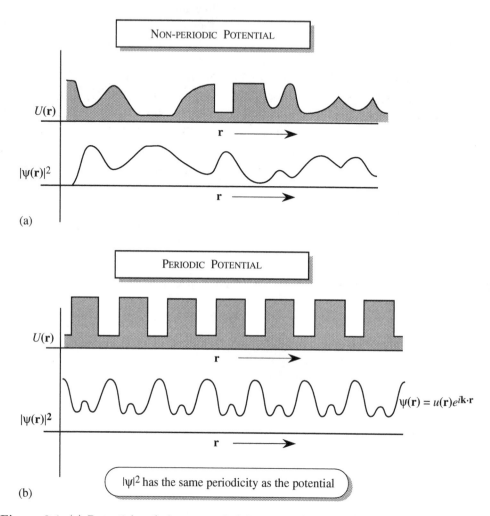

Figure 2.1: (a) Potential and electron probability value of a typical electronic wavefunction in a random material. (b) The effect of a periodic background potential on an electronic wavefunction. In the case of the periodic potential, $|\psi|^2$ has the same spatial periodicity as the potential. This puts a special constraint on $\psi(\mathbf{r})$ according to Bloch's theorem.

the product of a plane wave $e^{i\mathbf{k}\cdot\mathbf{r}}$ and a function $u_{\mathbf{k}}(\mathbf{r})$, which has the same periodicity as the periodic potential. Thus

$$\psi_{\mathbf{k}}(\mathbf{r}) = e^{i\mathbf{k}\cdot\mathbf{r}} u_{\mathbf{k}}(\mathbf{r}) \qquad (2.3)$$

is the form of the electronic function. The periodic part $u_{\mathbf{k}}(\mathbf{r})$ has the same periodicity as the crystal; i.e.,

$$u_{\mathbf{k}}(\mathbf{r}) = u_{\mathbf{k}}(\mathbf{r} + \mathbf{R}) \qquad (2.4)$$

The wavefunction has the property

$$
\begin{aligned}
\psi_{\mathbf{k}}(\mathbf{r} + \mathbf{R}) &= e^{i\mathbf{k}\cdot(\mathbf{r}+\mathbf{R})} u_{\mathbf{k}}(\mathbf{r} + \mathbf{R}) = e^{i\mathbf{k}\cdot\mathbf{r}} u_{\mathbf{k}}(\mathbf{r}) e^{i\mathbf{k}\cdot\mathbf{R}} \\
&= e^{i\mathbf{k}\cdot\mathbf{R}} \psi_{\mathbf{k}}(\mathbf{r})
\end{aligned}
\qquad (2.5)
$$

In Fig. 2.1b we show a typical wavefunction. The vector $\mathbf{k}$ used above is called $\mathbf{k}$-vector and plays an important role in the electronic properties of crystals.

2.2.1 Significance of the k-vector

An important implication of the Bloch theorem is that in the perfectly periodic background potential that the crystal presents, *the electron propagates without scattering.* The electronic state ($\sim \exp(i\mathbf{k}\cdot\mathbf{r})$) is an extended wave which occupies the entire crystal. We need to derive an equation of motion for the electrons which tells us how electrons will respond to external forces. If F_{ext} represents an external force applied on the electron and F_{int} represents the internal force due to atoms in the crystal we may write Newton's equation as

$$\frac{d\mathbf{p}}{dt} = \mathbf{F}_{ext} + \mathbf{F}_{int} \qquad (2.6)$$

However this equaiton is quite useless for a meaningful description of the electron because it includes the internal forces on the electron. We need a description which does *not* include the evaluation of the internal forces. We will now give a simple derivation for such an equation of motion. The equation of motion also provides a conceptual understanding of the vector $\mathbf{k}$ that has been introduced by the Bloch theorem.

Using the time-dependent Schrödinger equation the general solution for electrons is

$$\psi_R(\mathbf{r}, t) = u_k(\mathbf{r}) e^{i(k\cdot r - \omega t)} \qquad (2.7)$$

where the electron energy E is related to the frequency ω by

$$E = \hbar\omega \qquad (2.8)$$

The Bloch function is a plane wave which extends over all crystalline space. To define a localized electron we examine a wavepacket made up of wavefunctions near a particular k-value. We can define the group velocity of this wavepacket as

$$
\begin{aligned}
\mathbf{v}_g &= \frac{d\omega}{d\mathbf{k}} \\
\mathbf{v} &= \frac{1}{\hbar}\frac{dE}{d\mathbf{k}} \\
&= \frac{1}{\hbar}\nabla_{\mathbf{k}}E(\mathbf{k})
\end{aligned}
\tag{2.9}
$$

If we have an electric field $\mathbf{F}$ present, the work done on the electron during a time interval δt is

$$
\delta E = -e\mathbf{F}\cdot\mathbf{v}_g\delta t
$$

We may also write, in general,

$$
\begin{aligned}
\delta E &= \left(\frac{dE}{d\mathbf{k}}\right)\delta\mathbf{k} \\
&= \hbar\mathbf{v}_g\cdot\delta\mathbf{k}
\end{aligned}
$$

Comparing the two equations for δE, we get

$$
\delta\mathbf{k} = -\frac{e\mathbf{F}}{\hbar}\delta t
$$

giving us the relation

$$
\hbar\frac{d\mathbf{k}}{dt} = -e\mathbf{F}
\tag{2.10}
$$

Since $-e\mathbf{F}$ is the external force on the electron, we can generalize this equation and write

$$
\boxed{\hbar\frac{d\mathbf{k}}{dt} = \mathbf{F}_{ext}}
\tag{2.11}
$$

Eqn. 2.11 looks identical to Newton's second law of motion;

$$
\frac{d\mathbf{p}}{dt} = \mathbf{F}_{ext}
$$

in free space if we associate the quantity $\hbar\mathbf{k}$ with the momentum of the electron in the crystal. The term $\hbar\mathbf{k}$ responds to the external forces as if it is the momentum of the electron, although, as can be seen by comparing the true Newtons equation of motion, it is clear that $\hbar\mathbf{k}$ contains the effects of the internal crystal potentials and is therefore not the true electron momentum. The quantity $\hbar\mathbf{k}$ is called the crystal momentum. Once the E versus k relation is established, we can, for all practical purposes, forget about the background potential $U(\mathbf{r})$ and treat the electrons as if they are free and obey the effective Newtons equation of motion. This physical picture is summarized in Fig. 2.2.

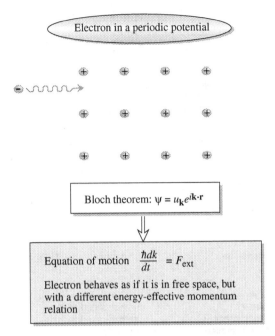

Figure 2.2: A physical description of electrons in a periodic potential. As shown the electrons can be treated as if they are in free space except that their energy-momentum relation is modified because of the potential.

2.3 METALS, INSULATORS, AND SEMICONDUCTORS

We know from atomic physics that bound electrons have discrete energy levels separated by forbidden energy regions. In solids the discrete levels broaden to form allowed bands which are separated by bandgaps. Once one knows the electronic spectra the important question is: Which of these allowed states are occupied by electrons and which are unoccupied? Two important situations arise when we examine the electron occupation of allowed bands: In one case we have a situation where an allowed band is completely filled with electrons, while the next allowed band is separated in energy by a gap E_g and is completely empty at 0 K. In a second case, the highest occupied band is only half full (or partially full). These cases are shown in Fig. 2.3.

At this point a very important concept needs to be introduced. *When an allowed band is completely filled with electrons, the electrons in the band cannot conduct any current.* This important concept is central to the special properties of metals and insulators. Being fermions the electrons cannot carry any net current in a filled band since an electron can only move into an empty state. One can imagine a net cancellation of the motion of electrons moving one way and those moving the other. Because of this effect, when we have a material in which a band is completely filled, while the next allowed band is separated in energy and empty, the material has, in principle, infinite resistivity and is called an *insulator* or a *semiconductor*. The material in which a band is only half full with electrons has a very low resistivity and is a *metal*.

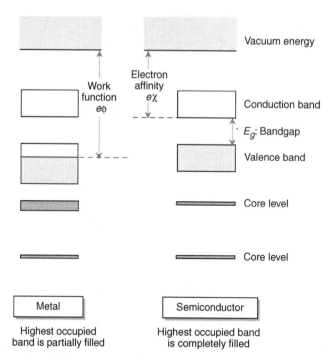

Figure 2.3: A schematic description of electron occupation of the bands in a metal and semi-conductor (or insulator). In a metal, the highest occupied band at 0 K is partially filled with electrons. Also shown is the metal work function. In a semiconductor at 0 K, the highest occupied band is completely filled with electrons and the next band is completely empty. The separation between the two bands is the bandgap E_g. The electron affinity and work function are also shown.

The band that is normally filled with electrons at 0 K in semiconductors is called the valence band, while the upper unfilled band is called the conduction band. The energy difference between the vacuum level and the highest occupied electronic state in a metal is called the metal work function. The energy between the vacuum level and the bottom of the conduction band is called the electron affinity. This is shown schematically in Fig. 2.3.

Metals have a very high conductivity because of the very large number of electrons that can participate in current transport. It is, however, difficult to alter the conductivity of metals in any simple manner as a result of this. On the other hand, semiconductors have zero conductivity at 0 K and quite low conductivity at finite temperatures, but it is possible to alter their conductivity by orders of magnitude. This is the key reason why semiconductors can be used for active devices.

As noted, semiconductors are defined as materials in which the valence band is full of electrons and the conduction band is empty at 0 K. At finite temperatures some of the electrons leave the valence band and occupy the conduction band. The valence band is then left with some unoccupied states. Let us consider the situation as shown

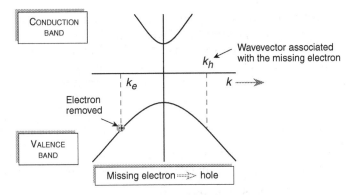

Figure 2.4: Illustration of the wavevector of the missing electron k_e. The wavevector is $-\mathbf{k}_e$, which is associated with the hole.

in Fig. 2.4, where an electron with momentum $\mathbf{k}_e$ is missing from the valence band.

When all of the valence band states are occupied, the sum over all wavevector states is zero; i.e.,

$$\sum \mathbf{k}_i = 0 = \sum_{\mathbf{k}_i \neq \mathbf{k}_e} \mathbf{k}_i + \mathbf{k}_e \tag{2.12}$$

This result is just an indication that there are as many positive k states occupied as there are negative ones. Now in the situation where the electron at wavevector $\mathbf{k}_e$ is missing, the total wavevector is

$$\sum_{\mathbf{k}_i \neq \mathbf{k}_e} \mathbf{k}_i = -\mathbf{k}_e \tag{2.13}$$

The missing state is called a hole and the wavevector of the system $-\mathbf{k}_e$ is attributed to it. It is important to note that the electron is missing from the state $\mathbf{k}_e$ and the momentum associated with the hole is at $-\mathbf{k}_e$. The position of the hole is depicted as that of the missing electron. But in reality the hole wavevector $\mathbf{k}_h$ is $-\mathbf{k}_e$, as shown in Fig. 2.4.

$$\mathbf{k}_h = -\mathbf{k}_e \tag{2.14}$$

Note that the hole is a representation for the valence band with a missing electron. As discussed earlier, if the electron is not missing the valence band electrons cannot carry any current. However, if an electron is missing the current flow is allowed. If an electric field is applied, all the electrons move in the direction opposite to the electric field. This results in the unoccupied state moving in the field direction. *The hole thus responds as if it has a positive charge.* It therefore responds to external electric and magnetic fields **F** and **B**, respectively, according to the equation of motion

$$\hbar \frac{d\mathbf{k}_h}{dt} = e\left[\mathbf{F} + \mathbf{v}_h \times \mathbf{B}\right] \tag{2.15}$$

where $\hbar \mathbf{k}_h$ and $\mathbf{v}_h$ are the momentum and velocity of the hole.

Thus the equation of motion of holes is that of particles with a *positive* charge e. The mass of the hole has a positive value, although the electron mass in its valence band is negative. When we discuss the conduction band properties of semiconductors or insulators we refer to electrons, but when we discuss the valence band properties, we refer to holes. This is because in the valence band only the missing electrons or holes lead to charge transport and current flow.

2.4   TIGHT BINDING METHOD

Before examining the various semiconductors it is extremely useful to examine the atomic structure of some of the elements which make up the various semiconductors.

IV Semiconductors

C $1s^2\, \underbrace{2s^2 2p^2}$

Si $1s^2 2s^2 2p^6\, \underbrace{3s^2 3p^2}$

Ge $1s^2 2s^2 2p^6 3s^2 3p^6 3d^{10}\, \underbrace{4s^2 4p^2}$

III–V Semiconductors

Ga $1s^2 2s^2 2p^6 3s^2 3p^6 3d^{10}\, \underbrace{4s^2 4p^1}$

As $1s^2 2s^2 2p^6 3s^2 3p^6 3d^{10}\, \underbrace{4s^2 4p^3}$

A very important conclusion can be drawn about the elements making up the semiconductors: The outermost valence electrons are made up of electrons in either the s-type or p-type orbitals. While this conclusion is strictly true for the elements in the atomic form, it turns out that even in the crystalline semiconductors the electrons in the valence and conductor band retain this s- or p-type character, even though they are "free" Bloch electrons. It is extremely important to appreciate this simple feature since it plays a key role in optical and transport processes in semiconductors.

We will now discuss several techniques for obtaining bandstructure of materials. As noted in the introduction there are methods which describe the entire bandstructure reasonably well and methods that are valid over a narrow energy range. The first technique we will discuss is the tight binding method.

The tight binding method (TBM) is an empirical technique, i.e., experimental inputs are used to fit the bandstructure. TBM uses atomic functions as a basis set for the Bloch functions. The periodic part of the Bloch function is represented by some combination of the atomic orbitals centered at the lattice points. If $\phi_n(\mathbf{r} - \boldsymbol{R})$ represents such an orbital centered at $\boldsymbol{R}$, we could write a Bloch function of the form:

$$\psi_{\mathbf{k}}(\mathbf{r}) = \sum_{\mathbf{R}_n} \phi_n(\mathbf{r} - \boldsymbol{R}) \exp(i\mathbf{k} \cdot \boldsymbol{R}_n) \qquad (2.16)$$

The periodic part of the Bloch function is expanded in terms of the atomic-like orbitals of the atoms of the unit cell (index n in the summation).

As noted earlier, the elements making up all the semiconductors of interest have the valence electrons described by s- or p-type atomic orbitals. The core electrons are usually not of interest. As the atoms of the elements making up the semiconductors are

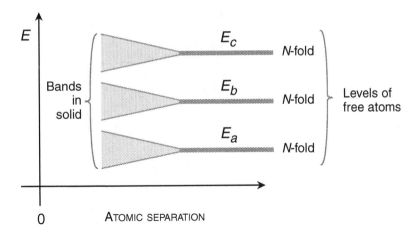

Figure 2.5: Atomic levels spreading into bands as the atoms come together. When the atoms are far apart, the electron levels are discrete and N-fold degenerate (N is the number of atoms). As the atoms are brought closer, the discrete levels form bands.

brought together to form the crystal, the electronic states are perturbed by the presence of neighboring atoms and discrete states broaden to form bands as shown in Fig. 2.5. While the original atomic functions describing the valence electrons are, of course, no longer eigenstates of the problem, they can be used as a good approximate set of basis states to describe the "crystalline" electrons. This is the motivation for the tight binding method. We develop a simple mathematical description for the tight binding method and then discuss some details of the application of the method to real semiconductors.

We assume the solution of the atomic problem

$$H_{\text{at}}\psi_n = E_n\psi_n \tag{2.17}$$

is already known for the atoms forming the crystalline material. This solution leads to the description of the electronic structure of the various atoms. One could construct a Bloch state given by

$$\psi_{\mathbf{k}}(\mathbf{r}) = \sum_{n,\mathbf{R}} e^{i\mathbf{k}\cdot\mathbf{R}}\psi_n(\mathbf{r} - \boldsymbol{R}) \tag{2.18}$$

but this state does not describe the new problem of the crystalline material where now we have

$$H_{\text{cryst}} = H_{\text{at}} + \Delta U(\mathbf{r}) \tag{2.19}$$

where $\Delta U(\mathbf{r})$ is the additional perturbation coming in, due to the interaction of neighboring atoms as shown in Fig. 2.6.

The new wavefunctions are now chosen as the more general wavefunction (they must, of course, satisfy Bloch's Theorem):

$$\Psi_{\mathbf{k}}(\mathbf{r}) = \sum_{\mathbf{R}} e^{i\mathbf{k}\cdot\mathbf{R}}\phi(\mathbf{r} - \boldsymbol{R}) \tag{2.20}$$

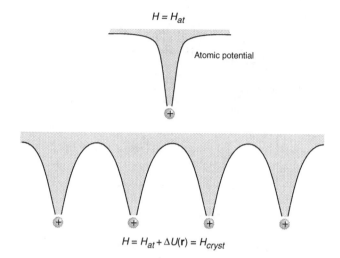

Figure 2.6: The effect of the neighboring atoms in a crystal is to alter the potential an electron experiences from that of an atomic potential (top figure) by an additional potential $\Delta U(\mathbf{r})$.

where $\phi(\mathbf{r})$ are not the atomic functions, but can be constructed out of the atomic functions. Expanding $\phi(\mathbf{r})$ in terms of the atomic eigenfunctions $\psi_n(\mathbf{r})$ we have

$$\phi(\mathbf{r}) = \sum_{n=1}^{N} b_n \psi_n(\mathbf{r}) \tag{2.21}$$

In the tight binding method, we will only include a finite number, N, of orbitals in terms of which $\phi(\mathbf{r})$ is described. The Schrödinger equation now involves the unknown coefficients b_n of Eqn. 2.21 and to solve for them we first derive a set of N–coupled equations, by using the orthonormal properties of the basis set ψ_n. The Schrödinger equation is now

$$H\Psi_{\mathbf{k}} = E(\mathbf{k})\Psi_{\mathbf{k}} \tag{2.22}$$

Using Eqns. 2.20 and 2.21 for the eigenfunction Ψ_k in Eqn. 2.22 and multiplying by $\psi_m^*(\mathbf{r})$ and integrating over space we get:

$$\int d^3r \, \psi_m^*(\mathbf{r}) \left\{ [H_{\text{at}} + \Delta U(\mathbf{r})] \sum_{\mathbf{R},n} b_n \, e^{i\mathbf{k}\cdot\mathbf{R}} \, \psi_n(\mathbf{r} - \mathbf{R}) \right.$$
$$\left. -E(\mathbf{k}) \sum_{\mathbf{R},n} b_n \, e^{i\mathbf{k}\cdot\mathbf{R}} \, \psi_n(\mathbf{r} - \mathbf{R}) \right\} = 0 \tag{2.23}$$

Since the atomic functions are orthogonal, we have

$$\int d^3r \, \psi_m^*(\mathbf{r}) \, \psi_n(\mathbf{r}) = \delta_{mn} \tag{2.24}$$

However, the atomic functions centered at different sites are not orthogonal, i.e.,

$$\int d^3r \, \psi_m^*(\mathbf{r}) \, \psi_n(\mathbf{r} - \mathbf{R}) \neq \delta_{mn} \text{ for } \mathbf{R} \neq 0 \tag{2.25}$$

In the summation over lattice vectors we separate the terms with $\mathbf{R} = 0$ and $\mathbf{R} \neq 0$ to get,

$$
\begin{aligned}
(E(\mathbf{k}) - E_m) b_m \\
= -(E(\mathbf{k}) - E_m) \sum_{n=1}^{N} \left(\sum_{\mathbf{R} \neq 0} \int \psi_m^*(\mathbf{r}) \, \psi_n(\mathbf{r} - \mathbf{R}) \, e^{i\mathbf{k} \cdot \mathbf{R}} \, d^3r \right) b_n \\
+ \sum_{n=1}^{N} \left(\int \psi_m^*(\mathbf{r}) \, \Delta U(\mathbf{r}) \, \psi_n(\mathbf{r}) \, d^3r \right) b_n \\
+ \sum_{n=1}^{N} \left(\sum_{\mathbf{R} \neq 0} \int \psi_m^*(\mathbf{r}) \, \Delta U(\mathbf{r}) \, \psi_n(\mathbf{r} - \mathbf{R}) \, e^{i\mathbf{k} \cdot \mathbf{R}} \, d^3r \right) b_n
\end{aligned}
\tag{2.26}
$$

Note that we have used the equality

$$
\int \psi_m^*(\mathbf{r}) \, H_{\text{at}} \, \psi_n(\mathbf{r}) \, d^3r = E_m \, \delta_{mn}
\tag{2.27}
$$

It may be pointed out that the atomic energies E_m may not correspond to the isolated atomic energies. This is due to the modifications arising from the neighboring atoms. As we will discuss later, the E_m are retained as fitting parameters. In tight binding methods, we also make the following approximation:

$$
\int \psi_m^*(\mathbf{r}) \, \psi_n(\mathbf{r} - \mathbf{R}) \, e^{i\mathbf{k} \cdot \mathbf{R}} \, d^3r \approx 0
\tag{2.28}
$$

This approximation assumes that there is negligible overlap between neighboring atomic functions, i.e, the atomic functions are tightly bound to the atoms.

The quantity

$$
\int \psi_m^*(\mathbf{r}) \, H \, \psi_n(\mathbf{r}) \, d^3r = E_m + \int \psi_m^*(\mathbf{r}) \, \Delta U(\mathbf{r}) \, \psi_m(\mathbf{r}) \, d^3r
\tag{2.29}
$$

is called the on-site matrix element. For most potentials, the on-site integral:

$$
\int \psi_m^*(\mathbf{r}) \, \Delta U(\mathbf{r}) \, \psi_n(\mathbf{r}) \, d^3r \text{ vanishes for } m \neq n.
\tag{2.30}
$$

The secular equation () for the eigenvalues $E(\mathbf{k})$ and eigenfunctions (whose coefficients are the b_m) is an $N \times N$ coupled set of equations whose solution is derived from an $N \times N$ secular equation. For each value of $\mathbf{k}$, there will be N solutions, which will provide the E vs. k relation or the bandstructure. It is illustrative to examine a simple s-band problem to develop an understanding of the method.

2.4.1 Bandstructure Arising From a Single Atomic s-Level

In semiconductors we have two atoms per basis and for each atom we need to include at least the outer shell s, p_x, p_y, p_z functions. Thus the secular equation for tight binding becomes quite difficult to solve. To get some physical insight into the problem we will solve a simple problem of one atom basis with only an s-function. Since there is only one atomic level, the coefficients $\{b_m\}$ are zero except for the s-level where $b_s = 1$. A single equation results in this case. Eqn. 2.26 becomes

$$
\begin{aligned}
E(\mathbf{k}) - E_s \;=\; & -(E(\mathbf{k}) - E_s) \sum_{\mathbf{R} \neq 0} \int \psi_s^*(\mathbf{r}) \, \psi_s(\mathbf{r} - \mathbf{R}) \, e^{i\mathbf{k} \cdot \mathbf{R}} \, d^3r \\
& + \int \psi_s^*(\mathbf{r}) \, \Delta U(\mathbf{r}) \, \psi_s(\mathbf{r}) \, d^3r \\
& + \sum_{\mathbf{R} \neq 0} \int \psi_s^*(\mathbf{r}) \, \Delta U(\mathbf{r}) \, \psi_s(\mathbf{r} - \mathbf{R}) \, e^{i\mathbf{k} \cdot \mathbf{R}} \, d^3r
\end{aligned}
\tag{2.31}
$$

Let us choose the following symbols for the integrals

$$
\begin{aligned}
\alpha(\mathbf{R}) \;&=\; \int \psi_s^*(\mathbf{r}) \, \psi_s(\mathbf{r} - \mathbf{R}) \, d^3r \\
\beta_s \;&=\; -\int \psi_s^*(\mathbf{r}) \, \Delta U(\mathbf{r}) \, \psi_s(\mathbf{r}) \, d^3r \\
\gamma(\mathbf{R}) \;&=\; -\int \psi_s^*(\mathbf{r}) \, \Delta U(\mathbf{r}) \, \psi_s(\mathbf{r} - \mathbf{R}) \, d^3r
\end{aligned}
\tag{2.32}
$$

As discussed before, in the tight binding method we choose $\alpha(\mathbf{R}) = 0$. This gives us

$$
E(\mathbf{k}) = E_s - \beta_s - \sum_{\mathbf{R}} \gamma(\mathbf{R}) \, e^{i\mathbf{k} \cdot \mathbf{R}}
\tag{2.33}
$$

The off-site integrals $\gamma(\mathbf{R})$ drop rapidly as the separation $\mathbf{R}$ increases. We will consider the case where we only have nearest neighbor interaction and all other off-site integrals are zero. Also, note that because of symmetry $\gamma(-\mathbf{R}) = \gamma(\mathbf{R})$. Let us solve the problem for the fcc lattice. The twelve nearest neighbors for an fcc point are at:

$$
\frac{a}{2}(\pm 1, \pm 1, 0); \frac{a}{2}(\pm 1, 0, \pm 1); \frac{a}{2}(0, \pm 1, \pm 1)
\tag{2.34}
$$

The energy equation then becomes

$$
\begin{aligned}
E(\mathbf{k}) \;=\; & E_s - \beta_s - \gamma \left[e^{i(k_x + k_y)a/2} + e^{i(k_x - k_y)a/2} \right. \\
& \left. + e^{i(-k_x + k_y)a/2} + e^{i(k_x - k_y)a/2} + \cdots \right] \\
\;=\; & E_s - \beta_s - 4\gamma \left[\cos\frac{k_x a}{2} \cos\frac{k_y a}{2} \right. \\
& \left. + \cos\frac{k_y a}{2} \cos\frac{k_z a}{2} + \cos\frac{k_z a}{2} \cos\frac{k_x a}{2} \right]
\end{aligned}
\tag{2.35}
$$

We will now examine this band in the first Brillouin zone of the fcc lattice. In Fig. 2.7 the Brillouin zone for the fcc structure is shown along with some of the high symmetry points and directions. We can see from Fig. 2.7 that there are six equivalent X-points and eight equivalent L-points.

If we examine the energy along various symmetry directions, we get (from Eqn. 2.35) the bandstructure:

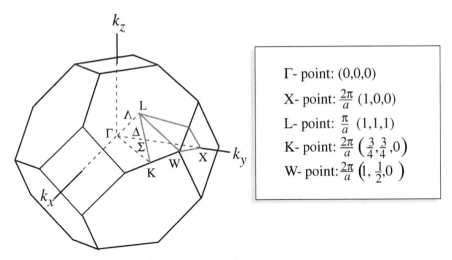

Figure 2.7: Brillouin zone and high symmetry points for the fcc lattice which forms the underlying Bravais lattice for diamond and zinc–blende structures.

Along ΓX $k_x = 2\pi\alpha/a$ $0 \leq \alpha \leq 1$

$k_y = k_z = 0$

$E(\mathbf{k}) = E_s - \beta_s - 4\gamma(1 + 2\cos\pi\alpha)$

Along ΓL $k_x = k_y = k_z = 2\pi\alpha/a$ $0 \leq \alpha \leq 1/2$

$E(\mathbf{k}) = E_s - \beta_s - 12\gamma\cos^2\pi\alpha)$

Along ΓK $k_x = k_y = 2\pi\alpha/a$ $0 \leq \alpha \leq 3/4$

$k_z = 0$

$E(\mathbf{k}) = E_s - \beta_s - 4\gamma(\cos^2\pi\alpha + 2\cos\pi\alpha)$

Bandstructure, i.e., the E vs. k relationship, is typically plotted in a form shown in Fig. 2.8. The result shown is for $\gamma = 1.0$ eV and $E_s + \beta = 0$. We see that the bandwidth of the allowed band is 16 γ. The stronger the overlap integral the wider the width of the allowed band. It is illustrative to examine the bandstructure near the γ point, where ka is small.

Putting $k_x^2 = k_y^2 = k_z^2 = k^2/3$, we have ($ka \ll 1$)

$$E(\mathbf{k}) = E_s - \beta_s - 12\gamma + \gamma k^2 a^2 \tag{2.36}$$

Comparing with the free electron problem solution

$$E(\boldsymbol{p}) = E_0 + \frac{p^2}{2m} \tag{2.37}$$

we write

$$E(\mathbf{k}) = E_s - \beta_s - 12\gamma + \frac{\hbar^2 k^2}{2m^*} \tag{2.38}$$

where we have defined an effective mass given by

$$m^* = \frac{\hbar^2}{2\gamma a^2}$$

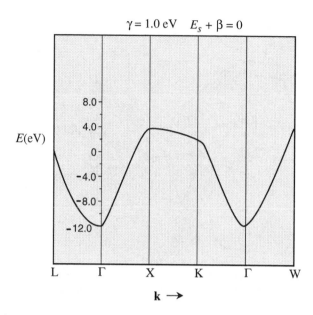

Figure 2.8: Bandstructure of the *s*-band model with parameters chosen as shown.

$$\gamma = \frac{\Delta W}{16} \tag{2.39}$$

If we choose $\gamma = 1.0$ eV and $a = 4$ Å we find that $m^* \approx 0.2m_0$. Thus, in this simple model the effective mass is determined by the parameters γ and a.

Note that the effective mass is positive at the bottom of the band, but has a negative value at the top of the band. It is important to note that a negative effective mass does not violate any physical principle. It simply means that the electron responds *as if* it has a negative mass, i.e., its energy decreases when a force is applied to it.

2.4.2   Bandstructure of Semiconductors

In 1954 Koster and Slater published their paper of the tight binding method for band-structure of materials like Si. This paper (*Physical Review*, volume 94, 1498, 1954) is an excellent reference on TBM. We will now discuss the TBM as applied to semiconductors. As we have discussed earlier, the atomic functions required to describe the outermost electrons in semiconductors are the s, p_x, p_y, and p_z type. It is possible to improve the technique by adding additional atomic levels. Since there are two atoms per basis in a semiconductor, we require eight functions to describe the central cell part of the Bloch functions. We choose a state of the form

$$\Psi(\mathbf{k}, \mathbf{r}) = \sum_{\mathbf{R}_i} \sum_{m=1}^{4} \sum_{j=1}^{2} C_{mj}(\mathbf{k})\, \phi_{mj}(\mathbf{r} - \mathbf{r}_j - \mathbf{R}_i) e^{i\mathbf{k}\cdot\mathbf{R}_i} \tag{2.40}$$

where the sum $\mathbf{R}_i$ is over unit cells, m are the different atomic functions ϕ_{mj} being used in the basis, and j are the atoms in each unit cell.

As described for the s-band case above we now cast the Schrödiner equation in the form of a secular determinant:

$$\left| < \phi_{m'j'} \, |H - E| \, \Psi(\mathbf{k}) > \right| = 0 \tag{2.41}$$

Here, H is the crystal Hamiltonian.

In principle, one can calculate the matrix elements in the secular determinant, Eqn. 2.41, from the first principles, by determining the crystal potential. This can be very difficult, however, because of the complexity of the problem. Slater and Koster were the first to advocate the use of the tight binding method as an empirical technique, instead of an ab initio technique. Measured or accurately calculated levels in the bandstructure can be used to fit the matrix elements. Once the elements are known the entire bandstructure is known.

In TBM using the sp^3 basis for zinc–blende crystals, there will be eight basis functions, an s and three p orbitals, p_x, p_y, and p_z, for each of the two atoms within the Wigner–Seitz cell. This approximation assumes that there is spin degeneracy in the bandstructure of the crystal. If the effects of the breaking of spin degeneracy are being considered, one would need a basis which includes the spin–up and spin–down variants of each orbital. That is, both spin–up and spin–down (i.e., $s \rightarrow s \uparrow$ and $s \downarrow$, etc.) orbitals for each atom in the Wigner–Seitz cell would need to be included in the basis. This results in a requirement of a sixteen function basis for zinc–blende semiconductors.

As discussed in the s-band case, there are a number of on–site and off–site overlap integrals that are retained as fitting parameters. The number of these parameters can reach 20 or more if the nearest and second nearest neighbor interactions are included. To write the matrix elements of the secular equation it is necessary to keep in mind the vectorial nature of the p-functions (i.e., a positive and negative lobe). The reader is urged to examine Koster and Slater's paper (referenced above) to see how each matrix element is set up.

The secular equation for real semiconductors is usually difficult to solve analytically. Matrix solving libraries are available to solve such eigenvalue problems. However, at certain high symmetry points (e.g., Γ-point, X-point, and L-point for the fcc lattice) it is possible to obtain analytical results.

As noted earlier TBM is usually used as an empirical fitting model. Experiments provide results on the relative positions of various high symmetry points. The overlap integrals are then adjusted to fit these measured points. Once a reasonable set of parameters is obtained the entire bandstructure can be plotted out. Several papers in the literature have provided sets of TBM parameters (e.g., see Talwar and Ting (1982).)

In Fig. 2.9 we show a typical TBM result calculated for GaAs. It is found that the bandedges for GaAs occur at the Γ point. The bottom of the conduction band is made from solely s-type states while the top of the valence band is made from only p-type states. Care has been taken to fit the bandgap and other high symmetry points in the Brillouin zone. However, it is known that the bandstructure of GaAs is quite different from this calculation especially at the top of the valence band. According to the tight binding method, the top of the valence band is 3-fold degenerate, corresponding to the degeneracy of (p_x, p_y, p_z). This degeneracy is 6-fold if spin is included. It is

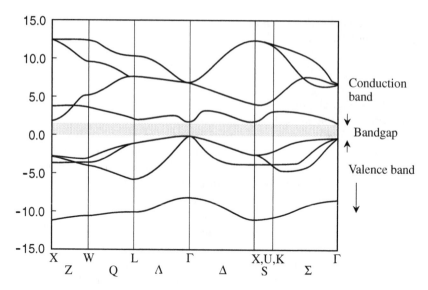

Figure 2.9: Calculated tight-binding bandstructure for GaAs without the effects of spin-orbit coupling. The method is not capable of giving an accurate valence bandedge description without the inclusion of spin-orbit coupling.

well known that in semiconductors, the top of the valence band is 2-fold (4-fold with spin) degenerate with another band (a 2-fold degenerate band) at a small energy below the valence bandedge. These effects can be incorporated only if relativistic effects are included in the problem. These effects are called the spin–orbit coupling.

2.5 SPIN–ORBIT COUPLING

In essentially all semiconductors it is found that the top of the valence band is made from primarily p-type states. As a result unless spin effects are included in bandstructure calculations, the description of the valence band is inaccurate. The spin provides the electron with a means to interact with the magnetic field produced through its orbital motion. An electron in the p-state has an orbital angular momentum of $\hbar$. Thus there is a strong interaction of the spin with the orbital motion of the electron. As noted above, the top of the valence bandedge states are primarily p-type. Thus the spin–orbit coupling has a strong effect there. While it is possible to calculate spin–orbit coupling in isolated atoms, it is difficult to do so in crystals. Thus, a general form of the interaction is assumed with a fitting parameter which is adjusted to experimentally observed effects. In most materials, the spin–orbit interaction is quite small and one adds its effect in a perturbative approach.

If we add the spin–orbit interaction energy to the previously discussed tight binding Hamiltonian, we have

$$H = H_{\mathrm{tb}} + H_{\mathrm{so}} \tag{2.42}$$

The matrix elements arising from the spin–orbit component of the Hamiltonian can

couple states of different spin. To calculate these terms, the spin–orbit interaction is
written as

$$H_{\text{so}} = \lambda \boldsymbol{L} \cdot \boldsymbol{S} \tag{2.43}$$

Here, $\boldsymbol{L}$ represents the operator for orbital angular momentum, $\boldsymbol{S}$ is the operator for
spin angular momentum, and we can treat λ as a constant. The addition of the spin and
orbital angular momentum is the total angular momentum, $\boldsymbol{J}$, which can be expressed
in the following form

$$
\begin{aligned}
\boldsymbol{J}^2 &= (\boldsymbol{L} + \boldsymbol{S})^2 \\
&= \boldsymbol{L}^2 + \boldsymbol{S}^2 + 2\boldsymbol{L} \cdot \boldsymbol{S}
\end{aligned} \tag{2.44}
$$

thus

$$
\begin{aligned}
<\boldsymbol{L} \cdot \boldsymbol{S}> &= \frac{1}{2} <\boldsymbol{J}^2 - \boldsymbol{L}^2 - \boldsymbol{S}^2> \\
&= \frac{\hbar^2}{2} [j(j+1) - l(l+1) - s(s+1)]
\end{aligned} \tag{2.45}
$$

Here, j, l, and s are the quantum numbers for the operators $\boldsymbol{J}$, $\boldsymbol{L}$, and $\boldsymbol{S}$ respectively.
This gives a straightforward technique for evaluating the spin–orbit interaction energy,
but it is only applicable to pure angular momentum states, that is, one needs to know
the total angular momentum of the states to which Eqn. 2.45 is applied. States like $p_x \uparrow$
are mixed states, that is they are made up of a combination of pure states. To determine
the spin–orbit interaction energy, the basis must first be decomposed into states of pure
angular momentum.

In this section we will examine the effect of spin–orbit interaction in some detail.
The reason for such a detailed examination is that later we will see that strain spittings
(e.g., in strained quantum wells) and optical selection rules are intimately tied to spin–
orbit coupling effects. To evaluate the effect of spin–orbit terms we need p_x, p_y, p_z states
in terms of pure angular momentum, $\phi_{1,1}, \phi_{1,-1}, \phi_{1,0}$. The relationship is

$$
\begin{aligned}
p_x &= \frac{1}{\sqrt{2}} (-\phi_{1,1} + \phi_{1,-1}) \\
p_y &= \frac{i}{\sqrt{2}} (\phi_{1,1} + \phi_{1,-1}) \\
p_z &= \phi_{1,0}
\end{aligned} \tag{2.46}
$$

The ϕ_{ij} are eigenfunctions of $\boldsymbol{L}^2$ and L_z with respective quantum numbers $l = i$ and
$l_z = j$ (e.g. $L^2\phi_{1,-1} = \hbar^2(1)(1+1)\phi_{1,-1} = 2\hbar^2\phi_{1,-1}$ and $L_z\phi_{1,-1} = -1\hbar\phi_{1,-1}$).

$$
\begin{aligned}
\phi_{1,\pm 1} &\equiv Y_{1,\pm 1}(\theta,\varphi) = \mp\sqrt{\tfrac{3}{8\pi}} \sin\theta e^{\pm i\varphi} \\
\phi_{1,0} &\equiv Y_{1,0}(\theta,\varphi) = \sqrt{\tfrac{3}{4\pi}} \cos\theta
\end{aligned} \tag{2.47}
$$

By using a natural extension of these equations, similar definitions for the spin–up and
spin–down p-states can be made as exemplified by

$$p_x \uparrow = \frac{1}{\sqrt{2}} (-\phi_{1,1} + \phi_{1,-1}) \uparrow \tag{2.48}$$

This formulation, however, is still in terms of mixed states. To decompose these mixed states into states of pure angular momentum, we must perform the addition of the spin and the orbital angular momentum to obtain the total angular momentum states. Applying the standard Clebsch–Gordan technique for the addition of angular momentum to the ϕ states plus the spin states yields the following six equations:

$$
\begin{aligned}
\Phi_{3/2,3/2} &= \phi_{1,1} \uparrow \\
&= \frac{-1}{\sqrt{2}}(p_x + ip_y) \uparrow \\
\Phi_{3/2,1/2} &= \frac{1}{\sqrt{3}}\phi_{1,1} \downarrow + \frac{\sqrt{2}}{\sqrt{3}}\phi_{1,0} \uparrow \\
&= \frac{-1}{\sqrt{6}}\left[(p_x + ip_y) \downarrow -2p_z \uparrow\right] \\
\Phi_{3/2,-1/2} &= \frac{\sqrt{2}}{\sqrt{3}}\phi_{1,0} \downarrow + \frac{1}{\sqrt{3}}\phi_{1,-1} \uparrow \\
&= \frac{1}{\sqrt{6}}\left[(p_x - ip_y) \uparrow +2p_z \downarrow\right] \\
\Phi_{3/2,-3/2} &= \phi_{1,-1} \downarrow \\
&= \frac{1}{\sqrt{2}}(p_x - ip_y) \downarrow \\
\Phi_{1/2,1/2} &= \frac{-1}{\sqrt{3}}\phi_{1,0} \uparrow + \frac{\sqrt{2}}{\sqrt{3}}\phi_{1,1} \downarrow \\
&= \frac{-1}{\sqrt{3}}\left[(p_x + ip_y) \downarrow +p_z \uparrow\right] \\
\Phi_{1/2,-1/2} &= \frac{-\sqrt{2}}{\sqrt{3}}\phi_{1,-1} \uparrow + \frac{1}{\sqrt{3}}\phi_{1,0} \downarrow \\
&= \frac{-1}{\sqrt{3}}\left[(p_x - ip_y) \uparrow -p_z \downarrow\right] \quad\quad (2.49)
\end{aligned}
$$

These six equations must be inverted to find the states like $\phi_{1,0} \uparrow$ in terms of the total angular momentum states. Once this is done, we can substitute back into the definitions for the basis states, exemplified by Eqn. 2.46, to get them in terms of the total angular momentum states. This procedure results in the following six equations:

$$
\begin{aligned}
p_x \uparrow &= \frac{1}{\sqrt{2}}\left[-\Phi_{3/2,3/2} + \frac{1}{\sqrt{3}}\Phi_{3/2,-1/2} - \frac{\sqrt{2}}{\sqrt{3}}\Phi_{1/2,-1/2}\right] \\
p_x \downarrow &= \frac{1}{\sqrt{2}}\left[-\frac{1}{\sqrt{3}}\Phi_{3/2,1/2} - \frac{\sqrt{2}}{\sqrt{3}}\Phi_{1/2,1/2} + \Phi_{3/2,-3/2}\right] \\
p_y \uparrow &= \frac{i}{\sqrt{2}}\left[\Phi_{3/2,3/2} + \frac{1}{\sqrt{3}}\Phi_{3/2,-1/2} - \frac{\sqrt{2}}{\sqrt{3}}\Phi_{1/2,-1/2}\right]
\end{aligned}
$$

$$p_y \downarrow \;=\; \frac{i}{\sqrt{2}}\left[\frac{1}{\sqrt{3}}\Phi_{3/2,1/2} + \frac{\sqrt{2}}{\sqrt{3}}\Phi_{1/2,1/2} + \Phi_{3/2,-3/2}\right]$$

$$p_z \uparrow \;=\; \frac{\sqrt{2}}{\sqrt{3}}\Phi_{3/2,1/2} - \frac{1}{\sqrt{3}}\Phi_{1/2,1/2}$$

$$p_z \downarrow \;=\; \frac{\sqrt{2}}{\sqrt{3}}\Phi_{3/2,-1/2} + \frac{1}{\sqrt{3}}\Phi_{1/2,-1/2} \tag{2.50}$$

The phases used in the above expressions of Φ_{j,m_j} in terms of $p_x \uparrow,\ldots,p_z \downarrow$, result from the use of the standard phase conventions in the derivation of the Clebsch–Gordan coefficients. However, the overall phase of a state is arbitrary and has no effect on the physical predictions. This admits the possiblity of other phase conventions for the expressions of Φ_{j,m_j} in terms of $p_x \uparrow,\ldots,p_z \downarrow$. One such convention which is in widespread use is that used by Luttinger and Kohn (1955):

$$\Phi_{3/2,3/2}^{\mathrm{LK}} \;=\; -\Phi_{3/2,3/2}^{\mathrm{CG}}$$

$$=\; \frac{1}{\sqrt{2}}(p_x + ip_y)\uparrow$$

$$\Phi_{3/2,1/2}^{\mathrm{LK}} \;=\; -i\Phi_{3/2,1/2}^{\mathrm{CG}}$$

$$=\; \frac{i}{\sqrt{6}}\left[(p_x + ip_y)\downarrow -2p_z\uparrow\right]$$

$$\Phi_{3/2,-1/2}^{\mathrm{LK}} \;=\; \Phi_{3/2,-1/2}^{\mathrm{CG}}$$

$$=\; \frac{1}{\sqrt{6}}\left[(p_x - ip_y)\uparrow +2p_z\downarrow\right]$$

$$\Phi_{3/2,-3/2}^{\mathrm{LK}} \;=\; i\Phi_{3/2,-3/2}^{\mathrm{CG}}$$

$$=\; \frac{i}{\sqrt{2}}(p_x - ip_y)\downarrow$$

$$\Phi_{1/2,1/2}^{\mathrm{LK}} \;=\; -\Phi_{1/2,1/2}^{\mathrm{CG}}$$

$$=\; \frac{1}{\sqrt{3}}\left[(p_x + ip_y)\downarrow +p_z\uparrow\right]$$

$$\Phi_{1/2,-1/2}^{\mathrm{LK}} \;=\; i\Phi_{1/2,-1/2}^{\mathrm{CG}}$$

$$=\; \frac{-i}{\sqrt{3}}\left[(p_x - ip_y)\uparrow -p_z\downarrow\right] \tag{2.51}$$

where the superscript LK identifies the states with the Luttinger–Kohn phase and the superscript CG identifies the states with the standard Clebsch–Gordan phase.

With these decompositions, the evaluation of terms like $\langle p_x \uparrow |H_{\mathrm{so}}|p_y \downarrow\rangle$ becomes straightforward. From Eqns. 2.43 and 2.45, we can write the spin–orbit Hamiltonian as:

$$H_{\mathrm{so}} = \frac{\lambda\hbar^2}{2}\left[j(j+1) - l(l+1) - s(s+1)\right] \tag{2.52}$$

For p-type electron orbitals, $l = 1$ and $s = 1/2$. j is given by the first index of Φ in the decompositions of the states, Eqn. 2.49. Because the pure states are orthogonal,

Semiconductor	Δ (eV)
Si	0.044
Ge	0.29
GaAs	0.35
InAs	0.41
InSb	0.82
InP	0.14
GaP	0.094

Table 2.1: Spin–orbit splitting for different semiconductors.

many of the terms will be zero. Evaluation of all possible terms gives nonzero results only in the following cases:

$$\langle p_x \uparrow |H_{so}|p_y \uparrow \rangle = -i\frac{\Delta}{3}$$

$$\langle p_x \uparrow |H_{so}|p_z \downarrow \rangle = \frac{\Delta}{3}$$

$$\langle p_y \uparrow |H_{so}|p_z \downarrow \rangle = -i\frac{\Delta}{3}$$

$$\langle p_x \downarrow |H_{so}|p_y \downarrow \rangle = i\frac{\Delta}{3}$$

$$\langle p_x \downarrow |H_{so}|p_z \uparrow \rangle = -\frac{\Delta}{3}$$

$$\langle p_y \downarrow |H_{so}|p_z \uparrow \rangle = -i\frac{\Delta}{3} \tag{2.53}$$

as well as the cases which are reflections about the diagonal of these terms which are the conjugates of the values given. The parameter Δ is the spin–orbit splitting $\Delta = \Delta_{so} = 3\lambda\hbar^2/2$. The values of Δ_{so} for several semiconductors is listed in Table 2.1.

If the spin–orbit effects are included in the bandstructure, the top of the valence band loses part of its degeneracy shown earlier in Fig. 2.10.

The effect is easier to demonstrate in the total angular momentum basis instead of the p_x, p_y, p_z basis. In the total angular momentum basis, the p-states (with spin) can be written as the six states (already discussed) $|j, m\rangle$ where j and m take the values: $|3/2, +3/2\rangle$; $|3/2, -3/2\rangle$; $|3/2, +1/2\rangle$; $|3/2, -1/2\rangle$; $|1/2, +1/2\rangle$; $|1/2, -1/2\rangle$. As can be seen from the equations for the spin–orbit perturbation, there is a splitting between the $j = 3/2$ states and the $j = 1/2$ states. This splitting in energy is simply Δ_{so}. The general valence bandedge of semiconductors then has a form shown in Fig. 2.10.

One has a doubly degenerate state (4-fold with spin) at the zone center and a split-off state (2-fold degenerate with spin). The degenerate states at the zone center have different curvatures and are called the light hole (LH) and the heavy hole (HH) state. Fig. 2.11 shows the bandstructure of GaAs with spin–orbit coupling. Contrast this to Fig. 2.9, where the spin–orbit coupling was ignored. Before ending this section

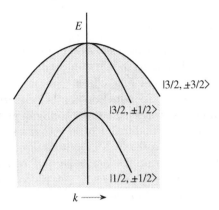

Figure 2.10: The general form of the valence bandstructure after including the effects of spin–orbit coupling. The states are described by pure angular momentum states only at $k = 0$. The splitting between the $j = 3/2$ and $j = 1/2$ states is Δ.

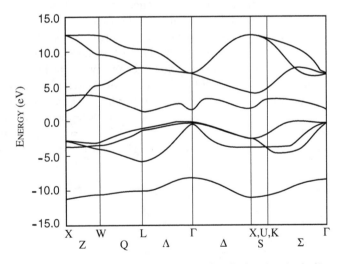

Figure 2.11: Calculated tight–binding bandstructure for GaAs after including spin–orbit coupling. The light-hole heavy-hole split-off band degeneracy is removed.

it is important, once more, to emphasize that spin–orbit coupling plays a key role in understanding strained structures and optical selection rules.

2.5.1   Symmetry of Bandedge States

In this section we will discuss the properties of electron near the conduction and valence bandedges. Bandedges play a dominant role in semiconductor physics and devices, since electrons and holes occupy states near bandedges.

It is useful to distinguish the conduction bandedge states for the direct bandgap materials such as GaAs and InAs, from the indirect bandgap materials such as Si and Ge. In direct gap semiconductors, the conduction band minima states occur at the Γ-point and have a central cell periodic part which is spherically symmetric. It is described as being made up of s-type states at the bandedge. Of course, as one examines the states away from the edge, there is an increasing p-type contribution that is mixed in the eigenfunction. Because of this predominant s-type nature of the wavefunction, very important selection rules for optical transitions arise. This will be discussed in detail later.

For indirect bandgap materials such as Si (with conduction bandedge near the X-point and a 6-fold degeneracy) and Ge (with an edge at the L-point) there is a strong anisotropy of the wavefunction. The anisotropy is described by appropriate combinations of the s, p_x, p_y, and p_z type functions. In Fig. 2.12 we show schematically the nature of the bandedge states.

The character of the valence bandedge states of most semiconductors is quite similar. The central cell part of the wavefunction is primarily p-type. This makes the spin–orbit interaction very important. In absence of this interaction, the top of the valence band is 3-fold degenerate (6-fold if spin degeneracy is included). However, in presence of the spin–orbit coupling, the degeneracy is lifted as shown in Fig. 2.12 leaving a 4-fold degeneracy and a 2-fold degeneracy. The 4-fold degenerate (at the top of the valence band) state consists of two heavy hole (HH) and two light hole (LH) bands while the other 2-fold bands are the split-off bands. Since optical transitions depend critically upon the nature of the hole states, it is useful to describe these states in terms of angular momentum states $\phi_{l,m}$ (l = total angular momentum, m = projection of the angular momentum along the z-axis) and spin state ($\uparrow = +1/2$, $\downarrow = -1/2$). The relationship between the total angular momentum states and the orbital angular momentum states has been discussed earlier and is the following.

Heavy hole states:

$$\Phi_{3/2,3/2} = \frac{-1}{\sqrt{2}} \left(|p_x\rangle + i|p_y\rangle \right) \uparrow$$

$$\Phi_{3/2,-3/2} = \frac{1}{\sqrt{2}} \left(|p_x\rangle - i|p_y\rangle \right) \downarrow \tag{2.54}$$

Light hole states:

$$\Phi_{3/2,1/2} = \frac{-1}{\sqrt{6}} \left[(|p_x\rangle + i|p_y\rangle) \downarrow -2|p_z\rangle \uparrow \right]$$

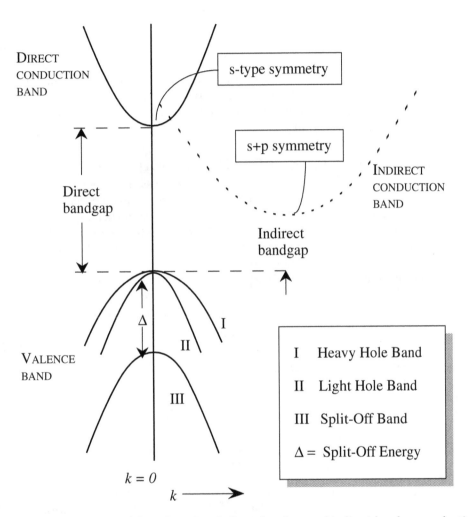

Figure 2.12: Schematic of the valence band, direct bandgap and indirect bandgap conduction bands. The conduction band of the direct gap semiconductor is shown in the solid line while the conduction band of the indirect semiconductor is shown in the dashed line.

$$\Phi_{3/2,-1/2} \quad = \quad \frac{1}{\sqrt{6}} \left[(|p_x\rangle - i|p_y\rangle) \uparrow + 2|p_z\rangle \downarrow \right] \tag{2.55}$$

Split–off hole states:

$$\Phi_{1/2,1/2} \quad = \quad \frac{-1}{\sqrt{3}} \left[(|p_x\rangle + i|p_y\rangle) \downarrow + |p_z\rangle \uparrow \right]$$

$$\Phi_{1/2,-1/2} \quad = \quad \frac{-1}{\sqrt{3}} \left[(|p_x\rangle - i|p_y\rangle) \uparrow + |p_z\rangle \downarrow \right] \tag{2.56}$$

This description of the hole states in terms of the total angular momentum states is extremely useful in calculating the optical properties of semiconductors. It is also useful to remember that the pure description of the hole states is strictly valid only at $\mathbf{k} = 0$ (center of the Brillouin zone) since at $\mathbf{k} \neq 0$ the HH and LH states mix strongly.

2.6   ORTHOGONALIZED PLANE WAVE METHOD

We know from quantum mechanics that we can solve the Schrödinger equation by expanding the eigenfunction in terms of a complete basis function and developing a matrix eigenvalue equation. In the tight binding method we have used the atomic functions as a basis set to describe the bandstructure. It is also possible to use a plane wave basis to do so. The plane wave basis is an attractive basis but has difficulty because too many plane waves are needed to describe the problem adequately. To express the Schrödinger equation in plane wave basis we need to use the reciprocal lattices vectors to expand the periodic potential.

The reciprocal lattice vectors $\mathbf{G}$ have the property that if $\mathbf{R}$ represent the general lattice vector then

$$e^{i\mathbf{G}\cdot\mathbf{R}} = 1 \tag{2.57}$$

and the potential is periodic in $\mathbf{R}$, i.e.,

$$U(r + \mathbf{R}) = U(r)$$

The periodic potential can, in general, be written as

$$V(\mathbf{r}) = \sum_{\mathbf{G}} V_{\mathbf{G}} e^{i\mathbf{G}\cdot\mathbf{r}} \tag{2.58}$$

If the potential is short-ranged, a large number of reciprocal lattice vectors are required to describe it. This makes the size of the secular equation to be solved correspondingly large. The orthogonalized plane wave (OPW) method is an approach to avoid having to deal with a very large number of plane wave states. The basic idea is that the valence and conduction band states are orthogonal to the core states of the crystal and this fact should be utilized in the selection of the plane waves. This simple imposition greatly simplifies the problem.

The general form of the Bloch function in k-space is

$$\Psi(\mathbf{k},\mathbf{r}) = \sum_m C_m(\mathbf{k}) \, e^{i(\mathbf{k}+\mathbf{G}_m)\cdot\mathbf{r}} \tag{2.59}$$

In general one should include a normalization factor $1/\sqrt{Nr_0}$ where N is the number of unit cells, and r_0 is the volume of the unit cell. The secular equation to be solved has the form

$$\left\| \langle \phi_{\mathbf{k},m}|H - E|\phi_{\mathbf{k},n}\rangle \right\| = 0 \tag{2.60}$$

where $\phi_{\mathbf{k},n}$, $\phi_{\mathbf{k},m}$ are the plane waves

$$\phi_{\mathbf{k},m} = \frac{1}{\sqrt{Nr_0}} e^{i(\mathbf{k}+\mathbf{G}_m)\cdot\mathbf{r}} \tag{2.61}$$

Instead of working with the general plane wave basis set, one uses the fact that the core states are known and the conduction and valence band states are orthogonal to them. To incorporate this property, one defines the orthogonal states

$$\chi_{\mathbf{k},m} = \phi_{\mathbf{k},m} - \sum_c \left\langle \psi_c|\phi_{\mathbf{k},m} \right\rangle \psi_c$$

where ψ_c are the core states. The Bloch functions are now expanded in the orthogonalized states χ and one gets a secular equation

$$\left\| < \chi_{\mathbf{k},m} |H - E| \chi_{\mathbf{k},n} > \right\| = 0 \tag{2.62}$$

This matrix equation only gives the eigenvalues corresponding to the valence and conduction bands, since the core states have been eliminated from the basis being used. This allows for easier convergence of the problem. To be able to solve the problem one needs three ingredients:

1. the form of the background atomic potentials which is then written as a Fourier series;

2. the form of the core states so that the orthogonal states can be determined. Usually the isolated atomic core states are used since the presence of other neighboring atoms is not expected to alter the core states;

3. the crystal structure which is, of course, known.

Once all this information is there, the solution of the problem is straightforward. Of course, detailed knowledge of the atomic potential in the crystal is difficult to obtain and one may have to resort to consulting the experimental results for high symmetry points and develop an approach similar to the tight-binding method where the matrix elements are adjusted to obtain good fits with some known points in the bandstructure.

2.7 PSEUDOPOTENTIAL METHOD

The pseudopotential is a powerful technique to solve for bandstructures of semiconductors and is often used as a benchmark for comparison of other techniques. We will describe the spirit of the method without going into its details. Like the orthogonal plane wave method, the pseudopotential method makes use of the information that the valence and conduction band states are orthogonal to the core states. However, this information is not just used in the construction of the Bloch states, but is included in an ingenious manner in the Hamiltonian itself. Thus the background periodic potential is replaced by a new "pseudopotential," which is obtained by subtracting out the effects of the core levels. The pseudopotential then has a smooth spatial dependence and yet includes all the relevant information to give the valence and conduction band bandstructure. The following formal equations define the procedure. The Schrödinger equation for the valence or conduction band states is

$$\left[\frac{p^2}{2m} + V(\mathbf{r}) \right] \Psi_v(\mathbf{k}, \mathbf{r}) = E_v \Psi_v(\mathbf{k}, \mathbf{r}) \tag{2.63}$$

with the orthogonality condition

$$\langle \Psi_c | \Psi_v \rangle = 0$$

where Ψ_c are the core states. This condition is explicitly incorporated into the definition of the valence band states by defining new states $\phi_v(\mathbf{k}, \mathbf{r})$ where

$$\Psi_v(\mathbf{k}, \mathbf{r}) = \phi_v(\mathbf{k}, \mathbf{r}) - \sum_c \langle \Psi_c | \phi_v \rangle \, \Psi_c \tag{2.64}$$

The equation for $\phi_v(\mathbf{k}, \mathbf{r})$ then becomes

$$\left[\frac{p^2}{2m} + V(\mathbf{r}) \right] \phi_v(\mathbf{k}, \mathbf{r}) - \sum_c [E_v(\mathbf{k}) - E_c] \langle \Psi_c | \phi_v \rangle \, \Psi_c = E_v(\mathbf{k}) \, \phi_v(\mathbf{k}, \mathbf{r})$$

where E_c are the known core level energies. The Schrödinger equation for $\phi_v(\mathbf{k}, \mathbf{r})$ has the same eigenvalues as the original equation for $\Psi_v(\mathbf{k}, \mathbf{r})$ together with the orthogonality condition, but with a new background potential. The original potential $V(\mathbf{r})$ is replaced by the operator

$$V(\mathbf{r})\Psi_v(\mathbf{k}, \mathbf{r}) \Rightarrow \left[V(\mathbf{r}) \, \phi_v(\mathbf{k}, \mathbf{r}) + \sum_c [E_v(\mathbf{k}) - E_c] \langle \Psi_c | \phi_v \rangle \, \Psi_c \right] = V_p \tag{2.65}$$

The new potential operator which involves subtraction of the core energies weighted with $\phi_v(\mathbf{k}, \mathbf{r})$ from the eigenvalues $E_v(\mathbf{k})$ is called the pseudopotential. Of course, there is no simplification as yet since the new equation is as difficult to solve as the starting equation. The pseudopotential V_p is a nonlocal eigenvalue dependent operator.

The problem is simplified due to the realization that the pseudopotential is much smoother as shown schematically in Fig. 2.13 than the original starting potential since the term $-E_c \langle \Psi_c | \phi_v \rangle$ is a position dependent term which subtracts the strong core effects near the atomic sites leaving the potential in the regions between atoms unchanged.

The pseudopotential is thus equivalent to a constant potential plus a weak background potential as far as the valence and conduction band states are concerned. The solution can then be expanded in terms of plane waves. Usually a few dozen plane wave states are found to be adequate for convergence.

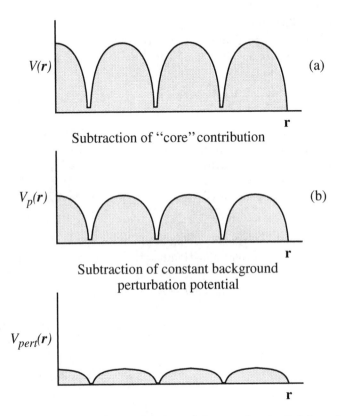

$V(r)$ (a)

r

Subtraction of "core" contribution

$V_p(r)$ (b)

r

Subtraction of constant background
perturbation potential

$V_{pert}(r)$

r

Figure 2.13: Schematic steps in the application of the pseudopotential formalism. The true potential $V(\mathbf{r})$ along with the orthogonality condition (a) is rewritten in terms of a pseudopotential (b) which is much smoother. The problem can then be solved perturbatively by separating the $[p^2/(2m) + V(0)]$ term in the Hamiltonian.

2.8 k · p METHOD

The **k · p** model has become one of the most widely used bandstructure models for describing not only 3–dimensional semiconductors, but lower dimensional systems such as a quantum wells, wires, and dots. It is quite accurate near the bandedges. In the **k · p** method, one starts with the known form of the bandstructure problem at the bandedges and using perturbation theory attempts to describe the bands away from the high symmetry points. Since for the central cell functions, we only expand around the high symmetry points in terms of known functions, the problem is considerably simplified, often leading to analytical results.

Let us consider a semiconductor with a bandedge at $\mathbf{k}_0$. We assume that the eigenvalues and Bloch functions are known for the bandedge; i.e., the equation

$$\left[\frac{p^2}{2m_0} + V(\mathbf{r}) \right] \psi_n(\mathbf{k}_0, \mathbf{r}) = E_n(\mathbf{k}_0)\, \psi_n(\mathbf{k}_0, \mathbf{r}) \tag{2.66}$$

is known. In most applications $\mathbf{k}_0$ is the Γ-point ($= [000]$) in the Brillouin zone. We can expand the general solutions away from the known $\mathbf{k} = \mathbf{k}_0$ solutions in the basis set $\exp[i(\mathbf{k} - \mathbf{k}_0) \cdot \mathbf{r}]$. Thus we may write

$$\psi(\mathbf{k}, \mathbf{r}) = \sum_n b_n(\mathbf{k})\, \psi_n(\mathbf{k}_0, \mathbf{r})\, e^{i(\mathbf{k} - \mathbf{k}_0) \cdot \mathbf{r}} \tag{2.67}$$

where b_n are the expansion coefficients, that are to be determined. This general approach is shown in Fig. 2.14. The secular equation has the usual form, formally represented by

$$\left\| \langle e^{i(\mathbf{k} - \mathbf{k}_0) \cdot \mathbf{r}} \psi_{n'}(\mathbf{k}_0, \mathbf{r}) | H - E | e^{i(\mathbf{k} - \mathbf{k}_0) \cdot \mathbf{r}} \psi_n(\mathbf{k}_0, \mathbf{r}) \rangle \right\| = 0 \tag{2.68}$$

A simple expansion allows us to rewrite this equation for just the central cell part of the Bloch states. Remembering that $\boldsymbol{p} = -i\hbar\nabla$ and

$$\nabla \left(e^{i(\mathbf{k} - \mathbf{k}_0) \cdot \mathbf{r}} \psi_n \right) = e^{i(\mathbf{k} - \mathbf{k}_0) \cdot \mathbf{r}} \left(\nabla + i(\mathbf{k} - \mathbf{k}_0) \right) \psi_n$$

$$\left[\frac{p^2}{2m_0} + V(\mathbf{r}) \right] e^{i(\mathbf{k} - \mathbf{k}_0) \cdot \mathbf{r}} \psi_n(\mathbf{k}_0, \mathbf{r})$$

$$= e^{i(\mathbf{k} - \mathbf{k}_0) \cdot \mathbf{r}} \left[\frac{\{\boldsymbol{p} + \hbar(\mathbf{k} - \mathbf{k}_0)\}^2}{2m_0} + V(\mathbf{r}) \right] \psi_n(\mathbf{k}_0, \mathbf{r})$$

$$= e^{i(\mathbf{k} - \mathbf{k}_0) \cdot \mathbf{r}} \left[\frac{\hbar^2}{2m_0}(\mathbf{k} - \mathbf{k}_0)^2 + \frac{\hbar}{m_0}(\mathbf{k} - \mathbf{k}_0) \cdot \boldsymbol{p} + E_n(\mathbf{k}_0) \right] \psi_n(\mathbf{k}_0, \mathbf{r}) \tag{2.69}$$

The eigenvalue determinant then becomes

$$\left\| \left\langle \left[\frac{\hbar^2}{2m_0}(\mathbf{k} - \mathbf{k}_0)^2 + E_n(\mathbf{k}_0) - E \right] \delta_{n'n} + \frac{\hbar}{m_0}(\mathbf{k} - \mathbf{k}_0) \cdot \boldsymbol{P}_{n'n}(\mathbf{k}_0) \right\rangle \right\| = 0 \tag{2.70}$$

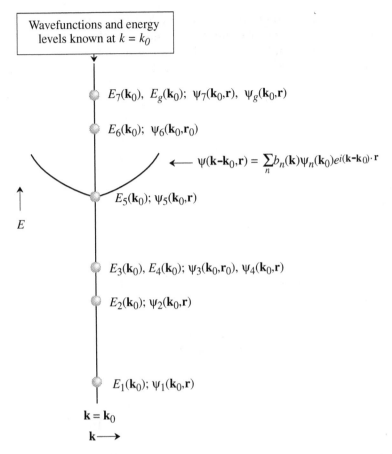

Figure 2.14: A schematic of the basis of the **k · p** method. The states away from $\mathbf{k}_0$ are expanded in terms of the known $\mathbf{k} = \mathbf{k}_0$ states.

where $\boldsymbol{P}_{n'n}$ is the momentum matrix element between the different bandedge states

$$\boldsymbol{P}_{n'n} = \int \psi_{n'}^*(\mathbf{k}_0, \mathbf{r}) \, \boldsymbol{p} \, \psi_n(\mathbf{k}_0, \mathbf{r}) \, d^3r \qquad (2.71)$$

If the momentum matrix elements are known, the eigenvalue problem can be easily solved for $E(\mathbf{k}_0)$ and $b_n(\mathbf{k})$. The integral in Eqn. 2.71 is nonzero only for certain symmetries of $\psi_{n'}(\mathbf{k}_0, \mathbf{r})$ and $\psi_n(\mathbf{k}_0, \mathbf{r})$. It is this reduction in the number of independent parameters, together with the fact that near a particular energy $E_n(\mathbf{k}_0)$, only bands which have the energy difference $E_{n'}(\mathbf{k}_0) - E_n(\mathbf{k}_0)$ small (as we see later in Eqns. 2.75 and 2.76), will contribute significantly, that makes the **k·p** method so attractive. Usually the matrix elements $\boldsymbol{P}_{n'n}$ are used as fitting parameters and adjusted to fit measured quantities such as carrier masses, etc.

Consider, for example, the **k · p** description of the nondegenerate bands (e.g.,

conduction bandedge or the split-off band in the valence band for the case of large spin–
orbit coupling). In this case one can use the perturbation theory to obtain the energy
and wavefunctions away from $\mathbf{k}_0$. For simplicity, let us assume $\mathbf{k}_0 = 0$. The Schrödinger
equation for the perturbation Hamiltonian is simply

$$(H_0 + H_1 + H_2)\, u_{n\mathbf{k}} = E_{n\mathbf{k}}\, u_{n\mathbf{k}} \tag{2.72}$$

where

$$H_0 = \frac{p^2}{2m_0} + V(r)$$

$$H_1 = \frac{\hbar}{m}\mathbf{k}\cdot\mathbf{p}$$

$$H_2 = \frac{\hbar^2 k^2}{2m_0}$$

In the perturbation approach, H_1 is a first order term in $\mathbf{k}$, and H_2 is a second
order term. The $u_{n\mathbf{k}}$ are the central cell part of the Bloch functions $\psi_n(\mathbf{k})$. To zero
order, we then have

$$u_{n\mathbf{k}} = u_{n0}$$

$$E_{n\mathbf{k}} = E_n(0) \tag{2.73}$$

To first order we have

$$u_{n\mathbf{k}} = u_{n0} + \frac{\hbar}{m_0}\sum_{n'\neq n}\mathbf{k}\cdot\frac{\langle n0|\mathbf{p}|n0\rangle}{E_n(0) - E_{n'}(0)}u_{n0}$$

$$E_{n\mathbf{k}} = E_n(0) + \frac{\hbar}{m_0}\mathbf{k}\cdot\langle n0|\mathbf{p}|n0\rangle \tag{2.74}$$

If the states $|n0\rangle$ or u_{n0} have inversion symmetry, the first order matrix element
is zero because $\mathbf{p}$ has an odd parity. This would occur in crystals with inversion symmetry
(e.g., Si, Ge, and C). For crystals lacking inversion symmetry, the functions $|n0 >$ may
not have a well-defined parity leading a small correction to energy proportional to $\mathbf{k}$.
This correction is usually very small and leads to a small band warping, i.e., the energy
is not an extrema at the high symmetry point.

To second order, the energy becomes

$$E_n(\mathbf{k}) = E_n(0) + \frac{\hbar^2 k^2}{2m_0} + \frac{\hbar^2}{m_0^2}\sum_{n'\neq n}\frac{|\mathbf{k}\cdot\langle n'0|\mathbf{p}|n0\rangle|^2}{E_n(0) - E_{n'}(0)} \tag{2.75}$$

This equation can be expressed in terms of an effective mass m^*

$$E_n(\mathbf{k}) = E_n(0) + \sum_{i,j}\frac{\hbar^2}{m_{i,j}^*}k_i\cdot k_j$$

where

$$\frac{m_0}{m^*_{i,j}} = \delta_{i,j} + \frac{2}{m_0} \sum_{n' \neq n} \frac{\langle n0|p_i|n'0\rangle \langle n'0|p_j|n0\rangle}{E_n(0) - E_{n'}(0)} \tag{2.76}$$

This equation is valid for the conduction bandedge and the split–off bands. For the conduction band, retaining only the valence bandedge bands in the summation, we get

$$E_c(\mathbf{k}) = E_c(0) + \frac{\hbar^2 k^2}{2m^*_c} \tag{2.77}$$

with

$$\frac{1}{m^*_c} = \frac{1}{m_0} + \frac{2p^2_{cv}}{m^2_0}\frac{1}{3}\left(\frac{2}{E_{g\Gamma}} + \frac{1}{E_{g\Gamma} + \Delta}\right)$$

where $E_{g\Gamma}$ is the gap at the zone center and Δ is the HH–SO band separation. The momentum matrix element is $\langle \sigma|P_x|p_x\rangle$ (σ is the s–function at the conduction bandedge and p_x is the p–function at the valence bandedge). For the SO band we have

$$\begin{aligned} E_{so} &= -\Delta - \frac{\hbar^2 k^2}{2m^*_{so}} \\ \frac{1}{m^*_{so}} &= \frac{-1}{m_0} + \frac{2p^2_{cv}}{3m^2(E_{g\Gamma} + \Delta)} \end{aligned} \tag{2.78}$$

Notice that in this simple model the SO band mass we have no contribution from the HH, LH bands in this simple treatment since the momentum matrix elements are zero, by symmetry.

According to Eqn. 2.76 we see that for the conduction band (for direct gap semiconductors) as the bandgap decreases, the carrier effective mass also decreases. This result holds quite well as can be seen by examining the effective masses of a range of semiconductors as shown in Fig. 2.15.

We have seen that at the top of the valence band we have the HH, LH degeneracy. As we move away from $k = 0$ there is a strong interaction between these states which causes a splitting between the two bands. The split–off band also can have an effect on the valence band states, since it is close to these HH, LH bands. If one ignores the effects of the conduction band in the determinant equation, one then gets a 6×6 eigenvalue secular equation. Symmetry considerations are then used to select the form and nonzero matrix elements of this matrix equation. A treatment which has proven very valuable is due to Kohn and Luttinger. The matrix equation, with the Luttinger

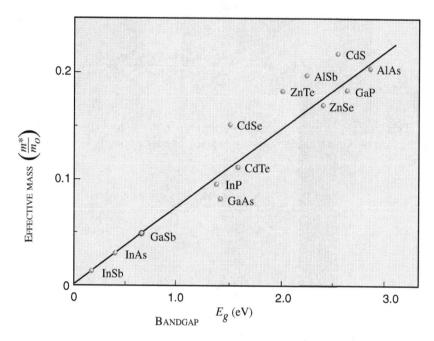

Figure 2.15: Electron effective mass, m^* as a function of the lowest–direct gap E_g for various III–V compounds.

and Kohn phases, is of the form:

j:	$3/2$	$3/2$	$3/2$	$3/2$	$1/2$	$1/2$
m_j:	$3/2$	$1/2$	$-1/2$	$-3/2$	$1/2$	$-1/2$

$$H = - \begin{bmatrix} H_{hh} & b & c & 0 & ib/\sqrt{2} & -i\sqrt{2}c \\ b^* & H_{lh} & 0 & c & -iq & i\sqrt{3}b/\sqrt{2} \\ c^* & 0 & H_{lh} & -b & -i\sqrt{3}b^*/\sqrt{2} & -iq \\ 0 & c^* & -b^* & H_{hh} & -i\sqrt{2}c^* & -ib^*/\sqrt{2} \\ -ib^*/\sqrt{2} & iq & i\sqrt{3}b/\sqrt{2} & i\sqrt{2}c & H_{so} & 0 \\ i\sqrt{2}c^* & -i\sqrt{3}b^*/\sqrt{2} & iq & ib/\sqrt{2} & 0 & H_{so} \end{bmatrix} \quad (2.79)$$

The elements in the Hamiltonian are given by

$$H_{hh} = \frac{\hbar^2}{2m_0} \left[(\gamma_1 + \gamma_2) \left(k_x^2 + k_y^2 \right) + (\gamma_1 - 2\gamma_2) k_z^2 \right]$$

$$H_{lh} = \frac{\hbar^2}{2m_0} \left[(\gamma_1 - \gamma_2) \left(k_x^2 + k_y^2 \right) + (\gamma_1 + 2\gamma_2) k_z^2 \right]$$

$$H_{so} = (H_{hh} + H_{lh})/2 + \Delta_0$$

$$b = \frac{-\sqrt{3}i\hbar^2}{m_0} \gamma_3 \left(k_x - ik_y \right) k_z$$

$$c = \frac{\sqrt{3}\hbar^2}{2m_0}\left[\gamma_2\left(k_x^2 - k_y^2\right) - 2i\gamma_3 k_x k_y\right]$$

$$q = (H_{hh} - H_{lh})/\sqrt{2} \tag{2.80}$$

In most semiconductors with large spin–orbit coupling, the 6×6 matrix equation can be separated into a 4×4 and a 2×2 equation. The appropriate equation defining the HH and LH states is then

$$-\begin{bmatrix} H_{hh} & b & c & 0 \\ b^* & H_{lh} & 0 & c \\ c^* & 0 & H_{lh} & -b \\ 0 & c^* & -b^* & H_{hh} \end{bmatrix}\begin{bmatrix} \Phi_{(3/2,3/2)} \\ \Phi_{(3/2,1/2)} \\ \Phi_{(3/2,-1/2)} \\ \Phi_{(3/2,-3/2)} \end{bmatrix} = E\begin{bmatrix} \Phi_{(3/2,3/2)} \\ \Phi_{(3/2,1/2)} \\ \Phi_{(3/2,-1/2)} \\ \Phi_{(3/2,-3/2)} \end{bmatrix} \tag{2.81}$$

where the coefficients H_{hh}, H_{lh}, c, b are the same as above and γ_1, γ_2 and γ_3 are the Kohn–Luttinger parameters. These can be obtained by fitting experimentally obtained hole masses. It may be noted that the 4×4 matrix equation can be reduced to two 2×2 matrix equations (Broido and Sham, 1985), and the E vs. k relations can be solved analytically. The 2×2 matrix is

$$\begin{vmatrix} H_{hh} & |b| - i|c| \\ (|b| - i|c|)^* & H_{lh} \end{vmatrix}\psi = E\psi \tag{2.82}$$

where ψ is a 1×2 vector.

Let us examine the momentum matrix element p_{cv}, in terms of which, the conduction band masses are expressed in the **k·p** formalism. We note that the conduction bandedge state for direct gap semiconductors has an s-type symmetry, and is denoted by $|+\sigma\rangle$ (spin up) and $|-\sigma\rangle$ (spin down).

The matrix elements of interest are of the form $\langle r|P|\sigma\rangle$; $r = x, y, z$. From parity considerations $\langle x|P_x|\sigma\rangle = \langle y|P_y|\sigma\rangle = \langle z|P_z|\sigma\rangle$ are the only nonzero matrix elements while the rest of the elements are zero.

The nonvanishing matrix elements are

$$\langle \pm 3/2|P_x| \pm \sigma\rangle = \frac{1}{\sqrt{2}}\langle x|P_x|\sigma\rangle$$

$$\langle \pm 1/2|P_x| \mp \sigma\rangle = \frac{1}{\sqrt{6}}\langle x|P_x|\sigma\rangle$$

$$\langle \pm 3/2|P_z| \pm \sigma\rangle = \frac{2}{\sqrt{6}}\langle x|P_x|\sigma\rangle \tag{2.83}$$

We define a quantity

$$E_p = \frac{2}{m_0}|\langle x|P_x|\sigma\rangle|^2$$

$$= \frac{2}{m_0}p_{cv}^2 \tag{2.84}$$

The following are the values of E_p for various semiconductors in (eV) (Lawaetz, 1971).

Devices	Region of Bandstructure of Importance Measured from Bandedge
• Transistors (Source Gate) Low Field Region	$\sim 2k_BT$ (~ 50 meV at $T = 300$ K)
• Transistors High Field Region (Gate-Drain Region)	~ 0.5 eV
• Power Transistors Avalanche Detectors and Other Devices Involving Breakdown	$\sim E_{\mathrm{gap}} \sim 1.0$ eV
• Lasers	$\sim 2 - 3k_BT$ ($\sim 50 - 75$ meV at $T = 300$ K)
• Detectors	Variable, depending upon photon wavelength

Table 2.2: Various electronic and optoelectronic devices and the range of bandstructure that is important for their performance.

$$
\begin{array}{ll}
\text{GaAs} & 25.7 \\
\text{InP} & 20.4 \\
\text{InAs} & 22.2 \\
\text{CdTe} & 20.7
\end{array}
$$

It is interesting to note that the matrix element is nearly the same for all semiconductors.

2.9   SELECTED BANDSTRUCTURES

We will now examine special features of some semiconductors. Of particular interest are the bandedge properties since they dominate the transport and optical properties. In this context, it is important to appreciate the range of energies away from the band-edges which control various physical properties of devices. These energies are shown in Table 2.2 for various kinds of electronic and optical devices. As can be seen, the region of interest varies depending upon the kind of devices one is interested in. Bandedge properties are often captured by density of states (number of allowed states per unit volume per energy interval). This concept is reviewed in Appendix C.

Silicon

Silicon forms the backbone of modern electronics industry. The bandstructure of silicon is shown in Fig. 2.16 and, as can be seen, it has an indirect bandgap. This fact greatly limits the applications of Si in optical devices, particularly for light emitting devices. The bottom of the conduction band in Si is at point ($\sim (2\pi/a)(0.85, 0.0)$ i.e., close

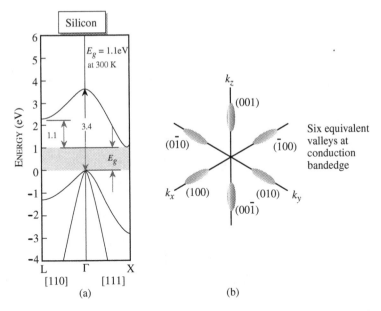

Figure 2.16: (a) Bandstructure of Si. (b) Constant energy ellipsoids for Si conduction band. There are six equivalent valley in Si at the bandedge.

to the X-point. There are six degenerate X-points and consequently six conduction bandedge valleys. The central cell part of the Bloch functions near the bandedge is a strong mixture of s and p_x functions along the x-axis (the longitudinal axis) and p_y and p_z along the transverse direction from the bandedge. The near bandedge bandstructure can be represented by ellipsoids of energy with simple E vs. k relations of the form (for examples for the [100] valley)

$$E(\mathbf{k}) = \frac{\hbar^2 k_x^2}{2m_l^*} + \frac{\hbar^2 \left(k_y^2 + k_z^2\right)}{2m_t^*} \tag{2.85}$$

where we have two masses, the longitudinal and transverse. The constant energy surfaces of Si are ellipsoids according to Eqn. 2.85. The six surfaces are shown in Fig. 2.16.

The longitudinal electron mass m_l^* is approximately 0.98 m_0, while the transverse mass is approximately 0.19 m_0.

The next valley in the conduction band is the L-point valley, which is about 1.1 eV above the bandedge. Above this is the Γ-point edge. The direct bandgap of Si is ~ 3.4 eV. This direct gap is quite important for optical transitions since, as we shall see later, the absorption coefficient for photons above this energy is very strong. It is important to note that due to the 6-fold degeneracy of the conduction bandedge, the electron transport in Si is quite poor. This is because of the very large density of states near the bandedge leading to a high scattering rate.

The top of the valence band has the typical features seen in all semiconductor valence bands. One has the HH, LH degeneracy at the zone edge. The split-off (SO)

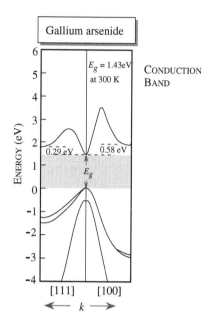

Figure 2.17: Bandstructure of GaAs. The bandgap at 0 K is 1.51 eV and at 300 K it is 1.43 eV. The bottom of the conduction band is at $k = (0,0,0)$, i.e., the Γ-point. The upper conduction band valleys are at the L-point.

band is also very close for Si since the split-off energy is only 44 meV. This is one of the smallest split off energies of any semiconductors.

GaAs

The near bandedge bandstructure of GaAs is shown in Fig. 2.17. The bandgap is direct, which is the chief attraction of GaAs. The direct bandgap ensures excellent optical properties of GaAs as well as superior electron transport in the conduction band. The bandstructure can be represented by the relation

$$E = \frac{\hbar^2 k^2}{2m^*} \tag{2.86}$$

with $m^* = 0.067m_0$. A better relationship is the nonparabolic approximation

$$E(1 + \alpha E) = \frac{\hbar^2 k^2}{2m^*} \tag{2.87}$$

with $\alpha = 0.67$ eV^{-1}.

For high electric field transport, it is important to note that the valleys above Γ point are the L-valleys. There are eight L-points, but since half of them are connected by a reciprocal lattice vector, there are four valleys. The separation $\Delta E_{\Gamma L}$ between the Γ and L minima is 0.29 eV. The L valley has a much larger effective mass than the Γ valley. For GaAs, $m_L^* \sim 0.25m_0$. This difference in masses is extremely important

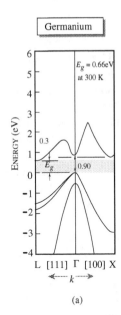
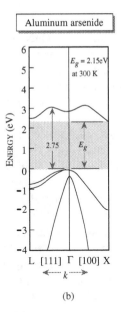

Figure 2.18: (a) Bandstructure of Ge. Like Si, Ge is an indirect semiconductor. The bottom of the conduction band occurs at the L-point. The hole properties of Ge are the best of any semiconductor with extremely low hole masses. (b) Bandstructure of AlAs. AlAs is an important III-V semiconductor because of its excellent lattice constant, matching GaAs. The material has indirect bandgap and is usually used in AlGaAs alloy for barrier materials in GaAs/AlGaAs heterostructures.

for high electric field transport and leads to negative differential resistance. Above the L-point in energy is the X-valley with $\Delta E_{\Gamma L} \sim 0.58$ eV. The mass of the electron in the X-valley is also quite large ($m_X^* \sim 0.6 m_0$). At high electric fields, electrons populate both the L and X valleys in addition to the Γ-valley, making these regions of bandstructure quite important.

The valence band of GaAs has the standard HH, LH, and SO bands. Due to the large spin–orbit splitting, for most purposes, the SO band does not play any role in electronic or optoelectronic properties.

The Kohn-Luttinger parameters for GaAs are

$$\gamma_1 = 6.85$$
$$\gamma_2 = 2.1$$
$$\gamma_3 = 2.9$$

leading to the density of states hole masses of $m_{HH}^* = 0.45 m_0$; $m_{LH}^* = 0.08 m_0$. As discussed in the case of Si, the effective masses and E vs. k relation for holes is highly anisotropic.

The bandstructures of Ge and AlAs, two other important semiconductors are shown in Fig. 2.18, along with brief comments about their important properties.

InN, GaN, and AlN

The III–V nitride family of GaN, InN, and AlN have become quite important due to progress in the ability to grow the semiconductor. The nitrides and their combinations, which have a wurtzite structure can provide bandgaps ranging from ~1.0 eV to over 6.0 eV. This large range is very useful for short wavelength light emitters (for blue light emission and for high resolution reading/writing applications in optoelectronics) and high power electronics. In Fig. 2.19 we show the bandstructure of InN, GaN, and AlN near the bandedges. Also shown is the Brillouin zone and the notations used for the high symmetry points.

It should be noted that it is difficult to obtain the bandgap of InN, since it is difficult to grow thick defect free layers due to substrate non-availability. It was thought (prior to ~2001) that the bandgap of InN was 1.9 eV. However, recent results have shown a bandgap closer to 0.9 eV.

Also important to note is that the bandgap of semiconductors generally decreases as temperature increases. The bandgap of GaAs, for example, is 1.51 eV at $T = 0K$ and 1.43 eV at room temperature. These changes have very important consequences for both electronic and optoelectronic devices. The temperature variation alters the laser frequency in solid state lasers, and alters the response of modulators and detectors. It also has effects on intrinsic carrier concentration in semiconductors. In Table 2.3 we show the temperature dependence of bandgaps of several semiconductors.

2.10   MOBILE CARRIERS: INTRINSIC CARRIERS

From our brief discussion of metals and semiconductors in Section 2.3, we see that in a metal, current flows because of the electrons present in the highest (partially) filled band. This is shown schematically in Fig. 2.20a. The density of such electrons is very high ($\sim 10^{23}$ cm^{-3}). In a semiconductor, on the other hand, no current flows if the valence band is filled with electrons and the conduction band is empty of electrons. However, if somehow empty states or holes are created in the valence band by removing electrons, current can flow through the holes. Similarly, if electrons are placed in the conduction band, these electrons can carry current. This is shown schematically in Fig. 2.20b. If the density of electrons in the conduction band is n and that of holes in the valence band is p, the total mobile carrier density is $n + p$.

In Appendix B we discuss the density of states of electrons near bandedges. The density of state $N(E)$ is a very important concept and gives us the number of allowed electronic states per unit volume per energy (units: cm^{-3} eV^{-1}). For 3–dimensional systems the density of states has the form

$$N(E) = \frac{\sqrt{2}\,(m_{dos}^*)^{3/2}\,(E - E_c)^{1/2}}{\pi^2 \hbar^3} \tag{2.88}$$

where m_{dos}^* is the density of states mass and E_c is the conduction bandedge. A similar expression exists for the valence band except the energy term is replaced by $(E_v - E)^{1/2}$ and the density of states exist below the valence bandedge E_v. In Fig. 2.21 we show a schematic view of the density of states.

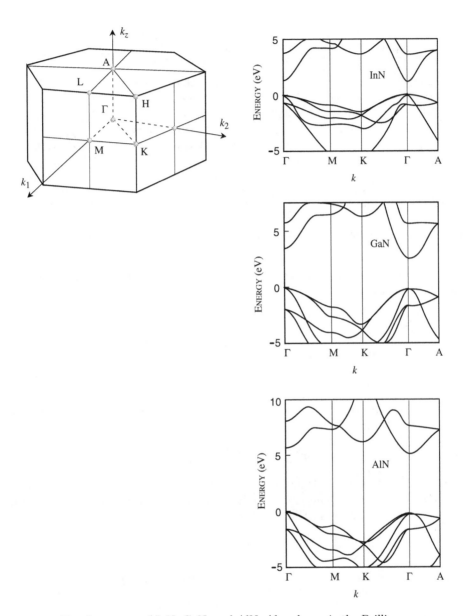

Figure 2.19: Bandstructure of InN, GaN, and AlN. Also shown is the Brillioun zone.

Compound	Type of Bandgap	Experimental Bandgap E_G (eV)		Temperature Dependence of Bandgap $E_G(T)$ (eV)
		0 K	300 K	
AlP	Indirect	2.52	2.45	$2.52 - 3.18 \times 10^{-4}T^2/(T+588)$
AlAs	Indirect	2.239	2.163	$2.239 - 6.0 \times 10^{-4}T^2/(T+408)$
AlSb	Indirect	1.687	1.58	$1.687 - 4.97 \times 10^{-4}T^2/(T+213)$
GaP	Indirect	2.338	2.261	$2.338 - 5.771 \times 10^{-4}T^2/(T+372)$
GaAs	Direct	1.519	1.424	$1.519 - 5.405 \times 10^{-4}T^2/(T+204)$
GaSb	Direct	0.810	0.726	$0.810 - 3.78 \times 10^{-4}T^2/(T+94)$
InP	Direct	1.421	1.351	$1.421 - 3.63 \times 10^{-4}T^2/(T+162)$
InAs	Direct	0.420	0.360	$0.420 - 2.50 \times 10^{-4}T^2/(T+75)$
InSb	Direct	0.236	0.172	$0.236 - 2.99 \times 10^{-4}T^2/(T+140)$

Table 2.3: Bandgaps of binary III–V compounds. (From Casey and Panish, 1978).

In direct gap semiconductors m^*_{dos} is just the effective mass for the conduction band. In indirect gap materials it is given by

$$m^*_{dos} = (m^*_1 m^*_2 m^*_3)^{1/3}$$

where $m^*_1 m^*_2 m^*_3$ are the effective masses along the three principle axes. For Si counting the 6 degenerate X–valleys we have

$$m^*_{dos} = 6^{2/3} \left(m_\ell m_t^2\right)^{1/3}$$

For the valence band we can write a simple expression for a density of states mass which includes the HH and LH bands

$$m^*_{dos} = \left(m_{hh}^{*3/2} + m_{\ell h}^{*3/2}\right)^{2/3}$$

In pure semiconductors electrons in the conduction band came from the valence band and $n = p = n_i = p_i$ where n_i and p_i are the intrinsic carrier concentrations. In general the electron density in the conduction band is

$$n = \int_{E_c}^{\infty} N_e(E) f(E) dE$$

$$n = \frac{1}{2\pi^2} \left(\frac{2m^*_e}{\hbar^2}\right)^{3/2} \int_{E_c}^{\infty} \frac{(E - E_c)^{1/2} dE}{\exp\left(\frac{E - E_F}{k_B T}\right) + 1} \tag{2.89}$$

In Fig. 2.21 we show how a change of temperature alters the shape of the Fermi function and alters the electron and hole densities. For small values of n (non-degenerate statistics where we can ignore the unity in the Fermi function) we get

$$n = N_c \exp\left[(E_F - E_c)/k_B T\right] \tag{2.90}$$

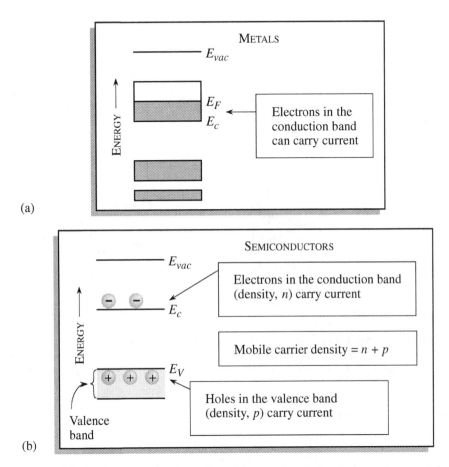

Figure 2.20: (a) A schematic showing allowed energy bands in electrons in a metal. The electrons occupying the highest partially occupied band are capable of carrying current. (b) A schematic showing the valence band and conduction band in a typical semiconductor. In semiconductors only electrons in the conduction band and holes in the valence band can carry current.

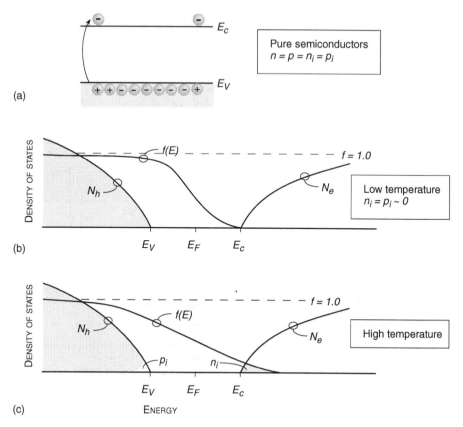

Figure 2.21: (a) A schematic showing that electron and hole densities are equal in a pure semiconductor. (b) Density of states and Fermi occupation function at low temperature. (c) Density of states and Fermi function at high temperatures when n_i and p_i became large.

where the effective density of states N_c is given by

$$N_c = 2 \left(\frac{m_e^* k_B T}{2\pi \hbar^2} \right)^{3/2} \tag{2.91}$$

Similar results arise for holes. We also obtain

$$np = 4 \left(\frac{k_B T}{2\pi \hbar^2} \right)^3 (m_e^* m_h^*)^{3/2} \exp \left(-E_g / k_B T \right) \tag{2.92}$$

We note that the product np is independent of the position of the Fermi level and is dependent only on the temperature and intrinsic properties of the semiconductor. This observation is called the law of mass action. If n increases, p must decrease, and vice versa. For the intrinsic case $n = n_i = p = p_i$, we have from the square root of the

MATERIAL	CONDUCTION BAND EFFECTIVE DENSITY (N_C)	VALENCE BAND EFFECTIVE DENSITY (N_V)	INTRINSIC CARRIER CONCENTRATION ($n_i = p_i$)
Si (300 K)	2.78 x 10^{19} cm⁻³	9.84 x 10^{18} cm⁻³	1.5 x 10^{10} cm⁻³
Ge (300 K)	1.04 x 10^{19} cm⁻³	6.0 x 10^{18} cm⁻³	2.33 x 10^{13} cm⁻³
GaAs (300 K)	4.45 x 10^{17} cm⁻³	7.72 x 10^{18} cm⁻³	1.84 x 10^{6} cm⁻³

Table 2.4: Effective densities and intrinsic carrier concentrations of Si, Ge, and GaAs. The numbers for intrinsic carrier densities are the accepted values even though they are smaller than the values obtained by using the equations derived in the text.

equation above,

$$n_i = p_i = 2 \left(\frac{k_B T}{2\pi\hbar^2} \right)^{3/2} (m_e^* m_h^*)^{3/4} \exp\left(-E_g/2k_B T \right)$$

$$E_{Fi} = \frac{E_c + E_v}{2} + \frac{3}{4} k_B T \ell n \left(m_h^*/m_e^* \right) \qquad (2.93)$$

Thus the Fermi level of an intrinsic material lies close to the midgap. Note that in calculating the density of states masses m_h^* and m_e^* the number of valleys and the sum of heavy and light hole states have to be included.

In Table 2.4 we show the effective densities and intrinsic carrier concentrations in Si, Ge, and GaAs. The values given are those accepted from experiments. These values are lower than the ones we get by using the equations derived in this section. The reason for this difference is due to inaccuracies in carrier masses and the approximate nature of the analytical expressions.

We note that the carrier concentration increases exponentially as the bandgap decreases. Results for the intrinsic carrier concentrations for some semiconductors are shown in Fig. 2.22. The strong temperature dependence and bandgap dependence of intrinsic carrier concentration can be seen from this figure. In electronic devices where current has to be modulated by some means, the concentration of intrinsic carriers is fixed by the temperature and therefore is detrimental to device performance. Once the intrinsic carrier concentration increases to $\sim 10^{15}$ cm⁻³, the material becomes unsuitable for electronic devices due to the high leakage current arising from the intrinsic carriers. A growing interest in high-bandgap semiconductors such as diamond (C), SiC, etc., is partly due to the potential applications of these materials for high-temperature devices where, due to their larger gap, the intrinsic carrier concentration remains low up to very high temperatures.

EXAMPLE 2.1 Calculate the effective density of states for the conduction and valence bands of GaAs and Si at 300 K. Let us start with the GaAs conduction-band case. The effective density of states is

$$N_c = 2 \left(\frac{m_e^* k_B T}{2\pi\hbar^2} \right)^{3/2}$$

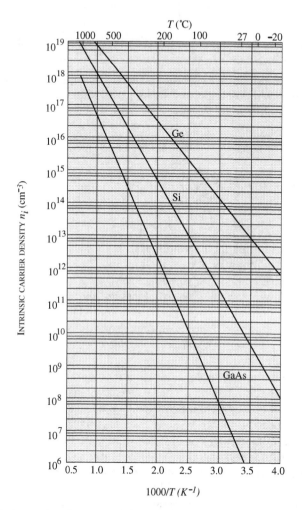

Figure 2.22: Intrinsic carrier densities of Ge, Si, and GaAs as a function of reciprocal temperature.

Note that at 300 K, $k_B T = 26$ meV $= 4 \times 10^{-21}$ J.

$$
\begin{aligned}
N_c &= 2 \left(\frac{0.067 \times 0.91 \times 10^{-30} \text{ (kg)} \times 4.16 \times 10^{-21} \text{ (J)}}{2 \times 3.1416 \times (1.05 \times 10^{-34} \text{ (Js)})^2} \right)^{3/2} \text{m}^{-3} \\
&= 4.45 \times 10^{23} \text{ m}^{-3} = 4.45 \times 10^{17} \text{ cm}^{-3}
\end{aligned}
$$

In silicon, the density of states mass is to be used in the effective density of states. This is given by

$$
m^*_{dos} = 6^{2/3} (0.98 \times 0.19 \times 0.19)^{1/3} \, m_0 = 1.08 \, m_0
$$

The effective density of states becomes

$$
\begin{aligned}
N_c &= 2 \left(\frac{m^*_{dos} k_B T}{2\pi\hbar^2} \right)^{3/2} \\
&= 2 \left(\frac{1.06 \times 0.91 \times 10^{-30} \text{ (kg)} \times 4.16 \times 10^{-21} \text{ (J)}}{2 \times 3.1416 \times (1.05 \times 10^{-34} \text{ (Js)})^2} \right)^{3/2} \text{m}^{-3} \\
&= 2.78 \times 10^{25} \text{ m}^{-3} = 2.78 \times 10^{19} \text{ cm}^{-3}
\end{aligned}
$$

One can see the large difference in the effective density between Si and GaAs.

In the case of the valence band, one has the heavy hole and light hole bands, both of which contribute to the effective density. The effective density is

$$
N_v = 2 \left(m_{hh}^{3/2} + m_{\ell h}^{3/2} \right) \left(\frac{k_B T}{2\pi\hbar^2} \right)^{3/2}
$$

For GaAs we use $m_{hh} = 0.45 m_0, m_{\ell h} = 0.08 m_0$ and for Si we use $m_{hh} = 0.5 m_0, m_{\ell h} = 0.15 m_0$, to get

$$
\begin{aligned}
N_v(\text{GaAs}) &= 7.72 \times 10^{18} \text{cm}^{-3} \\
N_v(\text{Si}) &= 9.84 \times 10^{18} \text{cm}^{-3}
\end{aligned}
$$

EXAMPLE 2.2 Calculate the position of the intrinsic Fermi level in Si at 300 K.

The density of states effective mass of the combined six valleys of silicon is

$$
m^*_{dos} = (6)^{2/3} \left(m^*_\ell \, m_t^2 \right)^{1/3} = 1.08 \, m_0
$$

The density of states mass for the valence band is 0.55 m_0. The intrinsic Fermi level is given by (referring to the valence bandedge energy as zero)

$$
\begin{aligned}
E_{Fi} &= \frac{E_g}{2} + \frac{3}{4} k_B T \ell n \left(\frac{m^*_h}{m^*_e} \right) = \frac{E_g}{2} + \frac{3}{4} (0.026) \ell n \left(\frac{0.55}{1.08} \right) \\
&= \frac{E_g}{2} - (0.0132 \text{ eV})
\end{aligned}
$$

The Fermi level is then 13.2 meV below the center of the mid-bandgap.

EXAMPLE 2.3 Calculate the intrinsic carrier concentration in InAs at 300 K and 600 K.

The bandgap of InAs is 0.35 eV and the electron mass is $0.027m_0$. The hole density of states mass is $0.4m_0$. The intrinsic concentration at 300 K is

$$
\begin{aligned}
n_i = p_i &= 2 \left(\frac{k_B T}{2\pi\hbar^2} \right)^{3/2} (m_e^* \, m_h^*)^{3/4} \exp \left(\frac{-E_g}{2k_B T} \right) \\
&= 2 \left(\frac{(0.026)(1.6 \times 10^{-19})}{2 \times 3.1416 \times (1.05 \times 10^{-34})^2} \right)^{3/2} \\
&\quad \left(0.027 \times 0.4 \times (0.91 \times 10^{-30})^2 \right)^{3/4} \exp \left(-\frac{0.35}{0.052} \right) \\
&= 1.025 \times 10^{21} \text{ m}^{-3} = 1.025 \times 10^{15} \text{cm}^{-3}
\end{aligned}
$$

The concentration at 600 K becomes

$$
n_i(600 \text{ K}) = 2.89 \times 10^{15} \text{cm}^{-3}
$$

2.11 DOPING: DONORS AND ACCEPTORS

In the last section of this chapter we will discuss electron and hole densities arising from doping. There are two kinds of dopants—donors, which donate an electron to the conduction band, and acceptors, which accept an electron from the valence band (thus creating a hole). To understand the donor (or acceptor) problem, we consider a donor atom on a crystal lattice site. The donor atom could be a pentavalent atom in silicon or a Si atom on a Ga site in GaAs. Thus it has one or two extra electrons in its outermost shell compared to the host atom it replaces. Focusing on the pentavalent atom in Si, four of the valence electrons of the donor atom behave as they would in a Si atom; the remaining fifth electron now sees a positively charged ion to which it is attracted, as shown in Fig. 2.23. The ion has a charge of unity and the attraction is simply a Coulombic attraction suppressed by the dielectric constant of the material. The attractive potential is (ϵ is the dielectric constant of the semiconductor, i.e., the product of ϵ_0 and the relative dielectric constant)

$$
U(r) = \frac{-e^2}{4\pi\epsilon r}
$$

This problem is now essentially the same as that of an electron in the hydrogen atom problem. The only difference is that the electron mass is m^* and the coulombic potential is reduced by ϵ_0/ϵ.

The lowest-energy solution for this problem is

$$
\begin{aligned}
E_d &= E_c - \frac{e^4 m_e^*}{2(4\pi\epsilon)^2 \hbar^2} \\
&= E_c - 13.6 \left(\frac{m^*}{m_o} \right) \left(\frac{\epsilon_o}{\epsilon} \right)^2 \text{ eV}
\end{aligned}
\tag{2.94}
$$

Note that in the hydrogen atom problem the electron level is measured from the vacuum energy level, which is taken as $E = 0$. In the donor problem, the energy level is measured from the bandedge and the ground state is shown schematically in Fig. 2.24.

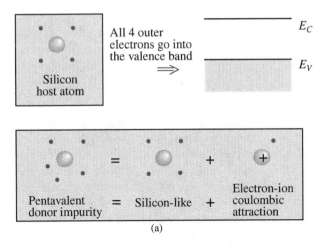

(a)

Figure 2.23: A schematic showing the approach to understanding donors in semiconductors. The donor problem is treated as the host atom problem together with a Coulombic interaction term. The silicon atom has four "free" electrons per atom. All four electrons are contributed to the valence band at 0 K. The dopant has five electrons out of which four are contributed to the valence band, while the fifth one can be used for increasing electrons in the conduction band.

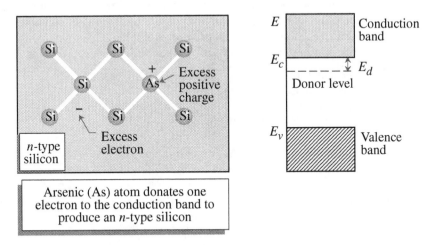

Figure 2.24: Charges associated with an arsenic impurity atom in silicon. Arsenic has five valence electrons, but silicon has only four valence electrons. Thus four electrons on the arsenic form tetrahedral covalent bonds similar to silicon, and the fifth electron is available for conduction. The arsenic atom is called a donor because when ionized it donates an electron to the conduction band.

The effective mass to be used is the conductivity effective mass m_σ^*, which tells us how electrons respond to external potentials. This mass is used for donor energies as well as for charge transport in an electric field. For direct bandgap materials like GaAs, this is simply the effective mass. For materials like Si the conductivity mass is

$$m_\sigma^* = 3 \left(\frac{2}{m_t^*} + \frac{1}{m_\ell^*} \right)^{-1} \tag{2.95}$$

According to the simple picture of the donor impurity discussed above, the donor energy levels depend only upon the host crystal (through ϵ and m^*) and *not* on the nature of the dopant. According to Eqn. 2.94, the donor energies for Ge, Si, and GaAs should be 0.006 V, 0.025, and 0.007 eV, respectively. However there is a small deviation from these numbers, depending upon the nature of the dopant. This difference occurs because of the simplicity of our model. In a real sense there is a small distortion of the atomic potential—the so called central cell correction—which modifies the donor energy as given by the simple model discussed here.

Another important class of intentional impurities is the acceptors. Just as donors are defect levels that are neutral when an electron occupies the defect level and positively charged when unoccupied, the acceptors are neutral when empty and negatively charged when occupied by an electron. The acceptor levels are produced when impurities that have a similar core potential as the atoms in the host lattice, but have one less electron in the outermost shell, are introduced in the crystal. Thus group III elements can form acceptors in Si or Ge, while Si could be an acceptor if it replaces As in GaAs.

From the discussion above, it is clear that while in group IV semiconductors, the donor- or acceptor-like nature of an impurity is unambiguous, in compound semiconductors, the dopants can be "amphoteric." For example, Si can act as a donor in GaAs if it replaces a Ga atom while it can act as an acceptor if it replaces an As atom.

EXAMPLE 2.4 Calculate the donor and acceptor level energies in GaAs and Si. The shallow level energies are in general given by

$$E_d = E_c - 13.6(\text{eV}) \times \frac{m^*/m_0}{(\epsilon/\epsilon_0)^2}$$

The conduction-band effective mass in GaAs is $0.067m_0$ and $\epsilon = 13.2\epsilon_0$, and we get for the donor level

$$E_d(\text{GaAs}) = E_c - 5.2 \text{ meV}$$

The problem in silicon is a bit more complicated since the effective mass is not so simple. For donors we need to use the conductivity mass, which is given by

$$m_\sigma^* = \frac{3m_0}{\left(\frac{1}{m_\ell^*} + \frac{2}{m_t^*} \right)}$$

where m_ℓ^* and m_t^* are the longitudinal and transverse masses for the silicon conduction band. Using $m_\ell^* = 0.98$ and $m_t^* = 0.2m_0$, we get

$$m_\sigma^* = 0.26m_0$$

Using $\epsilon = 11.9\epsilon_0$, we get

$$E_d(\text{Si}) = 25 \text{ meV}$$

The acceptor problem is much more complicated due to the degeneracy of the heavy hole and light hole band. The simple hydrogen atom problem does not give very accurate results. However, a reasonable approximation is obtained by using the heavy hole mass ($\sim 0.45m_0$ for GaAs, $\sim 0.5m_0$ for Si), and we get

$$\begin{aligned} E_a(\text{Si}) &= 48 \text{ meV} \\ E_a(\text{GaAs}) &\cong 36 \text{ meV} \end{aligned}$$

This is the energy above the valence bandedge. However, it must be noted that the use of the heavy hole mass is not strictly valid.

2.11.1 Carriers in Doped Semiconductors

As noted above, if a donor is placed in a crystal it can, in principle, provide an electron to the conduction band. However, whether this extra electron gives into the conduction band or stays bound to the donor depends upon temperature, donor binding energy and donor density. At very low temperatures, the donor electrons are tied to the donor sites and this effect is called carrier freezeout. At higher temperatures, however, the donor electron is "ionized" and resides in the conduction band as a free electron. Such electrons can carry current and modify the electronic properties of the semiconductor. The ionized donor atom is positively charged. Similarly, an ionized acceptor is negatively charged and contributes a hole to the valence band.

In general the relation between electron density and Fermi level, E_F, is given by Eqn. 2.89. While in general this relation is to be obtained numerically, a good approximation is given by the Joyce-Dixon approximation . According to this relation, we have

$$E_F = E_c + k_B T \left[\ell n \frac{n}{N_c} + \frac{1}{\sqrt{8}} \frac{n}{N_c} \right] = E_v - k_B T \left[\ell n \frac{p}{N_v} + \frac{1}{\sqrt{8}} \frac{p}{N_v} \right] \quad (2.96)$$

where the effective density of states N_c or N_v for the conduction or valence band is given by

$$N_c = 2 \left(\frac{m^* k_B T}{2\pi \hbar^2} \right)^{3/2}$$

This relation can be used to obtain the Fermi level if n is specified. Alternatively, it can be used to obtain n if E_F is known by solving for n iteratively. If the term $(n/\sqrt{8} \, N_c)$ is ignored the result corresponds to the Boltzmann approximation given in Eqn. 2.90.

EXAMPLE 2.5 A sample of GaAs has a free electron density of 10^{17} cm^{-3}. Calculate the position of the Fermi level using the Boltzmann approximation and the Joyce-Dixon approximation at 300 K.

In the Boltzmann approximation, the carrier concentration and the Fermi level are related by the following equation:

$$\begin{aligned} E_F &= E_c + k_B T \left[\ell n \frac{n}{N_c} \right] \\ &= E_c + 0.026 \left[\ell n \left(\frac{10^{17}}{4.45 \times 10^{17}} \right) \right] = E_c - 0.039 \text{ eV} \end{aligned}$$

The Fermi level is 39 meV below the conduction band. In the Joyce-Dixon approximation we have

$$
\begin{aligned}
E_F &= E_c + k_B T \left[ln\left(\frac{n}{N_c}\right) + \frac{1}{\sqrt{8}}\frac{n}{N_c} \right] \\
&= E_c + 0.026 \left[ln\left(\frac{10^{17}}{4.45 \times 10^{17}}\right) + \frac{10^{17}}{\sqrt{8}(4.45 \times 10^{17})} \right] \\
&= E_c - 0.039 + 0.002 = E_c - 0.037 \text{ eV}
\end{aligned}
$$

The error produced by using the Boltzmann approximation (compared to the more accurate Joyce-Dixon approximation) is 2 meV.

EXAMPLE 2.6 Assume that the Fermi level in silicon coincides with the conduction band-edge at 300 K. Calculate the electron carrier concentration using the Boltzmann approximation and the Joyce-Dixon approximation.

In the Boltzmann approximation, the carrier density is simply

$$
n = N_c = 2.78 \times 10^{19} \text{ cm}^{-3}
$$

According to the Joyce-Dixon approximation, the carrier density is obtained from the solution of the equation

$$
E_F = 0 = k_B T \left[ln\frac{n}{N_c} + \frac{n}{\sqrt{8}N_c} \right]
$$

Solving by trial and error, we get

$$
\frac{n}{N_c} = 0.76 \text{ or } n = 2.11 \times 10^{19} \text{ cm}^{-3}
$$

We see that the less accurate Boltzmann approximation gives a higher charge density.

2.11.2   Mobile Carrier Density and Carrier Freezeout

In the lowest energy state of the donor atom, the extra electron of the donor is localized at the donor site. Such an electron cannot carry any current and is not useful for changing the electronic properties of the semiconductor. At very low temperatures, the donor electrons are, indeed, tied to the donor sites and this effect is called carrier freeze-out. At higher temperatures, however, the donor electron is "ionized" and resides in the conduction band as a free electron as shown schematically in Fig. 2.25. Such electrons can, of course, carry current and modify the electronic properties of the semiconductor. The ionized donor atom is negatively charged and offers a scattering center for the free electrons. We will discuss the scattering in a later chapter.

The electron + donor system can be in one of the following states: *i*) the electron is free; *ii*) there is one electron attached to the donor with spin up or spin down; or *iii*) there are two electrons attached to a donor. While in principle the latter condition is allowed, the repulsion energy for two electrons on the same donor is so large that the third condition is not allowed. Thus, two electrons each with opposite spin cannot be bound to a donor.

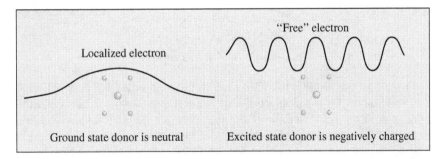

Figure 2.25: An electron band to a donor does not contribute to charge conduction. However, if the donor is "ionized," the electron becomes free and can contribute to the charge transport.

The occupation statistics for the donor system (an electron can occupy the donor state in two independent ways) gives the following occupation function:

$$f_{donor} = \frac{1}{\frac{1}{2}e^{(E_d - E_F)/k_B T} + 1}$$

Thus the number density of electrons bound to the donors, n_d is

$$n_d = \frac{N_d}{\frac{1}{2}e^{(E_d - E_F)/k_B T} + 1}$$

$$\approx 2N_d e^{-(E_d - E_F)/k_B T} \text{ for } (E_d - E_F)/k_B T \gg 1 \qquad (2.97)$$

Here N_d is the number density of the donor energies.

An analysis similar to the one for acceptors above gives (note that the acceptor levels are doubly degenerate, since the valence band is doubly degenerate. This allows a total of 4 possible ways in which a gole can be bound to an acceptor)

$$p_a = \frac{N_a}{\frac{1}{4}\exp\left(\frac{(E_F - E_a)}{k_B T}\right) + 1} \qquad (2.98)$$

where p_a is the density of holes trapped at acceptor levels.

2.11.3 Equilibrium Density of Carriers in Doped Semiconductors

In a doped semiconductor we may have donors, acceptors, or both. To find the electron and hole densities, we need to use the occupation functions for electrons, holes, and the dopants. We start with the charge neutrality condition

$$n_c + n_d = N_d - N_a + p_v + p_a \qquad (2.99)$$

where

$$n_c = \text{free electrons in the conduction band}$$

$$n_d \quad = \quad \text{electrons bound to the donors}$$
$$p_v \quad = \quad \text{free holes in the valence band}$$
$$p_a \quad = \quad \text{holes bound to the acceptors}$$

This equation along with the explicit form of n_c, n_d, p_v, and p_a, in terms of the Fermi level, allow us to calculate E_F at a given temperature. The general analysis requires numerical techniques. Essentially one chooses a Fermi level and adjusts it until the charge neutrality condition is satisfied. In case of low doping (low n, p densities) it is possible to get analytical results for electron and hole densities by ignoring the unity in the denominator of the occupation probability we can write (using Boltzmann approximation for free electrons)

$$n \quad = \quad N_c \, \exp \, - \left(\frac{E_c - E_F}{k_B T} \right)$$
$$n_d \quad = \quad \frac{N_d}{2} \, \exp \left(\frac{E_d - E_F}{k_B T} \right)$$

In this case we can write

$$\frac{n_d}{n + n_d} = \frac{1}{\frac{N_c}{2 N_d} \exp \left[-\frac{(E_c - E_d)}{k_B T} \right] + 1} \tag{2.100}$$

A similar treatment for p-type material with acceptor doping N_a gives the ratio

$$\frac{p_a}{p + p_a} = \frac{1}{\frac{N_v}{4 N_a} \exp \left[-\frac{(E_a - E_v)}{k_B T} \right] + 1} \tag{2.101}$$

In Fig. 2.25 we show how free electron density varies with temperature in a n-type silicon sample. As temperature increases, the fraction of "ionized" donors starts to increase until all of the donors are ionized and the free carrier density is equal to the donor density. This region is called the saturation region. Eventually, as the temperature is further raised, the carrier density starts to increase because of the intrinsic carrier density exceeding the donor density. At low temperatures the electrons are bound to the donors. This is the freezeout regime. Semiconductor devices usually operate in the saturation region where the mobile carrier density is essentially independent of temperature and is approximately equal to the doping density.

Semiconductor devices cannot operate in the high temperature intrinsic regime, since it is not possible to control intrinsic carrier density by applying an external bias. Thus devices cannot be "shut off" due to the leakage current from intrinsic carriers. High temperature electronics require large bandgap semiconductors for which the upper temperature limit is high.

EXAMPLE 2.7 Consider a silicon sample doped with phosphorus at a doping density of 10^{16} cm^{-3}. Calculate the fraction of ionized donors at 300 K. How does this fraction change if the doping density is 10^{18} cm^{-3}?

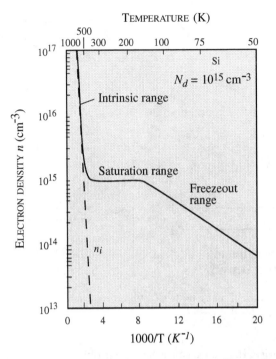

Figure 2.26: Electron density as a function of temperature for a Si sample with donor impurity concentration of 10^{15} cm^{-3}. It is preferable to operate devices in the saturation region where the free carrier density is approximately equal to the dopant density.

We have for silicon for $N_d = 10^{16}$ cm^{-3} (donor binding energy is 45 meV)

$$\frac{n_d}{n + n_d} = \frac{1}{\frac{2.8 \times 10^{19}}{2(10^{16})} \exp\left(-\frac{0.045}{0.026}\right) + 1} = 0.004$$

Thus n_d is only 0.4% of the total electron concentration and almost all the donors are ionized. For a doping level of 10^{18}, we get

$$\frac{n_d}{n + n_d} = 0.29$$

We see that in this case where the doping is heavy, only 71% of the dopants are ionized.

2.11.4 Heavily Doped Semiconductors

In the theory discussed so far, we have made several important assumptions that are valid only when the doping levels are low: i) we have assumed that the bandstructure of the host crystal is not seriously perturbed and the bandedge states are still described by simple parabolic bands; ii) the dopants are assumed to be independent of each other and their potential is thus a simple coulombic potential. These assumptions become invalid as the doping levels become higher. When the average spacing of the impurity atoms reaches ~ 100 Å, the potential seen by the impurity electron is influenced by the

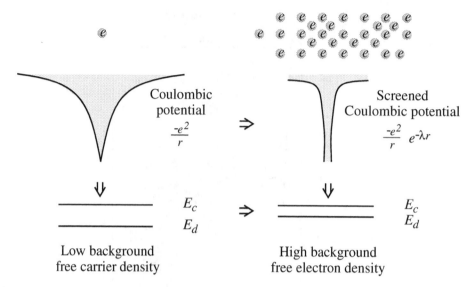

Figure 2.27: A schematic of many body effects arising from high doping levels. The free carrier affect the donor energy level E_d.

neighboring impurities. In a sense this is like the problem of electrons in atoms. When the atoms are far apart, we get discrete atomic levels. However, when the atomic separation reaches a few angstroms, as in a crystal, we get electronic bands. At high doping levels we get impurity bands. Several other important effects occur at high doping levels.

Screening of Impurity Potential
As the doping levels are increased, the background mobile electron density also increases. This background density adjusts itself in response to the impurity potential causing a screening of the potential. We will discuss this screening in Chapter 5, but in essence the $1/r$ long-range potential is reduced by an exponential factor in the simplest theory. This effect shown schematically in Fig. 2.27 lowers the binding energy of the electron to the donor.

The reduction in the donor ionization energy causes the donor level to move toward the conduction bandedge E_c as the donor concentration is gradually increased and ultimately merges into the conduction band. Measurements of the ionization energy of arsenic in germanium have shown that E_d can be expressed by the following empirical relation

$$E_d = E_{d0} \left[1 - (N_d/N_{\text{crit}})^{1/3} \right] \tag{2.102}$$

where E_{d0} is the ionization energy in a lightly doped crystal and $N_{\text{crit}} = 2 \times 10^{17}$ cm^{-3} is the critical donor concentration at which the ionization energy will vanish. From the known value of $E_{d0} = 0.0127$ eV for arsenic, and using Bohr's model for an electron attached to the donor atom, it can be shown that E_d drops to zero when the average spacing between the arsenic atoms is about $3r_0$, where r_0 is the ground state Bohr radius of the electron attached to the arsenic atom. Note that the electron donor interaction causes only a shift in the donor level toward E_c, but no change occurs in the position of bandedges E_c and E_v so that the energy gap E_g remains unchanged.

Electron–Electron Interaction

The bandstructure calculations we discussed in earlier chapters define the position of the valence and conduction bandedges for the situation where the valence band is full and the conduction band is empty. When a large number of electrons are introduced in the conduction band (or a large number of holes in the valence band), the electrons which are Fermions interact strongly with each other.

Electron-electron interaction results on a downward shift in the conduction band edge E_c. This shift is caused by electron exchange energy which evolves from the Pauli exclusion principle. When the electron concentration in the semiconductor becomes sufficiently large, their wavefunctions begin to overlap. Consequently, the Pauli exclusion principle becomes operative, and the electrons spread in their momenta in such a way that the overlapping of the individual electron wavefunctions is avoided. The Bloch states that we devised earlier are thus modified by the presence of the other electrons. In general, the electron-electron interaction can be represented by the usual Coulombic interaction and the exchange interaction. The latter comes about due to the constraint of the Pauli exclusion principle which forces any multiparticle electronic wavefunction to be antisymmetric in the exchange of two electrons. The average Coulombic energy between an electron with other electrons and with the background positive charge cancels out. The exchange term which "keeps the electrons away from each other," then lowers the energy of the system. This lowering has been studied in 3D systems in great detail and is an area of continuing research in sub-3D systems. For bulk GaAs the electron–electron interaction results in a lowering of the bandgap which is given by (H. Casey and F. Stern, 1976),

$$E_g = E_{g0} - 1.6 \times 10^{-8}(p^{1/3} + n^{1/3}) \text{ eV} \qquad (2.103)$$

where E_{g0} is the bandgap at zero doping and n and p are the electron and hole densities. As can be seen at background charge levels of $\sim 10^{18} \text{cm}^{-3}$, the bandgap can shrink by ~ 16 meV. In silicon the bandgap shrinkage is found to be

$$\Delta E_g \simeq -22.5 \left(\frac{N_d}{10^{18}} \frac{300}{T(k)} \right)^{1/2} \text{ meV} \qquad (2.104)$$

Band Tailing Effects

In addition to the band shrinkage that occurs when a semiconductor is heavily doped, another important effect occurs near the bandedge. This effect occurs because of the disordered arrangement of the impurity atoms on the host lattice. As shown schematically in Fig. 2.28, the randomly placed impurity atoms cause a random fluctuation in the effective bandedge.

Deep well regions as indicated in the figure are produced which lead to low-energy electronic states. The random nature of the potential fluctuation leads to a bandedge tail in the density of states. The underlying reasons behind the effect of disorder will be discussed in Chapter 8.

In view of the various effects discussed above, the "optical bandgap" of the semiconductor changes as the doping changes. The optical bandgap is defined from

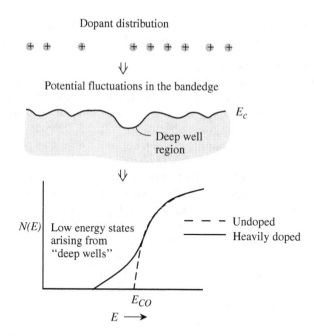

Figure 2.28: A schematic description showing how the randomly distributed impurities cause band tail states.

the optical absorption versus photon energy data and represents the energy separation where the valence and conduction band density of states are significant. In addition, one often defines artificially an electrical bandgap. This is defined from the np product discussed earlier. According to this definition

$$np = n_i^2$$

$$= C \exp\left(-\frac{E_g^{\text{elect}}}{k_B T}\right) \quad (2.105)$$

This definition has only a mathematical significance and does not represent any real physical property. The electrical bandgap is usually smaller than the optical bandgap.

2.12 TECHNOLOGY CHALLENGES

Bandstructure of semiconductors has a direct impact on electronic and optoelectronic devices. In information processing devices there are hosts of (sometimes conflicting) demands. In Table 2.5 we show how some of these demands can be met by choice of the correct bandstructure. As shown in this table, depending upon applications, we have to identify and choose semiconductors which have the desired bandstructure. This does not mean that it is simple to satisfy technology needs, since most semiconductor technologies are very immature. Only Si, GaAs, InP, InGaAsP, AlGaAs, and Ge have mature technologies.

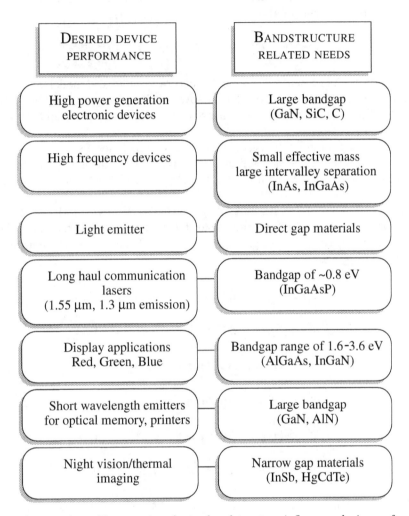

Table 2.5: An overview of how semiconductor bandstructure influences device performance.

2.13 PROBLEMS

2.1 Using the relation

$$\sum_{\mathbf{k}} e^{i\mathbf{k}\cdot\mathbf{R}} = N\delta_{\mathbf{R},0}$$

use the orthonormality of the Bloch functions to show that the Wannier functions $\phi_n(\mathbf{r})$ at different sites are orthogonal, i.e.

$$\int \phi_n^*(\mathbf{r} - \mathbf{R})\, \phi_{n'}(\mathbf{r} - \mathbf{R}')\, d^3r \propto \delta_{n,n'}\, \delta_{\mathbf{R},\mathbf{R}'}$$

2.2 In the text we had discussed the bandstructure of an s-band model in an fcc lattice. What is the bandstructure in a bcc lattice? Assume only nearest neighbor interaction $\gamma = 1.0$ eV. What is the effective mass near the bottom of the band and the top of the band? Assume a lattice constant of 4 Å.

2.3 If in the s-band model discussed in the text one includes a second neighbor interaction γ_2, what is the expression for the bandstructure? What is the electron effective mass at the bottom of the band if $\gamma_1 = 1.0$ eV; $\gamma_2 = 0.2$ eV? Assume a lattice constant of 4 Å.

2.4 Consider a cubic lattice with an s, p_x, p_y, p_z basis. Write down the 4×4 eigenvalue equation for the system. Assume only nearest neighbor interactions and make use of the following symmetries of the tight binding matrix elements (use the notation $s \rightarrow 1$, $p_x \rightarrow 2$, $p_y \rightarrow 3$, $p_z \rightarrow 4$)

$$\int d^3r\ \phi_1(x,y,z)\ \Delta U\ \phi_1(x-a,y,z) = V_{ss\sigma}$$

$$\int d^3r\ \phi_1(x,y,z)\ \Delta U\ \phi_2(x-a,y,z) = V_{sp\sigma}$$

$$\int d^3r\ \phi_1(x,y,z)\ \Delta U\ \phi_2(x,y-a,z) = 0$$

$$\int d^3r\ \phi_2(x,y,z)\ \Delta U\ \phi_1(x-a,y,z) = V_{sp\sigma}$$

$$\int d^3r\ \phi_2(x,y,z)\ \Delta U\ \phi_3(x,y-a,z) = V_{pp\pi}$$

$$\int d^3r\ \phi_2(x,y,z)\ \Delta U\ \phi_3(x-a,y,z) = 0$$

Show that the matrix elements in the 4×4 secular determinant are (E_{s0}, E_{p0} are on-site matrix elements),

$$
\begin{aligned}
H_{1,1} &= <s|H|s> \\
&= E_{s0} + 2V_{ss\sigma}[\cos k_x a + \cos k_y a + \cos k_z a] \\
H_{1,2} &= <p_x|H|s> \\
&= 2iV_{sp\sigma}\sin k_x a \\
H_{2,2} &= <p_x|H|p_x>
\end{aligned}
$$

$$= E_{p0} + 2V_{pp\sigma}\cos k_x a + 2V_{pp\pi}[\cos k_y a + \cos k_z a]$$

$$H_{3,2} = <p_y|H|p_x>$$

$$= 0.$$

Other elements are generated by cyclical permutations. What are the solutions of the problem at $\mathbf{k} = 0$?

2.5 Write out the matrix elements for the diamond structure using the tight binding approach. Without spin–orbit coupling, compare your results to the matrix elements given in the classic paper of Slater and Koster (1954), (see Reference section). Note that there is a misprint of the matrix elements in the original paper.

2.6 Include spin–orbit coupling in the tight binding matrix using an sp^3 basis per site. A good source of matrix elements for the tight binding for a variety of semiconductors is Talwar and Ting (1982).

2.7 In Si electron transport, the intervalley scattering is very important. Two kinds of intervalley scatterings are important: in the g–scattering an electron goes from one valley (say a [001] valley) to an opposite valley ([00$\bar{1}$]), while in an f–scattering, the electron goes to a perpendicular valley ([001] to [010], for example). The extra momentum for the transitions is provided by a phonon and may include a reciprocal lattice vector. Remembering that Si valleys are not precisely at the X–point, calculate the phonon vector which allows these scatterings.

2.8 Assuming k conservation, what is the phonon wavevector which can take an electron in the GaAs Γ–valley to the L–valley?

2.9 Using the 2×2 representation for the $k \cdot p$ model for the heavy-hole light-hole bands express the hole masses in terms of the Kohn-Luttinger parameters along the (100) and (110) k-direction.

2.10 Plot the conduction-band and valence-band density of states in Si, Ge, and GaAs. Use the following data:

$$\text{Si} : m^*_{dos}(c-\text{band}) = 1.08 \, m_0$$
$$m^*_{hh} = 0.49 \, m_0$$
$$m^*_{\ell h} = 0.16 \, m_0$$

$$\text{Ge} : m^*_{dos}(c-\text{band}) = 0.56 \, m_0$$
$$m^*_{hh} = 0.29 \, m_0$$
$$m^*_{\ell h} = 0.044 \, m_0$$

$$\text{GaAs} : m^*_e = m^*_{dos} 0.067 \, m_0$$
$$m^*_{hh} = 0.5 \, m_0$$
$$m^*_{\ell h} = 0.08 \, m_0$$

2.11 A conduction-band electron in silicon is in the (100) valley and has a k-vector of $2\pi/a(1.0, 0.1, 0.1)$. Calculate the energy of the electron measured form the conduction bandedge. Here a is the lattice constant of silicon.

2.12   The bandstructure of GaAs conduction-band electrons is given by a simple parabolic relation ($m^* = 0.067\ m_0$):

$$E = \frac{\hbar^2 k^2}{2m^*}$$

A better approximation is

$$E(1 + \alpha E) = \frac{\hbar^2 k^2}{2m^*}$$

where $\alpha = 0.67$ eV^{-1}. Calculate the density of states and the difference in the energy of an electron using the two expressions at $k = 0.01$ Å^{-1} and $k = 0.1$ Å^{-1}.

2.13   Estimate the intrinsic carrier concentration of diamond at 700 K (you can assume that the carrier masses are similar to those in Si). Compare the results with those for GaAs and Si. The result illustrates one reason why diamond is useful for high-temperature electronics.

2.14   An Si device is doped at 5×10^{16} cm^{-3}. Assume that the device can operate up to a temperature where the intrinsic carrier density is less than 10% of the total carrier density. What is the upper limit for the device operation?

2.15   Estimate the change in intrinsic carrier concentration with temperature, $d/n_i/dT$, for InAs, Si, and GaAs near room temperature.

2.16   Calculate and plot the position of the intrinsic Fermi level in Si between 77 K and 500 K.

2.17   A donor atom in a semiconductor has a donor energy of 0.045 eV below the conduction band. Assuming the simple hydrogenic model for donors, estimate the conduction bandedge mass.

2.18   Calculate the density of electrons in a silicon conduction band if the Fermi level is 0.1 eV below the conduction band at 300 K. Compare the results by using the Boltzmann approximation and the Joyce-Dixon approximation.

2.19   The electron density in s silicon sample at 300 K is 10^{16} cm^{-3}. Calculate $E_c - E_F$ and the hole density using the Boltzmann approximation.

2.20   A GaAs sample is dope n-type at 5×10^{17} cm^{-3}. What is the position of the Fermi level at 300 K?

2.21   Consider an n-type silicon with a donor energy 45 meV below the conduction band. The sample is doped at 10^{17} cm^{-3}. Calculate the temperature at which 90% of the donors are ionized.

2.22   Consider a GaAs sample doped at $N_d = 10^{17}$ cm^{-3}. The donor energy is 6 meV. Calculate the temperature at which 90% of the donors are ionized.

2.23   Calculate the n-type doping efficiency for phosphorous in silicon at room temperature for the following dopant densities: (a) $N_d = 10^{15}$ cm^{-3}; (b) $N_d = 10^{18}$ cm^{-3}; and (c) $N_d = 5 \times 10^{19}$ cm^{-3}. Assume that the donor level is 45 meV below the conduction band. Doping efficiency is the fraction of dopants that are ionized.

2.24   A 2 meV α-particle hits a semiconductor sample. The entire energy of the α-particle is used up in producing electron-hole pairs by knocking electrons from the valence to the conduction band. Estimate the number of electron-hole pairs produced in Si and GaAs.

2.25 Calculate the average separation of dopants in Si when the doping density is: (a) 10^{16} cm^{-3}; (b) 10^{18} cm^{-3}; and (c) 10^{20} cm^{-3}.

2.14 REFERENCES

- **General Bandstructure**

    - Bassani, F., in *Semiconductors and Semimetals* (edited by R. K. Willardson and A. C. Beer, Academic Press, New York, 1966), vol. 1, p. 21.
    - For general electronic properties a good reference is: Casey, H. C., Jr. and M. B. Panish, *Heterostructure Lasers, Part A, Fundamental Principles* and *Part B, Materials and Operating Characteristics* (Academic Press, New York, 1978).
    - Fletcher, G. C., *Electron Band Theory of Solids* (North-Holland, Amsterdam, 1971).
    - Harrison, W. A., *Electronic Structure and Properties of Solids* (W. H. Freeman, San Francisco, 1980).
    - Kane, E. O., *Semiconductors and Semimetals* (edited by R. K. Willardson and A. C. Beer, Academic Press, New York, 1966), vol. 1, p. 81.
    - Phillips, J. C., *Bonds and Bands in Semiconductors* (Academic Press, New York, 1973).

- **Tight Binding Method**

    - Bassani, F., in *Semiconductors and Semimetals* (edited by R. K. Willardson and A. C. Beer, Academic Press, New York, 1966), vol. 1, p. 21.
    - Chadi, D. and M. L. Cohen, Phys Stat. Sol. B, **68**, 404 (1975).
    - Jaffe, M. D., *Studies of the Electronic Properties and Applications of Coherently Strained Semiconductors*, Ph.D. Thesis, (The University of Michigan, Ann Arbor, 1989).
    - Slater, J. C. and G. F. Koster, Phys. Rev., **94**, 1498 (1954).
    - Talwar, D. N. and C. S. Ting, Phys. Rev. B, **25**, 2660 (1982).

- **Spin–Orbit Effects**

    - Chadi, D., Phys. Rev. B, **16**, 790 (1977).
    - Dresselhaus, D., A. F. Kip, and C. Kittel, Phys. Rev., **98**, 368 (1955).
    - Kane, E. O., in *Semiconductors and Semimetals* (edited by R. K. Willardson and A. C. Beer, Academic Press, New York, 1966), vol. 1, p. 81.
    - Luttinger, J. and W. Kohn, Phys. Rev., **97**, 869 (1955).
    - Sakurai, J. J., *Advanced Quantum Mechanics* (Addison–Wesley, New York, 1982).

− Slater, J. C., *Quantum Theory of Solids* (McGraw–Hill, New York, 1965), vol. 2.

- **OPW Method**

  − Woodruff, T. O., Solid State Physics, **4**, 367 (1957).

- **Pseudopotential Method**

  − Chelikowsky, J. R. and M. L. Cohen, Phys. Rev. B, **14**, 556 (1976).

  − Cohen, M. H. and V. Heine, Phys. Rev., **122**, 1821 (1961).

  − Harrison, W. A., *Pseudopotentials in the Theory of Metals* (W. A. Benjamin, New York, 1966).

  − Phillips, J. C. and L. Kleinmann, Phys. Rev., **116**, 287 (1959).

- **k · p Method**

  − Broido, D. A. and L. J. Sham, Phys. Rev. B, **31**, 888 (1985).

  − Hinckley, J. M., *The Effect of Strain in Pseudomorphic p-$Si_{1-x}Ge_x$: Physics and Modelling of the Valence Bandstructure and Hole Transport*, Ph.D. Thesis, (The University of Michigan, Ann Arbor, 1990).

  − Kane, E. O., J. Phys. Chem. Solids, **8**, 38 (1959).

  − Kane, E. O., in *Semiconductors and Semimetals* (edited by R. K. Willardson and A. C. Beer, Academic Press, New York, 1966), vol. 1, p. 81.

  − Lawaetz, P., Phys. Rev. B, **4**, 3460 (1971).

  − Luttinger, J., Phys. Rev., **102**, 1030 (1956).

  − Luttinger, J. and W. Kohn, Phys. Rev., **97**, 869 (1955).

- **General Semiconductors**

  − M.L. Cohen and J.R. Chelikowsky, *Electronic Structure and Optical Properties of Semiconductors*, Springer, Berlin (1988).

  − John H. Davies, *The Physics of Low Dimensional Semiconductors: An Introduction*, Cambridge University Press (1997).

  − S. Nakamura (Ed.) and S.F. Chichibu, *Nitride Semiconductors, Blue Lasers, and Light-Emitting Diodes*, Taylor and Francis (2000).

  − S. Nakamura, *Semiconductor Science and Technology*, **14**, 27 (1999).

  − *Properties of Group III–Nitrides*, EMIS Datareview Series, (ed. J.E. Idgar (INSPEC, London, 1994), vol. 11.

  − P.Y. Yu and M. Cardone, *Fundamentals of Semiconductors: Physics and Materials Properties*, Springer Verlag (2001).

Chapter

3

BANDSTRUCTURE
MODIFICATIONS

In the previous chapter we have seen how the intrinsic properties of a semiconductor as reflected by its chemical composition and crystalline structure lead to the unique electronic properties of the material. Can the bandstructure of a material be changed? The answer is yes, and the ability to tailor the bandstructure is a powerful tool. Novel devices can be conceived and designed for superior and tailorable performance. Also new physical effects can be observed. In this chapter we will establish the physical concepts which are responsible for bandstructure modifications. There are three widely used approaches for band tailoring (or engineering). These three approaches are shown in Fig. 3.1 and are:

1. Alloying of two or more semiconductors;

2. Use of heterostructures to cause quantum confinement; and

3. Use of built-in strain via lattice mismatched epitaxy.

These three concepts are increasingly being used for improved performance in electronic and optical devices.

3.1 BANDSTRUCTURE OF SEMICONDUCTOR ALLOYS

The easiest way to alter the electronic properties or to produce a material with new properties is based on making an alloy. Alloying of two materials is one of the oldest techniques to modify properties of materials, not only in semiconductors, but in metals and insulators as well.

The desire to form alloys in semiconductors is motivated by two objectives:

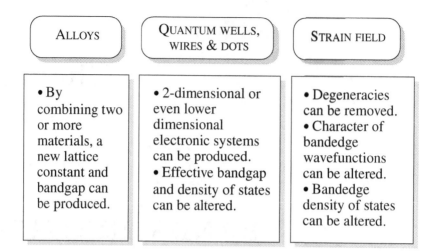

Figure 3.1: Approaches used to modify bandstructure of semiconductors.

1. Achieving a desired bandgap. This motivation drives a great deal of alloy studies in the laser/detector area. The bandgap essentially determines the energy of the light emitted and absorbed.

2. To create a material with a proper lattice constant to match or mismatch with an available substrate. For example, one uses $In_{0.53}Ga_{0.47}As$ alloy since it is lattice matched with InP substrates, which are easily available.

When an alloy $A_x B_{1-x}$ is produced by a random mixing of two (the concepts can be generalized to more than two components) the lattice constant of the alloy is given by Vegard's law:

$$a_{\text{alloy}} = xa_A + (1-x)a_B \tag{3.1}$$

Vegard's law is applicable to random alloys (i.e., there is no phase separation) and to components which have the same crystal structure. Lattice constants for some alloys are shown in Fig. 3.2, assuming Vegard's law. We have used the term random in describing alloys above without describing it properly. When an alloy $A_x B_{1-x}$ is formed, an important question is what is the arrangement of the atoms A and B in the alloy? One could have several extreme cases, namely:

1. A atoms are localized in one region while the B atoms are localized in another region. Such alloys are called *phase separated*.

2. The probability that an atom next to an A-type atom is A is x and the probability that it is B is $(1-x)$. Such alloys are called *random alloys*.

3. A and B atoms form a well-ordered periodic structure, leading to a *superlattice*.

In Fig. 3.3 we show a schematic of the three possibiities described above. Most semiconductor alloys used in the electronics/optoelectronics industry are grown with the intention of making perfectly random alloys.

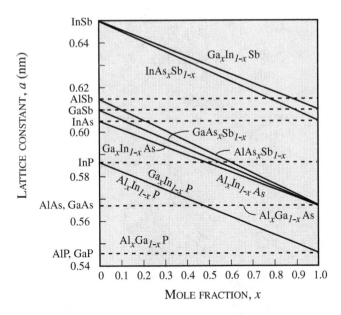

Figure 3.2: Lattice constant as a function of composition for ternary III-V crystalline solid solutions. Vegard's law is assumed to be obeyed in all cases. The dashed lines show regions where miscibility gaps are expected. (After Casey and Panish (1978).

Alloys have a well–defined crystal structure, but the randomness of atoms on the lattice sites means that other than the case of a superlattice (or ordered alloy) there is *no periodicity* in the background crystalline potential. This means that we can no longer use Bloch's theorem to describe the electronic wavefunctions. The electronic states, in general, are not the simple traveling waves but are much more complex with position dependent probabilities.

A simple approximation is usually made to study the bandstructure of alloys and is motivated by Fig. 3.4. The figure shows the atomic potentials distributed in real space. The randomness of the potentials seen by the electrons is obvious. However, in the virtual crystal approximation the random potential is replaced by an average periodic potential as shown

$$U_{\mathrm{av}}(\boldsymbol{r}) = xU_A(\boldsymbol{r}) + (1-x)U_B(\boldsymbol{r}) \tag{3.2}$$

For example, an implementation of this approach in the tight binding method involves taking the weighted average of the matrix elements. For direct bandgap materials (band-edges are at Γ–point or at $\boldsymbol{k} = 0$) this implies that the bandgaps are also weighted linearly, since the bandedges at $\boldsymbol{k} = 0$ are linear sums of the tight binding matrix elements.

$$E_g^{\mathrm{alloy}} = xE_g^{\mathrm{A}} + (1-x)E_g^{\mathrm{B}} \tag{3.3}$$

In most alloys, however, there is a bowing effect arising from the increasing disorder due to the alloying. An equation that is found to describe the bandgap of

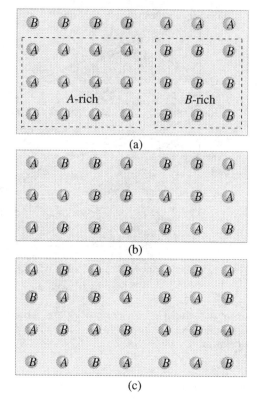

Figure 3.3: A schematic example of (a) a clustered, (b) a random, and (c) an ordered alloy.

alloys resonably well is

$$E_g^{\text{alloy}} = a + bx + cx^2 \tag{3.4}$$

where c is the bowing parameter. In Fig. 3.5 we show bandgaps of alloys made from various material combinations. The solid lines represent direct bandgap regions and the dotted lines the indirect gap regions.

In the virtual crystal approximation it can be easily seen that the bandedge masses scale as

$$\frac{1}{m_{\text{alloy}}^*} = \frac{x}{m_A^*} + \frac{1-x}{m_B^*} \tag{3.5}$$

since

$$
\begin{aligned}
E_{\text{alloy}}(\boldsymbol{k}) &= \frac{\hbar^2 k^2}{2m_{\text{alloy}}^*} \\
&= x\frac{\hbar^2 k^2}{2m_A^*} + (1-x)\frac{\hbar^2 k^2}{2m_B^*}
\end{aligned} \tag{3.6}
$$

Several important alloys have made extremely important contributions in electronics and optoelectronics. We will briefly discuss a few specific cases.

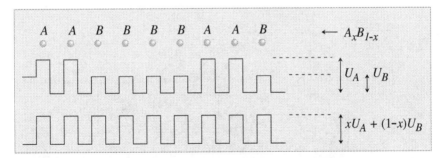

Figure 3.4: Motivation for the virtual crystal approximation. The uppermost section of this figure shows the arrangements of A and B type atoms on a lattice. Associated with these atoms are the corresponding short ranged atomic potentials, U_A and U_B. In the virtual crystal approximation it is assumed that the extended states see an "average" weighted potential $xU_A + (1 - x)U_B$ which is then considered to be periodic.

3.1.1   GaAs/AlAs Alloy

The AlGaAs system is one of the most important alloy system. AlAs and GaAs are nearly lattice matched so that the alloy can be grown on GaAs substrate without strain energy build-up. As a result AlGaAs has become an important component of high speed electronic devices (the modulation doped field effect transistors or MODFETs) and optoelectronic devices (modulators, detectors, and lasers). The alloy has a very interesting switching of the bandgap from direct to indirect. Fig. 3.6 shows the composition dependence of the conduction band valleys. Since the material becomes indirect for Al fraction above ∼35%, most device structures use a composition below this value.

3.1.2   InAs/GaAs Alloy

InAs and GaAs have a lattice mismatch of 7% and their alloys can span a bandgap range of 0.39 eV to 1.5 eV. However, because of the high strain it is not possible to grow all compositions, and the choice of a substrate is very important. Special properties of the $In_xGa_{1-x}As$ alloy have made it an active ingredient of very high speed transistors as well as for fiber optic communication lasers. A lattice matched composition often used is $In_{0.53}Ga_{0.47}As$ ($E_g = 0.88$ eV) which matches to InP. The alloy can then form a quantum well with InP ($E_g = 1.35$ eV) or $In_{0.52}Al_{0.48}As$ ($E_g = 1.45$ eV) as barriers. Other compositions that have become increasingly favorable are the lattice mismatched compositions of $In_xGa_{1-x}As$ (grown on GaAs) and $In_{0.53+x}Ga_{0.47-x}As$ (grown on InP). These "strained" structures have remarkable properties which are being exploited for high speed electronics and optoelectronics and will be discussed in detail in the next chapter. The $In_{0.53}Ga_{0.47}As$ alloy has found a niche in high speed electronics because of its small effective mass ($m_e^* = 0.04m_0$) and excellent low and high field transport properties. Fig. 3.7 shows some properties of this alloy.

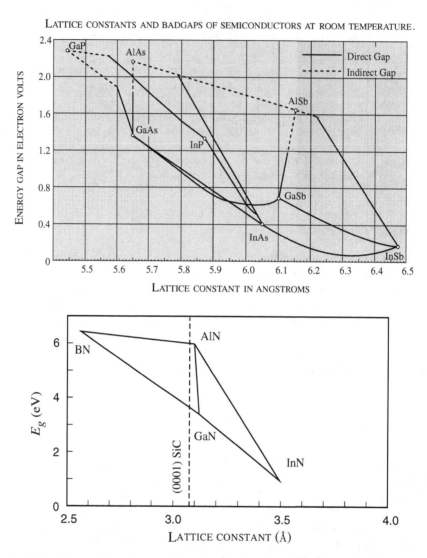

Figure 3.5: The range of bandgaps achievable by alloy formation in some III-V compound semiconductors. For the nitride system the lattice constant of SiC is given, since it is a useful substrate for nitrides. The bandgap value of InN has some uncertainty, since it is difficult to obtain high quality InN samples.

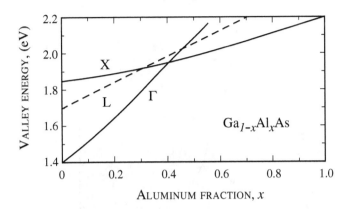

Figure 3.6: The variation of conduction band valleys in AlGaAs as a function of composition at 300 K.

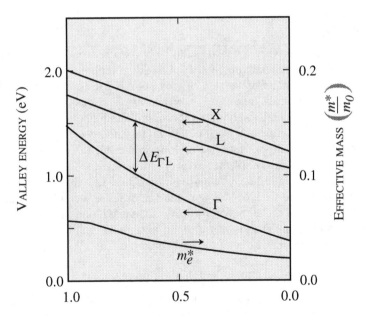

Figure 3.7: Variation of the principal gap energy levels in the alloy of InAs and GaAs. Also shown is the electron effective mass in the alloy.

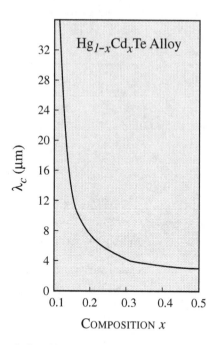

Figure 3.8: Cutoff wavelength for alloy HgCdTe as a function of composition.

3.1.3 HgTe/CdTe Alloy

The HgCdTe system is a very versatile alloy. The two components are very well lattice matched and span a bandgap of 0 to 1.5 eV, remaining direct throughout the composition range. The chief attraction of this alloy is for very small bandgap device applications. These primarily involve night vision applications, thermal imaging in industrial and medical applications, imaging through dense fog, etc. The alloy also has the advantage that it can be grown on both CdTe, ZnTe, and even Si and GaAs substrates. This allows for the possibility of integrating long wavelength detectors with high speed GaAs devices and high density silicon circuitry.

The semimetal HgTe has a "negative" bandgap, i.e., the Γ–point of the s-type states normally associated with conduction band in other semiconductors lies below the p-type Γ–point states. As the alloy composition is changed, one can obtain a wide range of bandgaps. The electron masses can also be extremely small which offer attractive possibilities of very high speed devices. The bandgap of the alloy $Hg_{1-x}Cd_x Te$ is given by the relation

$$E_g(x) = -0.3 + 1.9x \text{ eV} \tag{3.7}$$

In Fig. 3.8 we show the cutoff wavelength corresponding to the alloy bandgap as well as the carrier mass for the HgCdTe alloy. The cutoff wavelength is given by

$$\lambda_c = \frac{1.24 \ \mu m}{E_g(eV)} \tag{3.8}$$

3.1.4 Si/Ge Alloy

Silicon and Germanium have a lattice mismatch of ~4% so only thin layers can be grown on a silicon substrate. The alloy composition remains indirect, there is little motivation from optoelectronic device considerations to grow the alloy. However, there has been great interest in this alloy since it can be a component of Si–SiGe structures and allow heterostructure concepts to be realized in Si technology. Extremely high speed Si/SiGe transistors have been fabricated making this alloy a very important electronic material. The conduction band minima of Si is at X–point while that of Ge is at the L–point. The alloy retains an X–point character for the bandedge up to 85% Ge and then changes its character to L–like.

3.1.5 InN, GaN, AlN System

As can be seen from Fig. 3.5, alloys made from InN, GaN, and AlN can span a very large bandgap range. The alloys can be used for light emission in short wavelength regimes (e.g., for blue light emission), for light absorption (e.g., for solar blind detectors) and for high temperature/high power electronics where large bandgap materials are essential.

A key problem that needs to be surmounted in nitride technology is the lack of an appropriate substrate. Nitrides are usually grown on sapphire or SiC substrates by first depositing a GaN layer to create an effective GaN substrate. Although a variety of novel growth approaches have been explored to produce this effective GaN substrate, there is a high density of dislocations in the wafer.

In Chapter 1 we have discussed polarization effects in the nitride system. These effects play a strong role in electronic properties of nitride heterostructures and will be discussed later in this chapter.

EXAMPLE 3.1 Calculate the bandgap of $Al_{0.3}Ga_{0.7}As$ and $Al_{0.6}Ga_{0.4}As$. Use the virtual crystal approximation with the following values for the conduction band energies measured from the top of the valence band (at 300 K): 1.43 eV and 1.91 eV for the GaAs Γ-point and X-point, respectively. For AlAs the corresponding values are 2.75 eV and 2.15 eV. In the virtual crystal approximation one calculates the position of an energy point in an alloy by simply taking the *weighted average of the same energy points in the alloy components*.

For the $Al_{0.3}Ga_{0.7}As$ alloy, the Γ- and X-point energies are

$$E(\Gamma\text{-point}) = 2.75(0.3) + (1.43)(0.7) = 1.826 \text{ eV}$$
$$E(X\text{-point}) = 2.15(0.3) + (1.91)(0.7) = 1.982 \text{ eV}$$

For the $Al_{0.6}Ga_{0.4}As$ system, the energies are

$$E(\Gamma\text{-point}) = 2.75(0.6) + (1.43)(0.4) = 2.22 \text{ eV}$$
$$E(X\text{-point}) = 2.15(0.6) + (1.91)(0.4) = 2.05 \text{ eV}$$

We see that in $Al_{0.3}Ga_{0.7}As$, the lowest conduction band point is at the Γ-point and the bandgap is direct. However, for the $Al_{0.6}Ga_{0.4}As$, the lowest conduction band is at the X-point and the material is indirect.

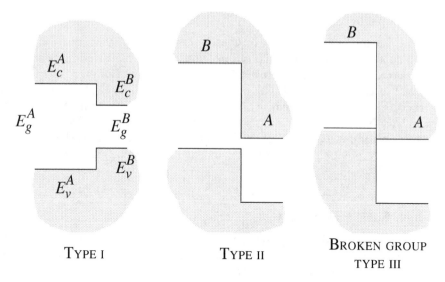

Figure 3.9: Various possible bandedge lineups in heterostructures.

3.2 BANDSTRUCTURE MODIFICATIONS BY HETEROSTRUCTURES

We have noted in Chapter 1 that it is possible to alter the chemical composition of a semiconductor structure in the growth direction by techniques such as MBE and MOCVD. By making heterostructures it is possible to confine electronic states and produce lower dimensional systems. Such lower dimensional systems are now widely exploited to make high performance optoelectronic and electronic devices.

Increased efforts are being dedicated to realizing confinement in two dimensions (quantum wires) and all three dimensions (quantum dots). While the epitaxial techniques for one-dimensional confinement are well-established and conceptually simple, confinement in other directions involves difficult and cumbersome epitaxial or processing technologies. In Chapter 1 we have discussed how strained epitaxy can lead to "self-assembled" quantum dots. This approach has led to a number of interesting devices based on quasi-0D systems. We will primarily focus on quantum wells, but will discuss some important issues in quantum dots as well.

One of the most important issues to be addressed when two semiconductors are brought together abruptly to form an interface is how the bandedges of the two materials line up at the interface. In principle, several possibilities could exist if a semiconductor A with bandgap E_g^A and edges E_v^A and E_c^A is grown with a semiconductor B with gap E_g^B and edges E_v^B and E_c^B. A number of possible band lineups can then arise, as shown in Fig. 3.9.

In semiconductor physics it is usually common to use the position of the electron affinity (or work function) to decide how conduction bands (or valence bands) of two materials line up when they form a heterostructure. However, band lineups based on electron affinity do not work in most cases when two different semiconductors form a heterostructure. This is because of subtle charge sharing effects that occur across atoms on the interface. There have been a number of theoretical studies that can predict general trends in how bands live up. However, the techniques are quite complex and heterostructure designs usually depend on experiments to provide line up information.

As seen in Fig. 3.9 it is possible to have a number of qualitatively different band line ups. The line up may be such that the conduction and valence bandedges of the smaller bandgap material are in the bandgap region of the larger bandgap material. In such a structure the lowest electron state and the highest valence band state exists in the same physical space (i.e., in the narrow gap material). Such heterostructures (called type I) are the most widely used ones. Material combinations such as GaAs/AlGaAs, InGaAs/InP, and GaN/AlGaN, etc., all have type I line up.

A different kind of band line up is produced when the lowest conduction band state is in one material and the highest valence band state is in the other region. If the lowest conduction bandedge is above the highest valence bandedge we have the type II heterostructure. In a type II heterostructure the "effective" bandgap of the structure (i.e., energy difference between the lowest conduction bandedge and highest valence bandedge) can be very small. These kinds of heterostructures are therefore quite useful for long wavelength optoelectronics. Of course, one has to recognize that since the electron and hole states are spatially separated optical transistors will be weak. Antimonide based structures (InSb, GaSb, etc.) are shown to have type II behavior. A system that has been widely studied is InAs/GaSb.

Finally it is possible to have a situation where, as shown on the right hand panel of Fig. 3.9, the conduction bandedge of material A is below the valence bandedge of material B. Such structures are called broken gap Type II (although other names have also been given to them).

Once the band line up is known one has to decide how the electronic states in the heterostructure should be described. An approach that has been quite successfully applied to heterostructures is the $\mathbf{k} \cdot \mathbf{p}$ approach. In its simplest form it is possible to represent the problem by the following simplification of the Schrödinger equation

$$\left[\frac{-\hbar^2}{2m_0} + V(r)\right]\psi(r) = E\psi \rightarrow \left[\frac{-\hbar^2}{2m^*} + E_{edge}\right]\phi = E\phi \tag{3.9}$$

Here the atomistic potential $V(r)$ is replaced by the bandedge energy E_{edge} and the effect of the background potential is contained in the effective mass. The state in the heterostructure has a central cell part which corresponds to that of the bandedge (see Section 2.5.1).

The simple effective mass approach outlined by Eqn. 3.9 works quite well for the conduction band and gives first order results for valence band for unstrained heterostructures. In general one may need to use a multiband $\mathbf{k} \cdot \mathbf{p}$ equation to describe heterostructure states.

3.2.1 Bandstructure in Quantum Wells

To calculate the bandstructure in quantum wells, we recall the nature of the energy levels and wavefunctions near the edges of the bandgap, since the quantum well problem is described in terms of these wavefunctions. We will discuss the case where the well region

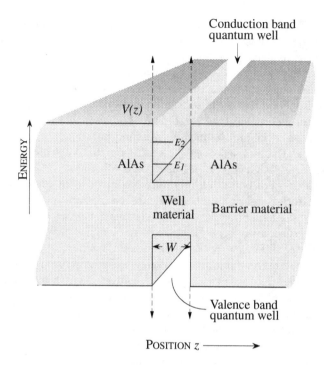

Figure 3.10: A schematic of a quantum well formed for the electron and holes in a heterostructure.

is made up of a direct bandgap material. In this case, the conduction band states are
s-type and valence band states are p-type. It is straightforward to extend these results
to other quantum wells. The simple quantum well structure such as shown in Fig. 3.10
is one of the most studied heterostructures.

The Schrödinger equation for the electron states in the quantum well can be
written in a simple approximation as

$$\left[-\frac{\hbar^2}{2m^*} \nabla^2 + V(z) \right] \Psi = E\Psi \tag{3.10}$$

where m^* is the effective mass of the electron. The wavefunction Ψ can be separated
into its z and ρ (in the x–y plane) dependence and the problem is much simplified

$$\Psi(x, y, z) = e^{ik_x \cdot x} \cdot e^{ik_y \cdot y} f(z) \tag{3.11}$$

where $f(z)$ satisfies

$$\left[-\frac{\hbar^2}{2m^*} \frac{\partial^2}{\partial z^2} + V(z) \right] f(z) = E_n f(z) \tag{3.12}$$

This simple 1D problem is solved in undergraduate quantum mechanics books. Assuming
an infinite barrier approximation, the values of $f(z)$ are (W is the well size)

$$f(z) \quad = \quad \cos \frac{\pi n z}{W}, \text{ if } n \text{ is even}$$

$$= \sin \frac{\pi n z}{W}, \text{ if } n \text{ is odd} \tag{3.13}$$

with energies

$$E_n = \frac{\pi^2 \hbar^2 n^2}{2m^* W^2} \tag{3.14}$$

The energy of the electron bands are then

$$E = E_n + \frac{\hbar^2 k_{\parallel}^2}{2m^*} \tag{3.15}$$

leading to subbands as shown in Fig. 3.11.

If the barrier potential V_c is not infinite, the wavefunction decays exponentially into the barrier region, and is a sine or cosine function in the well. By matching the wavefunction and its derivative at the boundaries one can show that the energy and the wavefunctions are given by the solution to the transcendental equations

$$\alpha \tan \frac{\alpha W}{2} = \beta$$

$$\alpha \cot \frac{\alpha W}{2} = -\beta \tag{3.16}$$

where

$$\alpha = \sqrt{\frac{2m^* E}{\hbar^2}}$$

$$\beta = \sqrt{\frac{2m^* (V_c - E)}{\hbar^2}}$$

These equations can be solved numerically. The solutions give the energy levels E_1, E_2, E_3 ... and the wavefunctions.

Each level E_1, E_2, etc., is actually a subband due to the electron energy in the x-y plane. As shown in Fig. 3.11 we have a series of subbands in the conduction and valence band. In the valence band we have a heavy hole. The subband structure has important consequences for optical and transport properties of heterostructures. An important manifestation of this subband structure is the density of states (DOS) of the electronic bands. The density of states figures importantly in both electrical and optical properties of any system.

The density of states in a quantum well is (see Appendix C)

• Conduction band

$$N(E) = \sum_i \frac{m^*}{\pi \hbar^2} \sigma(E - E_i) \tag{3.17}$$

where σ is the heavyside step function (unity if $E > E_i$; zero otherwise) and E_i are the subband energy levels.

• Valence band

$$N(E) = \sum_i \sum_{j=1}^{2} \frac{m_j^*}{\pi \hbar^2} \sigma(E_{ij} - E) \tag{3.18}$$

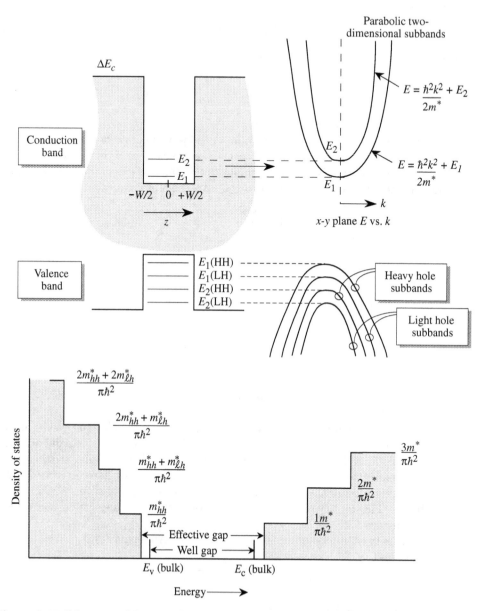

Figure 3.11: Schematic of density of states in a 3–, 2– and 1–dimensional system with parabolic energy momentum relations.

where i represents the subbands for the heavy hole ($j = 1$) and light holes ($j = 2$).

In the simple discussion for the quantum well structure, we have assumed that the conduction band state is a pure s-type state and we have used a very simple-minded effective mass theory to understand the quantum well bandstructure. A more sophisticated calculation can be done where one retains the full description of the bandstructure of the individual components (e.g., an eight–band model). When this is done, it turns out that the results for the conduction band states are not very much affected for unstrained structures. However, for highly strained systems one needs to do a more complicated study. Later we will discuss self-assembled quantum dots where an 8–band approach becomes necessary.

3.2.2 Valence Bandstructure in Quantum Wells

The description of the quantum well bandstructure and the density of states presented above is quite valid for electron states since, as noted earlier, the electron states in direct bandgap materials are adequately described by a single s-type band. However, the valence band states are formed from p-type states leading to HH and LH states. Unfortunately, while the HH(3/2,±3/2) and LH (3/2,±1/2) states are pure states at $\boldsymbol{k} = 0$, they strongly mix away from $\boldsymbol{k} = 0$. Thus, as far as subband level positions are concerned, the starting energies of the subbands can be solved for just as for the electron case, i.e., independently for the HH and LH. However, the dispersion relation for the hole states in the x–y plane is only approximately given by

$$E(\boldsymbol{k}) = E_{n,i} + \frac{\hbar^2 k^2}{2m_i^*} \tag{3.19}$$

where i denotes the HH or LH band. A better description of the hole states is given by solving the Kohn–Luttinger form of the Schrödinger equation

$$[H + V_p] \Psi = E\Psi \tag{3.20}$$

which for the HH, LH system is a 4×4 coupled equation given by the Kohn–Luttinger Hamiltonian discussed earlier. Treating the z component of the momentum as an operator ($k_z = -i(\partial/\partial z)$), we get the following new expressions for the matrix elements.

$$
\begin{aligned}
H_{hh} &= -\frac{\hbar^2}{2m_0} \left[(\gamma_1 + \gamma_2)\left(k_x^2 + k_y^2\right) - (\gamma_1 - 2\gamma_2)\frac{\partial^2}{\partial z^2} \right] + V_p(z) \\
H_{lh} &= -\frac{\hbar^2}{2m_0} \left[(\gamma_1 - \gamma_2)\left(k_x^2 + k_y^2\right) - (\gamma_1 + 2\gamma_2)\frac{\partial^2}{\partial z^2} \right] + V_p(z) \\
c &= \frac{\sqrt{3}\hbar^2}{2m_0} \left[\gamma_2 \left(k_x^2 - k_y^2\right) - 2i\gamma_3 k_x k_y \right] \\
b &= -i\frac{\sqrt{3}\hbar^2}{m_0} \left(-k_y - ik_x\right) \gamma_3 \frac{\partial}{\partial z}
\end{aligned}
\tag{3.21}
$$

where $V_p(z)$ represents the potential profile due to the quantum well. In the presence of biaxial strain, additional terms are to be included and will be discussed later. The general hole solutions can be written as

$$\Psi_h^m(\boldsymbol{k}_\|, z) = \sum_v g_m^v(z)\, U^v(\boldsymbol{r})\, e^{i\mathbf{k}_\| \cdot \boldsymbol{\rho}} \qquad (3.22)$$

where $g_m^v(z)$ is the z-dependent function arising from the confinement of the potential, v is the index representing the total angular momentum of the state, m index for each subband in the well, and $U^v(\boldsymbol{r})$ is the valence band central cell state for the v-spin component in the bulk material.

In the absence of the off-diagonal mixing terms in the Kohn–Luttinger Hamiltonian (the so-called diagonal approximation), the hole problem is as simple to solve as the electron problem. However, in real semiconductors, the off-diagonal mixing is quite strong and must be included for quantitative comparison to experiments.

A technique commonly employed to solve for the hole eigenfunctions and dispersion relations is to solve the problem variationally. Although this approach has been shown to work quite well, it has the usual disadvantages of variational techniques, viz. considerable insight is required to choose the form of the starting wave function and the techniques become increasingly complex for excited states. In fact, the hole dispersion relations need not be solved variationally. The equation can be written in a finite difference form as an eigenvalue problem and then solved by matrix solving techniques.

In Fig. 3.12 we show a typical valence bandstructure dispersion relation in a quantum well. Results are shown for the valence band of a 100 Å GaAs/Al$_{0.3}$Ga$_{0.7}$As quantum well. As can be seen the subbands are quite non-parabolic. The character of the hole states is represented by pure angular momentum states at the zone center, but there is a strong mixing of states as one proceeds away from the $k = 0$ point.

3.3   SUB–2-DIMENSIONAL SYSTEMS

Quantum well systems discussed in the previous section have found applications in a number of devices. Examples include high-speed optical modulators and low threshold lasers. Quantum wells are relatively easy to fabricate, since they require opening and closing of shutters or valves while the crystal is being grown. No etching/lithography processes are involved. It is even possible to grow quantum wells where they well region is just one monolayer.

If one examines the electronic properties of sub–2-dimensional systems it is clear that there are many advantages that can be exploited for superior device design. As a result there has been a great deal of activity to fabricate quantum wire (a quasi 1D system) and quantum dot (a quasi 0D) structures. The key difference between the fabrication of a quantum dot and a quantum wire (or dot) is that in a wire (or dot) material composition has to be altered not just in the growth direction, but in the plane normal to the growth direction. As a result the fabrication of sub–2D systems require a lot more than opening or closing of shutters in a growth chamber. As discussed in Chapter 1, there are two approaches to fabricating sub–2D systems: (i) etching/regrowth and (ii) self-assembly.

In the etching/regrowth approach a quantum well is grown and then etched leaving narrow regions of the "well" material exposed. The etched region is then filled

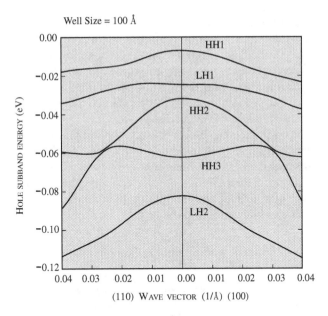

Figure 3.12: Hole dispersion relations in a 100 Å GaAs/Al$_{0.3}$Ga$_{0.7}$As quantum well assuming the same Kohn–Luttinger parameters in the well and barrier.

in by regrowth of a large bandgap (barrier) material to produce a sub–2D system. While this approach seems to simple in theory, in practice it has not proven to be very successful. The difficulty arises from the quality of the interface produced at the etched negative regrowth region. Due to impurities, broken or distorted bonds, etc., this interface is not defect-free. The defect states dominate the electronic properties of the system and the benefits of quantum effects are not realized. Of course, one has to realize that growth–etching–regrowth technology is improving and it may eventually be possible to fabricate sub–2D systems in this manner.

The self-assembly approach exploits thermodynamic and kinetic principles to grow sub–2D systems without any etching–regrowth. One example is the *V*-groove approach that has been used to grow quantum wires using the AlGaAs/GaAs system. A *V*-groove is etched into a substrate and an epilayer, say, GaAs is grown on the substrate. The initial buffer layer buries the initial damaged region. After that AlGaAs is grown on the *V*-groove. In MBE Al and Ga atoms impinge on the substrate randomly, but because the Ga-As bond is not as strong as the Al-As bond, the Ga atoms migrate a lot faster into the groove. This produces a Ga rich region at the groove which can then be covered with a barrier region to produce quantum wires.

An approach that has become quite successful in the fabrication of quantum dots is the use of strain. In Chapter 1, Section 1.4 we have discussed how the growth mode changes from monolayer-by-monolayer to island mode as strain is increased in the epilayer. This idea has been used to fabricate quantum dots from material systems like InGaAs/GaAs and SiGe/Si. In these self-assembled quantum dots no regrowth is needed and the structural quality of the system is very good.

The self-assembled quantum dots are under a very high degree of strain ($\epsilon \sim 2 - 3\%$). As a result the simple effective mass equation which allows one to describe the conduction band states and valence band states separately does not give accurate results. The energy levels are obtained by using bandstructure approaches that include the coupling between the valence and conduction bands. These include an 8-band $\mathbf{k} \cdot \mathbf{p}$ method and full band models. We will discuss results for electronic spectra in self-assembled dots after discussing the deformation potential theory.

The lower dimensional confinement problem can be solved in the same manner as the approach discussed for the 2–dimensional problem. The outcome is a series of subbands for the 1D problem and a series of δ-function density of states for the 0D problem. The density of states for the conduction band in the quantum wire is given by (a similar expression can be written for the valence band)

$$N(E) = \frac{\sqrt{2}m^{*1/2}}{\pi\hbar} \sum_i (E - E_i)^{-1/2} \tag{3.23}$$

where E_i are the subband levels. Note that there may several degenerate subbands depending upon the wire symmetry.

In Fig. 3.13 we show a schematic of the density of states in 3D, 2D, 1D, and 0D systems.

EXAMPLE 3.2 Using a simple infinite barrier approximation, calculate the "effective bandgap" of a 100 Å GaAs/AlAs quantum well. If there is a one-monolayer fluctuation in the well size, how much will the effective bandgap change? This example gives an idea of how stringent the control has to be in order to exploit heterostructures.

The confinement of the electron ($m^* = 0.067\ m_0$) pushes the effective conduction band up, and the confinement of electrons at the valence band push the effective edge down ($m_{hh}^* = 0.4\ m_0$). The change in the electron ground state is (using $n = 1$) 55.77 meV. The shift (downwards) in the valence band is (why do we only need to worry about the shift of the HH band to find the effective gap?) 9.34 meV. The net shift is 65.11 meV.

The effective bandgap is thus 65.11 meV larger than the bulk GaAs bandgap.

If the well size changes by one monolayer (e.g., goes from 100 Å to 102.86 Å), the change in the electron level is

$$\Delta E_c = E_e \left[1 - \frac{(100)^2}{(102.86)^2} \right] = E_c \times 0.055$$
$$= 3.06 \text{ meV}$$

The hole energy changes by

$$\Delta E_{hh} = E_{hh} \times 0.055 = 0.51 \text{ meV}$$

Thus, the bandgap changes by 3.56 meV for a one monolayer variation. In optical frequencies this represents a change of 0.86 Terrahertz, which is too large a shift for many optoelectronic applications.

EXAMPLE 3.3 Consider a 100 Å GaAs quantum well. The Fermi energy at 600 K is known to have a value

$$E_F = E_c(\text{GaAs}) + 0.15 \text{ eV}$$

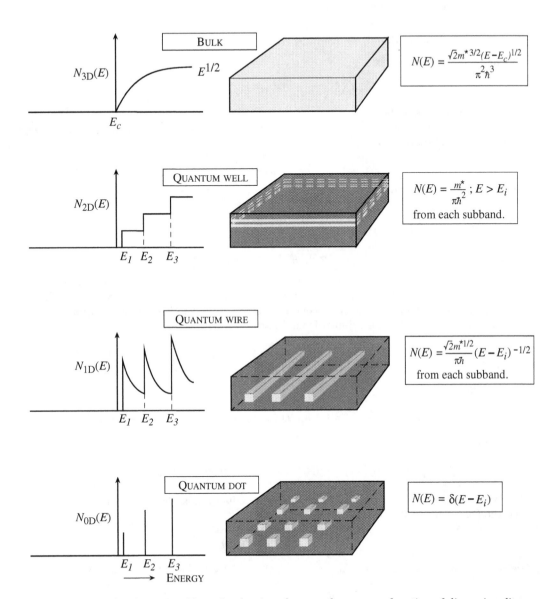

Figure 3.13: A schematic of how the density of states change as a function of dimensionality.

where $E_c(GaAs)$ is the bandedge of the GaAs conduction band.

- Calculate the electron density in the well in unites of cm^{-2}. You can use the infinite barrier model to get the subbband energies.

- Calculate the position of the Fermi level at 0 K.

The integral

$$\int \frac{dx}{e^x + 1} = \ell n(1 + e^{-x})$$

will be useful for this problem.

We first find the energies of the subband level sin the quantum well using the infinite barrier model. We have

$$E_1 = \frac{\hbar^2 \pi^2}{2m^* W^2}$$

Using $m^* = 0.067\ m_9$ and $W = 1.0 \times 10^{-8}$ m, we get

$$E_1 = 0.056 \text{ eV}$$

Also

$$E_2 = 0.222;\ E_3 = 0.504 \text{ eV; etc.}$$

We do not have to worry about the occupation of level 3 and higher, since their energies are well above the Fermi level. Note that the density of states of the first level starts at E_1, and is constant. The density of states of the second level starts at E_2 and is constant, etc. Using the integral given we have, for the 2-dimensional electron density,

$$n_{2D} = \frac{m^*}{\pi \hbar^2} k_B T \left(\ell n \left[+ \exp \left(\frac{E_F - E_1}{k_B T} \right) \right] + \ell n \left[1 + \exp \left(\frac{E_F - E_2}{k_B T} \right) \right] \right)$$

We have the following:

$$k_B T = 0.052 \text{ eV};\ E_F - E_1 = 0.15 - 0.052 = 0.094 \text{ eV};\ E_F - E_2 = -0.072 \text{ eV}$$

Using these values we get

$$n_{2D} = \left(2.82 \times 10^{13} \text{ cm}^{-2} \right) (0.052 \text{ eV}) (1.96 + 0.22) = 3.2 \times 10^{12} \text{ cm}^{-2}$$

Note that the contribution from the third band is negligible and is, therefore, not included. To find the Fermi level at 0 K we use the relation

$$n_{2D} = (E_F - E_1) N(E)$$

Using the value of n_{2D} calculated above and $N(E) = 2.82 \times 10^{13}$ eV^{-1} cm^{-2}, we get

$$E_F - E_1 = 0.113 \text{ eV}$$

or

$$E_F - E_c(GaAs) = 0.113 + 0.056 = 0.169 \text{ eV}$$

Note that the second subband has no electrons at 0 K, since the Fermi level is below the value of E_2.

3.4   STRAIN AND DEFORMATION POTENTIAL THEORY

The role of strain on electronic and optical properties of semiconductors has been studied for several decades. Such studies have been extremely important in developing a better theoretical understanding of the bandstructure of semiconductors by identifying symmetries of energy states involved in optically observed transitions. The studies have also been important since they allow determination of deformation potentials that are important in understanding electronic transport in presence of lattice scattering. Until recently, the strain in the semiconductor was introduced by external means, using sophisticated apparatus such as diamond anvil cells, and the experiments were done primarily to clarify the physics of semiconductors.

With the advent of strained heteroepitaxy it is now possible to incorporate strain into an epitaxial film. In fact, strain of a few percent can be built-in simply by growing a film on a mismatched substrate as discussed in Chapter 1, Section 1.4. In that section we have discussed the nature of strain tensor by epitaxy.

Once the strain tensor is known, we are ready to apply the deformation potential theory to calculate the effects of strain on various eigenstates in the Brillouin zone. The strain perturbation Hamiltonian is defined and its effects are calculated in the simple first order perturbation theory. In general we have

$$H_\epsilon^{\alpha\beta} = \sum_{ij} D_{ij}^{\alpha\beta} \epsilon_{ij} \tag{3.24}$$

where D_{ij} is the deformation potential operator which transforms under symmetry operations as a second rank tensor. $D_{ij}^{\alpha\beta}$ are the matrix elements of D_{ij}.

The deformation potential is not calculated in an ab initio manner, but is usually fitted to experimental results. As in the case of the force constants, the number of independent elements in the deformation potential can also be reduced based on the symmetry of the wavefunctions.

Let us consider the matrix elements of the strain tensor (Eqn. 3.29) which defines the perturbation Hamiltonian for the strain. We are interested in the $(\alpha\beta)$ element of the matrix where α and β represent the basic functions being used for the unperturbed crystal. The symmetry of the basis states is critical in simplifying the number of independent deformation potential elements. In Chapter 2, Fig. 2.12 we have shown a schematic of the nature of the symmetries encountered at the bandedges of direct and indirect semiconductors. Due to the importance of this central cell nature we reproduce a schematic in Fig. 3.14. We will discuss the deformation potentials realizing that we are primarily interested in seeing the effect of the strain on bandedge states.

Case 1: Let us first examine the nondegenerate Γ_2' state which represents the conduction bandedge of direct bandgap semiconductors. This state is an s-type state and has the full cubic symmetry associated with it. Let us perform the following two steps to establish the independent number of deformation potentials for this state:

Step 1: Rotate by 120° about (111). This causes the following transformation in the new axes (primes).

$$\begin{aligned} x' &= y \\ y' &= z \end{aligned}$$

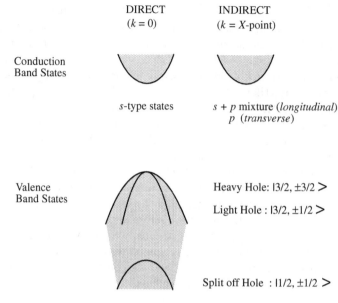

Figure 3.14: A schematic showing the nature of the central cell symmetry at the bandedges of direct and indirect semiconductors.

$$z' = x \tag{3.25}$$

Consequently, we have

$$\left. \begin{array}{rcl} D_{xx} & = & D_{z'z'} \\ D_{yy} & = & D_{x'x'} \\ D_{zz} & = & D_{y'y'} \end{array} \right\} \Rightarrow D_{xx} = D_{yy} = D_{zz} \tag{3.26}$$

$$\left. \begin{array}{rcl} D_{xy} & = & D_{z'x'} \\ D_{zx} & = & D_{y'z'} \\ D_{yz} & = & D_{x'y'} \end{array} \right\} \Rightarrow D_{xy} = D_{yz} = D_{zx} \tag{3.27}$$

Step 2: Rotate about (001) by 90° leading to the transformation

$$\begin{array}{rcl} x' & = & y \\ y' & = & -x \\ z' & = & z \end{array} \tag{3.28}$$

The deformation potentials transform as

$$\begin{array}{rcl} D_{xy} & = & -D_{y'x'} \\ & = & -D_{x'y'} \end{array} \tag{3.29}$$

Thus, $D_{xy} = 0$, and by symmetry, $D_{xy} = Dyz = D_{zx} = 0$.

Thus for Γ_2' states there is only one deformation potential and the effect of the strain is to produce a shift in energy.

$$
\begin{aligned}
\delta E^{(000)} &= H_\epsilon \\
&= D_{xx}(\epsilon_{xx} + \epsilon_{yy} + \epsilon_{zz})
\end{aligned} \tag{3.30}
$$

Conventionally we write

$$
D_{xx} = \Xi_d^{(000)} \tag{3.31}
$$

where $\Xi_d^{(000)}$ represents the dilation deformation potential for the conduction band (000) valley.

Case 2: States along the (100) direction in k-space or Δ_1 symmetry states. We again carry out the following symmetry transformation.

Step 1: Rotate by 90° about the (100) axis, producing the transformation

$$
\begin{aligned}
x' &= x \\
y' &= z \\
z' &= -y
\end{aligned} \tag{3.32}
$$

The deformation potentials transform as

$$
\left.
\begin{aligned}
D_{xx} &= D_{x'x'} \\
D_{yy} &= D_{z'z'} \\
D_{zz} &= D_{y'y'}
\end{aligned}
\right\} \Rightarrow D_{yy} = D_{zz}
$$
$$
\left.
\begin{aligned}
D_{xy} &= -D_{x'z'} \\
D_{zx} &= D_{x'y'}
\end{aligned}
\right\} \Rightarrow D_{xy} = -D_{x'y'}
$$
$$
D_{yz} = -D_{z'y'} \Rightarrow D_{yz} = 0 \tag{3.33}
$$

Thus,

$$
\delta E^{(100)} = D_{xx}\epsilon_{xx} + D_{yy}(\epsilon_{yy} + \epsilon_{zz}) \tag{3.34}
$$

Once again we write

$$
\begin{aligned}
D_{yy} &= \Xi_d^{(100)} \\
D_{xx} &= \Xi_d^{(100)} + \Xi_u^{(100)}
\end{aligned} \tag{3.35}
$$

where d and u represent dilation and uniaxial portions. This gives

$$
\delta E^{(100)} = \Xi_d^{(100)}(\epsilon_{xx} + \epsilon_{yy} + \epsilon_{zz}) + \Xi_u^{(100)}\epsilon_{xx} \tag{3.36}
$$

By symmetry we can write

$$
\delta E^{(010)} = \Xi_d^{(100)}(\epsilon_{xx} + \epsilon_{yy} + \epsilon_{zz}) + \Xi_u^{(100)}\epsilon_{yy} \tag{3.37}
$$

$$
\delta E^{(001)} = \Xi_d^{(100)}(\epsilon_{xx} + \epsilon_{yy} + \epsilon_{zz}) + \Xi_u^{(100)}\epsilon_{zz} \tag{3.38}
$$

We note that if the strain tensor is such that the diagonal elements are unequal (as is the case in strained epitaxy), the strain will split the degeneracy of the Δ_1 branches.

Case 3: The L_1 symmetry or states along (111) direction in k-space.
Step 1: Rotate by 120° about (111) axis causing the transformation

$$
\begin{aligned}
x' &= y \\
y' &= z \\
z' &= x
\end{aligned}
\tag{3.39}
$$

The deformation potentials transform as for the Γ_2' case giving

$$
D_{xx} = D_{yy} = D_{zz}
\tag{3.40}
$$

$$
D_{xy} = D_{yz} = D_{xz}
\tag{3.41}
$$

Thus we get a perturbation

$$
\delta E^{(111)} = D_{xx}(\epsilon_{xx} + \epsilon_{yy} + \epsilon_{zz}) + 2D_{xy}(\epsilon_{xy} + \epsilon_{yz} + \epsilon_{zx})
\tag{3.42}
$$

Conventionally, we write

$$
\begin{aligned}
D_{xx} &= \Xi_d^{(111)} + \frac{1}{3}\Xi_u^{(111)} \\
D_{xy} &= \frac{1}{3}\Xi_u^{(111)}
\end{aligned}
\tag{3.43}
$$

By similar transformation we find that

$$
\begin{aligned}
\delta E^{(11\bar{1})} &= D_{xx}(\epsilon_{xx} + \epsilon_{yy} + \epsilon_{zz}) + 2D_{xy}(\epsilon_{xy} - \epsilon_{yz} - \epsilon_{zx}) \\
\delta E^{(1\bar{1}1)} &= D_{xx}(\epsilon_{xx} + \epsilon_{yy} + \epsilon_{zz}) + 2D_{xy}(-\epsilon_{xy} - \epsilon_{yz} + \epsilon_{zx}) \\
\delta E^{(\bar{1}11)} &= D_{xx}(\epsilon_{xx} + \epsilon_{yy} + \epsilon_{zz}) + 2D_{xy}(-\epsilon_{xy} + \epsilon_{yz} - \epsilon_{zx})
\end{aligned}
\tag{3.44}
$$

Case 4: The triple degenerate states describing the valence bandedge.
The valence band states are defined (near the bandedge) by primarily p_x, p_y, p_z (denoted by x,y,z basis states. Consider a matrix element H_ϵ^{xx}.

$$
\begin{aligned}
H_\epsilon^{xx} &= \langle x|H_\epsilon|x\rangle \\
&= \sum_{ij} D_{ij}^{xx}\epsilon_{ij}
\end{aligned}
$$

Step 1: Rotate by 90° about (100). This causes the transformation

$$
\begin{aligned}
x' &= x \\
y' &= z \\
z' &= -y
\end{aligned}
\tag{3.45}
$$

Using arguments similar to those used before

$$
\begin{aligned}
D_{xx}^{xx} = D_{x'x'}^{x'x'} &= \ell \\
\left.\begin{aligned}
D_{yy}^{xx} &= D_{z'z'}^{x'x'} \\
D_{zz}^{xx} &= D_{y'y'}^{x'x'}
\end{aligned}\right\} &\Rightarrow D_{yy}^{xx} = D_{zz}^{xx} = m \\
\left.\begin{aligned}
D_{xy}^{xx} &= -D_{x'z'}^{x'x'} \\
D_{xz}^{xx} &= D_{x'y'}^{x'x'}
\end{aligned}\right\} &= 0 \\
D_{yz}^{xx} = -D_{z'y'}^{x'x'} &= 0
\end{aligned}
\tag{3.46}
$$

Also examining terms like H_ϵ^{xy}, we rotate by 90° about (001). Then it is easy to see that

$$\left.\begin{array}{rcl} D_{xx}^{xy} &=& -D_{y'y'}^{x'y'} \\ D_{yy}^{xy} &=& -D_{x'x'}^{x'y'} \end{array}\right\} \Rightarrow D_{xx}^{xy} = -D_{yy}^{xy} \tag{3.47}$$

These are zero if we consider a reflection through $x = y$.

$$D_{zz}^{xy} = -D_{z'z'}^{x'y'} = 0 \tag{3.48}$$

$$D_{zz}^{xy} = -D_{z'z'}^{x'y'} = 0 \tag{3.49}$$

$$\left.\begin{array}{rcl} D_{xz}^{xy} &=& D_{y'z'}^{x'y'} \\ D_{yz}^{xy} &=& -D_{x'z'}^{x'y'} \end{array}\right\} \Rightarrow D_{xz}^{xy} = D_{yz}^{xy} = 0 \tag{3.50}$$

Thus the unique deformation potentials are

$$\begin{array}{rcl} D_{xx}^{xx} &=& \ell \\ D_{yy}^{xx} &=& m \\ D_{xy}^{xy} &=& n \end{array} \tag{3.51}$$

We often use the related definition which is more useful for the angular momentum basis

$$\begin{array}{rcl} a &=& \dfrac{\ell + 2m}{3} \\[2mm] b &=& \dfrac{\ell - m}{3} \\[2mm] d &=& \dfrac{n}{\sqrt{3}} \end{array} \tag{3.52}$$

We have already discussed the strain tensor in epitaxial growth. For (001) growth which has been the main growth direction studied because of its compatibility with technology of processing we have

$$\begin{array}{rcl} \epsilon_{xx} = \epsilon_{yy} &=& \epsilon \\[2mm] \epsilon_{zz} &=& -\dfrac{2c_{12}}{c_{11}}\epsilon \end{array} \tag{3.53}$$

Let us evaluate the matrix elements in the angular momentum basis instead of the p_x, p_y, p_z basis, since the angular momentum basis is used to describe the heavy hole, light hole states (for a diagonal tensor produced by (001) epitaxy).

$$\begin{array}{rcl} \left\langle \dfrac{3}{2},\dfrac{3}{2} \Big| H_\epsilon \Big| \dfrac{3}{2},\dfrac{3}{2} \right\rangle &=& \dfrac{1}{2} \langle p_x - ip_y |H_\epsilon| p_x + ip_y \rangle \\[3mm] &=& \dfrac{1}{2}[\langle p_x |H_\epsilon| p_x \rangle + i\langle p_x |H_\epsilon| p_y \rangle \\[2mm] && -\, i\langle p_y |H_\epsilon| p_x \rangle + \langle p_y |H_\epsilon| p_y \rangle] \\[3mm] &=& \dfrac{1}{2}[\ell\epsilon_{xx} + m(\epsilon_{yy} + \epsilon_{zz}) + \ell\epsilon_{yy} + m(\epsilon_{zz} + \epsilon_{xx})] \\[3mm] &=& \dfrac{1}{2}[\epsilon_{xx}(\ell + m) + \epsilon_{yy}(\ell + m) + 2m\epsilon_{zz}] \end{array} \tag{3.54}$$

In terms of the deformation potentials, a, b, d,

$$
\begin{aligned}
m &= a - b \\
\ell &= a + 2b \\
n &= \sqrt{3}d
\end{aligned}
$$

(3.55)

$$
\begin{aligned}
\left\langle \frac{3}{2}, \frac{3}{2} \middle| H_\epsilon \middle| \frac{3}{2}, \frac{3}{2} \right\rangle &= \frac{1}{2}\left[\epsilon_{xx}(2a+b) + \epsilon_{yy}(2a+b) + \epsilon_{zz}(2a-2b)\right] \\
&= a(\epsilon_{xx} + \epsilon_{yy} + \epsilon_{zz}) + \frac{b}{2}(\epsilon_{xx} + \epsilon_{yy} - 2\epsilon_{zz})
\end{aligned}
$$

(3.56)

If we examine the case where growth is along the (001) direction, as noted earlier,

$$
\begin{aligned}
\epsilon_{xx} &= \epsilon_{yy} = \epsilon \\
\epsilon_{zz} &= -\frac{2c_{12}}{c_{11}}\epsilon
\end{aligned}
$$

(3.57)

we then obtain

$$
\left\langle \frac{3}{2}, \frac{3}{2} \middle| H_\epsilon \middle| \frac{3}{2}, \frac{3}{2} \right\rangle = 2a\frac{(c_{11} - c_{12})}{c_{11}}\epsilon + b\left(\frac{c_{11} + 2c_{12}}{c_{11}}\right)\epsilon
$$

(3.58)

Remember that in our definition

$$
\epsilon = \frac{a_S}{a_L} - 1
$$

(3.59)

In a similar manner it can be shown that

$$
\left\langle \frac{3}{2}, \pm\frac{1}{2} \middle| H_\epsilon \middle| \frac{3}{2}, \pm\frac{1}{2} \right\rangle = 2a\frac{(c_{11} - c_{12})}{c_{11}}\epsilon - b\left(\frac{c_{11} + 2c_{12}}{c_{11}}\right)\epsilon
$$

(3.60)

Restricting ourselves to the HH and LH states, the strain Hamiltonian can be written as (the state ordering is $|3/2, 3/2\rangle$, $|3/2, -1/2\rangle$, $|3/2, 1/2\rangle$, and $|3/2, -3/2\rangle$)

$$
H_\epsilon = \begin{bmatrix}
H_{hh}^\epsilon & H_{12}^\epsilon & H_{13}^\epsilon & 0 \\
H_{12}^{\epsilon*} & H_{lh}^\epsilon & 0 & H_{13}^\epsilon \\
H_{13}^{\epsilon*} & 0 & H_{lh}^\epsilon & -H_{12}^\epsilon \\
0 & H_{13}^{\epsilon*} & -H_{12}^{\epsilon*} & H_{hh}^\epsilon
\end{bmatrix}
$$

(3.61)

where the matrix elements are given by

$$
\begin{aligned}
H_{hh}^\epsilon &= a(\epsilon_{xx} + \epsilon_{yy} + \epsilon_{zz}) - b\left[\epsilon_{zz} - \frac{1}{2}(\epsilon_{xx} + \epsilon_{yy})\right] \\
H_{lh}^\epsilon &= a(\epsilon_{xx} + \epsilon_{yy} + \epsilon_{zz}) + b\left[\epsilon_{zz} - \frac{1}{2}(\epsilon_{xx} + \epsilon_{yy})\right] \\
H_{12}^\epsilon &= -d(\epsilon_{xz} - i\epsilon_{yz})
\end{aligned}
$$

$$= \frac{3}{2},\frac{3}{2} H_\epsilon \frac{3}{2},\frac{1}{2}$$

$$H_{13}^\epsilon = \frac{\sqrt{3}}{2} b(\epsilon_{yy} - \epsilon_{xx}) + id\epsilon_{xy}$$

$$= \frac{3}{2},\frac{3}{2} H_\epsilon \frac{3}{2},\frac{-1}{2} \tag{3.62}$$

Here, the quantities a, b, and d are valence band deformation potentials. As discussed earlier, strains achieved by lattice mismatched epitaxial growth along (001) direction can be characterized by $\epsilon_{xx} = \epsilon_{yy} = \epsilon$, and $\epsilon_{zz} = -(2c_{12}/c_{11})\epsilon$. All of the diagonal strain terms are zero. Using this information, we get

$$H_{hh}^\epsilon = \left[2a\left(\frac{c_{11} - c_{12}}{c_{11}}\right) + b\left(\frac{c_{11} + 2c_{12}}{c_{11}}\right)\right]\epsilon \tag{3.63}$$

$$H_{lh}^\epsilon = \left[2a\left(\frac{c_{11} - c_{12}}{c_{11}}\right) - b\left(\frac{c_{11} + 2c_{12}}{c_{11}}\right)\right]\epsilon \tag{3.64}$$

with all other terms zero. If the hole dispersion is to be described in a quantum well, the hole states $|m, \boldsymbol{k}\rangle$ can be written as

$$\langle \boldsymbol{r}_h|m, \boldsymbol{k}\rangle = \frac{e^{i\boldsymbol{k}\cdot\boldsymbol{\rho}_h}}{2\pi} \sum_v g_m^v(\boldsymbol{k}, z_h) \, U_0^v(\boldsymbol{r}_h) \tag{3.65}$$

where k is the in-plane two-dimensional wave vector, $\boldsymbol{\rho}_h$ is the in-plane radial coordinate, z_h is the coordinate in the growth direction, the U_0^v are the zone-center Bloch functions having spin symmetry v, and m is a subband index. The envelope functions $g_m^v(\boldsymbol{k}, z_h)$ and subband energies $E_m(\boldsymbol{k})$ satisfy the Kohn–Luttinger equation along with the strain effect

$$-\begin{bmatrix} H_{hh} - \frac{1}{2}\delta & b & c & 0 \\ b^* & H_{lh} + \frac{1}{2}\delta & 0 & c \\ c^* & 0 & H_{lh} + \frac{1}{2}\delta & -b \\ 0 & c^* & -b^* & H_{hh} - \frac{1}{2}\delta \end{bmatrix} \begin{bmatrix} g_m^{3/2,3/2}(\boldsymbol{k}, z_h) \\ g_m^{3/2,1/2}(\boldsymbol{k}, z_h) \\ g_m^{3/2,-1/2}(\boldsymbol{k}, z_h) \\ g_m^{3/2,-3/2}(\boldsymbol{k}, z_h) \end{bmatrix}$$

$$= E_m(\boldsymbol{k}) \begin{bmatrix} g_m^{3/2,3/2}(\boldsymbol{k}, z_h) \\ g_m^{3/2,1/2}(\boldsymbol{k}, z_h) \\ g_m^{3/2,-1/2}(\boldsymbol{k}, z_h) \\ g_m^{3/2,-3/2}(\boldsymbol{k}, z_h) \end{bmatrix} \tag{3.66}$$

The δ is the separation of the HH and LH states in a bulk material due to the strain and is given by Eqns. 3.63 and 3.64. For the $In_yGa_{1-y}As$ system it is given by $\delta = -5.966\epsilon$ eV. Note that the function $g_m^v(\boldsymbol{k}, z_h)$ depends on $\boldsymbol{k}$ as well as z_h and that the energy bands are not, in general, parabolic. The matrix elements in Eqn. 3.66 are given by

$$H_{hh} = \frac{\hbar^2}{2m_0}\left[(k_x^2 + k_y^2)(\gamma_1 + \gamma_2) - (\gamma_1 - 2\gamma_2)\frac{\partial^2}{\partial z_h^2}\right] - V^h(z_h)$$

$$H_{lh} = \frac{\hbar^2}{2m_0}\left[\left(k_x^2 + k_y^2\right)(\gamma_1 - \gamma_2) - (\gamma_1 + 2\gamma_2)\frac{\partial^2}{\partial z_h^2}\right] - V^h(z_h)$$

$$c = \frac{\sqrt{3}\hbar^2}{2m_0}\left[\gamma_2\left(k_x^2 - k_y^2\right) - 2i\gamma_3 k_x k_y\right]$$

$$b = -\frac{\sqrt{3}\hbar^2}{m_0}(k_x - ik_y)\gamma_3\frac{\partial}{\partial z_h}$$

where m_0 is the free-electron mass, V_h is the potential profile for the hole, and for GaAs $\gamma_1 = 6.85$, $\gamma_2 = 2.1$, and $\gamma_3 = 2.9$.

Equation 3.66 can be solved by writing it in finite-difference form and diagonalizing the resulting matrix (as described in Appendix C) to obtain the in-plane band structure. For GaAs, InAs type systems, the values of a and b are such that the net effect is that the $(3/2, \pm 3/2)$ state moves a total of 5.96ϵ eV and the $(3/2, \pm 1/2)$ state moves by energy equal to 12.0ϵ eV.

The relationships between the deformation potentials used in the discussion above and what is measured from a hydrostatic pressure measurement are

$$\Xi_d^{(000)} = \Xi_{hyd}^{(000)} + a$$

$$\Xi_d^{(100)} = \Xi_{hyd}^{(100)} - \frac{1}{3}\Xi_u^{(100)} + a$$

$$\Xi_d^{(111)} = \Xi_{hyd}^{(111)} - \frac{1}{3}\Xi_u^{(111)} + a \tag{3.67}$$

From hydrostatic compression

$$\delta E_g(\bar{k}) = \Xi_{hyd}^{(\bar{k})}(\epsilon_{xx} + \epsilon_{yy} + \epsilon_{zz})$$

$$\Xi_{hyd}^{(\bar{k})} = -\frac{1}{3}(c_{11} + 2c_{12})\frac{dE_g}{dP} \tag{3.68}$$

The elastic stiffness constants, c_{11}, c_{12}, and the hydrostatic pressure coefficient of the bandgap, dE_g/dP, are fairly accurately measured by various experiments and we thus get the values of the individual deformation potentials for different bands.

The effect of strain on bandstructure for both conduction band and valence band states is illustrated by examining the direct bandgap material $In_xGa_{1-x}As$ grown on GaAs and the indirect bandgap material Ge_xSi_{1-x} alloy grown on Si. For direct bandgap materials conduction bands, the strain tensor only moves the position of the bandedge and has a rather small effect on the carrier mass.

In Fig. 3.15 we show a schematic of how strain in a layer grown along the (001) direction influences the bandedges in a direct gap semiconductor. The conduction bandedge moves up or down with respect to its unstrained position as discussed earlier, but since it is a non-degenerate state there is no splitting. The valence bandedge is degenerate in the unstrained system. This degeneracy is lifted by quantum confinement even in an unstrained quantum well, but the splitting produced by quantum confinement is usually small (~ 10–15 meV). Under biaxial compressive strain the bandgap of the material increases and the HH and LH degeneracy is lifted. The splitting can easily

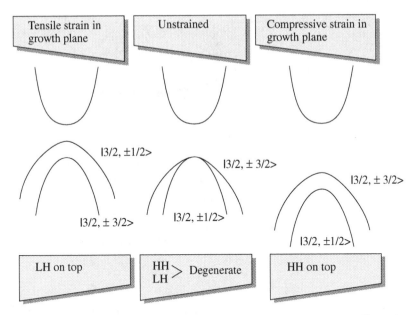

Figure 3.15: Effect of strain on bandedges of a direct bandgap material. The valence band degeneracy is lifted as shown.

approach 100 meV making strain an important resource to alter valence band density of states. Under biaxial compressive strain the HH state is above the LH state, while under biaxial tensile strain the LH state is above the HH state, as shown in Fig. 3.16.

In the case of the indirect bandgap $Si_{1-x}Ge_x$ alloy grown on Si, the conduction band also is significantly affected according to Eqns. 3.36 to 3.38. For (001) growth there is splitting in the 6 equivalent valleys. The results on the bandedge states are shown in Fig. 3.16. Note that the biaxial compressive strain causes a lowering of the four-fold in-plane valleys below the 2 two-fold out of plane valleys. We see that the bandgap of SiGe falls rapidly as Ge is added to Si. This makes the SiGe very useful for Si/SiGe heterostructure devices such as heterojunction bipolar transistors. the splitting of the HH, LH and SO bands also cause a sharp reduction in the density of states mass near the bandedge. The splitting of the conduction bandedge valleys also reduces the conduction band density of states in SiGe.

3.4.1 Strained Quantum Wells

Due to critical thickness-related issues most epitaxially grown strained layers form quantum wells. In these structures the effective bandgap and carrier masses are strongly dependent on strain. In our discussion so far we have mentioned that valence bandedge states are significantly altered by biaxial strain. Indeed, the bandedge density of states mass can be reduced by as much as a factor of 3 by incorporation of strain. To illustrate this we show results for the bandstructure and density of states mass in unstrained and strained quantum wells in Fig. 3.17. In Fig. 3.17a we show the in-plane dispersion of

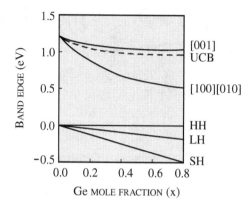

Figure 3.16: Splittings of the conduction band and valence band are shown as a function of alloy composition for $Si_{1-x}Ge_x$ grown on (001) Si. UCB: unstrained conduction band, HH: heavy hole, LH: light hole, SH: split-off hole.

the valence band subbands in a 100 Å $Al_{0.3}Ga_{0.7}As/GaAs$ quantum well. In Fig. 3.17b we show the bandstructure in a 100 Å $Al_{0.3}Ga_{0.7}As/In_{0.1}Ga_{0.9}As$ well. The addition of 10 % In in the well causes a biaxial compressive strain of 0.7 %. We can see the contrast between the results in Figs. 3.18a and 3.17b. The strain has clearly reduced the effective mass of the holes in the valence band.

In Fig. 3.17c we show results for the density of states mass in the HH, LH and SO bands as a function of strain in the 100 Å quantum well. Notice that both compressive and tensile strain cause a reduction in the bandedge hole masses.

We have discussed that in a quantum well the HH and LH states are not degenerate. The splitting is of the order of 10–15 meV, depending upon the well size and valence band discontinuity. We will see later in Chapter 9 that the HH and LH states have different interactions with polarized light. As a result the HH, LH splitting is quite important in determining the polarization response of quantum well devices.

It is interesting to note that the degeneracy lifting produced by quantum confinement of HH and LH states can be restored if the well is under tensile strain. This is illustrated schematically in Fig. 3.18.

It is quite evident from the discussions in this section that strain can significantly affect the bandstructure of semiconductors. The level of strain we have considered can be incorporated in the semiconductor reasonably easily through strained epitaxy. Degeneracy splittings in the valence band can be of the order of 100 meV, accompanied by large changes in band curvatures. It is important to appreciate that such large uniaxial strains are extremely difficult to obtain by external apparatus. It is important to identify whether or not the changes produced by the strain have any impact on the physical properties of the semiconductors. These consequences will be explored later in the chapters on transport and optical properties. The effects are, indeed, found to be significant.

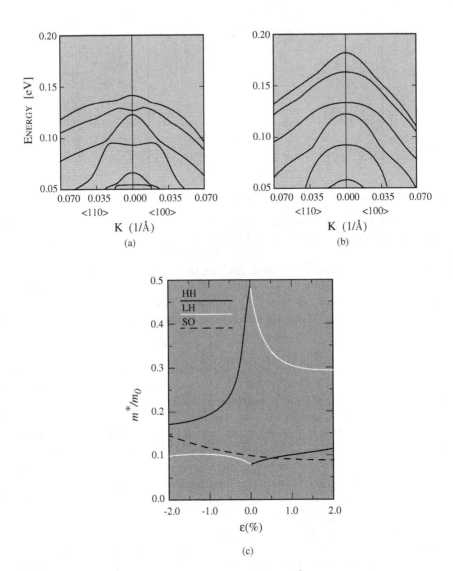

Figure 3.17: Hole subband dispersion curves for a 100 Å wide quantum well in the (a) $Al_{0.30}Ga_{0.70}As/GaAs$ material system and the (b) $Al_{0.30}Ga_{0.70}As/In_{0.1}Ga_{0.9}As$ system. (c) Change in the density of states mass at the valence bandedges as a function of strain.

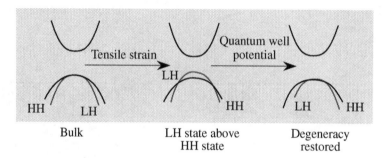

Figure 3.18: Effect of biaxial tensile strain and quantum confinement on conduction and valence band states. By a proper choice of tensile strain the heavy-hole light-hole states can be degenerate at $k = 0$ in a quantum well.

3.4.2   Self-Assembled Quantum Dots

In Chapter 1 we have discussed how under large strain epitaxy epilayers can grow in the island growth mode. This feature of epitaxy has been exploited to fabricate quasi-0-dimensional structures which are called self-assembled quantum dots. These dots are produced without any etching/regrowth making them very attractive for exploration of 0D physics as well as for devices. A number of material combinations have been grown as quantum dots using the strain epitaxy approach. The most widely studied system is the InGaAs/GaAs system where self-assembled dots appear when the In composition reaches ~ 0.4. InGaAs/GaAs quantum dots have been extensively studied and used to fabricate lasers and detectors.

Self-assembled dots usually grow as pyramidal dots although other "lens"-like shapes are also observed. The strain tensor in these dots was discussed in Chapter 2 and has quite large components. The strain between InAs and GaAs is 7 %—large enough that to fully understand the bandstructure of the dots one needs to include coupling between not only the HH, LH and SO bands, but also the conduction band states. This has led to the use of 8 band $k \cdot p$ method and other full bandstructure models.

In Fig. 3.19 we show electronic transitions in a InAs/GaAs dot with a base edge of 124 Å and a height of 62 Å. It can be seen that even though InAs has a bandgap of 0.35 eV, the bandgap of the dot is close to 1.1 eV! This large difference is primarily because of the large compressive strain in the dots.

The dimensions of the dots as well as their exact shape depends upon the strain value and to some extent the growth conditions. It is possible to have a dot density of $\sim 10^{11}$ cm^{-2} through self-assembly techniques. The ability to produce such a high density of 0D systems has naturally made this technique very attractive for device scientists. However, it must be recognized that there is considerable nonuniformity ($\sim 5\%$ in dot dimensions) in a sample. Also the placement of dots is rather random. As a result of the size fluctuations of the dots there are fluctuations in the transitional energies. For many applications such fluctuations cause deleterious effects and can even negate the potential benefits of quantum dots. Nevertheless, devices such as lasers and detectors have been successfully fabricated using self-assembled dots.

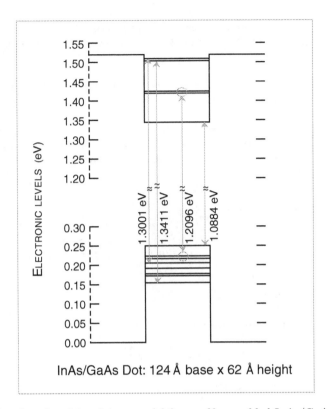

Figure 3.19: Results of an 8 band $k \cdot p$ model for a self-assembled InAs/GaAs quantum dot. Several transition energies are identified to give an idea of the energy ranges. The bandgap of unstrained InAs is only 0.35 eV.

EXAMPLE 3.4 Calculate the heavy hole-light hole splitting produced when $In_{0.2}Ga_{0.8}As$ is grown on a (001) GaAs substrate. Assume that the structure is pseudomorphic and the deformation potential is given by $b_d = -2.0$ eV. The force constants are $C_{11} = 11.5 \times 10^{11}$ dynes/cm^2; $C_{12} = 5.5 \times 10^{11}$ dynes/cm^2.

The splitting produced by a pseudomorphic strain is (in eV)

$$\Delta E_{hh} - \Delta E_{\ell h} = 2b_d \left(\frac{C_{11} + 2C_{12}}{C_{11}} \right) \epsilon$$
$$= -5.91\epsilon$$

The strain between $In_{0.2}Ga_{0.8}As$ and GaAs is

$$\epsilon = \frac{5.653 - 5.734}{5.653} = -0.014$$

Thus the splitting is

$$\Delta E_{hh} - \Delta E_{\ell h} = 0.085 \text{ eV}$$

EXAMPLE 3.5 Estimate the pseudomorphic strain needed in a 100 Å quantum well to cause a merger of the HH and LH states at the top of the valence band. Assume the following parameters:

$$
\begin{aligned}
b_d &= -2.0\ eV; C_{11} = 12.0 \times 10^{11}\ \text{dynes/cm}^2; C_{12} = 6 \times 10^{11}\ \text{dynes/cm}^2; \\
m^*_{hh} &= 0.45\ m_0; m^*_{\ell h} = 0.1\ m_0
\end{aligned}
$$

In this problem we have to ensure that the HH-LH splitting due to strain exactly balances the splitting due to the quantum confinement. The quantum confinement splitting is given approximately by (note that the hole mass is taken as positive, while the electron mass in the valence band is negative)

$$
\begin{aligned}
\Delta E_{hh} - \Delta E_{\ell h} &= \frac{-\hbar^2 \pi^2}{2W^2} \left(\frac{1}{m^*_{hh}} - \frac{1}{m^*_{\ell h}} \right) \\
&= \frac{(1.05 \times 10^{-34}\ \text{Js})^2 (\pi^2)}{2 \times (100 \times 10^{-10}\ \text{m})^2} \left[\frac{1}{(0.45 \times 9.1 \times 10^{-31}\ \text{kg})} - \frac{1}{(0.1 \times 9.1 \times 10^{-31}\ \text{kg})} \right] \\
&= 4.64 \times 10^{-21}\ \text{J} \\
&= 29\ \text{meV}
\end{aligned}
$$

The splitting produced by strain has to have an opposite effect, i.e.,

$$
\Delta E_{\ell h} - \Delta E_{hh} = -2b_d \left(\frac{C_{11} + C_{12}}{C_{11}} \right) \epsilon = 0.029\ \text{eV}
$$

Thus a tensile strain of 0.005 is needed.

3.5   POLAR HETEROSTRUCTURES

In Chapter 1, Section 1.6 we have discussed how in certain materials a net polarization can be present. This can be a result of spontaneous polarization, piezoelectric effect, or ferroelectric effect. In the fcc based semiconductors there is no spontaneous polarization, but there can be net polarization due to strain effects. An important result of polarization differences in heterostructures is that one can have large built-in electric fields at interfaces. A schematic example is shown in Fig. 3.20 where we show a nonpolar (undoped) quantum well and a polar (undoped) quantum well. The interface electric field is given by

$$
\mathbf{F} = \frac{\mathbf{P}}{\epsilon} \tag{3.69}
$$

where $\mathbf{P}$ is the net polarization of the heterostructure and ϵ is the dielectric constant.

The fixed polar charge at polar interfaces can be exploited for built-in field, band bending and creation of mobile charge. In principle, with proper structure design this fixed charge can be exploited to play the role of a sheet of dopants. As noted above, in the zinc blende semiconductors the piezoelectric effect controls the polar charge.

The general polarization developed due to strain is given by

$$
P_i = \sum_{k,l} e_{ikl}\ \epsilon_{kl} \tag{3.70}
$$

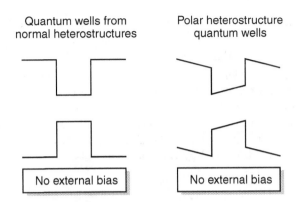

Quantum wells from normal heterostructures

Polar heterostructure quantum wells

No external bias

No external bias

Figure 3.20: A comparison of (undoped) nonpolar and polar quantum well band profiles in the absence of any external electric field.

It was shown by Nye (1957) that for zinc–blende structures, only one piezoelectric constant exists and in the reduced representation ($xx \Rightarrow 1$; $yy \Rightarrow 2$; $zz \Rightarrow 3$; $yz \Rightarrow 4$; $zx \Rightarrow 5$; $xy \Rightarrow 6$), e_{ikl} can be written as e_{im}(m = 1 to 6). The nonzero piezoelectric coefficients are

$$e_{14} = e_{25} = e_{36} \tag{3.71}$$

Thus, only for shear strain does one have a finite polarization. In Chapter 1, Table 1.5 we have given the piezoelectric coefficients for several semiconductors. In Chapter 1 we have seen that for (100) growth the strain tensor is diagonal and as a result there is no piezoelectric polarization for strained epitaxy along this growth. However, if growth is along other directions there are off-diagonal components in the strain tensor. For (111) growth of strained layers, one thus gets a strong dipole moment across a strained quantum well producing an electric field profile as shown in Fig. 3.20. The electric field is given by the equation

$$F = \sqrt{3}\frac{e_{14}\epsilon_{xy}}{\epsilon_s} \text{ V/m} \tag{3.72}$$

where e_{14} is the piezoelectric coefficient (usually in the range of ~ 0.1 C/m^2), ϵ_{xy} is the off-diagonal strain component, and ϵ_s is the dielectric constant of the semiconductor. A straightforward evaluation shows that a strain ϵ_{xy} of about 1% can easily produce an electric field of the range 10^5 V/cm.

In wurtzite structures (applicable for InN, GaN, and AlN) the growth is usually carried out along the c-axis, i.e., (0001) or (000$\bar{1}$) direction. In this case if the substrate is thick the inplane strain tensor is

$$\epsilon_{xx} = \epsilon_{yy} = \frac{a_3}{a_0} - 1 \tag{3.73}$$

where a_s is the substrate lattice constant and a_0 is the unstrained lattice constant of the growth material. For the growth direction the strain is given by

$$\epsilon_{zz} = -\frac{2c_{13}}{c_{33}}\left(\frac{a_s}{a_0} - 1\right) \tag{3.74}$$

These strains induce piezoelectric polarization $\mathbf{P}_{pz}$ in the strained layer which for the wurtzite system is

$$\mathbf{P}_{pz} = e_{33}\epsilon_{zz} + e_{31}\left(\epsilon_{xx} + \epsilon_{yy}\right) \tag{3.75}$$

where e_{33} and e_{31} are the piezoelectric constants. The direction of piezoelectric field is along (0001) direction, or Ga-faced direction according to the convention that the Ga atoms are on the top position of the bilayer. Table 1.5 shows the piezoelectric coefficients and the spontaneous polarization for AlN, GaN and InN. Typical values of c_{13} and c_{33} for the InGaN system are 109 Gpa and 355 Gpa respectively so that the ratio $2c_{13}/c_{33}$ is ~ 0.6. The electric field induced by the polarization is

$$F = \frac{P}{\epsilon_s} \tag{3.76}$$

When we examine the spontaneous polarization values for the nitrides we notice that InN and GaN have essentially the same values. As a result, in the InGaN/GaN structure the polar charges at the interface are due to the piezoelectric effect alone. On the other hand, for AlGaN films grown pseudomorphically on a GaN substrate the piezoelectric polarization and spontaneous polarization (difference between AlGaN and GaN values) are roughly equal.

When we compare the polarization values in the zinc blende semiconductors and wurtzite semiconductors we see that the wurtzite materials have an order of magnitude larger polarization. This results in large built-in interface charge and electric fields. Consider, for example, a case where the effective substrate is GaN and an $Al_xGa_{1-x}N$ overlayer is grown coherently. The polarization is found to have the value

$$\begin{aligned} \mathbf{P}(x) &= \mathbf{P}_{pz} + \mathbf{P}_{sp} \\ &= \left(-3.2x - 1.9x^2\right) \times 10^{-6} \ \mathrm{C/cm}^2 \\ &\quad 5.2 \times 10^{-6}x \ \mathrm{C/cm}^2 \end{aligned} \tag{3.77}$$

We see that in this system the effects arising from piezoelectric effect and spontaneous polarization mismathc are comparable. Note that the two effects cqan have opposite directions as well as depending on the surface termination conditions and the lattice mismatch between the overlayer and the effective substrate. The electric field associated with the polarization given above is

$$F(x) = \left(-9.5x - 2.1x^2\right) \ \mathrm{MV/cm} \tag{3.78}$$

We see that the built-in field and sheet charge values are very large. It is easy to produce fields around 10^6 V/cm and charge density around 10^{13} cm^2.

For the heterostructure $In_xGa_{1-x}N$ grown on a GaN effective substrate the polarization charge density and field values are

$$\begin{aligned} P(x) &= \left(14.1x + 4.9x^2\right) \times 10^{-6} \ \mathrm{C/cm}^2 \\ &\quad -0.3 \times 10^{-6}x \ \mathrm{C/cm}^2 \\ F(x) &= \left(15.6x + 5.5x^2\right) \ \mathrm{MV/cm} \end{aligned} \tag{3.79}$$

In this system the spontaneous polarization effect is negligible. However, the piezoelectric effect is very strong.

3.6 TECHNOLOGY ISSUES

At the end of Chapter 2, we examined some of the driving forces behind some of the technologies. The use of alloys and heterostructures adds a tremendous versatility to the available parameter space to exploit. Semiconductor alloys are already an integral part of many advanced technology systems. Consider the following examples.

- The HgCdTe alloy is the most important high-performance imaging material for long wavelength applications ($10 - 14$ μm). These applications include night vision, seeing through fog, thermal imaging of the human body parts for medical applications, and a host of special purpose applications involving thermal tracking.

- The AlGaAs alloy is an important ingredient in GaAs/AlGaAs heterostructure devices which drive a multitude of technologies including microwave circuits operating up to 100 GHz, lasers for local area networks, and compact disc players.

- InGaAs and InGaAsP alloy systems are active ingredients of MMICs operating above 100 GHz and long-haul optical communication lasers.

While alloys are important ingredients of many technologies, it must be emphasized again that alloys are not perfectly periodic structures. This results in random potential fluctuations which leads to an important scattering mechanism that limits certain performances. For example, the low temperature low field mobility is severely affected by alloy scattering as is the exciton linewidth of optical modulators (discussed in Chapter 10). The growth and fabrication issues in alloy systems are also sometimes serious due to miscibility gaps that may be present.

Technologies based on the heterostructure concepts are also evident in many high performance systems in use today. In Table 3.1 we provide an overview of some of the important technology needs that heterostructure-based structures have fulfilled. The most important heterostructure based devices are the MODFET, the HBT, and the quantum well laser. These devices are the driving force behind advances in MMICs operating above 100 GHz and low power optoelectronic systems. However, it must be emphasized that while the heterostructures can produce an extremely broad range of physical phenomena, not all of these effects can result in real devices. Over the last decade, hundreds of proposals, many seemingly very attractive, have been made for devices using heterostructure physics. Only a handful have made it into systems.

3.7 PROBLEMS

Section 3.1

3.1 Consider the alloy system $Si_{1-x}Ge_x$. Using the virtual crystal approximation, calculate the positions of the Γ, X, and L point energies in the lowest conduction band. Use the top of the valence band as a reference. At what Ge composition does the conduction bandedge change from X-like to L-like?

3.2 Using the bandstructure of GaAs and AlAs, calculate the conduction band minima at Γ, X, L points in $Al_xGa_{1-x}As$ alloy, as x varies from 0 to 1.

3.3 Glass fibers have a minimum loss window for light transmission at 1.3 μm and 1.55 μm. These wavelengths determine the laser materials of choice for long distance

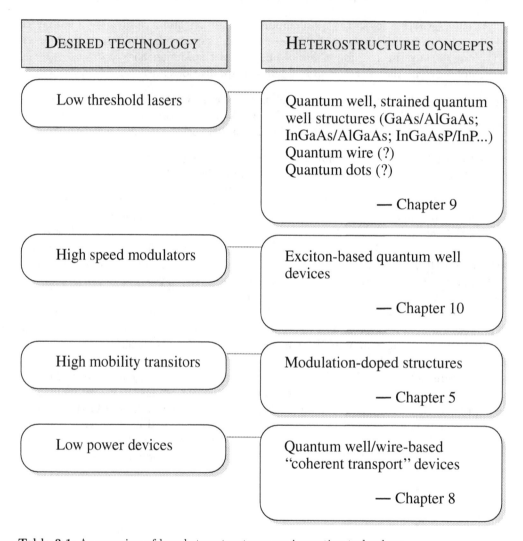

DESIRED TECHNOLOGY	HETEROSTRUCTURE CONCEPTS
Low threshold lasers	Quantum well, strained quantum well structures (GaAs/AlGaAs; InGaAs/AlGaAs; InGaAsP/InP...) Quantum wire (?) Quantum dots (?) — Chapter 9
High speed modulators	Exciton-based quantum well devices — Chapter 10
High mobility transitors	Modulation-doped structures — Chapter 5
Low power devices	Quantum well/wire-based "coherent transport" devices — Chapter 8

Table 3.1: An overview of how heterostructures are impacting technology.

optical communication. List the various semiconductor combinations that can be used for lasers and detectors in such a system. List only those materials that can be grown on GaAs or InP substrates with a strain of less than 2%. Using the virtual crystal approximation, estimate the electron effective masses of these materials.

Sections 3.2–3.3

3.4 Using Vegard's Law for the lattice constant of an alloy (i.e., lattice constant is the weighted average), find the bandgaps of alloys made in InAs, InP, GaAs, and GaP which can be lattice matched to InP.

3.5 For long haul optical communication, the optical transmission losses in a fiber dictate that the optical beam must have a wavelength of either 1.3 μm or 1.55 μm. Which alloy combinations lattice-matched to InP have a bandgap corresponding to these wavelengths?

3.6 Using the virtual crystal approximation, up to what Al composition does the alloy $Al_xGa_{1-x}As$ remain a direct gap semiconductor? What is the maximum bandgap achievable in the direct alloy?

3.7 Calculate the composition of $Hg_xCd_{1-x}Te$ which can be used for a night vision detector with bandgap corresponding to a photon energy of 0.1 eV. Bandgap of CdTe is 1.6 eV *and that of HgTe is -0.3 eV* at low temperters around 4 K.

3.8 Calculate the bandgap of the alloy InGaN needed to produce blue light and green light emission (do not include quantum or strain effects).

3.9 Consider the $GaAs/Al_xGa_{1-x}As$ ($0 \le x < 1$). Assume that the band discontinuity distribution between the conduction and valence band is given by

$$\Delta E_c = 0.6\Delta E_g \text{ (direct)}$$
$$\Delta E_v = 0.4\Delta E_g \text{ (direct)}$$

where ΔE_g(direct) is the bandgap discontinuity between AlGaAs (*even if the material is indirect*) and GaAs. Calculate the valence and conduction band offsets as a function of Al composition. Notice that the conduction band offset starts to decrease as Al composition increases beyond ∼0.4.

3.10 Show that the heterostructure $Al_xGa_{1-x}As/Al_yGa_{1-y}As$ ($x < y$) can be type II (i.e., valence band maxima is in $Al_xGa_{1-x}As$ and conduction band minima (X-type) is in $Al_yGa_{1-y}As$).

3.11 Calculate the first and second subband energy levels for the conduction band in a $GaAs/Al_{0.3}Ga_{0.7}As$ quantum well as a function of well size. Assume a 60:40 rule for ΔE_c:ΔE_v. Also, calculate the energy levels if an infinite barrier approximation was being used.

3.12 Calculate the first two valence subband levels for heavy hole and light hole states in a $GaAs/Al_{0.3}Ga_{0.7}As$ quantum well as a function of the well size. Note that to find the subband levels (at $k_\parallel = 0$), one does not need to include the coupling of the heavy hole and light hole states.

3.13 Using the eigenvalue method described in the Appendix, write the valence band 4×4 Schrödinger equation (using the Kohn–Luttinger formalism discussed in the text) as a difference equation. Write a computer program to solve this $4N \times 4N$ matrix equation where N is the number of grid points used to describe the region of the well and the barrier. Using the parameter values $\gamma_1 = 6.85$, $\gamma_2 = 2.1$, and $\gamma_3 = 2.9$, calculate the E vs. $k_\parallel$ diagram for the valence band in a 100 Å $GaAs/Al_{0.3}Ga_{0.7}As$ quantum well. You will need to find a proper mathematical library through your computer center to diagonalize the large matrix.

3.14 In the $In_{0.53}Ga_{0.47}As/InP$ system, 40% of the bandgap discontinuity is in the conduction band. Calculate the conduction and valence band discontinuities. Calculate the effective bandgap of a 100 Å quantum well. Use the infinite potential approximation and the finite potential approximation and compare the results.

3.15 Compare the electron and HH ground state energies in a $GaAs/Al_{0.3}Ga_{0.7}As$ well for infinite potential and finite potential models as a function of well size. The well size goes from 40 Å to 150 Å.

3.16 Show that in a parabolic potential well, the spacing between the energy levels is constant. In semiconductors, parabolic potential wells are often produced by using narrow square potential wells where the well to barrier width ratio gradually changes. Use the virtual crystal approximation to design a GaAs/AlAs parabolic well where the level spacing for the electron is approximately 10 meV. Hint: This is the "harmonic oscillator" problem.

3.17 Consider a 80 Å GaAs quantum well. The position of the Fermi level is given by

$$E_F = E_c + 0.1 \text{ eV}$$

where E_c is the conduction bandedge of bulk GaAs. Use the infinte barrier model.

- Calculate the subband energies and the electron density (cm^{-2}) in the first subband.

- Calculate the electron density (cm^{-2}) in the second subband.

- Plot the electron distribution ($cm^{-2}eV^{-1}$) versus energy in the range E_c to $E_c + 0.2$ eV.

3.18 Even in high quality glasses made for optical filters, small particulates of semiconductors (e.g., CdS) are suspended. The particle size can be altered by fabrication parameter variations from several hundred Angstroms to several tens of Angstroms in radius. Assuming an electron and hole mass of 0.07 m_0 and 0.2 m_0, calculate the effect on the bandgap of confinement in these "quantum dots" if the particle size is 20 Å, 100 Å and 500 Å in radius.

3.19 Consider a quantum dot of GaAs of dimensions $L \times L \times L$. Assume infinite potential barriers and calculate the separation of the ground and excited state energies in the conduction band as a function of L. If an energy separation of $k_B T$ is needed to observe 3-dimensional confinement effects, what is the maximum box size required to see these effects at 4 K and at 300 K, in the conduction band?

Sections 3.4–3.5

3.20 Using the deformation potentials provided in Appendix D, calculate the splitting of the 6-fold degenerate X-valleys when $Ge_{0.2}Si_{0.8}$ is grown on a (001) Si substrate pseudomorphically.

3.21 Calculate the splitting between the heavy hole and light hole states at the zone edge when $In_{0.2}Ga_{0.8}As$ is grown lattice matched to GaAs.

3.22 Consider a 100 Å $GaAsP/Al_{0.3}Ga_{0.7}As$ quantum well. At what composition of P will be zone center heavy-hole and light-hole states merge?

3.23 Consider a 100 Å $In_xGa_{1-x}As/Al_{0.3}GA_{0.7}As$ (001) quantum well structure. Calculate the heavy hole effective mass near the zone edge along the (100) and (110) direction as x goes from 0 to 0.3, by solving the Kohn–Luttinger equation. You may write the Kohn–Luttinger equation in the difference form and solve it by calling a matrix solving subroutine from your computer library.

3.24 A 150 Å $In_xGa_{1-x}As$ quantum well is to be designed so that the HH-LH separation is $3/2k_BT$ at 300 K. Calculate the In composition needed, assuming the infinite potential approximation.

3.25 Plot the electric field profile when a 100 Å /100 Å $GaAs/In_{0.2}Ga_{0.8}As$ multi-quantum well structure is grown on a (111) GaAs substrate. You can use the weighted average of e_{14} for the alloy.

3.8 REFERENCES

- **Semiconductor Alloys**

  - S. Adachi, *J. Appl. Phys.*, **58**, R–1 (1985).

  - For general alloy property values a good reference text is H. C. Casey, Jr. and M. B. Panish, *Heterostructure Lasers, Part A, Fundamental Principles, Part B, Materials and Operating Characteristics* (Academic Press, New York (1978).

  - For details of various methodologies beyond the virtual crystal approximation a good text is E. N. Economou, *Green's Functions in Quantum Physics* (Springer–Verlag, New York (1979).

  - D. L. Smith and T. C. McGill, *J. de Physique*, **45**, C5–509 (1984).

- **Heterojunctions: Band Offsets**

  - R. S. Bauer, P. Zurcher and H. W. Sang, *Appl. Phys. Lett.* 43, 663 (1983).

  - W. A. Harrison, *J. Vac. Sci. Technol.*, **14**, 1016 (1977).

  - H. Kroemer, in *Molecular Beam Epitaxy and Heterostructures*, edited by L. L. Chang and K. Ploog (NATO ASI Series E, No. 87, Martinus Nijhoff, Dordrecht, Netherlands (1985)).

  - J. W. Mayer and S. S. Lau, *Electronic Material Science: For Integrated Circuits in Si and GaAs* Macmillan, New York (1990).

  - A. G. Milnes and D. L. Feucht, *Heterojunctions and Metal Semiconductor Junctions* Academic Press, New York (1972).

  - J. Tersoff, *Phys. Rev. Lett.*, **56**, 2755 (1986).

– C. G. Van de Welle and R. M. Martin, *J. Vac. Sci. Technol.* B, **4**, 1055 (1986).

• **Bandstructure in Quantum Wells**

    – For the simple quantum well energy levels, any quantum mechanics book would do. An example is L. I. Schiff, *Quantum Mechanics* McGraw–Hill, New York (1968). For more detailed treatment of quantum wells, see the following references.

    – H. Akera, S. Wakahana and T. Ando, *Surf. Sci.*, **196**, 694 (1988).

    – E. Bangerk and G. Landwehr, Superlattices and Microstructures, **1**, 363 (1985).

    – Broido, D. A. and L. J. Sham, *Phys. Rev.* B, **31**, 888 (1985).

    – S. C. Hong, M. Jaffe and J. Singh, *IEEE J. Quant. Electron.*, **QE–23**, 2181 (1987).

    – G. D. Sanders and Y. C. Chang, *Phys. Rev.* B, **31**, 6892 (1985).

    – U. Ukenberg and M. Altarelli, *Phys. Rev.* B, **30**, 3569 (1984).

    – C. Weisbuch and B. Vinter, *Quantum Semiconductor Structures: Fundamentals and Applications*, Academic Press (1991).

• **Strain in Materials: General**

    – C. Kittel, *Introduction to Solid State Physics* John Wiley and Sons, New York (1971), 4th edition. New editions of this book do not have a chapter on this subject.

    – H. B. Huntington, *Elastic Constants of Crystals* Sol. St. Phys., **7**, 213 (1958).

    – J. F. Nye, *Physical Properties of Crystals: Their Representation by Tensors and Matrices* Clarendon Press, Oxford (1957).

• **Strain Tensor in Lattice Mismatched Epitaxy**

    – J. M. Hinckley and J. Singh, *Phys Rev.* B, **42**, 3546 (1990).

    – G.C. Osbourn, *J. Appl. Phys.*, **53**, 1586 (1982).

• **Deformation Potential Theory**

    – J. M. Hinckley, *The Effect of Strain in Pseudomorphic p-$Si_{1-x}Ge_x$: Physics and Modelling of the Valence Bandstructure and Hole Transport*, Ph.D. Thesis, (The University of Michigan, Ann Arbor, 1990).

    – Picus, G. E. and G. E. Bir, Sov. Phys. Sol. St., **1**, 1502 (1959).

    – Picus, G. L. and G. E. Bir, *Symmetry and Strain Induced Effects in Semiconductors* (John Wiley and Sons, New York, 1974).

- **Bandstructure Modification by Epitaxially Produced Strain**

  - Kato, H., N. Iguchi, S. Chika, M. Nakayama, and N. Sano, Jap. J . Appl. Phys., **25**, 1327 (1986).
  - Laurich, B. K., K. Elcess, C. G. Fonstad, J. G. Berry, C. Mailhiot and D. L. Smith, Phys. Rev. Lett., **62**, 649 (1989).
  - Mailhiot, C. and D. L. Smith, J. Vac. Sci. Technol. A, **5**, 2060 (1987).
  - Mailhiot, C. and D. L. Smith, Phys. Rev. B, **35**, 1242 (1987).
  - O'Reilly, E. P., Semicond. Sci. Technol., **4**, 121 (1989).
  - Osbourn, G. C., J. Appl. Phys., **53**, 1586 (1982).
  - People, R., IEEE J. Quant. Electron., **QE–22**, 1696 (1986).
  - See also the references in Chapter 1 on strain issues.

Chapter

4

TRANSPORT: GENERAL FORMALISM

4.1 INTRODUCTION

According to Bloch theorem even though an electron sees a complex background potential in a crystal, it *suffers no collision during its motion through the structure*. The motion of an electron in an electric field is described schematically in Figure 4.1. As shown in Fig. 4.1 electrons will simply follow a band to the Brillouin zone edge and then retrace their trajectory. This produces oscillations called Bloch oscillations which are discussed in Chapter 8. However, in real materials electrons are usually scattered by the presence of various imperfections in the crystal.

We will now calculate the effects of imperfections on the electron transport. It is important to remember that we use first order perturbation theory which gives us the Fermi golden rule. The rates given by the golden rule are calculated for Bloch states. Very often it is necessary to describe the electron by states that are well-defined in position space as well as momentum space. This is done by the wavepacket description. However, one ignores the uncertainty relation

$$\Delta x \Delta k \approx 1 \tag{4.1}$$

for the wavepacket. In this quasiclassical treatment the electron is then described as having a well-defined *position and momentum* while the *transition rates* are calculated using Bloch states. For most semiconductors this requires that the region of transport be larger than about 1000 Å. If the transit region is much smaller the use of a point particle description is inaccurate.

In this and the next few chapters we will develop the quasi-classical approach of electron (hole transport). In this approach wavefunctions for electrons are used (in

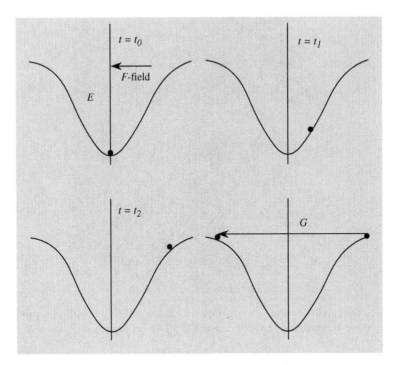

Figure 4.1: A schematic of how electrons move in absence of any imperfections in a crystal. An electron gains crystal momentum according to the equation $\hbar\ dk/dt = eF$. The electron climbs up the band until it reaches a zone edge and in the reduced zone scheme the electron appears as if it has been scattered by a reciprocal lattice vector.

the Fermi Golden rule or Born approximation) to calculate scattering rates, but then the electron is assumed to be a point particle that satisfies the modified Newton's equation in between scattering events. The approach used is shown schematically in Fig. 4.2 where an electron is shown moving in a band, suffering scattering at some time intervals. After scattering, the electron moves from one k–state to another. It is possible to loosely describe a "relaxation" time which represents some average time over which the particle scatters and loses coherence with its pre-scattering state.

In Fig. 4.3 we show an overall schematic of electron transport in real space. An electron's trajectory is considered to be made up of "free flight" periods where it moves without scattering and instantaneous scattering events in which the electron can change its momentum and/or energy. In this chapter we will discuss a generic approach used to calculate transport properties such as mobility or conductivity in a material.

4.2 BOLTZMANN TRANSPORT EQUATION

In this section we will establish the Boltzmann transport equation which describes the distribution function of electrons under an external perturbation in terms of the equilibrium distribution function (i.e., the Fermi–Dirac function). The distribution would

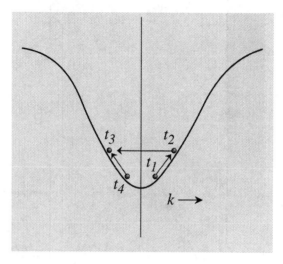

Figure 4.2: In the presence of scattering the electron moves in the band in a random manner. The presence of an electric field produces a net drift. The times $t_1, t_2, \ldots$ are times at which collisions occur.

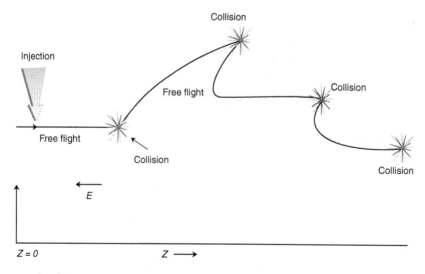

Figure 4.3: A schematic of a model used to understand carrier transport in semiconductors.

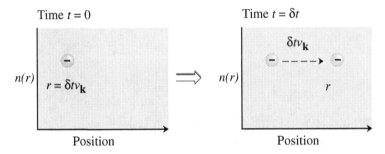

Figure 4.4: At time $t = 0$ particles at position $r - \delta t v_{\mathbf{k}}$ reach the position r at a later time δt. This simple concept is important in establishing the Boltzmann transport equation.

tell us how electrons are distributed in momentum space or k-space (and energy-space) and from this information all of the transport properties can be evaluated. We know that at equilibrium the distribution function is simply the Fermi-Dirac function

$$f(E) = \frac{1}{\exp\left(\dfrac{E - E_F}{k_B T}\right) + 1} \tag{4.2}$$

This distribution function describes the equilibrium electron gas and is independent of any collisions that may be present.

Let us denote by $f_{\mathbf{k}}(\boldsymbol{r})$ the local occupation of the electrons in state $\boldsymbol{k}$ in the neighborhood of $\boldsymbol{r}$. The Boltzmann approach begins with an attempt to determine how $f_{\mathbf{k}}(\boldsymbol{r})$ changes with time. Three possible reasons account for the change in the electron distribution in k-space and r-space:

1. Due to the motion of the electrons (diffusion), carriers will be moving into and out of any volume element around $\boldsymbol{r}$.

2. Due to the influence of external forces, electrons will be changing their momentum (or $\boldsymbol{k}$-value) according to $\hbar \, d\boldsymbol{k}/dt = \boldsymbol{F}_{ext}$.

3. Due to scattering processes, electrons will move from one $\boldsymbol{k}$-state to another.

We will now calculate these three individual changes by evaluating the partial time derivative of the function $f_{\mathbf{k}}(\boldsymbol{r})$ due to each source using a simple and intuitive approach.

4.2.1 Diffusion-Induced Evolution of $f_{\mathbf{k}}(\boldsymbol{r})$

If $\boldsymbol{v_k}$ is the velocity of a carrier in the state $\boldsymbol{k}$, in a time interval δt, the electron moves a distance $\delta t \boldsymbol{v_k}$. Thus the number of electrons in the neighborhood of $\boldsymbol{r}$ at time δt is equal to the number of carriers in the neighborhood of $\boldsymbol{r} - \delta t \, \boldsymbol{v_k}$ at time 0, as shown in Fig. 4.4.

We can thus define the following equality due to the diffusion

$$f_{\mathbf{k}}(\boldsymbol{r}, \delta t) = f_{\mathbf{k}}(\boldsymbol{r} - \delta t \, \boldsymbol{v_k}, 0) \tag{4.3}$$

or

$$f_{\mathbf{k}}(\mathbf{r}, 0) + \frac{\partial f_{\mathbf{k}}}{\partial t} \cdot \delta t = f_{\mathbf{k}}(\mathbf{r}, 0) - \frac{\partial f_{\mathbf{k}}}{\partial \mathbf{r}} \cdot \delta t \; \mathbf{v}_{\mathbf{k}}$$

$$\left. \frac{\partial f_{\mathbf{k}}}{\partial t} \right|_{\text{diff}} = -\frac{\partial f_{\mathbf{k}}}{\partial \mathbf{r}} \cdot \mathbf{v}_{\mathbf{k}} \qquad (4.4)$$

4.2.2   External Field-Induced Evolution of $f_{\mathbf{k}}(\mathbf{r})$

Next we calculate how the distribution function changes as a result of applied fields. The crystal momentum $\mathbf{k}$ of the electron evolves under the action of external forces according to Newton's equation of motion. For an electric and magnetic field ($\mathbf{F}$ and $\mathbf{B}$), the rate of change of $\mathbf{k}$ is given by

$$\dot{\mathbf{k}} = \frac{e}{\hbar} \left[\mathbf{F} + \mathbf{v}_{\mathbf{k}} \times \mathbf{B} \right] \qquad (4.5)$$

In analogy to the diffusion-induced changes, we can argue that particles at time $t = 0$ with momentum $\mathbf{k} - \dot{\mathbf{k}} \, \delta t$ will have momentum $\mathbf{k}$ at time δt and

$$f_{\mathbf{k}}(\mathbf{r}, \delta t) = f_{\mathbf{k} - \dot{\mathbf{k}} \delta t}(\mathbf{r}, 0) \qquad (4.6)$$

which leads to the equation

$$\left. \frac{\partial f_{\mathbf{k}}}{\partial t} \right|_{\text{ext. forces}} = -\dot{\mathbf{k}} \frac{\partial f_{\mathbf{k}}}{\partial \mathbf{k}}$$

$$= \frac{-e}{\hbar} \left[\mathbf{F} + \mathbf{v} \times \mathbf{B} \right] \cdot \frac{\partial f_{\mathbf{k}}}{\partial \mathbf{k}} \qquad (4.7)$$

4.2.3   Scattering-Induced Evolution of $f_{\mathbf{k}}(\mathbf{r})$

The electron distribution can also change due to scattering. Scattering processes take an electron from one state to another. We will assume that the scattering processes are *local* and *instantaneous* and change the state of the electron from $\mathbf{k}$ to $\mathbf{k}'$. Let $W(\mathbf{k}, \mathbf{k}')$ define the rate of scattering from the state $\mathbf{k}$ to $\mathbf{k}'$ if the state $\mathbf{k}$ is occupied and $\mathbf{k}'$ is empty. The rate of change of the distribution function $f_{\mathbf{k}}(\mathbf{r})$ due to scattering is

$$\left. \frac{\partial f_{\mathbf{k}}}{\partial t} \right)_{\text{scattering}} = \int \left[f_{\mathbf{k}'} \left(1 - f_{\mathbf{k}} \right) W(\mathbf{k}', \mathbf{k}) - f_{\mathbf{k}} \left(1 - f_{\mathbf{k}'} \right) W(\mathbf{k}, \mathbf{k}') \right] \frac{d^3 k'}{(2\pi)^3} \qquad (4.8)$$

The first term in the integral represents the rate at which electrons are coming from an occupied $\mathbf{k}'$-state (hence the factor $f_{\mathbf{k}'}$) to an unoccupied $\mathbf{k}$-state (hence the factor $(1 - f_{\mathbf{k}})$). The second term represents the loss term.

Under steady-state conditions, there will be no net change in the distribution function and the total sum of the partial derivative terms calculated above will be zero.

$$\left. \frac{\partial f_{\mathbf{k}}}{\partial t} \right)_{\text{scattering}} + \left. \frac{\partial f_{\mathbf{k}}}{\partial t} \right)_{\text{fields}} + \left. \frac{\partial f_{\mathbf{k}}}{\partial t} \right)_{\text{diffusion}} = 0 \qquad (4.9)$$

Let us define

$$g_{\mathbf{k}} = f_{\mathbf{k}} - f_{\mathbf{k}}^0 \tag{4.10}$$

where $f_{\mathbf{k}}^0$ is the equilibrium distribution.

Instead of calculating the distribution function $f_{\mathbf{k}}$, we will calculate $g_{\mathbf{k}}$, which represents the deviation of the distribution function from the equilibrium case.

Substituting for the partial time derivatives due to diffusion and external fields we get

$$-v_{\mathbf{k}} \cdot \nabla_r f_{\mathbf{k}} - \frac{e}{\hbar} \left(\boldsymbol{F} + \boldsymbol{v}_{\mathbf{k}} \times \boldsymbol{B} \right) \cdot \nabla_k f_{\mathbf{k}} = \left. \frac{-\partial f_{\mathbf{k}}}{\partial t} \right)_{\text{scattering}} \tag{4.11}$$

Substituting $f_{\mathbf{k}} = f_{\mathbf{k}}^0 + g_{\mathbf{k}}$

$$
\begin{aligned}
&-v_{\mathbf{k}} \cdot \nabla_r f_{\mathbf{k}}^0 - \tfrac{e}{\hbar} \left(\boldsymbol{F} + \boldsymbol{v}_{\mathbf{k}} \times \boldsymbol{B} \right) \nabla_k f_{\mathbf{k}}^0 \\
&= \left. -\frac{\partial f_{\mathbf{k}}}{\partial t} \right)_{\text{scattering}} + \boldsymbol{v}_{\mathbf{k}} \cdot \nabla_r g_{\mathbf{k}} + \tfrac{e}{\hbar} \left(\boldsymbol{F} + \boldsymbol{v}_{\mathbf{k}} \times \boldsymbol{B} \right) \cdot \nabla_k g_{\mathbf{k}}
\end{aligned} \tag{4.12}
$$

We note that the magnetic force term on the left-hand side of Eqn. 4.12 is proportional to

$$v_{\mathbf{k}} \cdot \frac{e}{\hbar} \left(\boldsymbol{v}_{\mathbf{k}} \times \boldsymbol{B} \right)$$

and is thus zero. We remind ourselves that

$$v_{\mathbf{k}} = \frac{1}{\hbar} \frac{\partial E_k}{\partial \boldsymbol{k}}$$

$$f_{\mathbf{k}}^0 = \frac{1}{\exp \left[\dfrac{E_{\mathbf{k}} - E_F}{k_B T} \right] + 1}$$

It is easy to see that

$$\nabla_r f^0 = \frac{\partial f^0}{\partial E_{\mathbf{k}}} \left[-\nabla E_F - \frac{(E_{\mathbf{k}} - E_F)}{T} \nabla T \right]$$

Also

$$
\begin{aligned}
\nabla_k f^0 &= \frac{\partial f^0}{\partial E_{\mathbf{k}}} \cdot \nabla_k E_k \\
&= \hbar v_{\mathbf{k}} \frac{\partial f^0}{\partial E_{\mathbf{k}}}
\end{aligned}
$$

Substituting these terms and retaining terms only to second-order in electric field (i.e., ignoring terms involving products $g_{\mathbf{k}} \cdot \boldsymbol{F}$), we get (from Eqn. 4.12) the Boltzmann equation

$$
\begin{aligned}
&-\frac{\partial f^0}{\partial E_{\mathbf{k}}} \cdot v_{\mathbf{k}} \cdot \left[-\frac{(E_{\mathbf{k}} - \mu)}{T} \nabla T + e\boldsymbol{F} - \nabla \mu \right] \\
&= \left. -\frac{\partial f}{\partial t} \right)_{\text{scattering}} + \boldsymbol{v}_{\mathbf{k}} \cdot \nabla_r g_{\mathbf{k}} + \tfrac{e}{\hbar} \left(\boldsymbol{v}_{\mathbf{k}} \times \boldsymbol{B} \right) \cdot \nabla_k g_{\mathbf{k}}
\end{aligned} \tag{4.13}
$$

We will now apply the Boltzmann equation to derive some simple expressions for conductivity, mobility, etc., in semiconductors. We will attempt to relate the microscopic scattering events to the measurable macroscopic transport properties. Let us consider the case where we have a uniform electric field $\boldsymbol{F}$ in an infinite system maintained at a uniform temperature.

The Boltzmann equation becomes (all gradients are zero)

$$-\frac{\partial f^0}{\partial E_{\mathbf{k}}} \boldsymbol{v_k} \cdot e\boldsymbol{F} = -\frac{\partial f_{\mathbf{k}}}{\partial t}\Bigg)_{\text{scattering}} \tag{4.14}$$

This equation, although it looks simple, is a very complex equation which can only be solved analytically under simplifying assumptions. We make an assumption that the scattering induced change in the distribution function is given by

$$-\frac{\partial f_{\mathbf{k}}}{\partial t}\Bigg)_{\text{scattering}} = -\frac{\partial g_{\mathbf{k}}}{\delta t} = \frac{g_{\mathbf{k}}}{\tau} \tag{4.15}$$

We have introduced a time constant called the relaxation time τ whose physical interpretation can be understood when we consider what happens when the external fields are switched off. The perturbation in the distribution function will decay according to the equation

$$\frac{-\partial g_{\mathbf{k}}}{\partial t} = \frac{g_{\mathbf{k}}}{\tau}$$

or

$$g_{\mathbf{k}}(t) = g_{\mathbf{k}}(0)e^{-t/\tau}$$

The time τ thus represents the time constant for relaxation of the perturbation as shown schematically in Fig. 4.5. The approximation which allows us to write such a simple relation is called the relaxation time approximation (RTA).

Using RTA we get, from Eqn. 4.14,

$$\begin{aligned}
g_{\mathbf{k}} &= -\frac{\partial f_{\mathbf{k}}}{\partial t}\Bigg)_{\text{scattering}} \cdot \tau \\
&= \frac{-\partial f^0}{\partial E_{\mathbf{k}}} \tau \boldsymbol{v_k} \cdot e\boldsymbol{F}
\end{aligned} \tag{4.16}$$

In RTA we lave shifted the problem of finding the distribution function to finding the relaxation time. We will discuss later how this time is calculated. The distribution function may be written as

$$f_{\mathbf{k}}(E) = f^0(E) - \left(\frac{\partial f^0_{\mathbf{k}}}{\partial E_{\mathbf{k}}}\right) e\tau \boldsymbol{v_k} \cdot \boldsymbol{F} = f^0(E - e\tau \boldsymbol{v_k} \cdot \boldsymbol{F}) \tag{4.17}$$

To obtain the k-space distribution function we note that

$$f_{\mathbf{k}} = f^0_{\mathbf{k}} - \left(\nabla_k f^0_{\mathbf{k}}\right) \cdot \frac{\partial k}{\partial E_{\mathbf{k}}} \cdot e\tau \boldsymbol{v_k} \cdot \boldsymbol{F}$$

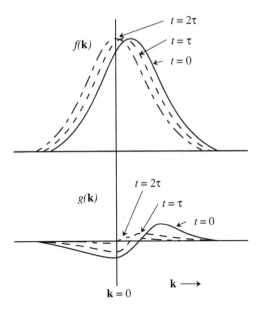

Figure 4.5: A schematic of the time evolution of the distribution function and the perturbation $g(k)$ when the external field is removed at time $t=0$.

Using the relation

$$\hbar \frac{\partial \boldsymbol{k}}{\partial E_{\mathbf{k}}} \cdot \boldsymbol{v}_{\mathbf{k}} = 1$$

We have

$$
\begin{aligned}
f_{\mathbf{k}} &= f_{\mathbf{k}}^0 - (\nabla_{\mathbf{k}} f_{\mathbf{k}}^0) \cdot \frac{e\tau \boldsymbol{F}}{\hbar} \\
&= f_{\mathbf{k}}^0 \left(\boldsymbol{k} - \frac{e\tau \boldsymbol{F}}{\hbar} \right)
\end{aligned}
\tag{4.18}
$$

This result, based on RTA and valid for low electric fields (typically fields of $\lesssim 1$ kV/cm) is extremely useful for transport studies. As we can see the distribution function $f_{\mathbf{k}}$ in the presence of an electric field is defined in terms of $f_{\mathbf{k}}^0$, the equilibrium function. One obtains $f_{\mathbf{k}}$ from $f_{\mathbf{k}}^0$ by shifting the original distribution function for $\boldsymbol{k}$ values parallel to the electric field by $e\tau \boldsymbol{F}/\hbar$. If the field is along the z-direction, only the distribution for k_z will shift. This is shown schematically in Fig. 4.6. In equilibrium there is a net cancellation between positive and negative momenta, but when a field is applied, there is a net shift in the electron momenta and velocities given by

$$
\begin{aligned}
\delta \boldsymbol{p} &= \hbar \delta \boldsymbol{k} = -e\tau \boldsymbol{F} \\
\delta \boldsymbol{v} &= -\frac{e\tau \boldsymbol{F}}{m^*}
\end{aligned}
\tag{4.19}
$$

This gives, for the mobility,

$$\mu = \frac{e\tau}{m^*} \tag{4.20}$$

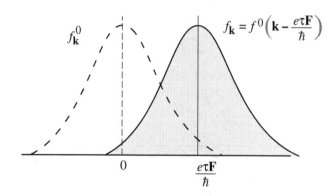

Figure 4.6: The displaced distribution function shows the effect of an applied electric field.

If the electron concentration is n, the current density is

$$
\begin{aligned}
\boldsymbol{J} &= ne\delta\boldsymbol{v} \\
&= \frac{ne^2\tau\boldsymbol{F}}{m^*}
\end{aligned}
$$

or the conductivity of the system is

$$
\sigma = \frac{ne^2\tau}{m^*} \tag{4.21}
$$

Transport properties such as mobility (and conductivity, diffusion constant, etc.,) are macroscopic quantities that are measured in the laboratory. The equation above allows us to relate a microscopic quantity τ to a macroscopic measurable quantity.

So far we have introduced the relaxation time τ, but not described how it is to be calculated. We will now relate it to the scattering rate $W(\boldsymbol{k}, \boldsymbol{k}')$, which can be calculated by using the Fermi golden rule.

Elastic Collisions

Elastic collisions represent scattering events in which the energy of the electrons remains unchanged after the collision. Impurity scattering and alloy scattering discussed in the next Chapter fall into this category. In the case of elastic scattering the principle of microscopic reversibility ensures that

$$
W(\boldsymbol{k}, \boldsymbol{k}') = W(\boldsymbol{k}', \boldsymbol{k}) \tag{4.22}
$$

i.e., the scattering rate from an initial state $\boldsymbol{k}$ to a final state $\boldsymbol{k}'$ is the same as that for the reverse process. The collision integral given by Eqn. 4.8 is now simplified as

$$
\begin{aligned}
\left.\frac{\partial f}{\partial t}\right)_{\text{scattering}} &= \int \left[f(\boldsymbol{k}') - f(\boldsymbol{k})\right] W(\boldsymbol{k}, \boldsymbol{k}') \frac{d^3\boldsymbol{k}'}{(2\pi)^3} \\
&= \int \left[g(\boldsymbol{k}') - g(\boldsymbol{k})\right] W(\boldsymbol{k}, \boldsymbol{k}') \frac{d^3\boldsymbol{k}'}{(2\pi)^3}
\end{aligned} \tag{4.23}
$$

Replacing the left hand side using the Boltzmann equation we get

$$\frac{-\partial f^0}{\partial E_{\mathbf{k}}} \mathbf{v_k} \cdot e\mathbf{F} = \int (g_{\mathbf{k}} - g_{\mathbf{k}'}) \; W(\mathbf{k}, \mathbf{k}') \frac{d^3 k'}{(2\pi)^3}$$

$$= \frac{-\partial f}{\partial t} \bigg)_{\text{scattering}}$$

The relaxation time was defined through

$$g_{\mathbf{k}} = \left(\frac{-\partial f^0}{\partial E} \right) e\mathbf{F} \cdot \mathbf{v_k} \cdot \tau$$

$$= \frac{-\partial f}{\partial t} \bigg)_{\text{scattering}} \cdot \tau$$

Substituting this value in the integral on the right-hand side, we get

$$\frac{-\partial f^0}{\partial E_{\mathbf{k}}} \mathbf{v_k} \cdot e\mathbf{F} = \frac{-\partial f^0}{\partial E_{\mathbf{k}}} e\tau \mathbf{F} \cdot \int (\mathbf{v_k} - \mathbf{v_{k'}}) \; W(\mathbf{k}, \mathbf{k}') \frac{d^3 k'}{(2\pi)^3}$$

or

$$\mathbf{v_k} \cdot \mathbf{F} = \tau \int (\mathbf{v_k} - \mathbf{v_{k'}}) \; W(\mathbf{k}, \mathbf{k}') \frac{d^3 k'}{(2\pi)^3} \cdot \mathbf{F}$$

and

$$\frac{1}{\tau} = \int W(\mathbf{k}, \mathbf{k}') \left[1 - \frac{\mathbf{v_{k'}} \cdot \mathbf{F}}{\mathbf{v_k} \cdot \mathbf{F}} \right] \frac{d^3 k'}{(2\pi)^3} \tag{4.24}$$

In general, this is a rather complex integral to solve. However, it becomes considerably simplified for certain simple cases. Consider, for example, the case of isotropic parabolic bands and elastic scattering. In Fig. 4.7 we show a geometry for the scattering process. We choose a coordinate axis where the initial momentum is along the z-axis and the applied electric field is in the y-z plane. The wavevector after scattering is given by $\mathbf{k}'$ represented by the angles α and ϕ. Assuming that the energy bands of the material is isotropic, $|\mathbf{v_k}| = |\mathbf{v_{k'}}|$. We thus get

$$\frac{\mathbf{v_{k'}} \cdot \mathbf{F}}{\mathbf{v_k} \cdot \mathbf{F}} = \frac{\cos \theta'}{\cos \theta} \tag{4.25}$$

We can easily see from Fig. 4.7 that

$$\cos \theta' = \sin \theta \sin \alpha \sin \phi + \cos \theta \cos \alpha$$

or

$$\frac{\cos \theta'}{\cos \theta} = \tan \theta \sin \alpha \sin \phi + \cos \alpha$$

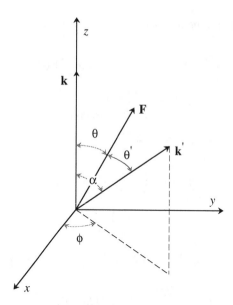

Figure 4.7: Coordinate system illustrating a scattering event for elastic scattering in an isotropic energy band.

When this term is integrated over ϕ to evaluate τ, the term involving $\sin\phi$ will integrate to zero for isotropic bands since $W(\boldsymbol{k},\boldsymbol{k}')$ does not have a ϕ dependence, only an α dependence. Thus

$$\frac{1}{\tau} = \int W(\boldsymbol{k},\boldsymbol{k}')\,(1-\cos\alpha)\,d^3k' \qquad (4.26)$$

This weighting factor $(1-\cos\alpha)$ confirms the intuitively apparent fact that large-angle scatterings are more important in determining transport properties than small-angle scatterings. Forward-angle scatterings $(\alpha = 0)$, in particular, have no detrimental effect on σ or μ for the case of elastic scattering.

Inelastic Collisions

In the case of inelastic scattering processes, the energy of the electron changes after scattering. Processes which cause inelastic scattering include lattice vibration (or phonon) scattering. For inelastic scattering we cannot assume that $W(\boldsymbol{k},\boldsymbol{k}') = W(\boldsymbol{k}',\boldsymbol{k})$. As a result, the collision integral cannot be simplified to give an analytic result for the relaxation time. However, if $f(E)$ is small, we can ignore second-order terms in f and we have

$$\left.\frac{\partial f}{\partial t}\right|_{\text{scattering}} = \int \left[g_{\boldsymbol{k}'} W(\boldsymbol{k}',\boldsymbol{k}) - g_{\boldsymbol{k}} W(\boldsymbol{k},\boldsymbol{k}') \right] \frac{d^3k'}{(2\pi)^3}$$

Under equilibrium we have

$$f^0_{\boldsymbol{k}'} W(\boldsymbol{k}',\boldsymbol{k}) = f^0_{\boldsymbol{k}} W(\boldsymbol{k},\boldsymbol{k}')$$

or

$$W(\boldsymbol{k}', \boldsymbol{k}) = \frac{f_{\mathbf{k}}^0}{f_{\mathbf{k}'}^0} W(\boldsymbol{k}, \boldsymbol{k}')$$

Assuming that this relation holds for scattering rates in the presence of the applied field, we have

$$\left. \frac{\partial f}{\partial t} \right|_{\text{scattering}} = \int W(\boldsymbol{k}, \boldsymbol{k}') \left[g_{\mathbf{k}'} \frac{f_{\mathbf{k}}^0}{f_{\mathbf{k}'}^0} - g_{\mathbf{k}} \right] \frac{d^3 \boldsymbol{k}'}{(2\pi)^3}$$

The relaxation time then becomes

$$\frac{1}{\tau} = \int W(\boldsymbol{k}, \boldsymbol{k}') \left[1 - \frac{g_{\mathbf{k}'}}{g_{\mathbf{k}}} \frac{f_{\mathbf{k}}^0}{f_{\mathbf{k}'}^0} \right] \frac{d^3 \boldsymbol{k}'}{(2\pi)^3} \tag{4.27}$$

The Boltzmann is usually solved iteratively using numerical techniques. We will discuss the approaches later in this chapter.

4.3 AVERAGING PROCEDURES

In our calculations we have assumed that the incident electron has a well-defined state. In general the electron gas will have an energy distribution and τ, in general, will depend upon the energy of the electron. Thus it is important to address the appropriate averaging procedure for τ to be used in macroscopic quantities such as mobility or conductivity.

Let us evaluate the average current in the system.

$$\boldsymbol{J} = \int e \, \boldsymbol{v}_{\mathbf{k}} \, g_{\mathbf{k}} \frac{d^3 k}{(2\pi)^3} \tag{4.28}$$

The perturbation in the distribution function is

$$\begin{aligned}
g_{\mathbf{k}} &= \frac{-\partial f^0}{\partial E_{\mathbf{k}}} \tau \boldsymbol{v}_{\mathbf{k}} \cdot e\boldsymbol{F} \\
&\approx \frac{f^0}{k_B T} \boldsymbol{v}_{\mathbf{k}} \cdot e\boldsymbol{F}
\end{aligned}$$

Substituting for $g_{\mathbf{k}}$ in the current equation, if we consider a field in the x-direction, the average current in the x-direction is given by

$$\langle J_x \rangle = \frac{e^2}{k_B T} \int \tau \, v_x^2 \, f^0 \frac{d^3 k}{(2\pi)^3} \, F_x \tag{4.29}$$

In the derivations below we will assume that the electric field is small so that the average energy of electrons is equal to the thermal energy $3/2k_B T$. Thus we can assume that $v_x^2 = v^2/3$, where $\boldsymbol{v}$ is the total velocity of the electron. Thus we get

$$\langle J_x \rangle = \frac{e^2}{3k_B T} \int \tau \, v^2 \, f^0(\boldsymbol{k}) \frac{d^3 k}{(2\pi)^3} \, F_x \tag{4.30}$$

Now we note that

$$\frac{1}{2}m^*\langle v^2\rangle = \frac{3}{2}k_B T$$

$$\Rightarrow k_B T = m^*\langle v^2\rangle/3$$

also

$$\langle v^2 \tau \rangle = \frac{\int v^2 \, \tau \, f^0(\boldsymbol{k}) \, d^3k/(2\pi)^3}{\int f^0(\boldsymbol{k}) \, d^3k/(2\pi)^3}$$

$$= \frac{\int v^2 \, \tau \, f^0(\boldsymbol{k}) \, d^3k/(2\pi)^3}{n}$$

Substituting in the right-hand side of Eqn. 4.30, we get (using $3k_B T = m \langle v^2\rangle$)

$$\langle J_x \rangle = \frac{ne^2}{m^*}\frac{\langle v^2\tau\rangle}{\langle v^2\rangle}F_x$$

$$= \frac{ne^2}{m^*}\frac{\langle E\tau\rangle}{\langle E\rangle}F_x \qquad (4.31)$$

Thus, for the purpose of transport, the proper averaging for the relaxation time is

$$\langle\langle\tau\rangle\rangle = \frac{\langle E\tau\rangle}{\langle E\rangle} \qquad (4.32)$$

Here the double brackets represent an averaging with respect to the perturbed distribution function while the single brackets represent averaging with the equilibrium distribution function. For calculations of low-field transport where the condition $v_x^2 = v^2/3$ is valid, one has to use the averaging procedure given by Eqn. 4.32 to calculate mobility or conductivity of the semiconductors.

In the next two chapters we will calculate scattering rates and relaxation times for several important scattering processes. For most scattering processes, one finds that it is possible to express the energy dependence of the relaxation time in the form

$$\tau(E) = \tau_0 \left(\frac{E}{k_B T}\right)^s \qquad (4.33)$$

where τ_0 is a constant and s is an exponent which is characteristic of the scattering process. When this form is used in the averaging we get, using a Boltzmann distribution for $f^0(\boldsymbol{k})$

$$\langle\langle\tau\rangle\rangle = \tau_0 \frac{\int_0^\infty [p^2/(2m^*k_B T)]^s \, \exp[-p^2/(2m^*k_B T)] \, p^4 \, dp}{\int_0^\infty \exp[-p^2/(2m^*k_B T)] \, p^4 \, dp}$$

where $\boldsymbol{p} = \hbar\boldsymbol{k}$ is the momentum of the electron.

Substituting $y = p^2/(2m^*k_B T)$, we get

$$\langle\langle\tau\rangle\rangle = \tau_0 \frac{\int_0^\infty y^{s+(3/2)}e^{-y}dy}{\int_0^\infty y^{3/2}e^{-y}dy} \qquad (4.34)$$

To evaluate this integral, we use Γ-functions which have the properties

$$
\begin{aligned}
\Gamma(n) &= (n-1)! \\
\Gamma(1/2) &= \sqrt{\pi} \\
\Gamma(n+1) &= n\,\Gamma(n)
\end{aligned}
$$
(4.35)

and have the integral value

$$
\Gamma(a) = \int_0^\infty y^{a-1} e^{-y} dy
$$

In terms of the Γ-functions we can then write

$$
\langle\langle \tau \rangle\rangle = \tau_0 \frac{\Gamma(s+5/2)}{\Gamma(5/2)}
$$
(4.36)

If a number of different scattering processes are participating in transport, the following approximate rule (Mathiesen's rule) may be used to calculate mobility:

$$
\frac{1}{\tau_{tot}} = \sum_i \frac{1}{\tau_i}
$$

$$
\frac{1}{\mu_{tot}} = \sum_i \frac{1}{\mu_i}
$$
(4.37)

where the sum is over all different scattering processes.

4.4 TRANSPORT IN A WEAK MAGNETIC FIELD: HALL MOBILITY

One of the most important transport characterization techniques for semiconductors is the Hall effect. This effect is used to obtain information on carrier type (electron or hole), carrier concentration, and mobility. We will discuss the theoretical basis of the Hall effect. The Hall effect experiment is shown schematically in Fig. 4.8. One applies an electric field F and a magnetic field B in the geometry shown in the figure. The conductivity of the sample is then measured as a function of the magnetic field. The electric field is maintained at a very low value (a few V/cm) and the analysis we will discuss below will also assume a low magnetic field. In Chapter 11 we will discuss how electrons respond if the magnetic field is high.

In the presence of a magnetic field the Boltzmann equation is

$$
e\left[F + v \times B\right] \cdot \nabla_p f = -\frac{(f - f^0)}{\tau}
$$
(4.38)

We have seen that

$$
\nabla_p f^0(E) = \frac{v \partial f^0}{\partial E}
$$

so that to zeroeth order the magnetic field term $[v \times B] \cdot \nabla_p f$ is zero if we replace f by f^0. We assume that the B-field is in the z-direction while the electric field is in the x-y plane. We write the distribution function in the form

$$
f = f^0 + a_1 v_x + a_2 v_y
$$
(4.39)

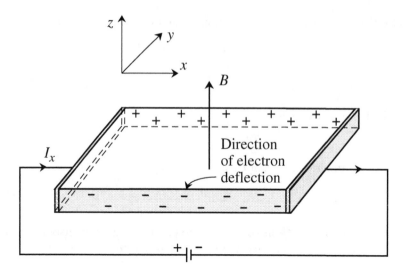

Figure 4.8: A rectangular Hall sample of an n-type semiconductor.

where a_1 and a_2 are to be determined. Substituting for f in the Boltzmann equation and equating the coefficients of v_x and v_y on the two sides we get

$$
\begin{aligned}
a_1 + \frac{e\tau B}{m^*}a_2 &= e\tau F_x \frac{\partial f^0}{\partial E} \\
-\frac{e\tau B}{m^*}a_1 + a_2 &= e\tau F_y \frac{\partial f^0}{\partial E}
\end{aligned}
\tag{4.40}
$$

This gives for a_1 and a_2

$$
\begin{aligned}
a_1 &= e\tau \frac{\partial f^0}{\partial E} \frac{F_x - \omega_c \tau F_y}{1 + (\omega_c \tau)^2} \\
a_2 &= e\tau \frac{\partial f^0}{\partial E} \frac{\omega_c \tau F_x + F_y}{1 + (\omega_c \tau)^2}
\end{aligned}
\tag{4.41}
$$

where ω_c is the cyclotron resonance frequency given by eB/m^*.

The current is given by

$$
\boldsymbol{J} = e \int \frac{d^3 k}{(2\pi)^3} \boldsymbol{v} \left(a_1 v_x + a_2 v_y \right)
$$

which can be written as

$$
J_i = \sigma_{ij} F_j
$$

We see that we have for our configuration

$$
\begin{aligned}
J_x &= \sigma_{xx} F_x - \sigma_{xy} F_y \\
J_y &= \sigma_{yx} F_x + \sigma_{yy} F_y
\end{aligned}
\tag{4.42}
$$

The conductivity tensor components are then

$$\sigma_{xx} = -\int \frac{\partial f^*}{\partial E} \frac{d^3k}{(2\pi)^3} \frac{e^2\tau}{1 + (\omega_c\tau)^2} v_x^2 \tag{4.43}$$

This is the same result as we had in the absence of a magnetic field except for the $1 + (\omega_c\tau)^2$ term in the denominator. If the B-field is small we simply have

$$\sigma_{xx} = \sigma_0 \tag{4.44}$$

where σ_0 is the conductivity in the absence of a magnetic field.

We also have the off-diagonal term given by

$$
\begin{aligned}
\sigma_{xy} &= -\int \frac{\partial f^0}{\partial E} \frac{d^3k}{(2\pi)^3} \frac{e^2\tau^2\omega_c}{1 + (\omega_c\tau)^2} v_x^2 \\
&= -\int \frac{\partial f^0}{\partial E} \frac{d^3k}{(2\pi)^3} \frac{e^3\tau^2}{m^*} v_x^2 \cdot \frac{B}{1 + (\omega_c\tau)^2}
\end{aligned} \tag{4.45}
$$

The term in the averaging is similar to what we had for the mobility in the absence of the magnetic field except that we are averaging τ^2 instead of τ. For very small B-fields (e.g., those employed in Hall effect experiments) we can ignore the B^2 and higher terms and the off-diagonal conductivity becomes

$$
\begin{aligned}
\sigma_{xy} &= \left[e \int \frac{d^3k}{(2\pi)^3} \tau^2 \frac{e^2}{m^*} \frac{\partial f^0}{\partial E} v_1^2 \right] B \\
&= \frac{e^3}{m^*} \int \frac{\tau^2 v^2}{3k_B T} \frac{d^3k}{(2\pi)^3} B \\
&= \frac{ne^3}{m^*} \langle\langle \tau^2 \rangle\rangle = ne\mu \frac{\langle\langle \tau^2 \rangle\rangle \mu}{\langle\langle \tau \rangle\rangle^2} B \\
&= \sigma_0 \mu_H B
\end{aligned} \tag{4.46}
$$

where μ and σ_0 are the mobility and conductivity in the absence of the magnetic field. The quantity μ_H is called the Hall mobility and is given by

$$
\begin{aligned}
\mu_H &= \frac{\langle\langle \tau^2 \rangle\rangle}{\langle\langle \tau \rangle\rangle^2} \mu \\
&= r_H \, \mu
\end{aligned} \tag{4.47}
$$

Once again

$$\langle\langle A \rangle\rangle \Rightarrow \langle EA \rangle$$

where the double-bracket averaging is over the actual perturbed distribution function and the single-bracket averaging is over the *equilibrium distribution function*. If we assume as before that the scattering time τ has an energy dependence

$$\tau = \tau_0 \left(\frac{E}{k_B T} \right)^s$$

it can be shown that

$$r_H = \frac{\Gamma\left(2s + 5/2\right)\Gamma\left(s/2\right)}{\left[\Gamma\left(s + 5/2\right)\right]^2} \tag{4.48}$$

As we can see, the Hall mobility can be quite different from the drift mobility, depending on which scattering mechanism dominates.

From the discussions above we see that in general, in the presence of both electric and magnetic fields, we have for very small B-fields

$$J_i = \sigma_{ij}\left(B\right)\boldsymbol{F}_j$$

where

$$[\sigma(B)] = \sigma_0 \begin{bmatrix} 1 & -\mu_H B_3 & \mu_H B_2 \\ \mu_H B_3 & 1 & -\mu_H B_1 \\ -\mu_H B_2 & \mu_H B_1 & 1 \end{bmatrix} \tag{4.49}$$

It must be noted that in general we have the $(1 + (\omega_c\tau)^2)$ term in the denominator, as seen in Eqns. 4.43 and 4.45. In Chapter 11 we will consider details of how Hall effect is used to get carrier concentration and Hall mobility information. As shown in this section Hall mobility is somewhat different from drift mobility.

4.5   SOLUTION OF THE BOLTZMANN TRANSPORT EQUATION

In the previous sections we have introduced the time τ which relates microscopic scattering to mobility and conductivity. It is straightforward to evaluate τ only for elastic scattering in parabolic bands. However, for most semiconductors, the scattering process is not always elastic or isotropic and the RTA itself fails at high electric fields. Numerous approaches have been developed to address the transport problem in such cases. We will discuss the iterative approach and the balance equation approach in this chapter. Another powerful technique, the Monte Carlo method, will be discussed in Chapter 7.

4.5.1   Iterative Approach

An approach which is quite successful for the solution of steady state, the homogeneous Boltzmann equation, is based on the iterative method. The equation to be addressed is

$$\begin{aligned} e\boldsymbol{F}\cdot\nabla_{\mathbf{p}}f &= \int \frac{d^3k^{'}}{(2\pi)^3}\,W(\boldsymbol{k}^{'},\boldsymbol{k})\,f(\boldsymbol{k}^{'}) - \int \frac{d^3k^{'}}{(2\pi)^3}\,W(\boldsymbol{k},\boldsymbol{k}^{'})\,f(\boldsymbol{k}) \\ &= I(\boldsymbol{k}) - \frac{f(\boldsymbol{k})}{\tau(\boldsymbol{k})} \end{aligned} \tag{4.50}$$

where we have defined

$$I(\boldsymbol{k}) = \int \frac{d^3k^{'}}{(2\pi)^3}\,W(\boldsymbol{k}^{'},\boldsymbol{k})\,f(\boldsymbol{k}^{'}) \tag{4.51}$$

$$\frac{1}{\tau(\boldsymbol{k})} = \int \frac{d^3k^{'}}{(2\pi)^3}\,W(\boldsymbol{k},\boldsymbol{k}^{'}) \tag{4.52}$$

We then obtain

$$f(\boldsymbol{k}) = \tau(\boldsymbol{k})\left[I(\boldsymbol{k}) - e\boldsymbol{F}\cdot\nabla_{\mathbf{p}}f\right] \tag{4.53}$$

This integral equation contains the unknown distribution function $f(\boldsymbol{k})$ on both sides of the equation. To solve this equation, one makes an intelligent guess for $f(\boldsymbol{k})$ and then calculates the next order $f(\boldsymbol{k})$ iteratively. If $f^n(\boldsymbol{k})$ is the nth iterative value for $f(\boldsymbol{k})$, the $n+1$ value is

$$f^{n+1}(\boldsymbol{k}) = \tau(\boldsymbol{k})\left[I^n(\boldsymbol{k}) - e\boldsymbol{F}\cdot\nabla_{\mathbf{p}}f^n\right] \tag{4.54}$$

It is interesting to note that if the starting distribution is chosen to be the equilibrium distribution, one gets upon the first iteration

$$f^2(\boldsymbol{k}) = \tau(\boldsymbol{k})\left[I^1(\boldsymbol{k}) - e\boldsymbol{F}\cdot\nabla_{\mathbf{p}}f^0\right]$$

and

$$I^1(\boldsymbol{k}) = \int \frac{d^3k}{(2\pi)^3}\, W(\boldsymbol{k}',\boldsymbol{k})\, f^0(\boldsymbol{k}')$$

which by detailed balance at equilibrium is equal to

$$
\begin{aligned}
I^1(\boldsymbol{k}) &= \int \frac{d^3k}{(2\pi)^3} W(\boldsymbol{k},\boldsymbol{k}')f^0(\boldsymbol{k})\\
&= \frac{f^0(\boldsymbol{k})}{\tau(\boldsymbol{k})}
\end{aligned}
$$

Thus

$$
\begin{aligned}
f^2(\boldsymbol{k}) &= f^0(\boldsymbol{k}) - e\tau\,\boldsymbol{F}\cdot\nabla_{\mathbf{p}}f^0\\
&= f^0(\boldsymbol{k} - \frac{e\tau\boldsymbol{F}}{\hbar})
\end{aligned}
$$

which is the result we obtained earlier. However, for more accurate results, one has to continue the iteration until the results for transport properties such as drift velocity or mobility converge.

4.6 BALANCE EQUATION: TRANSPORT PARAMETERS

As we have seen in our discussion so far, it is difficult to solve the Boltzmann equation except under very simplifying conditions, e.g., conditions of very low fields. For high fields or nonuniform fields one often has to use computer simulation techniques based on Monte Carlo methods which will be discussed in Chapter 7. However, it is useful to develop additional numerical approaches since Monte Carlo methods are very computer intensive. The Boltzmann equation can be written in the form of balance equations which prove to be very useful for such treatments.

Consider a general physical quantity $n_g(\boldsymbol{r},t)$ defined by the average value of a function $\theta(k)$

$$n_\theta(\boldsymbol{r},t) = \int \theta(\boldsymbol{k})\, f(\boldsymbol{r},\boldsymbol{k},t)\,\frac{d^3k}{(2\pi)^3} \tag{4.55}$$

By choosing various values of $\theta(\boldsymbol{k})$, one can get different physical quantities of interest. For example, $\theta(\boldsymbol{k})$ can be chosen to give particle density, momentum, energy, etc., of the system. The general balance equation is produced when we take the Boltzmann equation, multiply it by $\theta(\boldsymbol{k})$ and integrate over k-space. We get the equation

$$\int \theta(\boldsymbol{k}) \frac{\partial f}{\partial t} \frac{d^3k}{(2\pi)^3} + \int \theta(\boldsymbol{k})\,\boldsymbol{v} \cdot \nabla_{\mathbf{r}} f \; d^3k/(2\pi)^3 + \int \theta(\boldsymbol{k}) e\boldsymbol{F} \cdot \nabla_{\mathbf{p}} f \; d^3k/(2\pi)^3$$
$$= \int \theta(\boldsymbol{k}) \left. \frac{\partial f}{\partial t}\right|_{\text{coll.}} d^3k/(2\pi)^3 \tag{4.56}$$

From the definition of n_θ we have for the first term

$$\int \theta(\boldsymbol{k}) \frac{\partial f}{\partial t} \frac{d^3k}{(2\pi)^3} = \frac{\partial}{\partial t} n_g(\boldsymbol{r},t)$$

since $\theta(\boldsymbol{k})$ has no time dependence. The second term in Eqn. 4.56 becomes

$$\int \theta(\boldsymbol{k})\,\boldsymbol{v} \cdot \nabla_{\mathbf{r}} f \frac{d^3k}{(2\pi)^3} = \nabla \cdot \boldsymbol{F}_g(\boldsymbol{r},t)$$

where

$$\boldsymbol{F}_\theta(\boldsymbol{r},t) = \int \theta(\boldsymbol{k})\,\boldsymbol{v}\,f \frac{d^3k}{(2\pi)^3}$$

The function F_θ represents a flux associated with n_θ.

The electric field term of Eqn. 4.56 becomes

$$e\boldsymbol{F} \cdot \int \theta(\boldsymbol{k})\,\nabla_p f \frac{d^3k}{(2\pi)^3} = e\boldsymbol{F} \cdot \int \nabla_p(\theta f) \frac{d^3k}{(2\pi)^3} - e\boldsymbol{F} \cdot \int f\,\nabla_p \theta \frac{d^3k}{(2\pi)^3}$$

The first term can be represented by a surface integral which equals zero, since $f(\boldsymbol{k})$ goes to zero at large $\boldsymbol{k}$. We now define a generation term G_θ

$$G_\theta = -e\boldsymbol{F} \cdot \int \theta(\boldsymbol{k})\,\nabla_p f \frac{d^3k}{(2\pi)^3}$$
$$= e\boldsymbol{F} \cdot \int f\,\nabla_p \theta \frac{d^3k}{(2\pi)^3}$$

Finally coming to the collision term in the general balance equation, we note that the collisions are responsible for destroying momentum and can be physically represented by a recombination term R_θ.

$$R_\theta = -\int \theta(\boldsymbol{k}) \left.\frac{\partial f}{\partial t}\right|_{\text{coll.}} \frac{d^3k}{(2\pi)^3}$$
$$\equiv \left\langle\!\left\langle \frac{1}{\tau_\theta} \right\rangle\!\right\rangle \left[n_\theta(\boldsymbol{r},t) - n_\theta^0(\boldsymbol{r},t) \right]$$

where n_θ^0 is the equilibrium value of n evaluated with the Fermi function. The relation for R_θ defines a quantity $\langle\!\langle 1/\tau_\theta \rangle\!\rangle$ which represents a relaxation rate for the ensemble. Collecting these terms we get the final balance equation

$$\frac{\partial n_\theta(\boldsymbol{r},t)}{\partial t} = -\nabla \cdot \boldsymbol{F}_\theta + G_\theta - R_\theta \tag{4.57}$$

By substituting different physical quantities for θ, we get various balance equations. The equations that are important for most semiconductor devices are the equations pertaining to carrier density, momentum, and energy.

The balance equation for the carrier density is obtained by putting $\theta(\boldsymbol{k}) = 1$ so that $n_\theta = n$, the carrier density. The flux term is just the particle flux ($= \boldsymbol{J}/e$). The electric field term G_θ and the collision term R_θ are both just zero. The balance equation becomes

$$\frac{\partial n}{\partial t} = -\frac{1}{e}\nabla \cdot \boldsymbol{J} \tag{4.58}$$

which is just the current continuity equation.

To obtain the momentum balance equation, we use $\theta(\boldsymbol{k}) = p_z$ (for the z-component of the momentum). The quantity n_θ is now

$$
\begin{aligned}
n_\theta &= \int \frac{d^3k}{(2\pi)^3}\, p_z\, f \\
&= \langle p_z \rangle \\
&= n\, m^*\, v_{dz} \tag{4.59}
\end{aligned}
$$

where m^* is the carrier mass and v_{dz} is the z-component of the average velocity (i.e., the drift velocity). The flux term associated with the momentum is

$$\boldsymbol{F}_\theta = \int \frac{d^3k}{(2\pi)^3}\, \boldsymbol{v}\, p_z\, f$$

Denoting the x,y,z components of this flux by the index i, we can write

$$F_{\theta,i} = 2\, W_{iz} \tag{4.60}$$

where W_{iz} is an element of the kinetic energy tensor defined in Eqn. 4.62, below. The generation term from the electric field is

$$G_\theta = en\boldsymbol{F}_z$$

The momentum loss rate is ($p_z^0 = 0$)

$$R_\theta = \langle\langle \frac{1}{\tau_m} \rangle\rangle p_z$$

Collecting these terms we get the momentum balance equation

$$\frac{dp_z}{dt} = -\frac{\partial}{\partial x_i}(2\, W_{iz}) + ne\boldsymbol{F}_z - \langle\langle \frac{1}{\tau_m} \rangle\rangle\, p_z$$

or in general

$$\frac{\partial \boldsymbol{p}}{\partial t} = -2\, \nabla \cdot W + ne\boldsymbol{F} - \langle\langle \frac{1}{\tau_m} \rangle\rangle\, \boldsymbol{p} \tag{4.61}$$

where the i,j components of the kinetic energy tensor are

$$W_{ij} = \frac{1}{2} \int \frac{d^3k}{(2\pi)^3}\, v_i \cdot p_j\, f \tag{4.62}$$

and the derivative $\nabla \cdot W$ is the vector

$$\nabla \cdot W \equiv \sum_{ij} \frac{\partial}{\partial x_i} W_{ij}\, \hat{x}_j$$

If the bandstructure of the electron is parabolic

$$J = \frac{e\boldsymbol{p}}{m^*}$$

Thus, for such a case, the current density equation can be written as

$$\frac{\partial \boldsymbol{J}}{\partial t} = -2e\frac{\nabla \cdot W}{m^*} + \frac{e^2 n \boldsymbol{F}}{m^*} - \langle\langle \frac{1}{\tau_m}\rangle\rangle \boldsymbol{J} \tag{4.63}$$

If we define the mobility as

$$\mu = \frac{e}{m^* \,\langle\langle 1/\tau_m\rangle\rangle} \tag{4.64}$$

and assume that the current is not changing very rapidly during the time $1/\langle\langle 1/\tau_m\rangle\rangle$, we get

$$\boldsymbol{J} = ne\mu\boldsymbol{F} - 2\mu\nabla \cdot W \tag{4.65}$$

The kinetic energy contains terms due to the random motion of the electrons as well as the drift components. We assume that the drift components are negligible (valid at low fields), and that the kinetic energy tensor is diagonal. We then get

$$\begin{aligned} W_{ij} &= \frac{nk_B T_c}{2}\, \delta_{ij} \\ &= \frac{w}{3}\, \delta_{ij} \end{aligned}$$

where T_c defines the carrier temperature, and

$$w = \frac{3}{2}\, n\, k_B\, T_c \tag{4.66}$$

With this description of the kinetic energy we get,

$$\begin{aligned} \boldsymbol{J} &= ne\mu\boldsymbol{F} - \frac{2}{3}\, \mu\, \nabla\, w \\ &= ne\mu\boldsymbol{F} - eD\, \nabla n - e\, S\, \nabla T_c \end{aligned} \tag{4.67}$$

where D is the diffusion coefficient

$$D = \frac{k_B T_c}{e}\mu \tag{4.68}$$

and S is the Soret coefficient

$$S = \mu\frac{k_B}{e}n \tag{4.69}$$

This is the drift-diffusion equation and shows that a current may flow due to an electric field, a concentration gradient, or a temperature gradient.

Let us now go to the energy balance equation which is quite useful in understanding high field transport. Choosing $\theta(\boldsymbol{k}) = E(\boldsymbol{k})$ the particle energy, we get

$$\begin{aligned} n_\theta &= \int \frac{d^3k}{(2\pi)^3}\, E(\boldsymbol{k})\, f \\ &= w \end{aligned}$$

where w is defined in Eqn. 4.66.

The associated flux is

$$\begin{aligned} \boldsymbol{F}_\theta &= \int \frac{d^3k}{(2\pi)^3}\, \boldsymbol{v}\, E(\boldsymbol{k})\, f \\ &= \boldsymbol{F}_{Ess} \end{aligned}$$

which represents an energy flux. The energy supplied to the carriers is due to the electric field generation term

$$\begin{aligned} G_\theta &= e\boldsymbol{F} \cdot \int \frac{d^3k}{(2\pi)^3}\, \nabla_{\mathbf{p}} E(\boldsymbol{k})\, f \\ &= \boldsymbol{J} \cdot \boldsymbol{F} \end{aligned}$$

since $\nabla_{\mathbf{p}} E(\boldsymbol{k})$ is the carrier velocity. The energy is lost through the collision terms

$$R_\theta = \langle\langle \frac{1}{\tau_{Ess}} \rangle\rangle (w - w^0)$$

The energy balance equation is then

$$\frac{\partial w}{\partial t} = -\nabla \cdot \boldsymbol{F}_{Ess} + \boldsymbol{J} \cdot \boldsymbol{F} - \langle\langle \frac{1}{\tau_{Ess}} \rangle\rangle (w - w^0) \tag{4.70}$$

We had earlier mentioned the carrier temperature T_c in discussing the drift-diffusion equation. We will now formalize this concept. As the electric field is increased, the carriers gain energy from the field and a balance is established between this energy increase and the loss due to collisions. The distribution function describing the electrons is no longer the Fermi-Dirac function. It is sometimes useful to define this distribution function by a carrier temperature which is, in general, higher than the lattice temperature. In general, the carrier velocities can be written as

$$\boldsymbol{v} = \boldsymbol{v}_d + \boldsymbol{v}_r \tag{4.71}$$

where $\boldsymbol{v}_d$ is the drift velocity and $\boldsymbol{v}_r$ is the random velocity component. The kinetic energy is

$$w = \frac{1}{2}\, n\, m^*\, v_d^2 + \frac{1}{2}\, n\, m^*\, \langle v_r^2 \rangle \tag{4.72}$$

Note that there is no cross term since $\langle v_r \rangle = 0$. The first term in the energy, w, is the drift term, while the second term is due to the random thermal motion of the carriers,

though not at the lattice temperature, in general. We define the carrier temperature by
this thermal part of the kinetic energy

$$\frac{3}{2} n k_B T_c = \frac{1}{2} m^* n \langle v_r^2 \rangle$$

For simplicity we will assume that the carrier temperature is represented by a
diagonal tensor. The momentum balance equation can now be represented in terms of
a carrier temperature

$$\frac{\partial p_j}{\partial t} = -\frac{\partial(n m^* v_{di} v_{dj} + n k_B T_c)}{\partial x_i} + ne F_j - \langle\langle \frac{1}{\tau_m} \rangle\rangle p_j \tag{4.73}$$

Finally, to write the energy balance equation in terms of a carrier temperature, we
consider the energy flux

$$\boldsymbol{F}_{Ess} = \frac{n m^*}{2} \langle v^2 \boldsymbol{v} \rangle \tag{4.74}$$

Once again we separate the energy flux coming from the thermal motion by defining
    By the definition of the velocity $\boldsymbol{v}$, we get

$$\begin{aligned} \boldsymbol{F}_{Ess} &= \frac{n m^*}{2} \langle v^2 \rangle \boldsymbol{v}_d + \frac{n m^*}{2} \langle v^2 \boldsymbol{v}_r \rangle \\ &= w\boldsymbol{v}_d + \frac{n m^*}{2} \langle (v_d^2 + 2\boldsymbol{v}_d \cdot \boldsymbol{v}_r + v_r^2) \boldsymbol{v}_r \rangle \end{aligned} \tag{4.75}$$

The term to be averaged on the right-hand side has three terms, the first of which
averages to zero, the second is related to the carrier temperature, and the third is the
heat flux. Thus

$$\boldsymbol{F}_{Ess} = w\boldsymbol{v}_d + \boldsymbol{v}_d \cdot nk_B T_c + \boldsymbol{\theta}$$

$$\boldsymbol{\theta} = \frac{n m^*}{2} \langle v_r^2 \boldsymbol{v}_r \rangle \tag{4.76}$$

This term is nonzero if there is a nonzero gradient in carrier random motion, i.e., in
carrier temperature.
    If we write

$$\boldsymbol{\theta} = -H\nabla T_c \tag{4.77}$$

where H is the thermal conductivity, we get the following energy balance
equation

$$\begin{aligned} \frac{\partial w}{\partial t} &= \sum_i \left(-\frac{\partial}{\partial x_i} \left[(w + nk_B T_c)v_{di} - H\frac{\partial T_c}{\partial x_i} \right] + J_i \boldsymbol{F}_i \right) \\ &\quad - \langle\langle \frac{1}{\tau_{Ess}} \rangle\rangle (w - w^0) \end{aligned} \tag{4.78}$$

    The advantage of writing the balance equations in terms of the carrier temper-
atures is that the relationship between T_c and the applied electric field may be available
from either some simplifications or from Monte Carlo methods. Once this is known, the
balance equations can be solved.

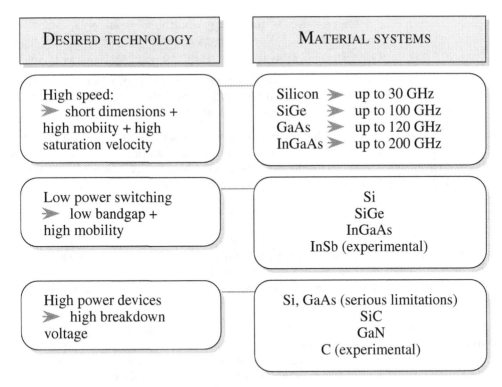

Table 4.1: An overview of materials used for various device needs.

4.7 TECHNOLOGY ISSUES

At present essentially all commercial semiconductor electronic devices are based on electron (hole) transport that involves scattering. Devices that involve "coherent" transport, i.e., transport without any scattering (discussed in Chapters 8 and 11) are still at various experimental stages. In electronic devices there are often conflicting needs: i) high speed devices that can switch at very high frequencies or amplify signals at very high frequencies. Speed is mainly controlled by electron (hole) transit time across critical device dimensions. Materials with high mobility or small carrier mass are desirable; ii) low power devices which require very small electrical power to switch. Usually this requires a small bandgap material where a small voltage charge translates into a large carrier density charge; and iii) high power devices which are needed for power switching or high power oscillators. These require large bandgap materials to avoid breakdown.

In Table 4.1 we give an overview on materials being considered for various needs. It has to be emphasized that the most important driving force for semiconductor electronic devices is device dimensions. In addition specific material parameters play a role. Let us briefly examine some important materials from the point of view of scattering processes.

Silicon: This is the dominant material used in electronics. Low-cost, high-yield, and reliable insulator (SiO_2) and reasonable performance have allowed Si to dominate electronics. Electron (and hole) mass in Si is quite large, resulting in relatively low mobility (~ 1400 cm^2 V·s in pure Si, $\sim$550 cm^2/V·s in NMOS channel). However, at high fields the velocity saturates at $\sim 10^7$ cm/s which is comparable to other materials, like GaAs.

SiGe: The SiGe alloy is used as the base region of a n(Si)-p(SiGe)-n(Si) bipolar transistor. Very high performance devices can be made on this heterostructure. The devices are useful for low power and high frequency applications.

GaAs: This material has a small electron effective mass and, as a result, high mobility. As we will see later in Chapter 7 it also shows negative differential resistance which can be exploited for microwave oscillations. GaAs is used by itself (GaAs MESFETs) and with AlGaAs (AlGaAs/GaAs MODFETs) to produce devices capable of operating up to 120 GHz.

InGaAs: Due to the very small electron mass in InAs, the InGaAs alloy is used for very high mobility devices. The devices are heterostructure devices (InP/InGaAs, In-AlAs/InGaAs) which can operate up to 200 GHz. Since there is no substrate for all InAs devices, such devices have not been successfully fabricated.

In addition to InAs, InSb is used in alloys for low power high speed devices. However, Sb-based devices are mostly experimental.

SiC, GaN: An important scattering mechanism limiting high power performance of semiconductors is impact ionization, which will be discussed in the next chapter. This scattering rate is suppressed for large bandgap semiconductors. As a result, materials like SiC and GaN (or AlGaN) are being used for high power applications.

4.8 PROBLEMS

4.1 Using the Maxwell-Boltzmann distribution function at equilibrium show that the average kinetic energy of an electron is given by $3k_BT/2$. Show that for a displaced Maxwell-Boltzmann distribution the average kinetic energy is given by

$$\frac{1}{2}m^*v_d^2 + \frac{3}{2}k_BT$$

where v_d is the drift velocity.

4.2 Show that for Boltzmann distribution for electrons, the average velocity is given by

$$\langle v \rangle = 2\left(\frac{2k_BT}{\pi m^*}\right)^{1/2}$$

For Si electron mobility is 1400 cm^2/V·s at 300 K.

- Calculate the mean free path.

- Calculate the energy gained by an electron within a mean free path if the applied field is kV/cm.

4.3 Consider a sample of GaAs with electron effective mass of 0.067 m_0. If an electric field of 1 kV/cm is applied, what is the drift velocity produced if *i)* $\tau = 10^{-13}$s; *ii)*

$\tau = 10^{-12}$s; or *iii)* $\tau = 10^{-11}$s ? How does the drift velocity compare to the average thermal speed of the electrons?

4.4 (1.) Plot the room temperature distribution function for electrons in GaAs ($f(E)$) when a field of 0.5, 1.0, 2.0 kV/cm is applied. Assume that $\tau = 10^{-12}$s, and assume a nondegenerate case.

(2.) Plot the k-space distribution function when an electric field $\boldsymbol{F} = \boldsymbol{F}_0 \hat{x}$ is applied to the sample of GaAs. Assume the field magnitudes and τ as given in part a.

4.5 In a semiconductor sample, the Hall probe region has a dimension of 0.5 cm by 0.25 cm by 0.05 cm thick. For an applied electric field of 1.0 V/cm, 20 mA current flows (through the long side) in the circuit. When a 10 kG magnetic field is applied, a Hall voltage of 10 mV is developed. What is the Hall mobility of the sample and what is the carrier density?

4.6 Show that the average energy gained between collisions is

$$\delta E_{\mathrm{av}} = \alpha m^* (\mu \boldsymbol{F})^2$$

where $\boldsymbol{F}$ is the applied electric field, and $\alpha \sim 1$. If the optical phonon energies in GaAs and Si are 36 meV and 47 meV, and the mobilities are 8000 cm^2/V-s and 1400 cm^2/V-s, respectively, what are the electric fields at which optical phonon emission can start? Note that because of the statistical nature of the scattering, a small fraction of the electrons can emit phonons at much lower electric fields.

4.7 Derive an expression for the Hall factor assuming a low carrier concentration of electrons and relaxation time given by

$$\tau = \tau_0 \left(\frac{E}{k_B T} \right)^{1/2}$$

If the Hall mobility in the sample is measured to be 5000 cm^{2}V·s, what is the drift mobility?

4.9 REFERENCES

- **General**

    - "Transport: The Boltzmann Equation," by E. Conwell, in *Handbook on Semiconductors* (North–Holland, New York, 1982), vol. 1.

    - J. Duderstadt and W. Martin, *Transport Theory* (John Wiley and Sons, New York, 1979).

    - D. K. Ferry, *Semiconductors* (Macmillan, New York) 1991.

    - S. Harris, *An Introduction to the Theory of the Boltzmann Equation* (Holt, Rinehart and Winston, New York, 1971).

    - J. R. Haynes and W. Shockley, Phys. Rev., **81**, 835 (1951).

    - M. Lundstrom, *Fundamentals of Carrier Transport* (Modular Series on Solid State Devices, edited by G. W. Neudeck and R. P. Pierre, Addison–Wesley, New York, 1990), vol. X.

- "Low Field Transport in Semiconductors," by D. S. Rode, in *Semiconductors and Semimetals* (edited by R. K. Willardson and A. C. Beer, Academic Press, New York, 1972), vol. 10.

- S. Wang, *Fundamentals of Semiconductor Theory and Device Physics*, Prentice Hall, New Jersey (1989).

Chapter

5

DEFECT AND CARRIER–CARRIER SCATTERING

In Chapter 4 we have derived a number of important mathematical relations necessary to calculate transport properties. A key ingredient of the theory is the scattering rate $W(\mathbf{k}, \mathbf{k}')$ which tells us how an electron in a state $\mathbf{k}$ scatters into the state $\mathbf{k}'$. We will now evaluate the scattering rates for a number of important scattering mechanisms. As noted in Chapter 4, the approach used by us is semiclassical—the electron is treated as a Bloch wave while calculating the scattering rate, but is otherwise treated as a particle. The Fermi golden rule is used to calculate the scattering rate.

In Fig. 5.1 we show an overview of how one goes about a transport calculation. Once the various imperfections in a material are identified the first and most important ingredient is an understanding of the scattering potential. This may seem like a simple problem, but is, in fact, one of the most difficult parts of the problem. Once the potential is known, one evaluates the scattering matrix element between the initial and final state of the electron. This effectively amounts to taking a Fourier transform of the potential since the initial and final states are essentially plane wave states. With the matrix element known one carries out an integral over all final states into which the electron can scatter and which are consistent with energy conservation. This kind of integral provides the various scattering times. Finally, one uses any of a variety of approaches to solve for the transport problem and obtain results for transport properties as outlined in Chapter 4.

In this chapter we will examine two kinds of scattering mechanisms. The first involves fixed defect centers such as ionized impurity atoms (donors, acceptors) and random crystal potential disorder arising in alloys. The second kind of scattering involves the scattering of electrons from other electrons (or holes). This can be an important

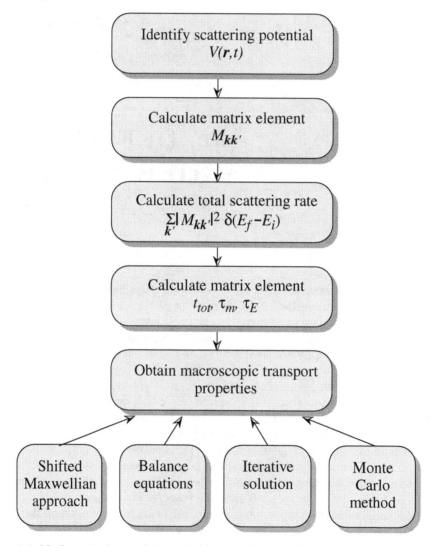

Figure 5.1: Mathematical steps in a typical transport calculation.

scattering mechanism in heavily doped semiconductors where there is a high density of background free carriers. This type of mechanism also gives rise to the important processes of impact ionization and Auger processes. The former is responsible for the breakdown of semiconductors at high electric fields while the latter is responsible for the nonradiative recombination of electrons and holes. Impact ionization is the key reason for limiting the high power performance of electronic devices. Similarly Auger scattering is the key reason why long wavelength light emission is very difficult in semiconductor lasers.

5.1 IONIZED IMPURITY SCATTERING

One of the most important scattering mechansims arises from ionized dopants in semiconductors. In a simplistic model we would expect that an ionized donor or acceptor will have a scattering potential given by

$$V(r) = \frac{e^2}{4\pi \epsilon r} \tag{5.1}$$

Such a potential is called the bare potential. However, we do not use this simple potential to describe scattering from donors or acceptors because of the "screening" of the potential by other free carriers. Free electrons (holes) respond to the bare potential by changing their local density. The variation produced in the electron density reduces the effect of the ionized impurity, particularly at distances that are far from the impurity. As a result the dielectric response of the material changes. To calculate the dielectric response we define the following:

- ρ_{ext}; ϕ_{ext}: the charge density and potential due to the external perturbation, e.g., the ionized impurity.

- ρ_{ind}; ϕ_{ind}: the charge density and potential due to the induced effects in the free carriers.

- ρ_{tot}; ϕ_{tot}: the total charge density and potentials.

We have

$$\phi_{tot} = \phi_{ext} + \phi_{ind}$$
$$\rho_{tot} = \rho_{ext} + \rho_{ind}$$

The dielectric response is defined by

$$\epsilon_{free}(\boldsymbol{k}) = \frac{\rho_{ext}}{\rho_{tot}}$$
$$= \frac{\phi_{ext}}{\phi_{tot}}$$
$$= 1 - \frac{\rho_{ind}}{\rho_{tot}} \tag{5.2}$$

since $\rho_{ext} = \rho_{tot} - \rho_{ind}$.

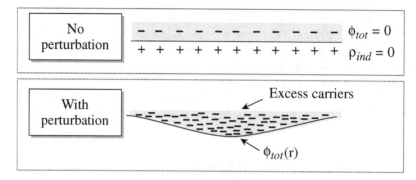

Figure 5.2: Effect of the impurity potential is to alter the uniform free charge by inducing charge. This in turn modifies the potential profile.

The dielectric response will be calculated in k-space due to the simplicity of the Poisson equation in k-space or Fourier space. Let us consider a particular Fourier component of the external potential

$$\rho_{\text{ext}}(\boldsymbol{k}) = \rho e^{i\mathbf{k}\cdot\mathbf{r}} \tag{5.3}$$

The Poisson equation is

$$\nabla^2 \phi = \frac{-\rho}{\epsilon} \tag{5.4}$$

and in k-space this equation becomes

$$\phi = \frac{\rho}{\epsilon k^2} \tag{5.5}$$

The free electron charge density will be assumed to be given by the Boltzmann statistics (i.e., we will deal with nondegenerate semiconductors). In the top panel of Fig. 5.2 we show that in the absence of any external perturbation there is local charge balance. The electron density is n_0 and is uniform. When a potential disturbance occurs as shown in the lower panel of Fig. 5.2, there will be an induced charge given by

$$\rho_{\text{ind}} = n_0 e - n_0 e \exp\left(\frac{e\phi_{\text{tot}}}{k_B T}\right) \tag{5.6}$$

where n_0 is the mean background carrier density. If the perturbation $\phi_{\text{tot}}(\boldsymbol{r})$ is small, we can linearize the exponential and get

$$
\begin{aligned}
\rho_{\text{ind}} &= \frac{-n_0 e^2}{k_B T}\phi_{\text{tot}} \\
&= \frac{-n_0 e^2}{k_B T}\cdot\frac{1}{\epsilon k^2}\rho_{\text{tot}}
\end{aligned} \tag{5.7}
$$

using Eqn. 5.5. Substituting from Eqn. 5.7 into Eqn. 5.2 we get

$$\epsilon_{\text{free}} = 1 + \frac{n_0 e^2}{\epsilon k_B T}\frac{1}{k^2}$$

$$= 1 + \frac{\lambda^2}{k^2} \tag{5.8}$$

where (note that ϵ is equal to $\epsilon_0 \epsilon_r$ where ϵ_r is the relative dielectric constant)

$$\lambda^2 = \frac{n_0 e^2}{\epsilon k_B T} \tag{5.9}$$

We now apply this formalism to the case where we have a bare potential which is Coulombic. In real space, the charge is

$$\rho_{\text{ext}}(\boldsymbol{r}) = q\delta(\boldsymbol{r}) \tag{5.10}$$

The potential is

$$\phi_{\text{ext}} = \frac{q}{4\pi\epsilon r} \tag{5.11}$$

We will express ϕ_{ext} in Fourier space by noting that

$$\delta(\boldsymbol{r}) = \frac{1}{(2\pi)^3} \int d^3 k\, e^{i\boldsymbol{k}\cdot\boldsymbol{r}} \tag{5.12}$$

i.e., the Fourier transform of $\delta(\boldsymbol{r})$ is unity. Thus

$$\rho_{\text{ext}}(\boldsymbol{k}) = q \tag{5.13}$$

and

$$\phi_{\text{ext}}(\boldsymbol{k}) = \frac{q}{\epsilon k^2} \tag{5.14}$$

Using our value of the dielectric constant, from Eqn. 5.8

$$\phi_{\text{tot}}(\boldsymbol{k}) = \frac{q}{\epsilon\left(k^2 + \lambda^2\right)} \tag{5.15}$$

The real space behavior of this function can be obtained by the Fourier transform:

$$
\begin{aligned}
\phi_{\text{tot}}(\boldsymbol{r}) &= \frac{q}{\epsilon(2\pi)^3} \int_0^\infty dk\, \frac{k^2}{k^2 + \lambda^2} \int_{-1}^1 d(\cos\theta) e^{ikr\cos\theta} \int_0^{2\pi} d\phi \\
&= \frac{q \cdot 2\pi}{\epsilon(2\pi)^3} \int_0^\infty dk\, \frac{k^2}{k^2 + \lambda^2} \left[\frac{1}{ikr} e^{ikr\cos\theta}\big|_{-1}^1 \right] \\
&= \frac{2q}{(2\pi)^2 \epsilon r} \int_0^\infty dk\, \frac{k \sin kr}{k^2 + \lambda^2} \\
&= \frac{q}{4\pi\epsilon r} e^{-\lambda r} \tag{5.16}
\end{aligned}
$$

This is the screened Coulombic potential which we will use for the calculations of the scattering rate in nondegenerate semiconductors. The effect of screening is to reduce the range of the potential from a $1/r$ variation to a $\exp(-\lambda r)/r$ variation as shown schematically in Fig. 5.3.

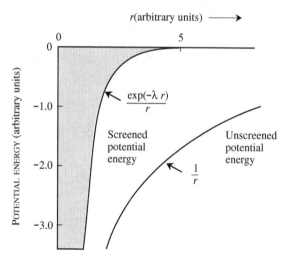

Figure 5.3: Comparison of screened and unscreened Coulomb potentials of a unit positive charge as seen by an electron. The screening length is λ^{-1}.

In doped semiconductors, a certain fraction of dopants are ionized. The ionized dopants have a charge Ze (usually $Z = 1$) which provides a scattering center for electrons and holes. To calculate the scattering rate we need to use the screened Coulombic potential and then use the sequence described by Fig. 5.1. We first calculate the matrix element for screened Coulombic potential

$$U(\boldsymbol{r}) = \frac{Ze^2 e^{-\lambda r}}{4\pi\epsilon r} \tag{5.17}$$

where Ze is the charge of the impurity. Using plane wave functions normalized to a volume V, the scattering matrix element is then

$$M_{\mathbf{kk'}} = \frac{Ze^2}{4\pi V\epsilon} \int e^{-i(\mathbf{k'}-\mathbf{k})\cdot\mathbf{r}} \frac{e^{-\lambda r}}{r} r^2 dr \sin\theta' d\theta' d\phi' \tag{5.18}$$

where $|\boldsymbol{k'}| = |\boldsymbol{k}|$ since the scattering is elastic. Then, as can be seen from Fig. 5.4

$$\left|\boldsymbol{k} - \boldsymbol{k'}\right| = 2k\sin(\theta/2) \tag{5.19}$$

where θ is the polar scattering angle.

$$
\begin{aligned}
M_{\mathbf{kk'}} &= \frac{Ze^2}{4\pi\epsilon V} 2\pi \int_0^\infty r\,dr \int_{-1}^1 d(\cos\theta') e^{-\lambda r} e^{-i|\mathbf{k'}-\mathbf{k}|r\cos\theta'} \\
&= \frac{Ze^2}{4\pi\epsilon V} 2\pi \int_0^\infty r\,dr e^{-\lambda r} \frac{1}{-i|\mathbf{k'}-\mathbf{k}|r} |e^{-i|\mathbf{k'}-\mathbf{k}|r\cos\theta'}|_1^{-1} \\
&= \frac{Ze^2}{4\pi\epsilon V} 2\pi \int_0^\infty dr\, e^{-\lambda r} \frac{1}{-i|\mathbf{k'}-\mathbf{k}|} \left[e^{i|\mathbf{k'}-\mathbf{k}|r} - e^{-i|\mathbf{k'}-\mathbf{k}|r} \right]
\end{aligned}
$$

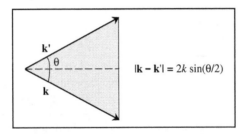

Figure 5.4: As a consequence of the elastic scattering, there is a simple relation between the magnitude of the scattered wavevector and the scattering angle q.

$$
= \frac{Ze^2}{4\pi\epsilon V} 2\pi \int_0^\infty dr \frac{1}{-i|\mathbf{k}'-\mathbf{k}|} \left[e^{(-\lambda+i|\mathbf{k}'-\mathbf{k}|)r} - e^{(-\lambda-i|\mathbf{k}'-\mathbf{k}|)r} \right]
$$

$$
= \frac{Ze^2}{4\pi\epsilon V} 2\pi \frac{1}{-i|\mathbf{k}'-\mathbf{k}|} \left[\frac{1}{-\lambda+i|\mathbf{k}'-\mathbf{k}|} - \frac{1}{-\lambda-i|\mathbf{k}'-\mathbf{k}|} \right]
$$

$$
= \frac{Ze^2}{4\pi\epsilon V} 2\pi \frac{2}{|\mathbf{k}'-\mathbf{k}|^2+\lambda^2}
$$

$$
= \frac{Ze^2}{V\epsilon} \frac{1}{4k^2\sin^2(\theta/2)+\lambda^2} \tag{5.20}
$$

The scattering rate is given by the Fermi golden rule

$$
W(\mathbf{k},\mathbf{k}') = \frac{2\pi}{\hbar} \left(\frac{Ze^2}{V\epsilon} \right)^2 \frac{\delta\left(E_\mathbf{k}-E_{\mathbf{k}'}\right)}{\left(4k^2\sin^2(\theta/2)+\lambda^2\right)^2} \tag{5.21}
$$

One can see that in the two extremes of no screening ($\lambda \to 0$) and strong screening ($\lambda \to \infty$), the rate becomes respectively

$$
W(\mathbf{k},\mathbf{k}') \propto \frac{1}{16k^4\sin^4(\theta/2)} \tag{5.22}
$$

and

$$
W(\mathbf{k},\mathbf{k}') \propto \frac{1}{\lambda^4} \tag{5.23}
$$

We see that for weak screening (low free carrier density) forward angle scattering dominates. However, we note that forward angle scattering ($\theta \sim 0$) is not as important in reducing mobility as large angle scattering. For high screening the scattering has less angular dependence as shown in Fig. 5.5. Using results from Chapter 4, Eqn. 4.26 we examine the relaxation time τ which is used to obtain the low field carrier mobility (using the normalized functions).

$$
\frac{1}{\tau} = \frac{V}{(2\pi)^3} \int (1-\cos\theta)\, W(\mathbf{k},\mathbf{k}')\, d^3k'
$$

$$
= \frac{2\pi}{\hbar} \left(\frac{Ze^2}{\epsilon} \right)^2 \frac{1}{V^2} \frac{V}{(2\pi)^3}
$$

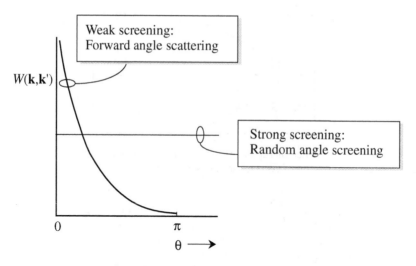

Figure 5.5: Angular dependence of the scattering by ionized impurities. The scattering has a strong forward angle preference.

$$\times \int (1 - \cos\theta) \frac{\delta \left(F_{\mathbf{k}} - F_{\mathbf{k}'} \right)}{\left(4k^2 \sin^2(\theta/2) + \lambda^2 \right)^2} k'^2 \, dk' \, \sin\theta \, d\theta \, d\phi$$

$$= \frac{1}{2\hbar} \left(\frac{Ze^2}{\epsilon} \right)^2 \frac{1}{V}$$

$$\times \int (1 - \cos\theta) \frac{\delta \left(F_{\mathbf{k}} - F_{\mathbf{k}'} \right)}{\left(4k^2 \sin^2(\theta/2) + \lambda^2 \right)^2} N(F_{\mathbf{k}'}) \, dF_{\mathbf{k}'} \, d(\cos\theta) \, d\phi$$

$$= \frac{1}{\hbar} \left(\frac{Ze^2}{\epsilon} \right)^2 \frac{1}{V} \frac{N(F_{\mathbf{k}})}{32k^4}$$

$$\times \int (1 - \cos\theta) \frac{1}{\left[\sin^2(\theta/2) + \left(\frac{\lambda}{2k} \right)^2 \right]^2} \, d(\cos\theta) \, d\phi$$

$$= F \int (1 - \cos\theta) \frac{1}{\left[\sin^2(\theta/2) + \left(\frac{\lambda}{2k} \right)^2 \right]^2} \, d(\cos\theta) \, d\phi \qquad (5.24)$$

with

$$F = \frac{1}{\hbar} \left(\frac{Ze^2}{\epsilon} \right)^2 \frac{1}{V} \frac{N(E_{\mathbf{k}})}{32k^4} \qquad (5.25)$$

Let $z = \cos\theta$, so that $\sin^2(\theta/2) = (1 - z)/2$.

$$\frac{1}{\tau} = 8\pi F \int_{-1}^{1} \frac{(1 - z) \, dz}{\left(1 + 2 \left(\frac{\lambda}{2k} \right)^2 - z \right)^2}$$

$$= 8\pi F \int_{-1}^{1} \frac{dz}{\left(1 + 2\left(\frac{\lambda}{2k}\right)^2 - z\right)^2} - \int_{-1}^{1} \frac{z\,dz}{\left(1 + 2\left(\frac{\lambda}{2k}\right)^2 - z\right)^2}$$

Upon integration this gives

$$\frac{1}{\tau} = \frac{\pi}{4\hbar} \left(\frac{4\pi Z e^2}{\epsilon}\right)^2 \frac{N(F_{\mathbf{k}})}{V k^4}$$

$$\times \left[\ln\left(1 + \left(\frac{2k}{\lambda}\right)^2\right) - \frac{1}{1 + \left(\frac{\lambda}{2k}\right)^2}\right]$$

$$N(F_{\mathbf{k}}) = \frac{m^{*3/2} E^{1/2}}{\sqrt{2}\pi^2 \hbar^3} \tag{5.26}$$

Spin degeneracy is not included, since the ionized impurity scattering cannot alter the spin of the electron. In terms of the electron energy, $F_{\mathbf{k}}$, we have

$$\frac{1}{\tau} = \frac{1}{V 16\sqrt{2\pi}} \left(\frac{Z e^2}{\epsilon}\right)^2 \frac{1}{m^{*1/2} E_{\mathbf{k}}^{3/2}}$$

$$\times \left[\ln\left(1 + \left(\frac{8m^* E_{\mathbf{k}}}{\hbar^2 \lambda}\right)^2\right) - \frac{1}{1 + \left(\frac{\hbar^2 \lambda}{8m^* E ss_{\mathbf{k}}}\right)^2}\right] \tag{5.27}$$

The average relaxation time is

$$\langle\langle\tau\rangle\rangle = \frac{\int_0^\infty F\, \tau_{\mathbf{k}}\, e^{-\beta E ss}\, dF}{\int_0^\infty F\, e^{-\beta E ss}\, dF}$$

To a good approximation, the effect of this averaging is essentially to replace $F_{\mathbf{k}}$ by $k_B T$ in the expression for $1/\tau$. An accurate determination of the average integral gives

$$\frac{1}{\langle\langle\tau\rangle\rangle} = \frac{1}{V 128\sqrt{2\pi}} \left(\frac{Z e^2}{\epsilon}\right)^2 \frac{1}{m^{*1/2} (k_B T)^{3/2}}$$

$$\times \left[\ln\left(1 + \left(\frac{8m^* k_B T}{\hbar^2 \lambda}\right)^2\right) - \frac{1}{1 + \left(\frac{\hbar^2 \lambda}{8m^* k_B T}\right)^2}\right] \tag{5.28}$$

The expression above gives the relaxation time from one scatterer in a volume V. If the density of ionized impurities is N_i, there are $N_i V$ impurities in the volume. We will assume that these impurities are randomly placed and are well separated from each other so that the scattering is incoherent from them. In this case according to quantum mechanics we simply sum the scattering rate from each scatterer. The total relaxation time is

$$\frac{1}{\langle\langle\tau\rangle\rangle} = N_i \frac{1}{128\sqrt{2\pi}} \left(\frac{Ze^2}{\epsilon}\right)^2 \frac{1}{m^{*1/2}(k_B T)^{3/2}}$$

$$\times \left[\ln\left(1 + \left(\frac{8m^* k_B T}{\hbar^2 \lambda}\right)^2\right) - \frac{1}{1 + \left(\frac{\hbar^2 \lambda}{8m^* k_B T}\right)^2} \right] \tag{5.29}$$

The mobility limited from ionized impurity scattering is

$$\mu = \frac{e\langle\langle\tau\rangle\rangle}{m^*}$$

The mobility limited by ionized dopant has the special feature that it decreases with temperature ($\mu \sim T^{3/2}$). This temperature dependence is quite unique to ionized impurity scattering. One can understand this behavior physically by saying that at higher temperatures, the electrons are traveling faster and are less affected by the ionized impurities.

The decrease in mobility as limited by ionized impurity scattering with decreasing temperature is a distinguishing feature which is not observed in other scattering processes. In the next chapter we will discuss scattering from lattice vibrations (phonon scattering) which causes mobility to decrease as temperature increases. Phonon scattering becomes very important at higher temperatures, while ionized impurity scattering becomes important at lower temperatures. Thus, for a semiconductor with some background doping, the mobility versus temperature relation has the form shown in curve (a) of Fig. 5.6. At low temperatures the mobility is low because of ionized impurity scattering. The mobility increases with temperature up to ~ 50 K or so, after which it starts decreasing because of phonon scattering effects. This behavior is in marked contrast to what happens in modulation doped semiconductor structures where ionized dopants are spatially separated from electrons by using heterostructures. Here, due to the remoteness of the dopants from the free electrons, the ionized impurity scattering is essentially absent. This leads to a mobility that continues to increase with decreasing temperature as shown in curve (b) of Fig. 5.6. By using modulation doping extremely high mobilities can be reached at low temperatures—a fact that is exploited for high performance electronic devices.

The formalism discussed above is applicable for low doping levels ($\lesssim 10^{18}$ cm^{-3}). The derivation for screening is for a nondegenerate semiconductor where the Boltzmann expression was sufficient to describe the carrier density variation in presence of the external potential (Eqn. 5.6). For heavily doped semiconductors with a high free carrier density, one may be in the degenerate limit where the Fermi level is in the band and one needs to use the proper distribution function to represent the fluctuations in the electron density. In the high degeneracy limit, this leads to a screening parameter λ given by

$$\lambda^2 = \frac{3n_0 e^2}{2\epsilon E_F} \tag{5.30}$$

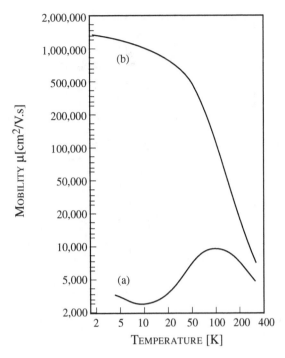

Figure 5.6: Typical measurements of electron mobility as a function of temperature in a uniformly doped GaAs with $N_D = 10^{17}$ cm^{-3}. The mobility drops at low temperature due to ionized impurity scattering becoming very strong. In contrast, the curve (b) shows the mobility in a modulation doped structure where ionized impurity is essentially eliminated.

where E_F is the Fermi level energy measured from the bandedge. The problem of scattering in heavily doped semiconductors is a lot more complex than the incoherent scattering model used here. In writing Eqn. 5.29, we made the assumption that each impurity center causes scattering independent of each other. While this is a reasonable assumption when the impurities are separated by a large distance (several hundred Angstroms), it is not a good approximation for inter-impurity separations less than approximately 100 Å. In general, the mobility falls faster with doping density than predicted by Eqn. 5.29 for heavily doped semiconductors due to the effects of multi-impurity scatterings. Fig. 5.7 shows the concentration dependence of mobility for both electrons and holes in Si and GaAs. In these results the scattering from lattice vibrations remains unchanged and limits the mobility at low impurity concentrations.

EXAMPLE 5.1 A sample of Si is doped with dopant A which gives one electron to the conduction band per dopant. Another sample of Si is doped with dopant B which gives two electrons per dopant. The electron density in both samples is found to be 10^{17} cm^{-3}. Calculate the mobility in the two samples at 300 K. Mobility in pure Si at 300 K is 1100 cm^2/V·s. Electron mass is 0.26 m_0.

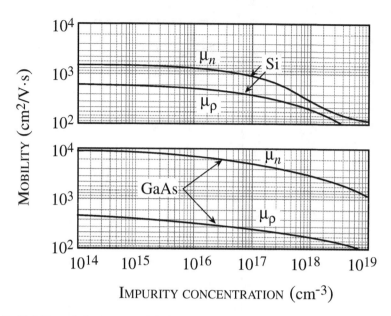

Figure 5.7: Mobility of electrons and holes in Ge, Si, and GaAs as 300 K versus impurity concentration. At doping of $\sim 10^{14}$ cm^{-3} the mobility is limited only by lattice vibration scattering.

In solving this problem we must note an important point. Both samples have the same free electron density. This means we have the following:

$$\text{Sample A}: \ N_i \ = \ 10^{17} \text{ cm}^{-3}; \ Z = 1$$
$$\text{Sample B}: \ N_i \ = \ 5 \times 10^{16} \text{ cm}^{-3}; \ Z = 2$$

where Z is the charge on the donor. Since the mobility goes as

$$\mu \propto Z^2 N_i$$

the impurity scattering limited mobility of sample A is twice that of sample B. The calculation for ionized impurity limited mobility gives

$$\mu_{ii}(\text{Sample A}) \ \sim \ 9500 \text{ cm}^2/V \cdot s$$
$$\mu_{ii}(\text{Sample B}) \ = \ \frac{\mu(\text{Sample A})}{2}$$

Using the mobility summing rate the total mobility is then

$$\mu(\text{Sample A}) \ = \ 985 \text{ cm}^2/V \cdot s$$
$$\mu(\text{Sample B}) \ = \ 893 \text{ cm}^2/V \cdot s$$

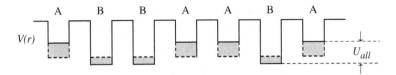

Figure 5.8: A schematic of the actual atomic potential (solid line) and the average virtual crystal potential (dashed line) of an A-B alloy. The shaded area shows the difference between the real potential and the virtual crystal approximation.

5.2   ALLOY SCATTERING

In Chapter 3 we have discussed how alloys can be used to tailor the bandgap of materials. As discussed there, alloys have no microscopic periodicity in the alloy crystal. For example, in the alloy $(AlAs)_x(GaAs)_{1-x}$ while the atom on the anion sublattice is always As, the atoms on the cation sublattice are randomly distributed between Ga and Al. In a random alloy the probability that a particular cation site is occupied by an Al atom is x and that it is occupied by a Ga atom is $(1 - x)$ regardless of the neighboring composition of the site. If, on the other hand, the alloy is clustered, there are regions in space where there is a higher than x or $(1 - x)$ probability of finding a particular kind of atom.

Due to the inherent disorder present in alloys, electrons and holes suffer scattering as they propagate through the material. Let us consider the scattering processes in a perfectly random alloy where the smallest physical size over which the crystal potential fluctuates randomly is the unit cell. An electron moving in the alloy A_xB_{1-x} will see a random potential schematically shown in the Fig. 5.8. The average potential and the average bandstructure of the alloy is described to the lowest order by the virtual crystal approximation. In this approximation, the averaging of the atomic potentials

$$\{M\}_{all} = x\{M\}_A + (1 - x)\{M\}_B \tag{5.31}$$

gives an average *periodic* potential represented by the dashed line in Fig. 5.8. To construct the scattering model the difference between the real potential and the assumed virtual crystal potential is represented within each unit cell by a highly localized potential. For example, for the A-type atoms, the difference is

$$
\begin{aligned}
E_{all} - E_A &= x E_A + (1 - x) E_B - E_A \\
&= (1 - x) [E_B - E_A] \\
&= (1 - x) U_{all} \tag{5.32}
\end{aligned}
$$

Similarly, for the B atom, the difference is

$$
\begin{aligned}
E_B - E_{all} &= x (E_B - E_A) \\
&= x U_{all} \tag{5.33}
\end{aligned}
$$

The quantities E_A, E_B, E_{all} are not well-defined and $(E_B - E_A)$, often called the alloy potential, is adjusted to fit experimental data.

The so-called "hard sphere" model for scattering is used to understand alloy scattering. The scattering potential is chosen to be of the form

$$\Delta U(\boldsymbol{r}) = \begin{array}{ll} U_0 & \text{for } |\boldsymbol{r}| \leq r_0 \\ = 0 & \text{for } |\boldsymbol{r}| > r_0 \end{array} \tag{5.34}$$

where r_0 is the interatomic distance and $U_0 = U_{\text{all}}$. If we use the Fermi golden rule to calculate the scattering rate, we have

$$W(\boldsymbol{k}) = \frac{2\pi}{\hbar} \sum_{\mathbf{k}'} |M_{\mathbf{kk}'}|^2 \delta(E_{\mathbf{k}} - E_{\mathbf{k}'})$$

and

$$M_{\mathbf{kk}'} = \int e^{i(\mathbf{k}-\mathbf{k}')\cdot\mathbf{r}} \, \Delta U(\boldsymbol{r}) \, d^3 r$$

We will now use the fact that the scattering potential only extends over a unit cell and over this small distance

$$e^{i(\mathbf{k}-\mathbf{k}')\cdot\mathbf{r}} \approx 1$$

Thus the matrix element has no k-dependence

$$M_{\mathbf{kk}'} = \frac{4\pi}{3} r_0^3 \, U_0 \tag{5.35}$$

and

$$\begin{aligned} W(\boldsymbol{k}) &= \frac{2\pi}{\hbar} \left(\frac{4\pi}{3} r_0^3 \, U_0 \right)^2 \frac{1}{(2\pi)^3} \int \delta(E_{\mathbf{k}} - E_{\mathbf{k}'}) \, d^3 k' \\ &= \frac{2\pi}{\hbar} \left(\frac{4\pi}{3} r_0^3 \, U_0 \right)^2 N(E_{\mathbf{k}}), \end{aligned}$$

In a fcc crystal of lattice constant a, we relate the extent of the potential r_0 to a by the relation

$$r_0 = \frac{\sqrt{3}}{4} a$$

This gives

$$\left(\frac{4\pi}{3} r_0^3 \right)^2 = \frac{3\pi^2}{16} V_0^2 \tag{5.36}$$

where $V_0 = a^3/4$ is the volume of the unit cell. We now obtain for the scattering rate

$$W(\boldsymbol{k}) = \frac{2\pi}{\hbar} \left(\frac{3\pi^2}{16} V_0^2 \right) U_0^2 \, N(E_{\mathbf{k}}) \tag{5.37}$$

We will now assume that all scattering centers cause incoherent scattering so that we can simply sum the scattering rates. For the A-type atoms, the scattering rate is (using $U_0 = (1-x)U_{all}$)

$$W_{\text{A}}(\boldsymbol{k}) = \frac{2\pi}{\hbar} \left(\frac{3\pi^2}{16} V_0^2 \right) (1-x)^2 \, U_{all}^2 \, N(E_{\mathbf{k}})$$

For the B-type atoms, the rate is (using $U_0 = xU_{all}$)

$$W_A(\mathbf{k}) = \frac{2\pi}{\hbar}\left(\frac{3\pi^2}{16}V_0^2\right)x^2\,U_{all}^2\,N(E_\mathbf{k})$$

There are x/V_0 A-type atoms and $(1-x)/V_0$ B-type atoms in the unit volume and under the assumption of incoherent scattering, the total scattering rate is

$$
\begin{aligned}
W_{\text{tot}} &= \frac{2\pi}{\hbar}\left(\frac{3\pi^2}{16}V_0\right)U_{all}^2\,N(E_\mathbf{k})\left[x\,(1-x)^2 + (1-x)\,x^2\right]\\
&= \frac{3\pi^3}{8\hbar}V_0\,U_{all}^2\,N(E_\mathbf{k})\,x\,(1-x) \qquad\qquad\qquad\qquad (5.38)
\end{aligned}
$$

Several important points are to be noted about the alloy scattering. The first is that the matrix element has no $\mathbf{k},\mathbf{k}'$ dependence, i.e., there is no angular dependence of the matrix element. If the density of states is isotropic, there will be no angular dependence of the scattering rate $W(\mathbf{k},\mathbf{k}')$. This is in contrast to the impurity scattering which had a strong forward angle scattering preference. After doing the proper ensemble averaging the relaxation time for the alloy scattering is

$$\frac{1}{\langle\langle\tau\rangle\rangle} = \frac{3\pi^3}{8\hbar}V_0 U_{all}^2 x(1-x)\frac{m^{*3/2}(k_BT)^{1/2}}{\sqrt{2}\pi^2\hbar^3}\frac{1}{0.75} \qquad (5.39)$$

according to which the mobility due to alloy scattering is

$$\mu_0 \propto T^{-1/2}$$

The temperature dependence of mobility is in contrast to the situation for the ionized impurity scattering. The quantity U_{all} is not known with any certainty from the scattering theory discussed here. Its value is usually obtained by carefully fitting the temperature dependent mobility data. The value of U_{all} is usually in the range of 0.5 eV.

In the discussions above we assumed that the alloy was cluster-free and the smallest region in which the disorder occurred was the unit cell. However, in some alloys one can expect alloy clustering as shown schematically in Fig. 5.9. Here we show a macroscopic alloy A_xB_{1-x} made up of microscopic regions with compositions $A_{x+\delta}B_{1-x-\delta}$ and $A_{x-\delta}B_{1-x+\delta}$. We can no longer calculate the total scattering rate assuming independent scattering from each unit cell. Atoms within a cluster size r_c will now scatter coherently making the problem more complex. One approximation for the alloy shown in Fig. 5.9 to is to use a scattering potential

$$
\begin{aligned}
\Delta U(\mathbf{r}) &= U_{all}\cdot\delta \text{ for } |\mathbf{r}| \le r_c\\
&= 0 \text{ for } |\mathbf{r}| > r_c \qquad\qquad\qquad\qquad\qquad (5.40)
\end{aligned}
$$

where we have used a potential with a smaller value but a larger spatial extent. Now in the matrix element evaluation we can no longer make the constant phase approximation, since r_c could be a large value. In fact, the matrix element now depends critically on the value of $|\mathbf{k}-\mathbf{k}'|$, reaching a maximum when $|\mathbf{k}-\mathbf{k}'| \approx 1/r_c$. The scattering rate

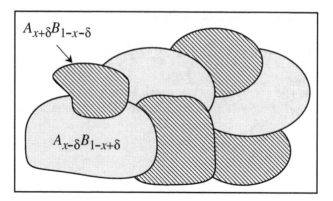

Figure 5.9: A clustered $A_x B_{1-x}$ alloy with A-rich and B-rich regions.

and mobility is, therefore, very much dependent upon the cluster size and temperature which determines the effective value of $\boldsymbol{k}$ and $\boldsymbol{k}'$.

In general, in the presence of clustering, mobility is reduced because of the coherent nature of scattering. For most applications in electronics and optoelectronics one therefore seeks alloys that are random.

EXAMPLE 5.2 Calculate the alloy scattering limited mobility in $Al_{0.3}Ga_{0.7}As$ at 77 K and at 300 K. Assume that the alloy scattering potential is 1.0 eV. The relaxation time at 300 K is $(m^* = 0.07\, m_0)$.

$$\frac{1}{\langle\langle r \rangle\rangle} = \frac{3\pi V_0 (U_{all})^2 x (1-x) m^{*3/2} (k_B T)^{1/2}}{8\sqrt{2}\hbar^4 (0.75)}$$

$$= 2.1 \times 10^{12}\ \text{s}$$

Here we have used $x = 0.3$, $V_0 = a^3/4$ with $a = 5.65$ Å.

The value of $\langle\langle r \rangle\rangle$ is 4.77×10^{-13} s. The mobility is then

$$\mu_{all}(300\ \text{K}) = 1.2 \times 10^4\ \text{cm}^2/\text{V}\cdot\text{s}$$

The mobility goes as $T^{-1/2}$ which gives

$$\mu_{all}(77\ \text{K}) = 2.36 \times 10^4\ \text{cm}^2/\text{V}\cdot\text{s}$$

5.3 NEUTRAL IMPURITY SCATTERING

Another source of scattering is the presence of neutral impurities and defects in semiconductors. Substitutional impurities and dopants that have not ionized can produce neutral impurities. Scattering from these impurities can be addressed using the same approach as used for alloy scatterings. The defect can be represented by a perturbation having a hard sphere like scattering potential. The scattering rate as in alloy scattering is

$$W(k) = \frac{2\pi}{\hbar} \left(\frac{4\pi}{3} r_0^3 U_0 \right)^2 N(E_{\mathbf{k}})$$

where U_0 is the scattering potential and r_0 is the radius describing the extent of the defect. The scattering time for mobility is simply

$$\frac{1}{\langle\langle\tau\rangle\rangle} = N_{\text{imp}} \cdot \frac{2\pi}{\hbar} \left(\frac{4\pi}{3} r_0^3 U_0\right)^2 \frac{m^{*3/2}(k_BT)^{1/2}}{\sqrt{2}\pi^2\hbar^3} \frac{1}{0.75} \tag{5.41}$$

Here N_{imp} is the density of the neutral impurities. Unless the neutral impurity density is very high ($\gtrsim 10^{18}$ cm^{-3}) there is little effect of this scattering (in comparison to other scattering processes) on mobility.

EXAMPLE 5.3 Consider a poor quality Si sample doped n-type with P at 10^{17} cm^{-3}. Assume all the donors are ionized. The sample also has 10^{17} cm^{-3} defects described by the scattering potential

$$
\begin{aligned}
V(r) &= 1.5 \text{ eV } \quad r \leq 10 \text{ Å} \\
&= 0 \text{ otherwise}
\end{aligned}
$$

i) Calculate the mobility of the sample limited by ionized impurity scattering.
ii) Calculate the mobility of the sample limited by defect scattering.
iii) What mobility will be measure in the lab?
Conductivity mass of electrons is 0.26 m_0. The mobility of electrons in pure Si is 1100 cm^2/V·s at 300 K.

The defect problem is similar to the alloy problem. The matrix element of scattering is

$$M_{kk'} = \frac{4\pi}{3} r_0^3 U_0$$

where $r_0 = 10$ Å and $U_0 = 1.5$ eV. The scattering time is

$$\frac{1}{\langle\langle\tau_{\text{defect}}\rangle\rangle} = N_{\text{defect}} \frac{2\pi}{\hbar} M_{kk'}^2 \frac{m^{3/2}(k_BT)^{1/2}}{\sqrt{2}\pi^2\hbar^3} \frac{1}{0.75}$$

Upon evaluating these terms we get

$$\frac{1}{\langle\langle\tau_{\text{defect}}\rangle\rangle} = 3.73 \times 10^{12} \text{ s}^{-1}$$

The defect limited mobility is then

$$\mu_{\text{defect}} = 0.18 \text{ m}^2/\text{V} \cdot \text{s} = 1800 \text{ cm}^2/\text{V} \cdot \text{s}$$

The ionized impurity limited mobility is found to be

$$\mu_{ii} = 9500 \text{ cm}^2/\text{V} \cdot \text{s}$$

The total mobility is given by the reciprocal relation and is

$$\mu_{tot} = 640 \text{ cm}^2/\text{V} \cdot \text{s}$$

5.4   INTERFACE ROUGHNESS SCATTERING

Interfaces between two materials are an important ingredient of modern electronic and optoelectronic devices. Depending upon the fabrication techniques, this interface has varying degrees of roughness. The interface roughness causes potential "bumps" in the path of the carriers, causing the carriers to scatter.

In Fig. 5.10 we show a schematic of the metal-oxide-semiconductor field effect transistor (MOSFET). In this device, a metal contact (the gate of the transistor) is isolated from the Si channel by a silicon oxide layer. The oxide layer is produced by oxidizing the silicon by exposing it to steam, and since SiO_x and Si have very different crystal structures, the Si/SiO_x interface has a roughness, schematically shown in Fig. 5.10c.

In an n-type MOSFET, electrons are induced in the channel in the "inversion" state as shown in Fig. 5.10. The electrons reside in a triangular quantum well with a wavefunction that goes to zero at the oxide side of the interface. A number of models have been suggested to account for the form of the interface roughness scattering potential. One model describes the potential as (r is the in-plane vector, z is the direction from the gate to the semiconductor)

$$
\begin{aligned}
\delta U(r, z) &= U(z + \Delta(r)) - U(z) \\
&\cong \Delta(r)\frac{\partial}{\partial z}U \\
&= e\tilde{E}(z)\Delta(r)
\end{aligned}
\tag{5.42}
$$

where $\tilde{E}(z)$ is the electric field in the inversion channel. The function $\Delta(r)$ is assumed to have the form

$$
\langle \Delta(r)\Delta(r' - r)\rangle = \Delta^2 \exp\left(-\frac{r^2}{\lambda^2}\right)
\tag{5.43}
$$

where Δ represents the height of the interface "islands" and λ represents their correlation length along the interface, as shown schematically in Fig. 5.10c.

The scattering matrix element is now

$$
M(k, k') = e\Delta(q) \int_0^\infty \tilde{E}(z) \mid \psi(z) \mid^2 dz
$$

where $\Delta(q)$ is the two-dimensional Fourier transform of $\Delta(r)$ and $\psi(z)$ is the wavefunction of the electrons in the inversion layer. If n_s is the electron sheet density and N_{depl} is the areal density of space charge (e.g., if there is any doping in the channel), then from Gauss's law we get

$$
\int \tilde{E}(z) \mid \psi(z) \mid^2 dz \simeq e\frac{N_{depl} + n_s/2}{\epsilon_s}
\tag{5.44}
$$

The factor $1/2$ multiplying n_s, is used because the interface field (which is of interest) is assumed to be about half of the average field in the channel. Under deep inversion conditions N_{depl} can be ignored in comparison to n_s.

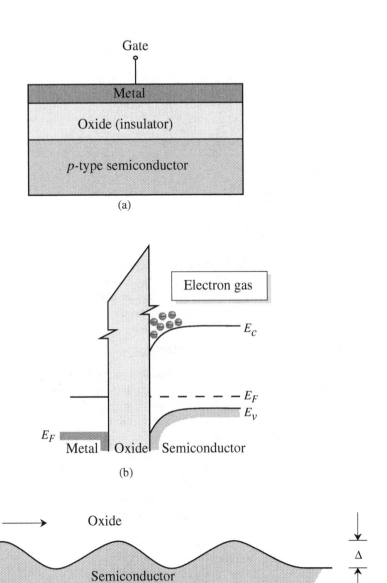

Figure 5.10: (a) A schematic of a metal-oxide-semiconductor structure; (b) a band profile of a MOSFET under inversion (ON) condition; and (c) a schematic of the roughness at the oxide-semiconductor interface.

For the surface roughness model assumed (A is area)

$$| \Delta(q) |^2 = \frac{\pi \Delta^2 \lambda^2}{A} \exp \left(-\frac{q^2 \lambda^2}{4} \right) \tag{5.45}$$

and the scattering rate becomes ($q = 2k \sin \theta/2$)

$$
\begin{aligned}
W(k) &= \frac{1}{A} \frac{2\pi}{\hbar} \frac{1}{4\pi^2} \int_0^{2\pi} d\theta \int_0^\infty q \, dq \, |M(k,k')|^2 \delta(E_k - E'_k) \\
&= \frac{m^*}{2\pi \hbar^3} \int_0^{2\pi} \frac{\pi \Delta^2 \lambda^2 e^4}{\epsilon_s^2} \left(N_{depl} + \frac{n_s}{2} \right)^2 \exp \left(-k^2 \lambda^2 \sin^2 \frac{\theta}{2} \right) d\theta \\
&= \frac{m^* \Delta^2 \lambda^2 e^4}{2\hbar^3 \epsilon_s^2} \left[I_0 \left(\frac{k^2 \lambda^2}{2} \right) - I_1 \left(\frac{k^2 \lambda^2}{2} \right) \right] \exp \left(\frac{-k^2 \lambda^2}{2} \right) \\
&\quad \left(N_{depl} + \frac{n_s}{2} \right)^2
\end{aligned}
\tag{5.46}
$$

where I_0 and I_1 are modified Bessel functions. The mobility, which is approximately obtained by replacing the energy (through the value of k) in the rate above by $k_B T$, is then inversely proportional to the square of the surface field or the charge density in the channel. This is borne out by experimental studies.

5.5 CARRIER–CARRIER SCATTERING

The various scattering sources discussed so far—ionized impurity, alloy disorder, neutral impurity and interface roughness—are all fixed in space and time. In these processes the electron energy remains unchanged after scattering because of the large mass of the scatterer. We will now discuss scattering of mobile carriers from other mobile carriers. Since electrons and holes are both charged particles there is Coulombic scattering between them. Also, there can be scattering between electrons themselves. The scattering between electrons is somewhat more complex due to the fact that they are identical Fermions. This introduces special features which we will discuss in this section. Both electron–hole and electron–electron scattering is quite important especially in materials where chrage densities reach or exceed 10^{18} cm^{-3}.

5.5.1 Electron–Hole Scattering

We have already discussed the scattering of electrons from the screened Coulombic potential of ionized dopants. Electrons can also scatter from holes and other electrons. This scattering mechanism is closely related in essence to the charged impurity scattering discussed earlier. We can approximate the interaction once again by a screened Coulombic potential between two point particles and obtain the rate in the Born approximation as done previously. As shown in Fig. 5.11, the interaction occurs as particle 1, with the initial wavevector $\boldsymbol{k}_1$, collides with particle 2, with initial wavevector $\boldsymbol{k}_2$. The collision causes a change in the wavevectors of both particles. Particle 1 leaves the interaction with a final wavevector $\boldsymbol{k}'_1$ and particle 2 leaves with a final wavevector $\boldsymbol{k}'_2$.

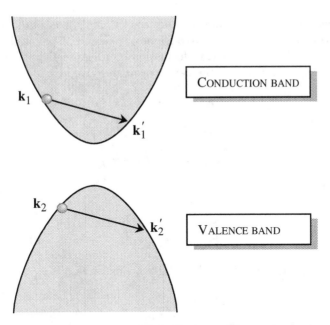

Figure 5.11: Scattering of an electron and hole. Each particle remains in the same band after scattering.

This interaction is described by the matrix element

$$
\begin{aligned}
\langle k_1' k_2' | eV | k_1 k_2 \rangle &= I(k_1, k_1') I(k_2, k_2') \\
&\times \frac{1}{V^2} \int\int d^3 r_1 d^3 r_2 e^{-i(\mathbf{k}_1' \cdot \mathbf{r}_1 + \mathbf{k}_2' \cdot \mathbf{r}_2)} \\
&\times \frac{e^2 \exp(-\lambda|\mathbf{r}_1 - \mathbf{r}_2|)}{4\pi\epsilon|\mathbf{r}_1 - \mathbf{r}_2|} e^{i(\mathbf{k}_1 \cdot \mathbf{r}_1 + \mathbf{k}_2 \cdot \mathbf{r}_2)}
\end{aligned}
\tag{5.47}
$$

where $I(k_1, k_1'), I(k_2, k_2')$ are the overlap integrals over the unit cell involving the cell-periodic parts of the Bloch functions

$$
\begin{aligned}
I(k_1, k_1')\, I(k_2, k_2') &= \int_{\text{cell}} u_{\mathbf{k}_1'}^*(\mathbf{r}_1)\, u_{\mathbf{k}_1}(\mathbf{r}_1)\, d^3 r_1 \\
&\times \int_{\text{cell}} u_{\mathbf{k}_2'}^*(\mathbf{r}_2)\, u_{\mathbf{k}_2}(\mathbf{r}_2)\, d^3 r_2
\end{aligned}
\tag{5.48}
$$

In our previous calculation we have not worried about the central cell part of the electron (hole) wavefunctions. This is reasonable, since the scattering carrier is in the same band (conduction band or valence band) and the overlap between the central cell parts is close to unity. For scattering events in which all wavevectors lie close to the bandedges, the integrals are usually assumed to be unity. In general, they are less than unity, especially for the hole states.

The interaction depends only on the separation distance of the particles, so it is convenient to transform the problem to a frame of reference in which the center of mass of the two particles is at rest. The transformation (nonrelativistic) to the center-of-mass frame is affected by converting to the new coordinates given by

$$
\begin{aligned}
\boldsymbol{K} &= \boldsymbol{k} - \boldsymbol{k}_{cm} \\
\boldsymbol{k}_{cm} &= \frac{1}{2}\left(\boldsymbol{k}_1 + \boldsymbol{k}_2\right) \\
\boldsymbol{K}_{12} &= \frac{1}{2}\left(\boldsymbol{k}_1 - \boldsymbol{k}_2\right)
\end{aligned}
$$

or

$$
\begin{aligned}
\boldsymbol{k}_1 &= \boldsymbol{k}_{cm} + \boldsymbol{K}_{12} \\
\boldsymbol{k}_2 &= \boldsymbol{k}_{cm} - \boldsymbol{K}_{12}
\end{aligned}
\tag{5.49}
$$

The corresponding transformation of spatial coordinates is

$$
\begin{aligned}
\boldsymbol{R} &= \boldsymbol{r} - \boldsymbol{r}_{cm} \\
\boldsymbol{r}_{cm} &= \frac{m_1^* \, \boldsymbol{r}_1 + m_2^* \, \boldsymbol{r}_2}{m_1^* + m_2^*} \\
\boldsymbol{r}_{12} &= \boldsymbol{r}_1 - \boldsymbol{r}_2
\end{aligned}
$$

leading to

$$
\begin{aligned}
\boldsymbol{r}_1 &= \boldsymbol{r}_{cm} + \frac{m_2^*}{m_1^* + m_2^*}\,\boldsymbol{r}_{12} \\
\boldsymbol{r}_2 &= \boldsymbol{r}_{cm} - \frac{m_1^*}{m_1^* + m_2^*}\,\boldsymbol{r}_{12}
\end{aligned}
\tag{5.50}
$$

The integral for the matrix element now splits into a product of two integrals, one over $\boldsymbol{r}_{cm}$ and the other over $\boldsymbol{r}_{12}$. The former gives unity and ensures the conservation of momentum. The integral over $\boldsymbol{r}_{12}$ is

$$
\begin{aligned}
\langle \boldsymbol{K}_{12}' \,|eV(\boldsymbol{r}_{12})|\, \boldsymbol{K}_{12} \rangle &= \frac{1}{V}\int \exp(-i\boldsymbol{K}_{12}' \cdot \boldsymbol{r}_{12}) \\
&\quad \times \frac{e^2 \exp(-\lambda r_{12})}{4\pi\epsilon\, r_{12}} \exp(i\boldsymbol{K}_{12} \cdot \boldsymbol{r}_{12})\, d^3 r_{12} \\
&= \frac{e^2}{\epsilon V}\, \frac{1}{\left|\boldsymbol{K}_{12}' - \boldsymbol{K}_{12}\right|^2 + \lambda^2}
\end{aligned}
\tag{5.51}
$$

as for the case of the ionized impurity scattering. The problem is exactly similar to the collision of a particle of mass μ^*, equal to the reduced mass of the two particles, with a fixed center. Thus one can use the calculations carried out for the ionized impurity scattering to obtain the angular dependence of the scattering rate.

The scattering angles θ in the center of mass system discussed above is related to the scattering angle in the laboratory system by the relation

$$\tan \theta_0 \frac{\sin \theta}{m_1^*/m_2^* + \cos \theta} \tag{5.52}$$

If the mass of the electron (m_1^*) is much smaller than that of the hole (m_2^*) the scattering angle in the laboratory frame and the center of mass frame is the same. In this case the holes cause the same scattering for electrons as ionized impurities (which are fixed center of essentially infinite mass).

In minority carrier transport, e.g., electrons moving in p-type semiconductors, the electron gets scattered from acceptors and holes with the result that if $m_h^* \gg m_e^*$, the scattering rate simply becomes *twice* that from the impurities alone. This approximation works quite well for most semiconductors.

5.5.2 Electron–Electron Scattering: Scattering of Identical Particles

In quantum mechanics there is an important distinction when scattering occurs between identical particles or distinguishable particles. In the calculation discussed so far (electron–impurity, electron–hole) electrons scatter from other particles that are distinguishable from them. We will now discuss electron–electron scattering which involves identical particles. In Fig. 5.12 we show two particles, A and B, scatter from each other and we measure their scattering rates by placing detectors. We will look for the angular distribution of the probability that some particle arrives at the detector D_1. We will work in the center of mass system and denote by $f(\theta)$ the amplitude that the particle A is scattered by angle θ. We have to consider the two possibilities, shown in the figure, where the particles are exchanged. The amplitude of scattering for the second case is $f(\pi - \theta)$. The following is observed experimentally.

- If the particles are distinguishable the probability of *some* particle appearing as D_1 is

$$|f(\theta)|^2 + |f(\pi - \theta)|^2 \tag{5.53}$$

- If the particles are indistinguishable and are bosons (e.g. α-particles, photons, mesons) the probability of one of the particles appearing is D_1 is

$$|f(\theta) + f(\pi - \theta)|^2 \tag{5.54}$$

- If the particles are identical but are fermions (e.g. electrons, neutrinoes, protons, neutrons) the probability is

$$|f(\theta) - f(\pi - \theta)|^2 \tag{5.55}$$

Note that if the two particles are electrons but have different spins, and the scattering is not supposed to alter the spin, then the probability is given by the *distinguishable* case.

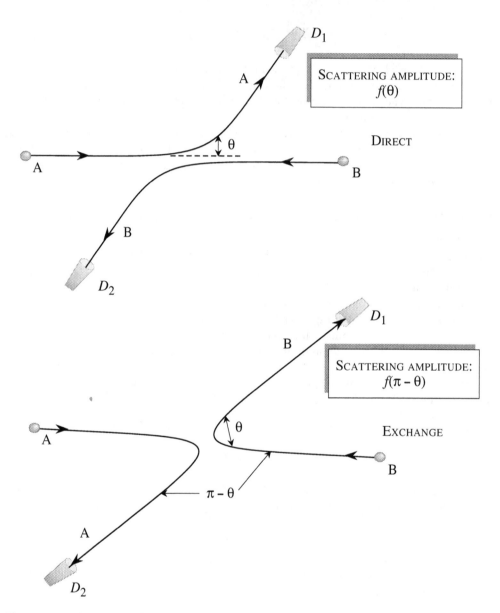

Figure 5.12: Scattering of two particles in the center of mass system. The detector D_1 is able to detect any particle being scattered at the angle.

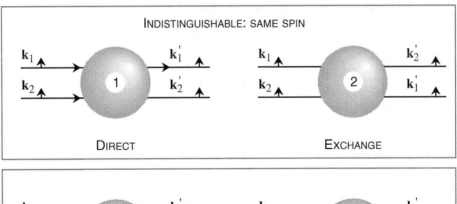

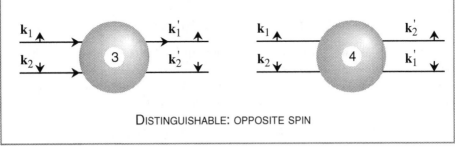

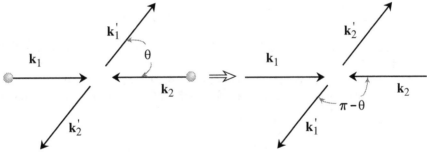

Figure 5.13: A schematic showing the direct and exchange scattering processes for identical particle scattering. When the spins of the two particles are the same, the two processes (1 and 2) are indistinguishable. When the spins are different, the two processes (3 and 4) are distinguishable.

As noted above, for e-e scattering we have to include the direct and exchange terms and add the scattering amplitudes with the prescription discussed above. In Fig. 5.13 we show the various possible scattering processes for both the indistinguishable (i.e., same spins) and distinguishable (opposite spins) cases.

We can essentially use the formalism we have already developed for the e-h collisions and apply it to the e-e case. The processes (1) and (2) are totally indistinguishable and will interfere. The matrix element we calculated earlier for e-h case is M_{12}

$$M_{12} \;=\; \langle \boldsymbol{K}'_{12} |V| \boldsymbol{K}_{12} \rangle$$

$$= \frac{e^2}{\epsilon V} \frac{1}{\left| \boldsymbol{K}'_{12} - \boldsymbol{K}_{12} \right|^2 + \lambda^2} \tag{5.56}$$

where

$$\left| \boldsymbol{K}'_{12} - \boldsymbol{K}_{12} \right| = 2K_{12} \sin(\theta/2)$$

The process (2) changes $\theta \to \pi - \theta$

$$M_{21} = \frac{e^2}{\epsilon V} \frac{1}{\left| \boldsymbol{K}'_{12} - \boldsymbol{K}_{12} \right|^2 + \lambda^2} \tag{5.57}$$

where

$$\left| \boldsymbol{K}'_{12} - \boldsymbol{K}_{12} \right| = 2K_{12} \cos(\theta/2)$$

Since these two processes are indistinguishable, they interfere destructively (since the electrons are fermions) at the amplitude level. The processes (3) and (4) are distinguishable and therefore do not interfere. One has to square and add the contributions separately The total matrix element squared is now

$$
\begin{aligned}
|M|^2 &= \frac{1}{2} \left[|M_{12}|^2 + |M_{21}|^2 + |M_{12} - M_{21}|^2 \right] \\
&= |M_{12}|^2 + |M_{21}|^2 - \frac{1}{2} \left(M_{21} M_{12}^* + M_{21}^* M_{12} \right)
\end{aligned}
\tag{5.58}
$$

The factor $1/2$ arises because in half the collisions the spins are aligned and in half they are opposed.

Following the arguments for the e-h case, or the ionized impurity case, we can write for the differential cross section in the center of mass frame (taking the overlap integrals to be unity)

$$
\begin{aligned}
\sigma(\theta) = \left(\frac{e^2}{16\pi\epsilon E_{12}} \right)^2 & \left[\frac{1}{\left\{ \sin^2(\theta/2) + \left(\frac{\lambda}{2K_{12}} \right)^2 \right\}^2} \right. \\
&+ \frac{1}{\left\{ \cos^2(\theta/2) + \left(\frac{\lambda}{2K_{12}} \right)^2 \right\}^2} \\
&\left. - \frac{1}{\left\{ \sin^2(\theta/2) + \left(\frac{\lambda}{2K_{12}} \right)^2 \right\} \left\{ \cos^2(\theta/2) + \left(\frac{\lambda}{2K_{12}} \right)^2 \right\}} \right]
\end{aligned}
\tag{5.59}
$$

where θ is the angle between $\boldsymbol{K}_{12}$ and $\boldsymbol{K}'_{12}$.

$$K_{12} = \left| \frac{\boldsymbol{K}_1 - \boldsymbol{K}_2}{2} \right|$$

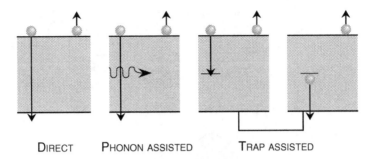

DIRECT   PHONON ASSISTED   TRAP ASSISTED

Figure 5.14: Various processes that contribute to Auger recombination. In high purity materials trap assisted processes are not important.

$$E_{12} = \frac{\hbar^2 K_{12}^2}{2\overline{m}^*}$$

$$\overline{m}^* = \frac{m^*}{2} \tag{5.60}$$

The total cross section is derived by integrating over all angles

$$\sigma = \left(\frac{e^2}{16\pi\epsilon E_{12}}\right)^2 8\pi \left[\frac{1}{(\lambda/2K_{12})^2\{1+(\lambda/2K_{12})^2\}}\right.$$
$$+ \left. \frac{1}{1+2(\lambda/2K_{12})^2}\ln\left\{\frac{(\lambda/2K_{12})^2}{1+(\lambda/2K_{12})^2}\right\}\right] \tag{5.61}$$

A momentum relaxation cross-section can be found by multiplying the first term in Eqn. 5.59 by $(1-\cos\theta)$, second term by $(1+\cos\theta)$, and the third term by $(1-\cos\theta)(1+\cos\theta)$. With $x = (\hbar^2\lambda^2)/(2\overline{m}^* E_{12})$

$$\frac{1}{\tau_m} = \frac{e^4}{8\pi\epsilon^2 V\sqrt{2\overline{m}^*}E_{12}^{3/2}}\left[\ln\left(1+\frac{4}{x}\right)-\frac{1}{1+x}-\frac{\pi}{2}\left\{\frac{1}{1+x\left(1+\frac{x}{4}\right)}\right\}\right] \tag{5.62}$$

Electron–electron scattering rate is usually too small for carrier concentrations of less than 10^{17} cm^{-3}.

5.6 AUGER PROCESSES AND IMPACT IONIZATION

In the scattering processes discussed so far the electron (hole) remains in the same band (conduction or valence) after scattering. The number of free carriers is unchanged as a result of scattering. In Auger processes and in impact ionization carriers scatter across bands. This can increase or decrease the total density of free carriers in a material. Fig. 5.14 shows a schematic of several possible Auger processes. For example, in the direct process an electron recombines with a hole and the extra energy is absorbed by another electron. In this process, the electron hole recombination does not produce a photon as would be the case for radiative transitions. Such processes are very important in narrow bandgap lasers where this process causes carrier recombination without producing useful

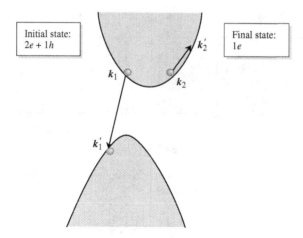

Figure 5.15: A schematic of the states of the electrons before and after an Auger scattering. The reverse of this process is the impact ionization process. This particular process is called CHCC (Conduction-Heavy hole-Conduction-Conduction).

photons. In addition to the processes shown for electrons, we have the processes for holes (where the energy of recombination is transferred to a hole). These processes are all mediated by the Coulombic interactions and involve e-e scattering. In high purity materials only the direct processes are of significance.

Impact ionization (which is the reverse of the Auger process) occurs in the presence of high electric fields. Under very high electric fields, electrons gain energy larger than the bandgap of the semiconductor. Thus a high energy electron in the conduction band scatters from an electron in the valence band. The second electron is raised to the conduction band, resulting in two electrons in the conduction band and a hole in the valence band. This causes carrier multiplication and the current in the semiconductor increases dramatically. This results in the breakdown of the semiconductor and limits the high performance behavior of electronic devices. Impact ionization is also exploited in avalanche photodetectors for very high gains.

Let us discuss the direct Auger process. The matrix element M_{12} for the scattering in the direct process is as before (see Fig. 5.15)

$$
\begin{aligned}
M_{12} \;=\; & \frac{1}{V^2} \int\!\!\int d^3r_1\, d^3r_2\, \frac{e^2 \exp(-\lambda|\boldsymbol{r}_1 - \boldsymbol{r}_2|)}{4\pi\epsilon|\boldsymbol{r}_1 - \boldsymbol{r}_2|} \\
& \times\; u^*_{v\mathbf{k}'_1}(\boldsymbol{r}_1)\, \exp(-i\boldsymbol{k}'_1\cdot\boldsymbol{r}_1)\, u^*_{c\mathbf{k}'_2}(\boldsymbol{r}_2)\, \exp(-i\boldsymbol{k}'_2\cdot\boldsymbol{r}_2) \\
& \times\; u^*_{c\mathbf{k}_1}(\boldsymbol{r}_1)\, \exp(i\boldsymbol{k}_1\cdot\boldsymbol{r}_1)\, u^*_{c\mathbf{k}_2}(\boldsymbol{r}_2)\, \exp(i\boldsymbol{k}_2\cdot\boldsymbol{r}_2)
\end{aligned}
\tag{5.63}
$$

Transforming to the center of mass system gives us the conservation of momentum

$$
\boldsymbol{k}_1 + \boldsymbol{k}_2 = \boldsymbol{k}'_1 + \boldsymbol{k}'_2
\tag{5.64}
$$

As before, we have

$$M_{12} = \left(\frac{e^2}{\epsilon V}\right) \frac{I(\boldsymbol{k}_1, \boldsymbol{k}_1^{'})\, I(\boldsymbol{k}_2, \boldsymbol{k}_2^{'})}{\left|\boldsymbol{k}_1^{'} - \boldsymbol{k}_1\right|^2 + \lambda^2} \tag{5.65}$$

The relevant overlap integrals are

$$I(\boldsymbol{k}_1, \boldsymbol{k}_1^{'}) = \int_{\text{cell}} d^3r_1\, u_{v\boldsymbol{k}_1^{'}}^{*}(\boldsymbol{r}_1)\, u_{c\boldsymbol{k}_1}^{*}(\boldsymbol{r}_1)$$

$$I(\boldsymbol{k}_2, \boldsymbol{k}_2^{'}) = \int_{\text{cell}} d^3r_2\, u_{c\boldsymbol{k}_2^{'}}^{*}(\boldsymbol{r}_2)\, u_{c\boldsymbol{k}_2}^{*}(\boldsymbol{r}_2)$$

Here we have used the fact that

$$K_{12} = \frac{1}{2}\left|\boldsymbol{k}_1 - \boldsymbol{k}_2\right|$$

so that due to momentum conservation

$$\left|\boldsymbol{K}_{12}^{'} - \boldsymbol{K}_{12}\right|^2 = \left|\boldsymbol{k}_1^{'} - \boldsymbol{k}_1\right|^2$$

For the exchange process

$$M_{21} = \left(\frac{e^2}{\epsilon V}\right) \frac{I(\boldsymbol{k}_1, \boldsymbol{k}_2^{'})I(\boldsymbol{k}_2, \boldsymbol{k}_1^{'})}{\left|\boldsymbol{k}_1^{'} - \boldsymbol{k}_2\right|^2 + \lambda^2}$$

and

$$I(\boldsymbol{k}_1, \boldsymbol{k}_2^{'}) = \int_{\text{cell}} d^3r_1 u_{c\boldsymbol{k}_2^{'}}^{*}(\boldsymbol{r}_1)u_{c\boldsymbol{k}_1}(\boldsymbol{r}_1)$$

$$I(\boldsymbol{k}_2, \boldsymbol{k}_1^{'}) = \int_{\text{cell}} d^3r_2 u_{v\boldsymbol{k}_1^{'}}^{*}(\boldsymbol{r}_2)u_{c\boldsymbol{k}_2}(\boldsymbol{r}_2)$$

As discussed for the case of the e-e scattering we have to consider the four processes of Fig. 5.13 which give us the total matrix element squared

$$|M|^2 = \left[|M_{12}|^2 + |M_{21}|^2 + |M_{12} - M_{21}|^2\right]$$

To calculate the Auger rates we must discuss the occupation statistics of the various electrons and hole states involved in the process. We need to weigh the rate with the probability that state $\boldsymbol{k}_2$ is full, $\boldsymbol{k}_1^{'}$ is empty and $\boldsymbol{k}_1$ is full. In general we have to use the Fermi–Dirac function to describe the occupation. This requires numerical evaluation of the scattering rate. However, if we assume nondegenerate statistics we can obtain analytical expressions. We have

$$P(\boldsymbol{k}_1, \boldsymbol{k}_2, \boldsymbol{k}_1^{'}) = f(\boldsymbol{k}_1)\, f(\boldsymbol{k}_2)\left(1 - f(\boldsymbol{k}_1^{'})\right) \tag{5.66}$$

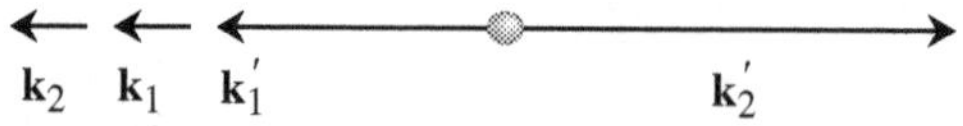

Figure 5.16: Procedure for finding the maximum in probability for Auger rates. This procedure is also used to find the threshold energy for the impact ionization process to occur. The threshold is reached when the high energy electron k_2' has its wavevector lined up opposite to those of the low energy electrons as shown.

where

$$
\begin{aligned}
f(\mathbf{k}_1) &= \frac{n}{N_c} \exp\left(\frac{-E_{c\mathbf{k}_1}}{k_B T}\right) \\
f(\mathbf{k}_2) &= \frac{n}{N_c} \exp\left(\frac{-E_{c\mathbf{k}_2}}{k_B T}\right) \\
1 - f(\mathbf{k}_1') &= \frac{p}{N_v} \exp\left(\frac{-E_{v\mathbf{k}_1'}}{k_B T}\right)
\end{aligned}
\tag{5.67}
$$

Here n and p are the electron and hole carrier densities and N_c, N_v are the conduction and valence band effective density of states.

This gives the total probability factor

$$
\begin{aligned}
P(\mathbf{k}_1, \mathbf{k}_2, \mathbf{k}_1') &= \frac{np}{N_c N_v} \frac{n}{N_c} \exp\left(-\frac{E_{c\mathbf{k}_2} + E_{v\mathbf{k}_1'} + E_{c\mathbf{k}_1}}{k_B T}\right) \\
&\approx \frac{n}{N_c} \exp\left(-\frac{E_g + E_{c\mathbf{k}_2} + E_{v\mathbf{k}_1'} + E_{c\mathbf{k}_1}}{k_B T}\right)
\end{aligned}
\tag{5.68}
$$

We have assumed that the state $\mathbf{k}_2'$ is always available since it is a high energy electron state in the conduction band.

It is useful to examine the energy at which this term maximizes. This involves finding the extremum of the expression (in the parabolic band approximation)

$$
\frac{k_1^2}{2m_c^*} + \frac{k_2^2}{2m_c^*} + \frac{k_1'^2}{2m_v^*} = \frac{1}{2m_c^*}\left[k_1^2 + k_2^2 + \mu k_1'^2\right]
\tag{5.69}
$$

where

$$
\mu = \frac{m_c^*}{m_v^*}
$$

The probability factor will maximize for the lowest energy values of $\mathbf{k}_1$, $\mathbf{k}_1'$ and $\mathbf{k}_2$ which are consistent with energy and momentum conservation. Since $\mathbf{k}_2'$ is the largest vector, we line up $\mathbf{k}_1$, $\mathbf{k}_1'$, and $\mathbf{k}_2$ with $\mathbf{k}_2'$ in the opposite direction as shown in Fig. 5.16. Thus we choose

$$
\mathbf{k}_1 + \mathbf{k}_1' + \mathbf{k}_2 = -\mathbf{k}_2'
\tag{5.70}
$$

We also write

$$
\begin{aligned}
\boldsymbol{k}_1 &= a\boldsymbol{k}'_1 \\
\boldsymbol{k}_2 &= b\boldsymbol{k}'_1
\end{aligned}
\tag{5.71}
$$

We now have from conservation of energy

$$
k'^2_2 = (a^2 + b^2 + \mu)k'^2_1 + K^2_g
\tag{5.72}
$$

where

$$
E_g = \frac{\hbar^2 K^2_g}{2m^*_c}
$$

Also the conservation of momentum gives us

$$
\boldsymbol{k}'_2 = (a + b + 1)\boldsymbol{k}'_2
$$

or

$$
k'^2_2 = (a^2 + b^2 + 1 + 2ab + 2a + 2b)k'^2_1
\tag{5.73}
$$

Eliminating k'^2_2 from Eqns. 5.72 and 5.73 we get

$$
k'^2_1(1 + 2ab + 2a + 2b - \mu) = K^2_g
\tag{5.74}
$$

The quantity to be minimized for maximum Auger rate or the impact ionization threshold is

$$
k^2_1 + k^2_2 + \mu k'^2_1 = k'^2_1(a^2 + b^2 + \mu)
$$

Substituting for k'^2_1 from Eqn. 5.74 we get

$$
(a^2 + b^2 + \mu)k'^2_1 = \frac{a^2 + b^2 + \mu}{1 + 2ab + 2a + 2b - \mu}K^2_g
$$

This quantity minimizes when

$$
a = b = \mu
\tag{5.75}
$$

This gives us the energy values

$$
\begin{aligned}
E_{c\boldsymbol{k}_1} &= E_{c\boldsymbol{k}_2} \\
&= \mu E_{v\boldsymbol{k}'_1} \\
&= \left(\frac{\mu^2}{1 + 3\mu + 2\mu^2}\right)E_g
\end{aligned}
\tag{5.76}
$$

The maximum probability function is now

$$
P(\boldsymbol{k}_1, \boldsymbol{k}_2, \boldsymbol{k}'_1) = \frac{n}{N_c}\exp\left(-\frac{1 + 2\mu}{1 + \mu}\frac{E_g}{k_B T}\right)
\tag{5.77}
$$

and the energy of the high energy electron is

$$E_{ck_2'} = \frac{1 + 2\mu}{1 + \mu} E_g \tag{5.78}$$

If $\mu \ll 1$, we have the approximation

$$E_{ck_2'} \approx (1 + \mu)E_g \tag{5.79}$$

The value for $E_{ck_2'}$ represents the threshold for the inverse process of impact ionization.

According to this (very approximate) expression the Auger process starts when the initial carriers have a certain minimum energy. This is because both momentum and energy have to be conserved in the scattering (in general, momentum has to be conserved within a reciprocal lattice vector). As a result of the threshold energy, Auger rates have a strong dependence on temperature and bandgap.

To solve the general integral for the Auger rates, one needs to evaluate the multiple integral (these are rates for a particular electron at k_1; the total rate over all electrons will involve a further integral over k_1, as well)

$$
\begin{aligned}
W_{\text{Auger}} \;=\; & 2 \left(\frac{2\pi}{\hbar} \right) \left(\frac{e^2}{\epsilon} \right)^2 \frac{1}{(2\pi)^9} \\
\times \;& \int d^3 k_2 \int d^3 k_1' \int d^3 k_2' \, |M|^2 \, P(\boldsymbol{k}_1, \boldsymbol{k}_2, \boldsymbol{k}_1') \\
\times \;& \delta(E_{ck_1} + E_{ck_2} - E_{vk_1'} - E_{ck_2'}) \tag{5.80}
\end{aligned}
$$

The matrix element $|M|^2$ has been discussed before and for most purposes it is adequate to use $\lambda = 0$. In general, one has to explicitly evaluate the overlap integrals by using an accurate bandstructure description. Typical results for such a calculation are shown in Fig. 5.17 where we show the Auger recombination rate for the narrow bandgap material $In_{0.53}Ga_{0.47}As$ ($E_g \approx 0.8$ eV) which is widely used for long-distance optical communication lasers. The Auger rate is approximately proportional to n^3 (for most lasers $n = p$), and is often written in the form

$$
\begin{aligned}
W \;=\; & R_{\text{Auger}} \\
=\; & Fn^3 \tag{5.81}
\end{aligned}
$$

where F is the Auger coefficient.

The Auger rates increase exponentially as the bandgap is decreased. They also increase exponentially as the temperature increases. These are direct results of the energy and momentum conservation constraints and the carrier statistics. Auger processes are more or less unimportant in semiconductors with bandgaps larger than approximately 1.5 eV (e.g., GaAs, AlGaAs, InP). However, they become quite important in narrow bandgap materials such as $In_{0.53}Ga_{0.47}As$ ($E = 0.8$ eV) and HgCdTe ($E < 0.5$ eV), and are thus a serious hindrance for the development of long wavelength lasers. Table 5.1 shows the values of Auger coefficients for some semiconductors.

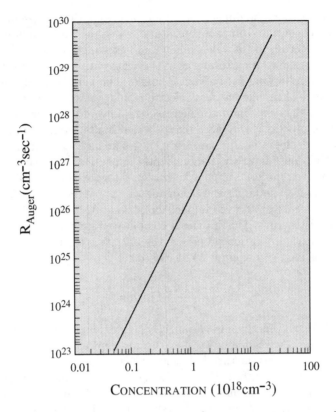

Figure 5.17: Auger rates calculated for $In_{0.53}Ga_{0.47}As$ at room temperature. The important process involves the final state with one hole in the split off band. The term CHHS (Conduction-Heavy hole-Heavy hole-Split off hole) is used for such events.

MATERIAL	BANDGAP	AUGER COEFFICIENT $(cm^6 s^{-1})$
InGaAs	0.8 eV	10^{28} at 300 K
HgCdTe	0.2 eV	10^{27} at 77 K
GaInAsP	0.8 eV	$\sim 6 \times 10^{28}$ at 300 K
GaInAsP	0.7 eV	$\sim 1.2 \times 10^{27}$ at 300 K
GaInAsSb	0.4 eV	$\sim 6 \times 10^{27}$ at 300 K

Table 5.1: Auger coefficients of some semiconductors. There is a large uncertainty in these values.

As noted earlier, the impact ionization process is the inverse of the Auger process. In an electron initiated impact ionization process a "hot" (high energy) electron scatters with an electron in the valence band and exchanges energy with it to knock it into the conduction band. As a result, after the scattering we have two electrons in the conduction band and a hole in the valence band. The net result of impact ionization is that the number of mobile carriers (electrons and holes) increases. As a result the current flowing in the semiconductor increases rapidly resulting in "breakdown." The impact ionization process is the main reason which limits the high power performance of semiconductor devices. The process only occurs at high applied electric fields since the initial electron must have an energy slightly larger than the bandgap of the material. Thus, for a given material impact ionization does not stand until the electric field reaches critical value. The larger the bandgap, the higher the critical field.

The full calculation for the impact ionization rates also involves the knowledge of the entire bandstructure. This implies a complicated numerical solution.

In the impact ionization processes, the initiating electron (hole) energy has to be larger than the bandgap energy. As discussed for the Auger processes the threshold energy is

$$E_{\mathbf{k}1} = \frac{1+2\mu}{1+\mu} E_g; \quad \mu = \frac{m_c^*}{m_v^*} \tag{5.82}$$

Thus the initial electron energy must be slightly larger than the bandgap of the material. The threshold energy expression given above is only approximate. Since bands are quite anisotropic at high energies the threshold energy has a strong angular dependence. Once the electron (hole) energy exceeds the threshold energy the impact ionization scattering rate becomes very strong reaching values of $\sim 10^{12}$ s^{-1} rapidly. As a result carrier multiplication occurs rapidly beyond a threshold applied electric field.

As noted earlier the detailed calculation for the impact ionization has to be done numerically due to the complexity of the final state density of states and the bandstructure. However, a simple expression has been derived by Ridley (1982) for parabolic bands

$$W_{\text{imp}} = 4.139 \times 10^{16}$$

$$\times \left\{ \frac{4\sqrt{m_c^* m_v^*}}{m_0} \left(\frac{m_c^*}{m_0} + \mu \right) \left(\frac{\epsilon_0}{\epsilon} \right)^2 \left[\frac{E_{\mathbf{ck}_2'}}{E_g} - (1+\mu) \right] \right\} \text{ s}^{-1} \tag{5.83}$$

In Fig. 5.18 we show the calculated impact ionization rates for electrons in GaAs assuming simple isotropic electron and hole masses and a non-parabolic relation for GaAs electrons

$$(1 + \alpha E)E = \frac{\hbar h^2}{2m^*} \tag{5.84}$$

where $\alpha = 0.67$ eV^{-1}.

The impact ionization process becomes important in semiconductor devices when they are operated at high electric fields (several hundred kV/cm). It causes breakdown in semiconductor structures and thus limits the high power application of transistors. The impact ionization process (or avalanche process) is, however, exploited in

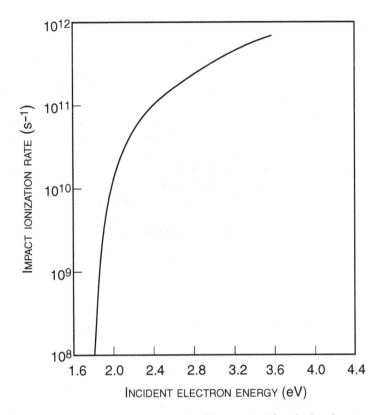

Figure 5.18: Impact ionization rates for GaAs ($E_g = 1.4$ eV) calculated using an isotropic bandstructure.

avalanche detectors where a photo-generated carrier causes carrier multiplication by impact ionization and thus provides a high gain detector.

5.7 PROBLEMS

5.1 Calculate and plot the screening length λ^{-1} as a function of free carrier density from $n_{\text{free}} = 1 \times 10^{14}$ cm^{-3} to 1×10^{18} cm^{-3} at 77 K and 300 K for GaAs and Si.

5.2 Plot the angular dependence of the scattering rate (like Fig. 5.5) due to ionized impurities when the background ionized donors are 10^{15} cm^{-3} and 10^{17} cm^{-3} at room temperature for GaAs. Assume an electron energy of $k_B T$.

5.3 Plot the bare and screened Coulombic potential (i.e., $V(r)$ vs. r) arising from a singly charged donor in Si at carrier concentrations of i) 10^{14} cm^{-3}; ii) 10^{16} cm^{-3}; and iii) 10^{18} cm^{-3}. Plot the results at 300 K and 77 K.

5.4 Plot the scattering rate versus scattering angle for electrons with energy equal to $3/2k_B T$ in Si doped at i) 10^{17} cm^{-3} and ii) 10^{18} cm^{-3}. Calculate the results at 77 K and 300 K.

5.5 Calculate the ionized impurity limited mobility ($N_D = 10^{16}$ cm^{-3}; 10^{17} cm^{-3}) in

GaAs from 77 K to 300 K.

5.6 If the measured room temperature mobility of electrons in GaAs doped n-type at 5×10^{17} cm^{-3} is 3500 cm^{2}V^{-1} s^{-1} calculate the relaxation time for phonon scattering.

5.7 Calculate the alloy scattering limited mobility in In$_{0.53}$Ga$_{0.47}$As as a function of temperature from 77 K to 400 K. Assume an alloy scattering potential of 1.0 eV.

5.8 Assume that the holes are much heavier than electrons in GaAs. Calculate the room temperature minority carrier mobility of electrons moving in a p-type base of an HBT where the base acceptor doping level is 10^{17} cm^{-3}. Remember that the electrons will scatter from the holes as well as the acceptors.

5.9 Assume that the scattering potential of neutral, or un-ionized impurities, in a semiconductor can be represented by a form similar to that used for the alloy scattering potential. What is the room temperature electron mobility in GaAs, due to 10^{15} neutral impurities per cm^3, each having a potential of 1.0 eV and a radial extent of 10 Å?

5.10 In a clustered alloy, the scattering potential is represented by

$$
\begin{aligned}
V(r) &= V_0 \quad \text{for } r \le r_c \\
&= 0 \quad\; \text{for } r > r_c,
\end{aligned}
$$

where r_c is the cluster radius. Explain why the temperature dependence of the alloy scattering limited mobility shows a peak at high temperatures.

5.11 In the text, when considering impurity scattering, we considered each scatterer to be independent. It is found experimentally that at high doping, the mobility is much lower than the theoretical value. Explain this qualitatively.

5.12 In our discussions of impact ionization, we continued to use the Fermi golden rule or Born approximation for the scattering rates. At the high energies encountered in impact ionization, the total scattering rates may approach 10^{13} s^{-1} or even 10^{14} s^{-1}. Discuss the effects this may have in terms of the energy conservation rule used in the scattering rate derivation.

5.13 In the lucky drift model for impact ionization, it is assumed that some lucky carriers are accelerated ballistically (i.e., without scattering) to energies above threshold, causing impact ionization. Assume that the average relaxation time is 0.01 ps. At approximately what electric fields will 0.1% of the electrons acquire threshold energy in GaAs?

5.14 Calculate the energy dependence of the impact ionization rate in GaAs and InAs.

5.15 Show that while in a 3-dimensional and 2-dimensional system, it is possible for electron–electron scattering to randomize the energy distribution of hot electrons, in a strictly 1-dimensional system, this is not possible.

5.8 REFERENCES

- **Screening and Dielectric Constant**

    – N. W. Ashcroft and N. D. Mermin, *Solid State Physics* Holt, Rinehart and Winston, New York (1976).

    – C. Kittel, *Introduction to Solid State Physics* John Wiley and Sons, New York (1986).

– J. M. Ziman, *Principles of the Theory of Solids* Cambridge University Press, Cambridge (1972).

• **Electron–Ionized Impurity Scattering: Theory**

– H. Brooks, *Phys. Rev.*, **83**, 879 (1951).

– E. M. Conwell and V. Weisskopf, *Phys. Rev.*, **77**, 388 (1950).

– M. Lundstrom, *Fundamentals of Carrier Transport* (Modular Series on Solid State Devices, edited by G. W. Neudeck and R. F. Pierret), Addison-Wesley, Reading (1990), vol. X.

– B. K. Ridley, *Quantum Processes in Semiconductors* Clarendon Press, Oxford (1982).

– N. Takimoto, *J. Phys. Soc. Japan*, **14**, 1142 (1959).

• **Impurity Dependence of Mobility: Experiments**

– D. A. Anderson, N. Aspley, P. Davies and P. L. Giles, *J. Appl. Phys.*, **58**, 3059 (1985).

– H. C. Casey, Jr. and M. B. Panish, *Heterostructure Lasers* Academic Press, New York (1978).

– "Modulation Doping of Semiconductor Heterostructures," by A. C. Gossard, in *Proceeding of NATO School of MBE and Heterostructures* at Erice, Italy, Nijhoft, Holland (1983).

– M. B. Prince, *Phys. Rev.*, **92**, 681 (1953).

– S. M. Sze, *Physics of Semiconductor Devices* John Wiley and Sons, New York (1981).

• **Alloy Scattering**

– D. K. Ferry, *Semiconductors* Macmillan, New York (1991).

– J. W. Harrison, and J. Hauser, *Phys. Rev. B*, **1**, 3351 (1970).

– L. Makowski and M. Glicksman, *J. Phys. Chem. Sol.*, **34**, 487 (1976).

• **Consequences of Alloy Clustering**

– J. M. Cowley, *Phys. Rev.*, **77**, 669 (1950)

– G. L. Hall, *Phys. Rev.*, **116**, 604 (1959).

– W. P. Hong, P. Bhattacharya and J. Singh, *Appl. Phys. Lett.*, **50**, 618 (1987).

– W. P. Hong, A. Chin, N. Debbar, J. M. Hinckley, P. Bhattacharya and J. Singh, *J. Vac. Sci. Technol. B*, **5**, 800 (1987).

– J. Marsh, *Appl. Phys. Lett.*, **4**, 732 (1982).

– L. Nordheim, *Ann. Physik*, **9**, 607 (1931).

- **Electron–Electron Scattering**

  - R. P. Feynman, R. B. Leighton and M. Sands, *The Feynman Lectures on Physics* Addison-Wesley, Reading (1965). Volume III has an excellent discussion on identical particle scattering.

  - D. Pines, *Rev. Mod. Phys.*, **28**, 184 (1956).

  - B. K. Ridley, *Quantum Processes in Semiconductors* Clarendon Press, Oxford (1982).

- **Auger Processes**

  - A. R. Beattie and P. T. Landsberg, *Proc. Roy. Soc.* A, **249**, 16 (1958).

  - A. R. Beattie, *J. Phys. Chem. Solids*, **49**, 589 (1988).

  - A. Haug, *J. Phys. Chem. Solids*, **49**, 599 (1988).

  - I. Suemune, *Appl. Phys. Lett.*, **55**, 2579 (1989).

- **Impact Ionization**

  - D. K. Ferry, *Semiconductors* Macmillan, New York (1991).

  - K. Hess, *Advanced Theory of Semiconductor Devices* Prentice-Hall, Englewood Cliffs (1988).

  - E. O. Kane, *Phys. Rev.*, **159**, 624 (1967). B. K. Ridley, *Quantum Processes in Semiconductors* Clarendon Press, Oxford (1982).

  - D. J. Robbins, *Phys. Status Solid* B, **97**, 9 (1980).

  - H. Shichijo and K. Hess, *Phys. Rev.* B, **23**, 4197 (1981).

Chapter

6

LATTICE VIBRATIONS: PHONON SCATTERING

In a crystalline material atoms vibrate about the rigid lattice sites and one of the most important scattering mechanisms for mobile carriers in semiconductors is due to these vibrations. In our discussions for the bandstructure we assumed that the background potential is periodic, and does not have any time dependence. In actual materials the background ions forming the crystal are not fixed rigidly but vibrate. The vibration increases as the temperature is increased. To understand the properties of electrons in a vibrating structure we use an approach shown schematically in Fig. 6.1. Scattering will occur due to the potential disturbances by the lattice vibration. Before we can answer the question regarding how lattice vibrations cause scattering, we must understand some basic properties of these vibrations. Once we understand the nature of the lattice vibrations we can begin to examine how the potential fluctuations arising from these vibrations cause scattering.

6.1 LATTICE VIBRATIONS

In Chapter 1 we have discussed how atoms are arranged in a crystalline material. The reason a particular crystal structure is chosen by a material has to do with the minimum energy of the system. As atoms are brought together to form a crystal, there is an attractive potential that tends to bring the atoms closer and a repulsive potential which tends to keep them apart. The attractive interaction is due to a variety of different causes including Van der Waals forces (resulting from the dipole moment created when

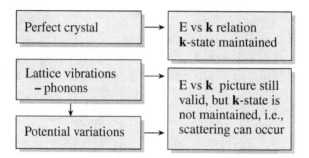

Figure 6.1: The effect of imperfections caused by either lattice vibrations or other potential fluctuations lead to scattering of electrons.

an atoms' electron cloud is disturbed by the presence of another atom), ionic bonding where electrons are transferred from one atom to another and covalent bonding where electrons are shared between atoms. When atoms are very close to each other there is a strong repulsion between the electrons on neighboring atoms. As a result the overall energy-configuration profile for the system has a schematic form, shown in Fig. 6.2. The total energy of the system is minimum when the atomic spacing becomes R_0 as shown in the figure.

In general we can expand the crystal binding energy around the point R_0 as follows:

$$U(R) = U(R_0) + \left(\frac{dU}{dR}\right)_{R_0} \Delta R + \frac{1}{2}\left(\frac{d^2U}{dR^2}\right)_{R_0} \Delta R^2 + \dots \tag{6.1}$$

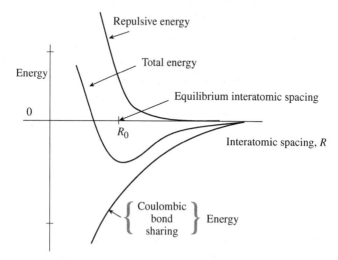

Figure 6.2: General form of the binding energy versus atomic distance of a crystal. In the case of most semiconductors, the long range attraction is due to either electrostatic interactions of the ions or the bond sharing energy of the covalent bond.

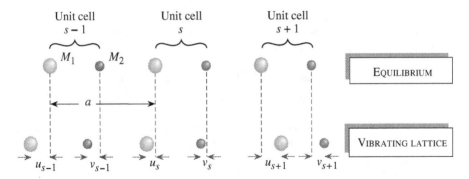

Figure 6.3: Vibrations in a crystal with two atoms per unit cell with masses M_1, M_2 connected by force constant C between adjacent planes.

The second term is zero since R_0 is the equilibrium interatomic separation. Retaining terms to the second order in ΔR (this is called the harmonic approximation),we get

$$U(R) = U(R_0) + \frac{1}{2}C(\Delta R)^2 \tag{6.2}$$

where

$$C = \frac{\partial^2 U}{\partial R^2} \tag{6.3}$$

is the force constant of the material. The restoring force is then

$$\text{Force} = -C\Delta R \tag{6.4}$$

Due to this restoring force the atoms in the crystal vibrate as a particle attached to a spring would do. We will now discuss such vibrations for semiconductors. Let us consider a diatomic lattice (two atoms per basis) as shown in Fig. 6.3. The atoms are at an equilibrium position around which they vibrate. There is a restoring force (let us assume this force is between the nearest neighbors only). We assume that the atoms have masses M_1 and M_2.

If u_s and v_s represent the displacements of the two kinds of atoms of the unit cell s (see Fig. 6.3), we get the following equations of motion for the atoms in the unit cell s:

$$M_1\frac{d^2u_s}{dt^2} = C(v_s + v_{s-1} - 2u_s) \tag{6.5}$$

$$M_2\frac{d^2v_s}{dt^2} = C(u_{s+1} + u_s - 2v_s) \tag{6.6}$$

We look for solutions of the traveling wave form, but with different amplitudes u and v on alternating planes

$$u_s = u\exp(iska)\exp(-i\omega t)$$
$$v_s = v\exp(iska)\exp(-i\omega t) \tag{6.7}$$

We note that a is the distance between nearest identical planes and not nearest planes, i.e., it is the minimum distance of periodicity in the crystal as shown in Fig. 6.3. Eqn. 6.7, when substituted in Eqns. 6.5 and 6.6 gives

$$-\omega^2 M_1 u = Cv\left[1 + \exp(-ika)\right] - 2Cu \tag{6.8}$$

$$-\omega^2 M_2 v = Cu\left[\exp(-ika) + 1\right] - 2Cv \tag{6.9}$$

These are coupled eigenvalue equations which can be solved by the matrix method. The equations can be written as the matrix vector product

$$\begin{vmatrix} -\omega^2 M_1 + 2C & -C\left[1 + \exp(-ika)\right] \\ -C\left[\exp(-ika) + 1\right] & -\omega^2 M_2 + 2C \end{vmatrix} \begin{vmatrix} u \\ v \end{vmatrix} = 0$$

Equating the determinant to zero, we get

$$\left|2C - M_1\omega^2 \quad -C[1 + \exp(-ika)] - C[1 + \exp(ika)] \quad 2C - M_2\omega^2\right| = 0 \tag{6.10}$$

or

$$M_1 M_2 \omega^4 - 2C(M_1 + M_2)\omega^2 + 2C^2(1 - \cos ka) = 0 \tag{6.11}$$

This gives the solution

$$\omega^2 = \frac{2C(M_1 + M_2) \pm [4C^2(M_1 + M_2)^2 - 8C^2(1 - \cos ka)M_1 M_2]^{1/2}}{2M_1 M_2} \tag{6.12}$$

It is useful to examine the results at two limiting cases. For small k, we get the two solutions

$$\omega^2 \approx 2C\left(\frac{1}{M_1} + \frac{1}{M_2}\right) \tag{6.13}$$

and

$$\omega^2 \approx \frac{C/2}{M_1 + M_2} k^2 a^2 \tag{6.14}$$

Near $k = \pi/a$ we get (beyond this value the solutions repeat)

$$\omega^2 = 2C/M_2$$
$$\omega^2 = 2C/M_1 \tag{6.15}$$

The general dependence of ω on k is shown in Fig. 6.4. Two branches of lattice vibrations can be observed in the results. The lower branch, which is called the acoustic branch, has the property that as for the monatomic lattice, ω goes to zero as k goes to zero. The upper branch, called the optical branch, has a finite ω even at $k = 0$.

The acoustical branch represents the propagation of sound waves in the crystal. The sound velocity is

$$v_s = \frac{d\omega}{dk} = \sqrt{\frac{C}{M_{av}}}\, a \tag{6.16}$$

where M_{av} is the average mass of the two atoms.

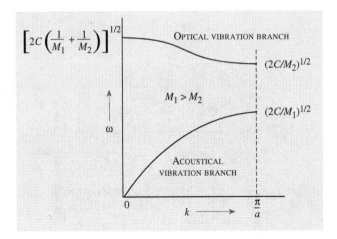

Figure 6.4: Optical and acoustical branches of the dispersion relation for a diatomic linear lattice

It is important to examine the eigenfunctions (i.e., u_s), for the optical branch and the acoustic branch of the dispersion relation. For $k = 0$, for the optical branch, we have, after substituting

$$\omega^2 = 2C \left(\frac{1}{M_1} + \frac{1}{M_2} \right) \tag{6.17}$$

in the equation of motion (say, Eqn. 6.8)

$$u = \frac{-M_2}{M_1} v \tag{6.18}$$

The two atoms vibrate against each other, but their center of mass is fixed. If we examine the acoustic branch, we get $u = v$ in the long wavelength limit. In Fig. 6.5a we show the different nature of vibration of the acoustic and optical mode.

Note that for each wavevector, k, there will be a longitudinal mode and two transverse modes. The frequencies of these modes will, in general, be different since the restoring force will be different. When optical vibration takes place in ionic materials like GaAs, polarization fields are set up that vibrate as well. These fields are important for longitudinal vibration, but not for translational vibration. As a result, in longitudinal vibrations there is an additional restoring force due to the long-range polarization. In Fig. 6.5b we show the lattice vibration frequency wavevector relation for GaAs. Notice that the longitudinal optical mode frequency is higher than that of the transverse mode frequency.

Phonons: Quantization of Lattice Vibrations
In the previous discussions we evaluated the dispersion relation ω vs. k for a set of coupled harmonic oscillator equations. If we consider a single harmonic oscillator problem

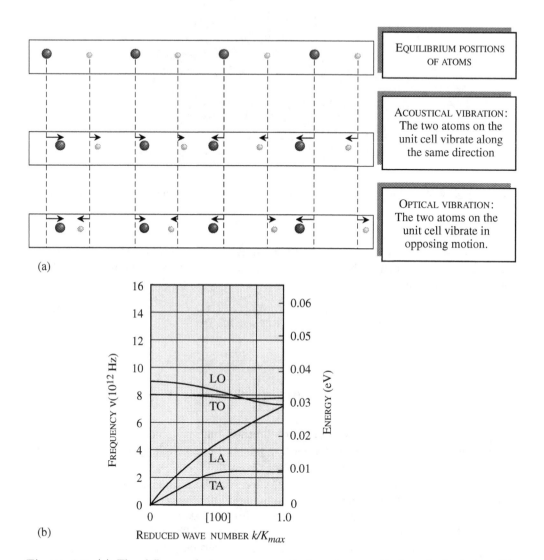

(a)

(b)

Figure 6.5: (a) The difference between as acoustical mode an optical mode is shown. (b) Phonon dispersion relation in GaAs. The longitudinal (LO, LA) and transverse (TO, TA) optical and acoustical modes are shown.

in quantum mechanics with the Hamiltonian

$$H = \frac{P^2}{2m} + \frac{1}{2}Cx^2 \tag{6.19}$$

the energy of the vibrating particles is quantized and is given by

$$\epsilon_n = \left(n + \frac{1}{2}\right)\hbar\omega \tag{6.20}$$

where $n = 0, 1, 2 \ldots$.

The frequency ω is just the classical frequency $(C/M)^{1/2}$. In classical physics the energy of the oscillator can be continuous and corresponds to a continuously increasing amplitude of vibration. The quantum oscillator has a minimum energy $\hbar\omega/2$ and the energy changes in steps of $\hbar\omega$. In the language of second quantization one says that the number n of Eqn. 6.20 represents the number of "particles" in the system or the occupation number of the system. One uses the term *phonon* to describe the lattice vibrations once they are treated as particles. For a single oscillator the frequency ω is fixed, but if we have a series of coupled oscillators as is the case for the atoms in a crystal, the frequency varies and we can introduce the phase determining vector $\mathbf{k}$ and get an ω vs. $\mathbf{k}$ relation of the form we derived. However, at each frequency $\omega_{\mathbf{k}}$ one can solve the harmonic oscillator problem in quantum mechanics and find that the energy is quantized and given by

$$\epsilon_{\mathbf{k}} = \left(n_{\mathbf{k}} + \frac{1}{2}\right)\hbar\omega_{\mathbf{k}} \tag{6.21}$$

In the context of lattice vibrations the number $n_{\mathbf{k}}$ denotes the number of phonons in the mode $\omega_{\mathbf{k}}$. To find out how many phonons are in a given mode one needs to define the statistics for phonons which we will discuss later, in Section 9.7.

The wave vector $\mathbf{k}$ can take on the values given by

$$|\mathbf{k}| = \frac{2\pi n}{Na}; \text{ for } n = 0, \pm 1, \ldots, \pm\frac{N-1}{2}, \frac{N}{2} \tag{6.22}$$

where N is the number of unit cells in the system. This leads to $3N$ (N longitudinal and $2N$ transverse) modes of vibration for the system. Each such mode can have a number n for its occupation number as given by the phonon statistics.

6.2 PHONON STATISTICS

We have briefly discussed how the lattice vibration can be represented by particles called phonons. How many phonons are present in a given mode $\omega_{\mathbf{k}}$ at temperature T? As in the case of electrons, to get this information we need to know the distribution function. The phonons are characterized as Bosons, i.e., particles which can "share" the same quantum state. Their occupation number is given by Bose–Einstein statistics which is valid at thermal equilibrium.

The phonon number is given by

$$\langle n_\omega \rangle = \frac{1}{\exp\frac{\hbar\omega}{k_B T} - 1} \tag{6.23}$$

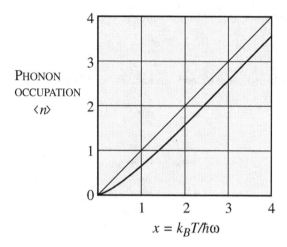

PHONON
OCCUPATION
$\langle n \rangle$

$x = k_B T/\hbar\omega$

Figure 6.6: Plot of the Bose–Einstein distribution function. At high temperatures the occupancy of a state is approximately linear in the temperature. The upper straight line is a classical limit.

In Fig. 6.6 we show a plot of this distribution function. As can be seen, unlike the Fermi–Dirac function, for occupancy of electron states, phonon occupancy can be larger than unity. The higher the temperature, the larger the vibration of the lattice atoms and larger the value of $\langle n_\omega \rangle$. It is also important to note that at low temperatures, the occupancy of the optical phonons is going to be very small since the optical phonons have a large energy for any value of k. On the other hand, the acoustic phonons exist with very small $\hbar\omega$ (at low k values) and are thus present even at very low temperatures. Thus at low temperatures, the optical phonons do not play as important a role. One can see that in the limit

$$\hbar\omega \ll k_B T$$
$$\langle n \rangle \approx \frac{k_B T}{\hbar\omega} \tag{6.24}$$

The total energy of the lattice vibrations is given by (ignoring the zero point energy)

$$U = \sum_{\mathbf{k},\rho} \langle n_{\mathbf{k},\rho} \rangle \, \hbar\omega_{\mathbf{k},\rho} \tag{6.25}$$

where $\mathbf{k}$ is the wavevector and ρ represents the polarization of the mode.

6.2.1 Conservation Laws in Scattering of Particles Involving Phonons

A phonon with a wave vector $\mathbf{k}$ will interact with particles such as electrons and photons and change their momentum as if its momentum was $\hbar\mathbf{k}$. Remember that in the case of electrons, the relevant electron momentum is $\hbar\mathbf{k}$ (wave vector) and not the true momentum of the electron. The phonons actually do not carry any momentum, they just behave as if they had a momentum. The actual physical momentum of the lattice vibrations is zero. Physically, this is obvious since the atoms are moving against each other, the crystal as a whole is not moving. Mathematically this can be seen as follows. The momentum of the crystal is

$$p = M\frac{d}{dt}\sum_s u_s \tag{6.26}$$

Due to the nature of the solution for the vibration u_s, this quantity is zero as one expects. When we discuss electron-phonon interactions in Section 6.5, we will see that one has, in general, the conservation laws (for first order phonon scattering)

$$k_i = k_f + q \tag{6.27}$$

$$E_i = E_f \pm \hbar w_q \tag{6.28}$$

where k_i and k_f are the initial and final electron momenta, q is the phonon momenta, E_i, E_f, $\hbar\omega_q$ are the corresponding energies. In more complex scattering problems where more phonons are involved, the conservation laws get appropriately modified.

EXAMPLE 6.1 In GaN the optical phonon energy is 93 meV. Calculate the optical phonon occupation at 77 K and 300 K.

The occupation at 77 K ($k_BT = 6.6$ meV) is

$$n(\omega_{op}) = \frac{1}{\exp\left(\dfrac{93}{6.66}\right) - 1} = 9 \times 10^{-7}$$

In all practical purposes no optical phonon modes are excited at this temperature. By contrast, in GaAs ($\hbar\omega_{op} = 36$ meV) the occupation at 77 K is 4.5×10^{-3}. We will see later in this chapter that scattering due to phonon absorption is proportional to $n(\omega)$. Thus phonon absorption plays no role in transport of carriers in GaN at 77 K. We will also see that phonon emission scattering is proportional to $(n(\omega) + 1)$. Thus electrons can scatter by emitting phonons even if $n(\omega) = 0$. However, they need an initial energy of $\hbar\omega_{op}$ to emit a phonon. At 300 K

$$n(\omega_{op}) = 0.029$$

6.3 POLAR OPTICAL PHONONS

In our discussions of optical phonons, we have ignored the fact that in some semiconductors, the atoms carry positive and negative charges (the anions and cations). This is not true only of group IV semiconductors like Si, Ge, C, etc. Due to this ionic nature of compound semiconductors, there is an additional restoring force due to the long-range polarization fields that are produced in the lattice vibrations. These polarization fields are only produced in the longitudinal modes and not in the transverse modes as can be seen from Fig. 6.7 . Due to this additional restoring force, there is a difference between the longitudinal and transverse frequencies. We will examine these differences in

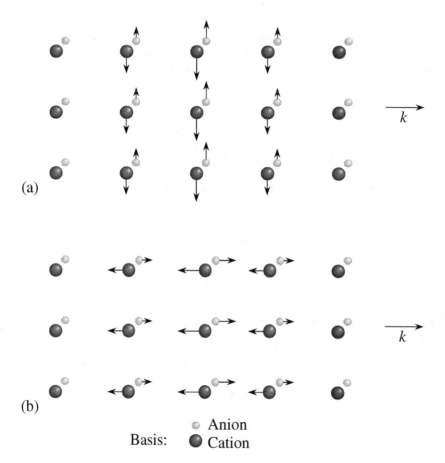

Basis: ◔ Anion
        ● Cation

Figure 6.7: Optical modes of vibration of an ionic crystal. (a) During transverse modes, the vibrations do not produce any polarization effects. (b) Long-range electric fields due to polarization are produced in longitudinal modes.

two approaches. The first one is extremely simplistic and only brings out the physical concepts. We will work in the relative vector

$$u_r = u - v \tag{6.29}$$

which represents the relative displacement between the positively and negatively charged ion lattices. The equation of motion for the transverse vibrations which do not produce any polarizations is

$$M\ddot{u}_r + M\omega_t^2 u_r = 0 \tag{6.30}$$

where M is the reduced mass of the two atoms and ω_t is the transverse optical phonon frequency. In a longitudinal vibration mode, an addition electric field is produced due to the polarization produced by the vibrations. The equation of motion is given by the restoring force, which is the sum of the force in Eqn. 6.30 and the electric field force. The equation for the longitudinal vibration is

$$-\omega_l^2 M u_r = -\omega_t^2 M u_r + F_i e^* \tag{6.31}$$

when F_i is the internal electric field due to the polarization, and e^* is an effective electronic charge per ion. If n is the number of unit cells per unit volume, the polarization is

$$P = ne^* u_r \tag{6.32}$$

and the electric field is (ϵ_0 is the free space dielectric constant)

$$
\begin{aligned}
F_i &= \frac{-P}{\epsilon_0} \\
&= \frac{-ne^* u_r}{\epsilon_0}
\end{aligned} \tag{6.33}
$$

The longitudinal frequency is then

$$\omega_l^2 = \omega_t^2 + \frac{ne^{*2}}{M\epsilon_0}$$

If the material has a relative static dielectric constant of ϵ_{rel} ($= \epsilon_s/\epsilon_0$) the result becomes

$$\omega_\ell^2 = \omega_t^2 + \frac{n\epsilon_{rel}e^{*2}}{m\epsilon_0} \tag{6.34}$$

The effective charge, e^*, will be discussed in Section 6.9, when we discuss polar optical phonon scattering. The longitudinal optical phonon frequency is higher than the transverse phonons at small k. Figs. 6.8 through 6.11 show some phonon dispersion curves for several semiconductors. For group IV semiconductors, there is, of course, no splitting at $k = 0$ between the longitudinal and transverse phonon frequencies. However, for III-V compounds there is an important difference arising from the ionicity of the crystal. An important point to note is that the optical phonons have little dispersion near $k = 0$, i.e., the bands are almost flat unlike the acoustic phonons. When we discuss scattering of electrons by phonons we will make the assumption that the optical phonons are dispersionless.

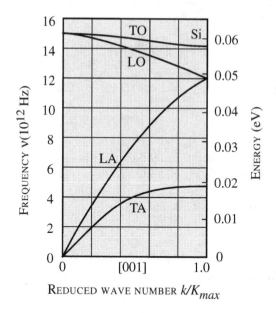

Figure 6.8: Phonon spectra of Si.

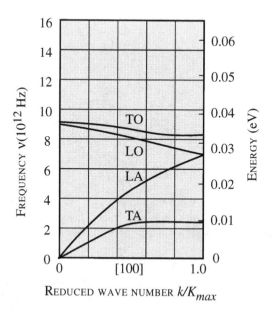

Figure 6.9: Phonon spectra in Ge.

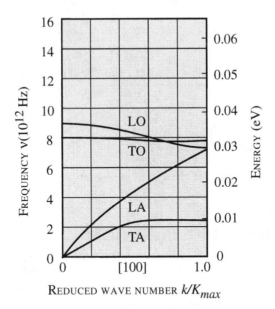

Figure 6.10: Phonon spectra of GaAs.

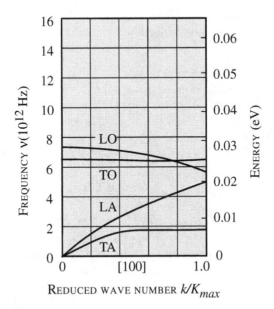

Figure 6.11: Phonon spectra in InAs.

EXAMPLE 6.2 Using results for phonon spectra for InAs shown, estimate the effective charge on In and As atoms.

The reduced mass of InAs is 7.57×10^{-26} kg. Using the values for ω_ℓ, ω_t, n and ϵ_0 we get ($\epsilon_{rel} \sim 14$)

$$
\begin{aligned}
e^{*2} &= \frac{\left(3.2 \times 10^{26} \text{ s}^{-2}\right)\left(7.57 \times 10^{-26} \text{ kg}\right)\left(8.84 \times 10^{12} \text{ F/m}\right)}{1.8 \times 10^{28} \text{ m}^{-3} \times 14} \\
&= 8.57 \times 10^{-40} \text{ C}^2
\end{aligned}
$$

or

$$
e^* = 2.9 \times 10^{-20} \text{ C} = 0.18 \, e
$$

InAs is quite ionic, i.e., a fairly large charge transfer occurs between In and As atoms. It is not as ionic as NaCl where $e^* = e$.

6.4 PHONONS IN HETEROSTRUCTURES

We have noticed that the phonon problem is quite similar to the electron problem, since both involve solutions of differential equations in a periodic potential. In Chapter 3 we discussed the electronic bandstructure in heterostructures. The concepts of quantum wells, superlattices, etc., have important consequences on the phonon dispersion just like they do for electronic spectra. The main difference between the two cases is qualitative. For electrons, the band offsets in semiconductors are such that heterostructure effects start becoming important when quantum well dimensions approach the electron de Broglie wavelength which is ~ 100 Å. Equivalent length scales for phonons are a few monolayers. As a result there are phonon modes associated with interfaces and narrow period superlattices that are quite important.

Confined Optical Phonons

In narrow period superlattices phonon modes can be confined within a single layer just as electron waves can be confined in a quantum well. For example, the optical phonon dispersion curves of GaAs and AlAs do not overlap and, therefore, the optical modes of one material cannot propagate into the other. These modes are confined within their respective layers. The picture of confined phonons is similar to the case of a particle in an infinite square well. The allowed wave vectors of these modes are:

$$
|k_n| = \frac{n}{2d} \; ; \text{ for } n = 1, 2, 3, \ldots \tag{6.35}
$$

where d is a thickness of the GaAs or AlAs layer. The energy of these modes has been found to follow the dispersion curve of the bulk material fairly well. In other words, $E(\boldsymbol{k})$ is nearly the same whether one is looking at the continuous values of $\boldsymbol{k}$ for, say, GaAs or the discrete values of $\boldsymbol{k}$ dictated by Eqn. 6.35.

Interface Phonons

In heterostructures there are also modes in the optical region that have their largest amplitude at the interfaces between two semiconductors. These interface modes carry a macroscopic electric field and can be discussed in terms of a simple electrostatic model.

Theoretical studies have shown that the combined effects of interface and confined phonons in thin structures on electron scattering is quite close to the scattering calculated by ignoring the finite size phonon effects.

6.5 PHONON SCATTERING: GENERAL FORMALISM

In our discussion of electronic bandstructure, we assumed a time-independent background periodic potential which led to our bandstructure picture. The problem of lattice vibrations, on the other hand, led us to the phonons and their dispersion. The ability to separate in the first order the lattice vibrations from the electronic spectra is dependent upon the so called adiabatic approximation which can be applied to a time-dependent problem. This approximation is applicable when the time variation of the Hamiltonian is slow enough that the problem can be treated as a stationary state (time-independent) problem at each instant in time. Of course, as we have seen from our discussions on phonon frequencies, the term slow variation is only relative, since the atoms move at frequencies of several terrahertz (10^{12} Hz)! Nevertheless, since the electron masses are so much smaller than the atomic masses, the adiabatic approximation works quite well. Thus, we are able to treat the problem in two steps:

1. Electronic states in a perfect lattice (bandstructure).

2. Interaction of electrons with lattice vibrations (phonon scattering).

In the previous sections we have seen that in semiconductors which have two atoms per basis, we have acoustic phonons which have an essentially linear ω vs. k relation near $k = 0$ and optical phonons which have essentially no dispersion (i.e., variation of ω) near $k = 0$. For both acoustic and optical phonons we have longitudinal and transverse modes of vibration. The motion of the lattice atoms produces strain in the crystal which according to the deformation potential theory produces perturbation in the electronic states. This relationship was examined in detail in our chapter on bandstructure modification by strain (Chapter 3). The effect of optical phonon strain can be understood on the basis of a similar formalism.

We have seen that there are two kinds of phonons. In acoustic phonons atoms vibrate essentially in phase with each other as shown in Fig. 6.5a. Strain energy is thus caused by the *derivative of the displacement* of atoms. In optical phonons atoms vibrate against each other and therefore the energy perturbation is proportional to the atomic displacement. In polar materials, longitudinal optical vibrations cause vibration of the ions with effective charge e^*. This causes additional scattering.

The electrons see the energy fluctuations produced by the phonons as shown schematically in Fig. 6.12 and scatter from them.

We will first develop a general formalism for scattering rate and then apply it to the specific case of acoustic optical or polar optical phonons.

In general, the electron-phonon interaction will depend directly upon the displacement $\boldsymbol{u}$ of the atoms in the crystal. For example, as we shall see later, the form of the perturbation potential is

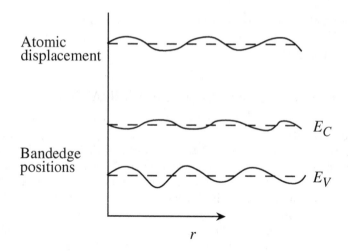

Figure 6.12: A schematic showing the effect of atomic displacement on bandedge energy levels in real space.

Acoustic phonons $U_{AP} \sim D \partial u / \partial x$
Optical phonons $U_{OP} \sim D_0 u$
Polar optical phonons $U_{OP} \sim e^* u$

where e^* is an effective charge which we will evaluate later and was discussed in Section 6.3 in connection with longitudinal polar optical phonons.

To understand scattering of electrons from lattice vibrations we invoke "second quantization" according to which these classical vibrations are to be represented by "particles" or phonons. In Fig. 6.13 we show a schematic of the "first" and "second" quantization whereby the wave-particle duality is established. According to the second quantization approach, a harmonic oscillator of classical frequency ω can have energies given by

$$E(\hbar\omega) = \left(n + \frac{1}{2}\right)\hbar\omega \tag{6.36}$$

where n is the number of quanta (of energy $\hbar\omega$) in the system. The number operator has the following properties:

$$n = a^\dagger a$$
$$\langle n|a^\dagger|n-1\rangle = \sqrt{n} = \langle n-1|a|n\rangle \tag{6.37}$$

Here $|n\rangle$ is the eigenstate corresponding to the number operator n and energy $(n + 1/2)\hbar\omega$, $a^\dagger$ is called the creation operator and a the destruction operator. In terms of the creation and destruction operators the displacement u and momentum of the oscillator are given by

$$u = \left(\frac{\hbar}{2M\omega}\right)^{1/2}\left(a^\dagger + a\right)$$

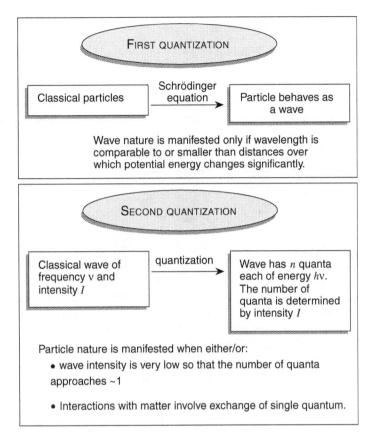

Figure 6.13: A conceptual picture of first and second quantization. The second quantization allows us to describe classical waves as quantum mechanical particles.

$$p = i \left(\frac{M\hbar\omega}{2} \right)^{1/2} \left(a^\dagger - a \right) \tag{6.38}$$

where M is the mass of the oscillator. According to quantum mechanics the vibrations of the atoms are described in terms of creation and destruction operators and in terms of the phonon occupation number. As shown in Fig. 6.13 the creation operator in the electron-phonon interaction would then lead to scattering processes in which a phonon is created after scattering (phonon-emission process). The destruction operator, on the other hand, would lead to processes where a phonon is destroyed (phonon absorption) in the scattering.

In general, the lattice vibrations represent coupled oscillators, and the displacement u representing the deformation of the unit cell or of the two atoms in the unit cell, must be represented in "normal coordinates" where the different modes are uncoupled. The conversion of real displacement to the normal coordinate description involves a simple transformation, and the normal modes can then be described by harmonic oscillator quantum mechanics. In general

$$\boldsymbol{u} = \frac{1}{\sqrt{N}} \sum_{\boldsymbol{q}} [\theta_{\mathbf{qb}} \, \boldsymbol{b}_{\mathbf{q}} \, \exp(i\boldsymbol{q} \cdot \boldsymbol{R}) + \text{c.c.}] \tag{6.39}$$

where c.c. is the complex conjugate of the first term, N is the number of unit cells, $\theta_{\mathbf{qb}}$ are the normal coordinate displacements for wavevector $\boldsymbol{q}$ and polarization $\boldsymbol{b}$, and $\boldsymbol{b}_{\mathbf{q}}$ is the polarization vector.

The normal coordinate displacement can now be written in terms of the phonon creation and destruction operators $a^{\dagger}$ and a as discussed above

$$\theta_{\mathbf{qb}} = \sqrt{\frac{\hbar}{2\,m\,\omega_{\mathbf{qb}}}} \left(a^{\dagger}_{-\mathbf{q}} + a_{\mathbf{q}} \right) \tag{6.40}$$

In the scattering problem of interest we consider the initial state of the electron-phonon system where the electron has a state $|\boldsymbol{k}\rangle$ and the phonons are described by the product state $\prod |n_{\mathbf{qb}}(\theta)\rangle$ which describes the phonon distribution. The product state essentially describes the phonons in all of the possible modes. After scattering, the electron is in the state $|\boldsymbol{k}'\rangle$ and the phonons are described by a new product state where the number of phonons may have changed due to scattering. We thus have the initial and final states of the electron- phonon system written as

$$\begin{aligned}
\Psi_i &= \psi_{\mathbf{k}}(\boldsymbol{r}) \prod_{\mathbf{qb}} |n_{\mathbf{qb}}(\theta)\rangle \\
\Psi_f &= \psi_{\mathbf{k}'}(\boldsymbol{r}) \prod_{\mathbf{qb}} |n'_{\mathbf{qb}}(\theta)\rangle
\end{aligned} \tag{6.41}$$

To keep the matrix element calculations clear, we will examine the phonon part and the electronic part of the matrix element separately for the sake of clarity. The phonon part is for a particular normal mode $\boldsymbol{q}$ and polarization $\boldsymbol{b}$

$$\prod_{\mathbf{q}'',\mathbf{b}''} \prod_{\mathbf{q}',\mathbf{b}'} \langle n_{\mathbf{q}''\mathbf{b}''} | \theta_{\mathbf{qb}} | n_{\mathbf{q}'\mathbf{b}'} \rangle \tag{6.42}$$

The quantum mechanics of the harmonic oscillator tells us that θ consists of terms $(a^{\dagger}_{-\mathbf{q}} + a_{\mathbf{q}})$ (as seen by Eqn. 6.40) and these operators will only mix states according to the following rules

$$\begin{aligned}
\langle n_{\mathbf{qb}} - 1 | a_{\mathbf{q}} | n_{\mathbf{qb}} \rangle &= \sqrt{n_{\mathbf{qb}}} \\
\langle n_{-\mathbf{qb}} + 1 | a^{\dagger}_{-\mathbf{q}} | n_{-\mathbf{qb}} \rangle &= \sqrt{n_{\mathbf{qb}} + 1}
\end{aligned} \tag{6.43}$$

The matrix element involving $\theta_{\mathbf{qb}}$ will only involve the above states resulting in a term

$$\sqrt{\frac{\hbar}{2\,m\,\omega_{\mathbf{qb}}}} \, \delta_{\mathbf{q},\mathbf{q}',\mathbf{q}''} \, \delta_{\mathbf{b},\mathbf{b}',\mathbf{b}''} \left[\sqrt{n_{\mathbf{qb}}} \, \delta_{n'-1,n} + \sqrt{n_{\mathbf{qb}} + 1} \, \delta_{n'+1,n} \right] \tag{6.44}$$

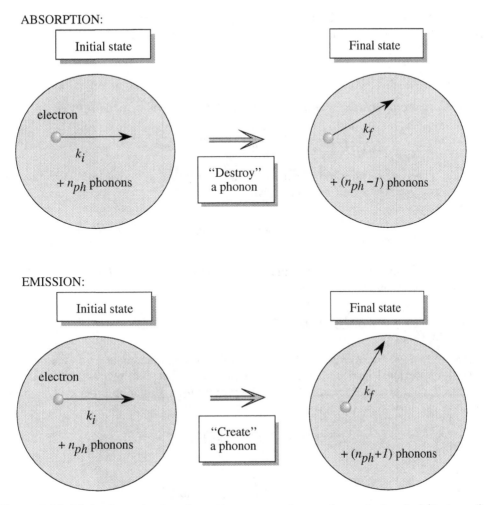

ABSORPTION:

Initial state

electron

k_i

$+ n_{ph}$ phonons

"Destroy" a phonon

Final state

k_f

$+ (n_{ph} - 1)$ phonons

EMISSION:

Initial state

electron

k_i

$+ n_{ph}$ phonons

"Create" a phonon

Final state

k_f

$+ (n_{ph} + 1)$ phonons

Figure 6.14: (a) A schematic of an absorption process where a phonon is absorbed (destroyed) and the energy and momentum of the electron are altered and (b) the emission of a phonon where a phonon is created.

The two processes, schematically shown in Fig. 6.14, represent the "destruction" and "creation" of a phonon. The first term in the brackets involves removing a phonon from the initial state during scattering and is the phonon absorption term. The second term involves adding an extra phonon to the final state and is the phonon emission term. Note that if there are no initial phonons in a system the absorption is zero. On the other hand, the emission process involves the $(n_{-\mathbf{qb}} + 1)$ term and is nonzero even if there are no initial phonons present. At equilibrium the phonon occupation is simply given by the phonon statistics discussed in Section 6.3.

We will now consider the electronic part of the matrix element. If we write the general electron-phonon interaction as

$$H_{ep} = \sum_{q} \left[H_{qb}\,\theta_{qb}\,e^{iq\cdot R} + \text{c.c.} \right] \tag{6.45}$$

then the electronic part of the matrix element for a particular mode (qb) is

$$\frac{1}{V} \int u^*_{n'k'}\, e^{-ik\cdot r}\, H_{qb}(r)\, e^{\pm iq\cdot R}\, u_{nk}\, e^{ik\cdot r}\, d^3r \tag{6.46}$$

The r-space interaction can be broken up into a sum over all lattice sites R, and an integral over a unit cell. This gives

$$\frac{1}{V} \int_{\text{cell}} \Psi^*_{n'k'}\, H_{qb}\, \Psi^*_{nk}\, d^3r \sum_{R} \exp\left[i(k \pm q - k')\cdot R \right] \tag{6.47}$$

We assume that over the small size of the unit cell $\exp[i(k - k')\cdot r] \approx 1$. The lattice sum is given by

$$\sum_{R} \exp\left[i(k \pm q - k')\cdot R \right] = \delta_{k\pm q-k',G} \tag{6.48}$$

where G is a reciprocal vector. This essentially says that momentum is conserved to within a reciprocal lattice vector in a crystal. In a scattering process, if $G = 0$ the process is the normal process. If $G = 0$, the process is an Umklapp process. The matrix element can now be written as

$$M^{\text{electronic}}_{k,k'} = \frac{1}{V}\, C_{qb}\, G(k, k')\, \delta_{k\pm q-k',G} \tag{6.49}$$

where

$$C_{qb}\, G(k, k') = \int_{\text{cell}} \Psi^*_{n'k'}\, H_{qb}\, \Psi^*_{nk}\, d^3r \tag{6.50}$$

In most phonon scattering problems H_{qb} does not vary over the unit cell and the equality in Eqns. 6.50 can be written as

$$C_{qb}G(k, k') = H_{qb}G(k, k') \tag{6.51}$$

where

$$G(k, k') = \frac{1}{2} \sum_{\mu,\mu'} I_{\mu,\mu'}(k, k')$$

$$I_{\mu,\mu'}(k, k') = \langle n'k'\mu' | nk\mu \rangle$$

$$n \quad \text{is the band index}$$

$$k \quad \text{is the electron wavevector}$$

$$\mu \quad \text{is the electron spin.}$$

i.e., $G(\boldsymbol{k}, \boldsymbol{k}')$ represents the overlap of the cell periodic part of the initial and final electron states. As we have discussed in the chapter on bandstructure, the bottom of conduction band for direct bandgap material is essentially s-type in which case the overlap integral for scattering within the conduction band would be unity. However, because of the small p-type mixture, the integral is given by (Fawcett et al., 1970)

$$G(\boldsymbol{k}, \boldsymbol{k}') = \frac{\left\{\sqrt{1 + \alpha E_{\boldsymbol{k}'}}\, \sqrt{1 + \alpha E_{\boldsymbol{k}}} + \alpha\, \sqrt{E_{\boldsymbol{k}'} E_{\boldsymbol{k}}}\, \cos\theta_{\boldsymbol{k}}\right\}^2}{(1 + 2\alpha E_{\boldsymbol{k}'})(1 + 2\alpha E_{\boldsymbol{k}})} \qquad (6.52)$$

where α is the non-parabolicity factor given by $\hbar^2 k^2/(2m^*) = E(1 + \alpha E)$, and $\theta_{\boldsymbol{k}}$ is the angle between $\boldsymbol{k}$ and $\boldsymbol{k}'$. The hole bands are p-type in nature with strong angular dependence of the overlap integral. For scattering within the same bands (e.g. heavy hole to heavy hole), the integral is approximately

$$G(\boldsymbol{k}, \boldsymbol{k}') = \frac{1}{4}(1 + 3\cos^2\theta_{\boldsymbol{k}}) \qquad (6.53)$$

while for interband scattering (e.g. heavy hole to light hole)

$$G(\boldsymbol{k}, \boldsymbol{k}') = \frac{3}{4}\sin^2\theta_{\boldsymbol{k}} \qquad (6.54)$$

The electron-phonon matrix element becomes for normal processes

$$|\langle f|H_{\mathrm{ep}}|i\rangle|^2 = \frac{\hbar}{2NM}\frac{C_{\mathbf{qb}}^2\, G(\boldsymbol{k}, \boldsymbol{k}')}{\omega_{\mathbf{qb}}}\left(n(\omega_{\mathbf{qb}}) + \frac{1}{2} \mp \frac{1}{2}\right)\delta_{\mathbf{k}\pm\mathbf{q}-\mathbf{k}',0} \qquad (6.55)$$

where the upper sign is for phonon absorption and the power sign is for phonon emission. The scattering rate, according to the golden rule, is then

$$\begin{aligned} W(\boldsymbol{k}) = \quad &\frac{1}{8\pi^2 NM}\int \frac{C_{\mathbf{qb}}^2\, G(\boldsymbol{k}, \boldsymbol{k}')}{\omega_{\mathbf{qb}}}\left(n(\omega_{\mathbf{qb}}) + \frac{1}{2} \mp \frac{1}{2}\right) \\ &\times \delta_{\mathbf{k}\pm\mathbf{q}-\mathbf{k}',0}\, \delta\left(E_{\boldsymbol{k}'} - E_{\boldsymbol{k}} \mp \hbar\omega_{\mathbf{qb}}\right)\, d^3 k' \end{aligned} \qquad (6.56)$$

6.6 LIMITS ON PHONON WAVEVECTORS

In a phonon scattering the electron's energy and momentum are altered. It is important to note that while the electrons and phonons have similar wavevectors, the electron energy is much larger than the phonon energy due to the very different dispersion relations. It is important to consider the consequences of the energy and momentum conservation.

In normal processes involving phonon scattering, both energy and momentum are conserved

$$E(\boldsymbol{k}') = E(\boldsymbol{k}) \pm \hbar\omega(\boldsymbol{q}) \qquad (6.57)$$

For parabolic electron bands we have

$$\frac{\hbar^2 k'^2}{2m^*} = \frac{\hbar^2 k^2}{2m^*} \pm \hbar\omega(\boldsymbol{q}) \qquad (6.58)$$

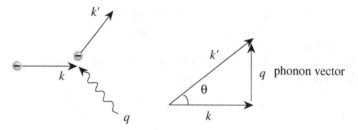

Figure 6.15: Wavevectors of the electron and phonon in a scattering event.

and

$$\boldsymbol{k}' = \boldsymbol{k} \pm \boldsymbol{q} \tag{6.59}$$

If θ is the scattering angle, we have, as shown in Fig. 6.15,

$$
\begin{aligned}
k'^2 &= k^2 + q^2 \pm 2\boldsymbol{k} \cdot \boldsymbol{q} \\
&= k^2 + q^2 \pm 2kq \cos \theta
\end{aligned}
\tag{6.60}
$$

From this and Eqn. 6.58 we have

$$
\begin{aligned}
\frac{\hbar^2 k'^2}{2m^*} &= \frac{\hbar^2 k^2}{2m^*} + \frac{\hbar^2 q^2}{2m^*} \pm \frac{\hbar^2 kq \cos \theta}{m^*} \\
&= \frac{\hbar^2 k^2}{2m^*} \pm \hbar \omega
\end{aligned}
$$

thus

$$\frac{\hbar^2 q^2}{2m^*} = \mp \frac{\hbar^2 kq \cos \theta}{m^*} \pm \hbar \omega \tag{6.61}$$

or

$$
\begin{aligned}
\hbar q &= \hbar k \left[\mp 2 \cos \theta \pm \frac{2\omega m^*}{\hbar k q} \right] \\
&= \hbar k \left[\mp 2 \cos \theta \pm \frac{2\omega}{v(\boldsymbol{k}) q} \right]
\end{aligned}
\tag{6.62}
$$

where $v(\boldsymbol{k}) = \hbar k / m^*$ is the electron velocity. The value of $\cos \theta$ is restricted between ± 1 and this imposes limits on the wavevectors of the phonons involved in the scattering. It is useful to consider the restriction arising from the momentum and energy conservation relations for various kinds of phonons.

6.6.1 Intravalley Acoustic Phonon Scattering

As we have seen in our discussion of Section 6.1, acoustic phonons have small energies for small values of k. Energy momentum conservation requires that the phonons involved in scattering electrons have small phonon wavevector. For acoustic phonons $\omega/q = v_s$, the sound velocity, and Eqn. 6.62 gives

$$\hbar q = 2\hbar k \left[\mp \cos \theta \pm \frac{v_s}{v(\boldsymbol{k})} \right] \tag{6.63}$$

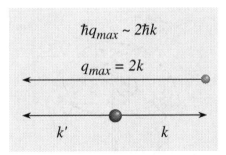

Figure 6.16: A scattering event in which the electron is scattered backwards. This defines the maximum limit for the phonon vector.

The upper limit on the phonon vector involved in a scattering is given by $\theta = \pi$ for phonon absorption and by $\theta = 0$ for phonon emission. The upper limits are then

$$\hbar q_{\max} = 2\hbar k \left[1 \pm \frac{v_s}{v(\boldsymbol{k})} \right] \tag{6.64}$$

since $v_s \sim 10^5$ cm/sec and typically the electron velocity, $v(\boldsymbol{k}) \sim 10^6 - 10^7$ cm/sec. This corresponds to a backward scattering of the electron as shown in Fig. 6.16. The maximum acoustic phonon wavevectors are also close to the zone center and energy change produced by these phonons is

$$\begin{aligned} \Delta \boldsymbol{E}_{\max} &= \hbar \omega_{\max} \\ &\sim \hbar q_{\max} v_s \\ &\sim 10^{-4} \text{ eV} \end{aligned} \tag{6.65}$$

Since the energy change is so small, one usually takes the acoustic phonon scattering to be elastic since $\Delta \boldsymbol{E}_{\max}$ is so much smaller than the electron energy. Only at very low temperatures does one need to be concerned about the inelastic nature of acoustic phonon scattering.

6.6.2 Intravalley Optical Phonon Scattering

We have seen from Section 6.1 that for optical phonons the energy is essentially independent of k near the zone center. We will use the constant phonon frequency model for our scattering calculations. For constant frequency ω_0, we can solve for q in Eqn. 6.62 to get

$$\hbar q = \hbar k \left[-\cos \theta + \sqrt{\cos^2 \theta \pm \frac{\hbar \omega_0}{E(\boldsymbol{k})}} \right] \tag{6.66}$$

The maximum phonon vector is then, $(\cos \theta = -1)$

$$\hbar q_{\max} = \hbar k \left[1 + \sqrt{1 \pm \frac{\hbar \omega_0}{E(\boldsymbol{k})}} \right] \tag{6.67}$$

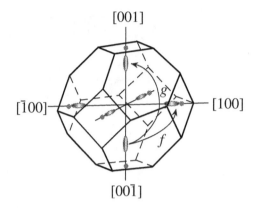

Figure 6.17: Bandstructure of Si and constant energy ellipsoids for Si conduction band. There are six equivalent valleys in Si at the bandedge. The g and f intervalley scattering processes are shown.

We note from this equation that an electron cannot emit an optical phonon unless its energy is larger than $\hbar\omega_0$. Also only zone center phonons are important in scattering.

6.6.3 Intervalley Phonon Scattering

Our bandstructure studies have shown that there are a number of "valleys" and "bands" in the E vs. k relationship. Electrons can scatter from one valley into another. A variety of important intervalley or interband scattering processes are induced by phonons. Fig. 6.17 shows the important processes in Silicon electron transport. where we have 6 equivalent X-valleys. Scattering between opposite valleys, e.g., $< 100 >$ to $< \bar{1}00 >$ is called a g-process and that between non-opposite valleys is called an f-process (e.g., between $< 100 >$ and $< 010 >$). Once again, we must maintain momentum conservation to *within a reciprocal lattice vector*. In Si scattering, the intervalley scattering involves an Umklapp process. The reciprocal lattice vector involved in the g-process is G_{100} and for an f-process it is G_{111}. It must be remembered that the minimum of the conduction band in Si is *not* at the zone edge (X-point) but is only 85% to the zone edge. Thus, an additional phonon of wavevector 0.3 times the X-point zone edge value $2\pi/a(100)$ is required for a g-process. For the f-process one needs a phonon with the wave vector along the Σ_1 symmetry line (11° off a $< 100 >$ direction) and with the zone edge magnitude is required.

In GaAs, as shown in Fig. 6.18, the lowest valley for electron transport is the Γ-valley and the next valley is the L-valley with a separation of 0.36 eV. The X-valleys are slightly higher in energy at ~ 0.5 eV above the Γ-point. Energy conservation ensures that an electron cannot scatter from the Γ- valley to the L-valley unless the electron energy is equal to at least $E_i = \Delta E_{\Gamma L} - \hbar\omega$, in which case the electron can absorb a phonon and move to the upper valley. The momentum of the phonon has to be close to the zone edge value to provide the electron the momentum difference between the Γ and the L-valleys. Scattering from the Γ-valley to the *L*-valley plays a very important role

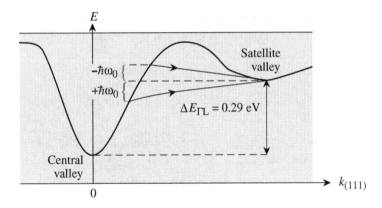

Figure 6.18: Scattering process involving intervalley scattering in GaAs. Electrons in Γ-valley can scatter into the L-valley (or even the X-valley) by absorbing or emitting an appropriate phonon. This scattering process is important at high electric fields and leads to the negative differential resistance in GaAs.

in GaAs electron transport. The scattering process also takes the electron from a low mass Γ-valley ($m^* \approx 0.067\ m_0$) to a high mass L-valley ($m^* \approx 0.35\ m_0$). This results in a negative differential mobility in the high field transport of direct gap materials like GaAs, InGaAs, etc.

Holes can also scatter from one band to another as shown in Fig. 6.19. The interband scatterings are important at both low and high electric fields, because of the degeneracy of the HH and LH bands at the zone center. In most semiconductors, the SO band is separated from the HH and LH bands by an energy of several hundred meV's, so that scattering into the SO band does not occur until high electric fields. However, for Si, this is not the case since the SO band is only 44 meV away from the top of the valence band. This is one of the reasons that hole transport in Si is comparatively poor.

6.7 ACOUSTIC PHONON SCATTERING

We will apply the general formalisms developed so far to specific phonon scattering processes. We start with acoustic phonons which is an almost elastic scattering process. We will assume that the overlap integral involving the central cell function is unity. In

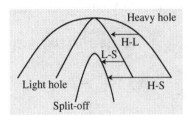

Figure 6.19: Important interband scattering between the hole bands.

the case of acoustic phonons, the electronic energy perturbation is related to the strain in the crystal ∇u and is given by the deformation potential theory as

$$
\begin{aligned}
H_{ep} &= D\frac{\partial u}{\partial r} \\
&= \frac{1}{\sqrt{N}}\sum_{\mathbf{q}}[i\,\theta_{\mathbf{qb}}\,D\,b_{\mathbf{q}}\cdot q\,\exp(iq\cdot R)+\text{c.c.}]
\end{aligned}
\tag{6.68}
$$

where D is the deformation potential for the particular valley of interest. Thus, for acoustic phonons we have from Eqn. 6.50

$$
C_{\mathbf{qb}}^2 = D^2 q^2
$$

Also for acoustic phonons the phonon dispersion relation is simply

$$
\omega_{\mathbf{q}} = v_s q
$$

where v_s is the sound velocity. In addition since the low energy phonons will dominate the scattering process

$$
\begin{aligned}
n(\omega_{\mathbf{q}}) &= \frac{1}{\exp\left(\hbar\omega_{\mathbf{q}}/k_B T\right)-1} \\
&\approx \frac{k_B T}{\hbar\omega_{\mathbf{q}}}
\end{aligned}
$$

Finally, assuming that the overlap integral is unity, we get for the scattering rate

$$
\begin{aligned}
W(\mathbf{k}) &= \frac{V}{8\pi^2 NM} \\
&\times \int d^3 k' \frac{D^2 q^2}{\omega_{\mathbf{q}}}\,\delta(\mathbf{k}\pm\mathbf{q}-\mathbf{k}')\,\delta(E_{\mathbf{k}'}-E_{\mathbf{k}}\mp\hbar\omega_{\mathbf{q}}) \\
&\times \left(n+\frac{1}{2}\mp\frac{1}{2}\right)
\end{aligned}
\tag{6.69}
$$

Note that $M = \rho V/N$, where ρ is the mass density. We also ignore $1/2$ in comparison to the occupation number n, which is a good approximation except at temperatures below ~ 50 K.

With these approximations, we find that the acoustic phonon scattering rate is

$$
W(\mathbf{k}) = \frac{D^2 k_B T}{8\pi^2\hbar\rho v_s^2}\int d^3 k'\,\delta(\mathbf{k}\pm\mathbf{q}-\mathbf{k}')\,\delta(E_{\mathbf{k}'}-E_{\mathbf{k}}\mp\hbar\omega_{\mathbf{q}})
\tag{6.70}
$$

Ignoring $\hbar\omega$ in the energy δ function (i.e., assuming elastic scattering) and using the definition of the density of states

$$
\frac{1}{8\pi^3}\int d^3 k = N(E_{\mathbf{k}})
$$

we get

$$W(\boldsymbol{k}) = \frac{\pi \, D^2 \, k_B T \, N(\boldsymbol{E_k})}{\hbar \rho v_s^2}$$

The total scattering is the sum of emission and absorption rates which for acoustic scattering (under our assumptions) are the same. Thus, the total rate is we get

$$W_{\mathrm{ac}}(\boldsymbol{k}) = \frac{2\pi \, D^2 \, k_B T \, N(\boldsymbol{E_k})}{\hbar \rho v_s^2} \qquad (6.71)$$

The acoustic phonon scattering is proportional to the temperature as one may expect.

In Fig. 6.20 we show a plot of acoustic phonon scattering rates as a function of energy for GaAs. We see that the scattering rates are of the order of $\sim 2 \times 10^{12}$ s^{-1}. Acoustic phonon scattering is usually the dominant phonon scattering at low temperatures and low fields, since the system does not have enough energy to excite optical phonons.

EXAMPLE 6.3 Calculate the 300 K acoustic phonon scattering rate for an electron in GaAs if the initial electron energy is 0.1 eV. The relevant material parameters are

$$
\begin{aligned}
D &= 7.8 \text{ eV} \\
\rho &= 5.37 \text{ g/cm}^3 \\
v_s &= 5.22 \times 10^5 \text{ cm/s}
\end{aligned}
$$

The scattering rate is

$$
\begin{aligned}
W_{ac} &= \frac{\sqrt{2}\,(m^*)^{3/2}\,k_B T\,D^2\,\sqrt{E}}{\pi\,\rho v_s^2 \hbar^4} \\[2mm]
&= \frac{\sqrt{2}\left(0.067 \times 9.1 \times 10^{-31} \text{ kg}\right)^{3/2}\left(0.026 \times 1.6 \times 10^{-19} \text{ J}\right)}{\pi\left(5.37 \times 10^3 \text{ kg/m}^3\right)\left(5.22 \times 10^3 \text{ m/s}\right)^2} \\[2mm]
&\quad \times \frac{\left(7.8 \times 1.6 \times 10^{-19} \text{ J}\right)^2\left(0.1 \times 1.6 \times 10^{-19} \text{ J}\right)^{1/2}}{\left(1.05 \times 10^{-34} \text{ Js}\right)^4} \\[2mm]
&= 3.13 \times 10^{11} \text{ s}^{-1}
\end{aligned}
$$

On an average the electron will scatter every 3.2 ps.

6.8 OPTICAL PHONONS: DEFORMATION POTENTIAL SCATTERING

In optical phonons the two atoms of the basis vibrate against each other. This produces strain and, if the two atoms have a charge e^*, a polar field. Both of these perturbations cause scattering. The strain produced causes fluctuations in the electron energy and is responsible for deformation potential scatterings. The scattering potential for the optical phonons is

$$H_{\mathrm{ep}} = D_0 \cdot \boldsymbol{u}$$

where D_0 is the optical deformation potential, and $\boldsymbol{u}$ is the relative displacement of the two atoms in the basis.

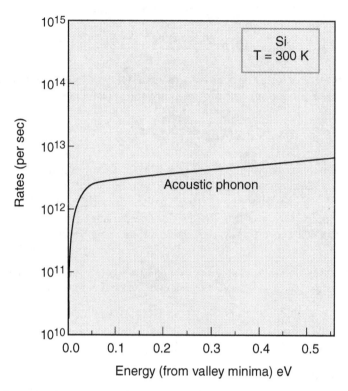

Figure 6.20: Acoustic phonon scattering rates in silicon as a function of energy.

Once again, expanding in terms of the normal coordinates, the interaction Hamiltonian is

$$H_{ep} = \frac{1}{\sqrt{N}} \sum_{\mathbf{q}} \{\theta_{\mathbf{qb}} \exp(i\mathbf{q} \cdot \mathbf{R}) + \text{c.c.}\} D_0 \cdot \mathbf{b_q} \qquad (6.72)$$

and the appropriate coupling coefficient is

$$C_{\mathbf{q}}^2 = D_0^2$$

where $\mathbf{b_q}$ represents the polarization vector. As noted earlier we will assume that optical phonon have a fixed frequency ω_0 which is the zone center frequency. For diamond structure the transverse and longitudinal phonons have the same frequency at the zone center. This is not the case for the zinc-blende structures, but the differences are quite small and can be ignored. With this dispersionless ω approximation, the energy conserving δ-function in the Fermi golden rule for the scattering rate (Eqn. 6.56) becomes independent of the scattering angle and one is simply left with an integration over the final energy states. This simply gives the final state density of states. Thus, the scattering rate becomes

$$W(\mathbf{k}) = \frac{\pi D_0^2}{\rho \omega_0} \left[n(\omega_0) N(\mathbf{E_k} + \hbar\omega_0) \right] \qquad \text{(absorption)}$$

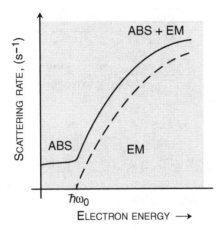

Figure 6.21: A schematic of the inelastic phonon scattering (e.g., optical phonon scattering–deformation type or polar type). The absorption (ABS) process can occur at low energies, but the emission process (EM) can only start once the electron has an energy equal to $\hbar\omega_0$.

$$= \frac{\pi D_0^2}{\rho\omega_0} \left[(n(\omega_0) + 1)\, N(E_\mathbf{k} - \hbar\omega_0)\right] \quad \text{(emission)} \tag{6.73}$$

where $N(E_\mathbf{k})$ is the density of states (without spin degeneracy). We have also assumed that $G(\mathbf{k}, \mathbf{k'})$, the overlap integral, is unity. This approximation also has to be removed for the hole bands.

In Fig. 6.21 we show a typical form of the scattering of electrons from optical phonons. The scattering rates have two distinct regions. At low electron energies the electrons scatter only by absorption of the phonons. This scattering rate is proportional to $n(\omega_0)$, the occupation probability, and is thus very sensitive to the lattice temperature. At higher electron energy, the scattering is dominated by the phonon emission process with the factor $(n(\omega_0) + 1)$. Phonons can be emitted even if $n(\omega_0)$ is zero, provided the starting electron has an energy at least equal to the phonon energy $\hbar\omega_0$.

EXAMPLE 6.4 Calculate the deformation optical phonon absorption for an electron with the following parameters:

$$
\begin{aligned}
E &= 0.1 \text{ eV} \\
D &= 1.0 \times 10^9 \text{ eV/cm} \\
\rho &= 2.33 \text{ g/cm}^3 \\
\omega_0 &= 9.8 \times 10^{13} \text{ rad/s} \\
T &= 300 \text{ K} \\
m^* &= 0.26\, m_0
\end{aligned}
$$

The phonon occupation at 300 K is 0.092. The absorption rate is

$$W_{abs} = \frac{D_0^2 n(\omega_0)(m^*)^{3/2}(E_f)^{1/2}}{\rho\omega_0\sqrt{2}\pi\hbar^3}$$

$$= \frac{\left(10^{11} \times 1.6 \times 10^{-19} \text{ J/m}\right)^2 (0.092)}{\left(2.33 \times 10^3 \text{ kg/m}^3\right) \left(9.8 \times 10^{13} \text{ s}^{-1}\right)}$$

$$\times \frac{\left(0.26 \times 9.1 \times 10^{-31} \text{ kg}\right)^{3/2} \left(0.164 \times 1.6 \times 10^{-19} \text{ J}\right)^{1/2}}{\left(\sqrt{2}\pi\right) \left(1.05 \times 10^{-34} \text{ Js}\right)^3}$$

$$= 3.7 \times 10^{12} \text{ s}^{-1}$$

6.9   OPTICAL PHONONS: POLAR SCATTERING

In addition to the strain energy perturbation produced by optical phonons, in polar materials there is a polarization related perturbation. As shown in Fig. 6.7 in polar materials such as GaAs, InAs, etc., when the cation and anion vibrate against each other in a longitudinal optical phonon mode, a polarization field is created. This causes a strong perturbation for the electrons resulting in the polar optical phonon scattering. The dipole produced by this perturbation is

$$\delta \boldsymbol{p} = e^* \boldsymbol{u} \tag{6.74}$$

where e^* is an effective charge in the cation or anion. We will first relate e^* to physically observable properties of the semiconductor. The dielectric constant of the semiconductor has a contribution from the electronic levels (the dominant contribution) and the lattice dipole moments. At low frequencies both of these are important. However, at high frequencies only the electronic contribution is present, since the lattice response is too slow. The polarization of the medium is given by

$$\boldsymbol{p}_{\text{tot}}(0) = \left(\frac{\epsilon_s - \epsilon_0}{\epsilon_s}\right) \boldsymbol{D} \quad \text{(low frequency)}$$

$$\boldsymbol{p}_{\text{tot}}(\infty) = \left(\frac{\epsilon_\infty - \epsilon_0}{\epsilon_\infty}\right) \boldsymbol{D} \quad \text{(high frequency)} \tag{6.75}$$

where ϵ_s is the low frequency (static) permitivity, ϵ_∞ is the high frequency permitivity, and $\boldsymbol{D}$ is the displacement vector. The difference in polarization is due to the lattice dipoles, which is then given by

$$\boldsymbol{p}_{\text{lattice}} = \boldsymbol{p}_{\text{tot}}(0) - \boldsymbol{p}_{\text{tot}}(\infty)$$

$$= \epsilon_0 \left(\frac{1}{\epsilon_\infty} - \frac{1}{\epsilon_s}\right) \boldsymbol{D} \tag{6.76}$$

The lattice vibrations causing the dipole satisfy the equation

$$\bar{m} \left(\frac{\partial^2 \boldsymbol{u}}{\partial t^2} + \omega_0^2 \boldsymbol{u}\right) = \boldsymbol{F} \tag{6.77}$$

where $\boldsymbol{F}$ is the applied force and $\bar{m}$ is the reduced mass. The polarization is given by

$$\boldsymbol{p}_{\text{lattice}} = \frac{e^* \boldsymbol{u}}{V_0} \tag{6.78}$$

where V_0 is the unit cell volume and e^* is the effective charge. The equation for the polarization is then (force $= e^* D/\epsilon_0$);

$$\frac{\bar{m} V_0}{e^*} \left(\frac{\partial^2 p}{\partial t^2} + \omega_0^2 p \right) = \frac{e^* D}{\epsilon_0} \tag{6.79}$$

The value of the polarization is for static case (no time dependence)

$$p_{\text{lattice}} = \frac{e^{*2} D}{\bar{m} \, V_0 \, \omega_0^2} \tag{6.80}$$

Equating these two values from Eqns. 6.78 and 6.80, we get

$$e^{*2} = \bar{m} \, V_0 \, \omega_0^2 \epsilon_0^2 \left(\frac{1}{\epsilon_\infty} - \frac{1}{\epsilon_s} \right) \tag{6.81}$$

ϵ_∞ and ϵ_s can be measured experimentally so that e^* can be evaluated.

We are now ready to calculate the scattering rate from the polar optical phonons. Only longitudinal modes will result in such a polarization fields as discussed in Section 6.3. Firstly we consider the interaction Hamiltonian. The basic energy interaction energy is

$$H_{\text{ep}} = \int \rho(R) \, \phi(R) \, d^3 R$$

where $\rho(R)$ $(= \nabla \cdot D)$ is the electronic charge density and $\phi(R)$ is the electric potential due to the polarization in the unit cell at R. The polarization, as before, is

$$p(R) = \frac{e^* u(R)}{V_0}$$

To calculate the perturbation we evaluate the integral

$$\begin{aligned} H_{\text{ep}} &= \int \{ \nabla \cdot D(R) \} \, \phi(R) \, d^3 R \\ &= - \int D(R) \cdot \nabla \phi(R) \, d^3 R \\ &= \int D(R) \cdot F(R) \, d^3 R \end{aligned} \tag{6.82}$$

where $D(R)$ is the electric displacement associated with the charge and $F(R)$ is the field associated with the polarization.

The electric displacement at a point R due to an electron at a point r is

$$D(R) = \frac{-1}{4\pi} \nabla \left(\frac{e}{|r - R|} \right)$$

In the presence of screening, the displacement is suppressed by a screening factor $\exp(-\lambda |r - R|)$ as discussed in the case of ionized impurity scattering. Thus, we have

$$D(R) = \frac{-1}{4\pi} \nabla \left\{ \frac{e}{|r - R|} e^{-\lambda |r - R|} \right\} \tag{6.83}$$

Substituting the normal coordinate form for the lattice displacement

$$u(\boldsymbol{R}) = \frac{1}{\sqrt{N}} \sum_{\mathbf{q},\mathbf{b}} \{\theta_{\mathbf{q},\mathbf{b}} \, b_{\mathbf{q}} \, \exp(i\boldsymbol{q} \cdot \boldsymbol{R}) + c.c.\} \tag{6.84}$$

and using the expression Eqn. 6.83 for the displacement we get (remember that $\boldsymbol{F}(\boldsymbol{R}) = -\boldsymbol{P}/\epsilon_0$)

$$H_{ep} = \frac{1}{\sqrt{N}} \frac{ee^*}{V_0 \epsilon_0} \sum_{\mathbf{q},\mathbf{b}} \int \left\{ \nabla \left[\frac{\exp(-\lambda \, |\boldsymbol{r} - \boldsymbol{R}|)}{|\boldsymbol{r} - \boldsymbol{R}|} \right] e^{i\mathbf{q}\cdot\mathbf{R}} \, \theta_{\mathbf{q},\mathbf{b}} \, b_{\mathbf{q}} + c.c. \right\} d^3 R$$

The spatial integral has the same form as the integral solved by us in ionized impurity scattering.

The resulting value for the electron-polar optical phonon interaction is

$$H_{ep} = -\frac{1}{\sqrt{N}\epsilon_0} \frac{ee^*}{V_0} \sum_{\mathbf{q}} \frac{q}{q^2 + \lambda^2} \{i\,\theta_{\mathbf{q}} \, \exp(i\boldsymbol{q} \cdot \boldsymbol{r}) + c.c.\} \tag{6.85}$$

The scattering rate then becomes from Eqns. 6.50 and 6.56

$$
\begin{aligned}
W(\boldsymbol{k}) &= \frac{V_0}{8\pi^2 \bar{m} \omega_0} \left(\frac{ee^*}{\epsilon_0 V_0} \right)^2 \int_0^{q_{max}} \int_{-1}^1 \int_0^{2\pi} \frac{q^4}{(q^2 + \lambda^2)^2} \delta_{\mathbf{k}\pm\mathbf{q}+\mathbf{k}',0} \\
&\quad \times \left(n\left(\omega_0 \right) + \frac{1}{2} \mp \frac{1}{2} \right) \delta \left(E_{\mathbf{k}'} - E_{\mathbf{k}} \mp \hbar\omega_0 \right) \, dq \, d(\cos\theta) \, d\phi
\end{aligned} \tag{6.86}
$$

The scattering rate depends explicitly on q making this integral more difficult than the ones encountered in the case of deformation scattering. The scattering rate integral becomes for parabolic bands

$$
\begin{aligned}
W(\boldsymbol{k}) &= \frac{V_0}{4\pi \, \bar{m} \, \omega_0} \left(\frac{ee^*}{\epsilon_0 V_0} \right)^2 \frac{1}{\hbar v} \left[n(\omega_0) \int_{q_{min,a}}^{q_{max,a}} \frac{q^3}{(q^2 + \lambda^2)^2} \, dq \right. \\
&\quad \left. + (n(\omega_0) + 1) \int_{q_{min,e}}^{q_{max,e}} \frac{q^3}{(q^2 + \lambda^2)^2} \, dq \right]
\end{aligned}
$$

where v is the electron velocity and q_{min} and q_{max} are the limits on the phonon wavevector. We will now proceed assuming no screening effect ($\lambda = 0$), in which case the integrals are simple

$$
\begin{aligned}
W(\boldsymbol{k}) &= \frac{V_0}{2\pi \, \bar{m} \, \hbar \, \omega_0 \, v} \left(\frac{ee^*}{\epsilon_0 V_0} \right)^2 \\
&\quad \times \left[n(\omega_0) \ln \left(\frac{q_{max}}{q_{min}} \right) + (n(\omega_0) + 1) \ln \left(\frac{q_{max}}{q_{min}} \right) \right]
\end{aligned} \tag{6.87}
$$

$$q_{max} = k \left(1 + \sqrt{1 \pm \frac{\hbar\omega_0}{E(k)}} \right)$$

and

$$q_{\text{min}} = k \left(\mp 1 \pm \sqrt{1 \pm \frac{\hbar \omega_0}{E(k)}} \right)$$

where the upper sign corresponds to phonon emission ($q_{\text{max,e}}$, $q_{\text{min,e}}$) and the lower sign corresponds to phonon absorption ($q_{\text{max,a}}$, $q_{\text{min,a}}$).

Using the identity

$$\sinh^{-1}(x) = \ln \left[x + \sqrt{1 + x^2} \right]$$

we get for the final scattering rate

$$
W(k) = \frac{V_0}{2\pi \bar{m} \hbar \omega_0 v} \left(\frac{ee^*}{V_0 \epsilon_0} \right)^2 \left[n(\omega_0) \sinh^{-1} \left(\frac{E_k}{\hbar \omega_0} \right)^{1/2} \right.
$$
$$
\left. + (n(\omega_0) + 1) \sinh^{-1} \left(\frac{E_k}{\hbar \omega_0} - 1 \right)^{1/2} \right] \tag{6.88}
$$

The emission rate is zero unless $E_k > \hbar \omega_0$. Substituting for the value of effective charge we get

$$
W(k) = \frac{e^2 \omega_0}{2\pi \hbar v} \left(\frac{1}{\epsilon_\infty} - \frac{1}{\epsilon_s} \right) \left[n(\omega_0) \sinh^{-1} \left(\frac{E_k}{\hbar \omega_0} \right)^{1/2} \right.
$$
$$
\left. + (n(\omega_0) + 1) \sinh^{-1} \left(\frac{E_k}{\hbar \omega_0} - 1 \right)^{1/2} \right] \tag{6.89}
$$

In Fig. 6.22 we show a plot of the polar optical phonon scattering for electrons in GaAs at room temperature. This scattering is not very important at low temperature since the occupation probability $n(\omega_0)$ is small, but is a dominant scattering mechanism in compound semiconductors at room temperature. In the presence of an electric field electrons have higher energies and polar optical phonon emission becomes a dominant scattering mechanism even at low temperatures.

To evaluate the momentum relaxation time for the polar optical phonons, one has to weigh each scattering by the change in momentum $\mp (q/k) \cos \theta$

$$\frac{q}{k} \cos \theta = \cos \theta \sqrt{\cos^2 \theta + \left(\frac{k'^2}{k^2} - 1 \right)} - \cos^2 \theta \quad \text{(absorption)}$$

$$\frac{-q}{k} \cos \theta = \pm \cos \theta \sqrt{\cos^2 \theta - \left(1 - \frac{k'^2}{k^2} \right)} - \cos^2 \theta \quad \text{(emission)}$$

This leads to the following expression

$$\frac{1}{\tau_m} = \frac{e^2 \omega_0}{(4\pi)^2 \hbar v} \left(\frac{1}{\epsilon_\infty} - \frac{1}{\epsilon_s} \right)$$

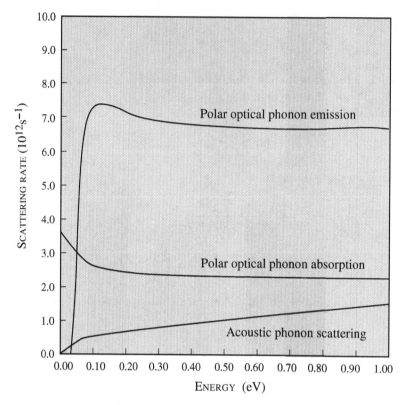

Figure 6.22: A comparison of the polar optical phonon scattering and acoustic phonon scattering in GaAs at room temperature.

$$
\times \; \left(n(\omega_0) \left(1 + \frac{\hbar\omega_0}{E_\mathbf{k}} \right)^{1/2} + (n(\omega_0) + 1) \left(1 - \frac{\hbar\omega_0}{E_\mathbf{k}} \right)^{1/2} \right.
$$

$$
+ \; \frac{\hbar\omega_0}{E_\mathbf{k}} \left[-n(\omega_0) \; \sinh^{-1} \left(\frac{E_\mathbf{k}}{\hbar\omega_0} \right)^{1/2} \right.
$$

$$
\left. \left. + \; (n(\omega_0) + 1) \; \sinh^{-1} \left(\frac{E_\mathbf{k}}{\hbar\omega_0} - 1 \right)^{1/2} \right] \right) \tag{6.90}
$$

Once again the emission process occurs only if $E_n > \hbar\omega_0$.

It should be noted that as with optical phonons, acoustic vibrations also cause polar vibrations. An acoustic vibration produces a polar charge fluctuation called the piezoelectric effect. In Chapter 1, Section 1.7 we have discussed this effect for semiconductors. Piezoelectric scattering is much weaker than polar optical phonon scattering and becomes significant only at very low temperatures in very pure materials.

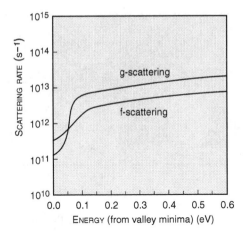

Figure 6.23: The g and f intervalley scattering rates in Si at 300 K. The nature of the g and f scattering are shown in Fig. 6.17.

6.10 INTERVALLEY SCATTERING

We will now discuss scattering of carriers between various valleys in the bandstructure. In Section 6.6.2 we have discussed how phonons can cause intervalley scattering. In Figs. 6.17 and 6.18 we have shown the various intervalley processes that can occur in direct and indirect bandgap materials. The mathematical treatment of intervalley scattering is done in a very simple phenomenological manner by simply postulating a deformation potential-like interaction

$$H_{\mathrm{ep}} = D_{if}\, u \tag{6.91}$$

where D_{if} is the intervalley deformation potential that scatters the electron from the valley i to f. In a sense this choice is like that for a deformation optical phonon scattering. In general, the scattering could involve an acoustic phonon, but one still chooses the expression in given above and simply redefines D_{ij}. The value of D_{ij} is usually found by fitting to experimental data. By analogy with the deformation optical phonon scattering discussed in Section 6.8, we have, for the scattering rate

$$W(\boldsymbol{k}) = \frac{\pi D_{ij}^2 Z_f}{\rho\, \omega_{ij}} \left(n(\omega_{ij}) + \frac{1}{2} \mp \frac{1}{2} \right) N\left(\boldsymbol{E} \pm \hbar\omega_{ij} - \Delta \boldsymbol{E}_{fi} \right) \tag{6.92}$$

where $N(\boldsymbol{E})$ is the density of states in the final valley (*without* spin degeneracy), Z_f is the number of equivalent final valleys, $\Delta \boldsymbol{E}_{fi}$ is the energy separation between the initial and final valley ($= 0$ for Si and 0.3 eV for Γ to L valley transfers and -0.3 eV for L to Γ valley transfers in GaAs). The frequency of the phonons responsible for scattering is ω_{ij}. In Figs. 6.23 and 6.24 we plot some typical results for the intervalley scattering in Si and GaAs.

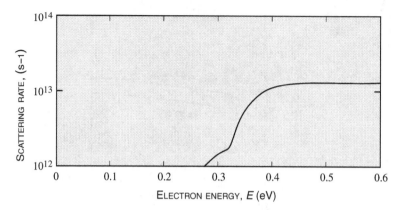

Figure 6.24: Intervalley scattering rates for electrons in Γ-valley scattering into the L-valley in GaAs at 300 K.

6.11   ELECTRON–PLASMON SCATTERING

We will now briefly review scattering of electrons from plasmons—a quantum description of free charge excitations or plasmons. The mathematics of this scattering is similar to that of polar optical phonon scattering. Plasmons involve the collective vibration of the free electrons against the fixed background of positive charges. Fig. 6.25 shows such a vibration. Just like the longitudinal polar optical phonons, these vibrations create a long-range electric field from which the electrons can scatter. This scattering rate is given by the same expressions derived by us for polar optical phonon scattering except that we do not need to worry about the effective charge e^*. We simply need to find the frequency of the plasmons.

The electron gas moves as a whole with respect to the positive ion background,

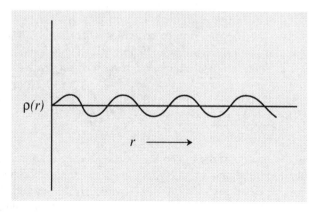

Figure 6.25: A schematic of the local charge density $\rho(r)$ produced in the collective oscillation of the free charge against the fixed positive background charge.

the displacement $\boldsymbol{u}$ creates an electric field

$$F = \frac{ne\boldsymbol{u}}{\epsilon} \tag{6.93}$$

where n is the electron density. This creates a restoring force on the electrons. The equation of motion then becomes

$$m^* \frac{d^2\boldsymbol{u}}{dt^2} = -e\boldsymbol{F}$$
$$= -\frac{ne^2\boldsymbol{u}}{\epsilon} \tag{6.94}$$

This gives a plasma frequency ω_p, where

$$\omega_p = \left(\frac{ne^2}{\epsilon m^*}\right)^{1/2} \tag{6.95}$$

The scattering from these plasmon fields has the same form as that of the polar optical phonon scattering. The scattering rate is

$$W_p(\boldsymbol{k}) = \frac{e^2 \omega_p \left(N(\omega_p) + \frac{1}{2} \mp \frac{1}{2}\right)}{4\pi\epsilon\hbar v} \ln \left[\frac{q_c/(2k)}{\mp 1 \pm \sqrt{1 \pm (\hbar\omega_p/\boldsymbol{E_k})}} \right] \tag{6.96}$$

Here q_c is either the maximum value of the phonon vector given by Eqn. 6.64 or is the inverse of the screening length (λ^{-1}), whichever is smaller. For plasma oscillations to be sustained, we must have $\omega_p\tau \gg 1$, where τ is the relaxation time for the scattering electrons. This tells us that the plasma oscillations can only be sustained for the carrier concentration $n > 10^{17}$ cm^{-3} since $\tau \sim 10^{-13}$ s. In Fig. 6.26 we show typical values of plasma scattering at room temperature for GaAs with a carrier concentration of 10^{17} cm^{-3}. As can be seen from these rates, this scattering is quite strong at such high carrier concentrations.

6.12 TECHNOLOGY ISSUES

Phonon scattering is intimately tied to the performance of semiconductor electronic and optoelectronic devices. Other scattering processes (ionized impurity, defect-related scattering, carrier-carrier scattering) can be eliminated by proper choices of material systems. Phonon scattering, on the other hand, is present in perfect material. Although it can be suppressed at low temperatures, it is still effective in limiting "coherent" electron devices of the type discussed in Chapters 8 and 11.

Phonon scattering is usually detrimental to high speed device performance. In Table 6.1 we show how low field mobility and high field mobilities are influenced by various phonon scattering processes. It is interesting to note that low field mobility is greatly improved by cooling a device. This is because optical phonon occupation essentially goes to zero and electrons (at low fields) do not have enough energy to emit optical phonons. On the other hand, at high fields there is only a small improvement

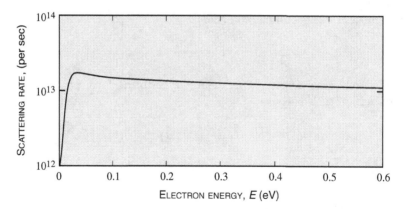

Figure 6.26: Electron-plasmon scattering rate versus energy for electrons in GaAs at room temperature. The electron density is $n_0 = 1.0 \times 10^{17}$ cm^{-3}.

in the carrier velocity, since optical phonon emission is proportional to $(n(\omega) + 1)$ and even at low temperature it can be very strong.

Phonon scattering is an important (usually the most dominant) mechanism of energy loss for electrons and holes. This does have some beneficial effects on devices, as shown in Table 6.1. In semiconductor lasers (discussed in Chapter 9) electrons and holes are injected into an active region with high energies. These carriers have to rapidly lose their energy to "thermalize" to the bandedges from where they recombine to produce photons. It is thus essential that the phonon emission process be very rapid.

In impact ionization we have seen in Chapter 5 that electrons (holes) must reach a certain energy threshold before carrier multiplication can start. In the absence of phonon emission this threshold will be reached at very low electric fields and devices will breakdown. Thus phonon emission helps in high power devices.

6.13   PROBLEMS

6.1   Consider a crystal which can be represented by a simple linear chain as far as the acoustic phonon spectra is concerned. If the sound velocity in this material is 3×10^5 cm/s, what is the phonon frequency at the zone edge ($a = 5.6$ Å)?

6.2   Calculate the Debye frequency for GaAs where the sound velocity is 5.6×10^5 cm/s. Assume that the volume of the unit cell is 4.39×10^{-23} cm^{-3}.

6.3   The optical phonon energies of GaAs and AlAs are 36 meV and 50 meV respectively at the zone center. What is the occupation probability of these optical phonons at 77 K and 300 K?

6.4   Consider a crystal of GaAs in which the sound velocity is 5.6×10^5 cm/s and the optical phonon energy is 36 meV and is assumed to have no dispersion. Using the Debye model for the acoustic phonon energy and the Einstein model for the optical phonons, calculate the lattice vibration energy per cm^3 at 77 K, 300 K and 1000 K. If the Ga–As bond energy is 1 eV, compare the phonon energy with the crystal binding energy per cm^3.

Table 6.1: A schematic of how phonon scattering influences transport and device parameters.

6.5 In Si electron transport, the intervalley scattering is very important. Two kinds of intervalley scatterings are important: in the g–scattering an electron goes from one valley (say a [001] valley) to an opposite valley ([00$\bar{1}$]), while in an f–scattering, the electron goes to a perpendicular valley ([001] to [010], for example). The extra momentum for the transitions is provided by a phonon and may include a reciprocal lattice vector. Remembering that Si valleys are not precisely at the X–point, calculate the phonon vectors which allow these scatterings.

6.6 Assuming k conservation, what is the phonon wavevector which can take an electron in the GaAs from the Γ–valley to the L–valley?

6.7 In the text, we solved the problem of phonon dispersion in a two atom basis system assuming that the force constants are the same between the M_1–M_2 bonds and the M_2–M_1 bonds. This is true for longitudinal modes, but not for transverse modes. Assume two force constants f_1 and f_2 and show that the phonon solutions are given by the equation

$$\omega^4 - \frac{M_1 + M_2}{M_1 M_2}(f_1 + f_2)\omega^2 + \frac{2f_1 f_2}{M_1 M_2}(1 - \cos\frac{ka}{2}) = 0$$

and

$$\frac{u_2}{u_1} = \frac{(f_1 + f_2 \exp(ika/2))/\sqrt{M_1 M_2}}{\dfrac{f_1 + f_2}{M_2} - \omega^2}$$

6.8 The optical phonon zone center energy for GaAs is 36 meV and for AlAs it is 50 meV. Calculate the force constants for the two materials, assuming a simple linear chain model. Based on these results, what are the sound velocities in the two semiconductors?

6.9 Based on a simple linear chain model, calculate the phonon dispersion in a 2 monolayer/2 monolayer GaAs/AlAs superlattice. Use the information obtained from the previous problem.

6.10 Estimate the rms fluctuation of the atoms in an fcc structure with a bulk modulus of 7.5×10^{11} erg cm^{-2} at a temperature of 300 K. Assume that each unit cell has a thermal energy of $\sim k_B T/2$.

6.11 Calculate and plot the total scattering rate for acoustic phonons as a function of electron energy in GaAs and Si at 77 K and 300 K. Plot the results for electron energies between 0.0 and 0.5 eV.

6.12 Calculate and plot the total scattering rate for polar optical phonons in GaAs for electron energy between 0 and 300 meV, at 77 K and 300 K. Calculate the results separately for the absorption and emission of the optical phonon.

6.13 Calculate the optical phonon scattering rates in Si at 77 K and 300 K as a function of electron energy between 0 and 300 meV.

6.14 Using momentum conservation considerations (i.e., momentum is conserved to within a reciprocal vector), identify the phonon vectors necessary for intervalley scattering in (a) GaAs for Γ-X and Γ-L scattering; (b) in Si for g and f scattering and in (c) in Ge for L-X scattering.

6.15 Calculate the Γ →L and L→ Γ intervalley scattering rate for GaAs electrons as a function of energy at 77 K and 300 K.

6.16 Calculate the f- and g-type scattering rates for Si as a function of electron energy at 77 K and 300 K.

6.17 Using GaAs and $In_{0.53}Ga_{0.47}As$ bandstructure plots, estimate the magnitude of the phonon vectors needed to cause Γ-L and Γ-X intervalley scatterings.

6.18 In the text, the optical phonon density is assumed to be given by the thermodynamic lattice temperature. At high fields, the electrons emit optical phonons at a very high rate, causing "hot phonon" related effects. Estimate the optical phonon generation rate in GaAs at electron energies of 200 meV. If the optical phonons cannot dissipate rapidly, the phonon occupancy $n(\omega_0)$ becomes large. What effect will this have in terms of hot carrier relaxation?

6.19 Calculate the energy dependence of f- and g-type scattering rates in Si. If an alloy $Si_{1-x}Ge_x$ is grown on Si(001), there is a splitting of the four in-plane valleys ([100], [$\bar{1}$00], [010], [0$\bar{1}$0]) and the two out of plane valleys. What would the f- and g-type scattering rates be as a function of energy for the $Si_{0.8}Ge_{0.2}$ alloy? You can use the results in Chapter 3 for the splitting values.

6.20 The static and high frequency relative dielectric constants in GaAs are 13.2 and 10.9, respectively. The transverse optical phonon frequency is $\omega_t = 5.4 \times 10^{13}$ rad s^{-1} and the ionic masses are $M_{Ga} = 69.7\ M$, $M_{As} = 74.9\ M$ ($M = 1.67 \times 10^{-24}$ gm). What is the ionic charge on the Ga atom?

6.21 Calculate and plot the energy dependence of acoustic and polar optical phonon scattering rates in GaAs at 77 K and at 300 K. It is observed experimentally that when temperature is lowered, the low field mobility improves but the high field mobility remains almost unaffected. Do your results explain this behavior?

6.22 In many experiments, electron (and hole) carrier distributions are specially prepared by laser pulses. Thus, at time zero, the electrons have a well-defined energy. In such experiments, the time- dependent optical spectroscopy of such monoenergetic particles show "phonon replicas." Discuss the source of such phonon replicas.

6.14 REFERENCES

- **Crystal Bonding**

  - N. W. Ashcroft and N. D. Mermin, *Solid State Physics* (Holt, Rinehart and Winston, New York, 1976).

  - C. Kittel, *Introduction to Solid State Physics* John Wiley and Sons, New York (1986).

  - L. Pauling, *The Nature of the Chemical Bond* Cornell (1960).

  - J. C. Phillips, *Bonds and Bands in Semiconductors* Academic Press, New York (1973).

- **Lattice Vibrations in Crystals**

  - N. W. Ashcroft and N. D. Mermin, *Solid State Physics* Holt, Rinehart and Winston, New York (1976).

  - M. Born and K. Huang, *Dynamical Theory of Crystal Lattices* Oxford University Press, London (1954).

– W. Cochran, *Dynamics of Atoms in Crystals* Crane, Russak, New York (1973).

- **Quantization of Lattice Vibrations**

  – Most quantum mechanics books have detailed discussions on the second quantization procedure which allows one to treat classical fields as particles. An example is Schiff, L. I., *Quantum Mechanics* McGraw Hill, New York (1968).

- **Polar Optical Phonons**

  – C. Kittel, *Quantum Theory of Solids* John Wiley and Sons, New York (1987).

  – J. M. Ziman, *Principles of the Theory of Solids* Cambridge (1972).

- **Phonon Dispersion: Measurements and Calculations**

  – N. W. Ashcroft and N. D. Mermin, *Solid State Physics* Holt, Rinehart and Winston, New York (1976).

  – B. N. Brockhouse, *Phys. Rev. Lett.*, **2**, 256 (1959).

  – B. N. Brockhouse and P. Iyengar, *Phys. Rev.*, **111**, 747 (1958).

  – Charles, R., *Phys. Rev. B*, **22**, 4804 (1980).

  – Landolt-Bornstein, *Semiconductors* (edited by O. Madelung, M. Schultz, and H. Weiss, Springer–Verlag, Berlin) (1982).

  – Martin, D. H., *Adv. in Physics*, **14**, 39 (1965).

  – "Vibration Spectra of Solids," by S. S. Mitra, in *Solid State Physics*, **13**, (1962).

  – J. L. T. Waugh and G. Dolling, *Phys. Rev.*, **132**, 2410 (1963).

  – G. B. Wright, editor, *Scattering Spectra of Solids* Springer–Verlag, New York (1969).

- **Phonons in Heterostructures**

  – C. Colvard, T. A. Gant, M. V. Klein, R. Merlin, R. Fischer, H. Morkoc and A. C. Gossard, *Phys. Rev. B*, **31**, 2080 (1985).

  – K. Huang and B. F. Zhu, *Phys. Rev. B*, **38**, 13377 (1988).

  – B. K. Ridley, *Phys. Rev. B*, **39**, 5282 (1989).

  – S. M. Rytov, *Sov. Phys.* JETP, **2**, 466 (1956).

  – N. Sawaki, *J. Phys.* C, **19**, 4965 (1986).

  – M. A. Stroscio, K. W. Kim and M. A. Littlejohn, in *Physical Concepts of Materials for Novel Optoelectronic Device Applications II: Device Physics and Applications* (edited by M. Razeghi, SPIE vol. 1362, 1990), p. 566.

  – S. Tamura and J. P. Wolfe, *Phys. Rev. B*, **35**, 2528 (1987).

  – R. Tsu and S. S. Jha, *Appl. Phys. Lett.*, **20**, 16 (1972).

- **General**

  - "High Field Transport in Semiconductors," by E. M. Conwell, in *Solid State Physics* (edited by F. Seitz, D. Turnbull and H. Ehrenreich, Academic Press, New York) (1967), vol. Suppl. 9.

  - W. Fawcett, A. D. Boardman and S. Swain, *J. Phys. Chem. Solids*, **31**, 1963 (1970).

  - D. K. Ferry, *Semiconductors* Macmillan, New York (1991).

  - H. Frohlich, *Proc. Roy Soc.* A, **160**, 230 (1937).

  - J. M. Hinckley, *The Effect of Strain on Pseudomorphic p- $Si_x Ge_{1-x}$: Physics and Modeling of the Valence Bandstructure and Hole Transport*, Ph.D. Thesis, (The University of Michigan, Ann Arbor, 1990).

  - M. Lundstrom, *Fundamentals of Carrier Transport* (Modular Series on Solid State Devices, edited by G. W. Neudeck and R. F. Pierret, Addison-Wesley, Reading, 1990), vol. X.

  - B. K. Ridley, *Quantum Processes in Semiconductors* Clarendon Press, Oxford (1982).

  - "Low Field Electron Transport," by D. L. Rode, in *Semiconductors and Semimetals* (edited by R. K. Willardson and A. C. Beer, Academic Press, New York, 1975), vol. 10.

  - J. M. Ziman, *Electrons and Phonons* Clarendon Press, Oxford (1980).

Chapter

7

VELOCITY-FIELD
RELATIONS
IN SEMICONDUCTORS

In this chapter we will examine how the various scattering processes discussed in the last two chapters influence transport properties of electrons and holes. In steady state the response of the free carriers to an external electric field is represented by the velocity versus electric field relationship. This relationship is vital to the understanding of electronic devices. There are three regions of the electric field which are important in charge transport:

1. The low electric field region where the velocity-field relation is linear and is defined by the mobility μ through the relation

$$v = \mu \mathbf{F}$$

   It is usually possible to develop analytic formalisms for this region based on the Boltzmann transport equation.

2. A higher electric field (usually $\mathbf{F} \geq 1$ kV/cm), where the v-$\mathbf{F}$ relation is no longer linear. To understand the transport in this region one usually requires numerical methods including those based on computer simulations.

3. Finally, at extremely high electric fields ($\mathbf{F} \geq 10^5$ V/cm), the semiconductor "breaks down" either due to impact ionization or due to electrons tunneling from the valence band to the conduction band.

   The steady state v-$\mathbf{F}$ relationships require an electron to undergo several (usually at least several tens) collisions before reaching steady state. Most collision times are of the order of a picosecond, and since electron velocities are of the order of 10^7 cm/s, it takes an electron several hundred Angstroms of travel before scattering occurs. Transport in regions which are submicron in length is therefore not described by the usual v-$\mathbf{F}$ curves. This region of transport is described by the transient velocity-field curves and for extremely short distances the electron moves "ballistically" (i.e., without scattering). While several numerical approaches have been developed to describe this region, it is best described the computer simulation techniques based on Monte Carlo methods.

7.1 LOW FIELD TRANSPORT

In Chapter 4 we have discussed how it is possible to obtain mobility through relaxation time calculations. In general we have for the mobility

$$\mu = \frac{e\tau}{m^*} \tag{7.1}$$

At low fields, the first order solution of the Boltzmann equation is quite valid and the distribution function in presence of an electric field is simply

$$f(\mathbf{k}) = f^0 \left(\mathbf{k} + \frac{e\tau\mathbf{F}}{\hbar} \right) \tag{7.2}$$

In case the carrier density is small, the distribution function is adequately given by a displaced Maxwellian function and as discussed in Chapter 4, the relaxation time is simply

$$\ll \tau \gg = \frac{\langle E\tau \rangle}{\langle E \rangle} \tag{7.3}$$

In most cases, one does not have a single scattering process present during carrier transport. If the various scattering processes are independent of each other, the total scattering rate is just the sum of the individual scattering rates. Thus, we have

$$\frac{1}{\tau_{\text{tot}}} = \sum_i \frac{1}{\tau_i} \tag{7.4}$$

Now, *if* the various scattering rates have the *same* energy dependence, then when one carries out the averaging the final mobility is simply given by

$$\frac{1}{\mu_{\text{tot}}} = \sum \frac{1}{\mu_i} \tag{7.5}$$

This is known as the Mathieson's rule and is strictly valid only when all the scattering mechanisms have the same energy dependence. As we have seen in the previous chapters, this is usually not true, so Eqn. 7.6 is not strictly valid. However, Mathieson's rule is widely used because of its ease of application and reasonable accuracy.

In Chapter 4 we showed that if a scattering mechanism has an energy dependence given by

$$\tau(E) = \tau_0 \left(\frac{E}{k_B T} \right)^s$$

the averaging gives

$$\left\langle\left\langle \frac{1}{\tau} \right\rangle\right\rangle = \frac{1}{\tau_0} \frac{\Gamma(5/2)}{\Gamma(s+5/2)} \tag{7.6}$$

In Chapter 4 we have calculated the averaged relaxation time for ionized impurity scattering. If N_i is the ionized impurity density (assuming a single charge per impurity), the mobility is

$$\mu_I = \frac{128\sqrt{2\pi}\epsilon^2\epsilon_0^2(k_BT)^{3/2}}{e^3\sqrt{m^*}N_i\left[\ln(1+\gamma_m^2) - \gamma_m^2/(1+\gamma_m^2)\right]} \tag{7.7}$$

where

$$\gamma_m = \frac{2}{\hbar\lambda}\sqrt{6m^*k_BT} \tag{}$$

The mobility increases with temperature, a distinguishing signature of the ionized impurity scattering.

In the case of acoustic phonon scattering we have from Chapter 6

$$\tau(E) = \tau_0\left(\frac{E}{k_BT}\right)^{-1/2} \tag{7.8}$$

with

$$\tau_0 = \frac{2\pi\hbar^4 c_\ell}{D^2}(2m^*k_BT)^{-3/2} \tag{7.9}$$

The mobility then becomes

$$\begin{aligned}
\mu_{AP} &= \frac{e\tau_0}{m^*}\frac{\Gamma(2)}{\Gamma(5/2)} \\
&= \frac{2\sqrt{2\pi}\,e\,\hbar^4\,c_\ell}{3D^2\,(m^*)^{5/2}\,(k_BT)^{3/2}}
\end{aligned} \tag{7.10}$$

The mobility limited by acoustic phonons decreases with temperature and has a characteristic $T^{-3/2}$ dependence. Mobilities limited by scattering from optical phonons can be calculated numerically or by the use of the Monte Carlo method described later.

Low Field Mobility in Silicon

Si has a rather poor low field mobility as compared to the mobilities in direct bandgap materials. However, the high field velocity is quite good. In Fig. 7.1. we show typical mobilities for Si as a function of temperature and doping. For high quality samples, the mobility continues to increase as the temperature decreases since ionized impurity scattering is absent and the acoustic phonon scattering has a $T^{-3/2}$ behavior. However, for doped samples the mobility shows a peak at low temperatures and then decreases.

Low Field Mobility in GaAs

Electrons in GaAs have a superior low field mobility as compared to the case in Si. This is mainly due to the lower density of states at the bandedge. Room temperature mobilities in high quality GaAs samples are ~ 8500 cm^{2}V^{-1}s^{-1} compared to only $\sim$ 1500 cm^{2}V^{-1}s^{-1} for Si. Fig. 7.2 shows a typical plot of mobility versus temperature for

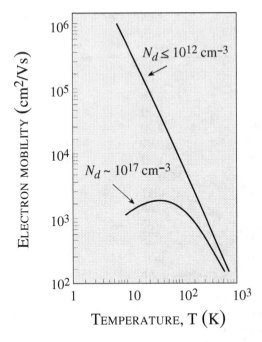

Figure 7.1: Low-field mobility of electrons in silicon as a function of temperature.

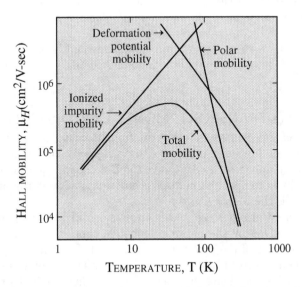

Figure 7.2: Low-field Hall mobility of electrons in GaAs as a function of temperature. The material has an ionized impurity concentration of 7×10^{13} cm^{-3}.

Semiconductor	Bandgap (eV) 300 K	Bandgap (eV) 0 K	Mobility at 300 K (cm^2/V-s) Elec.	Mobility at 300 K (cm^2/V-s) Holes
C	5.47	5.48	1800	1200
GaN	3.4	3.5	1400	350
Ge	0.66	0.74	3900	1900
Si	1.12	1.17	1500	450
α-SiC	3.00	3.30	400	50
GaSb	0.72	0.81	5000	850
GaAs	1.42	1.52	8500	400
GaP	2.26	2.34	110	75
InSb	0.17	0.23	80000	1250
InAs	0.36	0.42	33000	460
InP	1.35	1.42	4600	150
CdTe	1.48	1.61	1050	100
PbTe	0.31	0.19	6000	4000
In$_{0.53}$Ga$_{0.47}$As	0.8	0.88	11000	400

Table 7.1: Bandgaps along with electron and hole mobilities in several semiconductors. Properties of large bandgap materials (C, GaN, SiC) are continuously changing (mobility is improving) due to progress in crystal growth. Zero temperature bandgap is extrapolated.

GaAs. Also shown are the relative contributions of ionized impurity, acoustic phonons and polar optical phonons to the mobility. The polar optical phonon scattering is the dominant scattering for temperatures above 100 K.

In Table 7.1 we show mobilities for electrons and holes for a number of semiconductors. These results are given for "pure" materials and are limited by phonon scattering only.

7.2   HIGH FIELD TRANSPORT: MONTE CARLO SIMULATION

At low fields the energy gained by electrons from the field is small compared to the thermal energy at equilibrium. As the field is increased, the carrier energy increases and the simple approximations used to solve the Boltzmann equation do not hold. As a result, one needs fairly complex numerical techniques to address the problem. One very versatile technique is the Monte Carlo method.

In the Monte Carlo process, the electron is considered as a point particle whose scattering rates are given by the Fermi golden rule expressions. The Monte Carlo approach comes in because of the probabilistic nature of the transport phenomenon. The Monte Carlo method involves carrying out a computer simulation to represent as closely as possible the actual physical phenomena occurring during the carrier transport. Carrier transport in Monte Carlo techniques is viewed as a series of free flight and scattering events. The scattering events are treated as instantaneous as shown in Fig. 7.3. The simulation involves the following stages:

1. Particle Injection into the Region of Study: The particles under study are injected into the region (say from a contact) with a prechosen distribution of carrier momenta.

2. Free Flight of the Carrier: This is an important component of the Monte Carlo method. In the Monte Carlo approach, the scattering event is considered to be instantaneous and between the scattering process the electron simply moves in the electric field according to the "free particle" equation of motion.

3. Scattering Event: A specific prescription is used in the Monte Carlo method to determine the time between scattering events. At the end of a free flight a scattering occurs which alters the flight pattern of the electron.

4. Selection of the Scattering Event: Since a number of scattering processes will be simultaneously present one has to decide which event was responsible for the scattering that occurred. This choice is again based on the Monte Carlo method.

5. State of the Electron After Scattering: Finally, one uses the Monte Carlo method to determine the momentum of the electron immediately after the collision has occurred. Remember that the collision duration is assumed to be zero. To determine the final state after collision one needs detailed information on the scattering process.

Once the final state is known, the procedure is simply repeated with a new free flight. In Fig. 7.4 we provide a general flowchart for the Monte Carlo computer simulation program.

7.2.1 Simulation of Probability Functions by Random Numbers

The Monte Carlo approach depends upon the generation of "random" numbers which are cleverly used to simulate random physical events. Since the random numbers are actually generated by a computer program, they are not really random since the entire sequence of numbers is predictable. Nevertheless, if the random number generation code is good, there should be little correlation between the random numbers.

If $P(\beta)$ is a probability distribution of some variable β in a range a to b, the question one asks in the Monte Carlo methods is the following: What are the different values of β if the events are chosen randomly? For example, $P(\beta)$ may be the probability that an electron scatters by an angle β. In this case we would be interested in finding the scattering angle during successive random collisions. This choice is made by generating a random number which has a uniform distribution, say between 0 and 1. As shown in Fig. 7.5 we are then interested in mapping the probability function for the random number $P_u(R)$ to the probability function $P(\beta)$. The mapping is given by the equation

$$\int_0^{R_n} P_u(R)dR = \int_0^{\beta_n} P(\beta)d\beta \qquad (7.11)$$

where R_n is the random number generated in the nth try. The left-hand side of Eqn. 7.11 is simply R_n. In many cases the right-hand side can be expressed analytically in terms

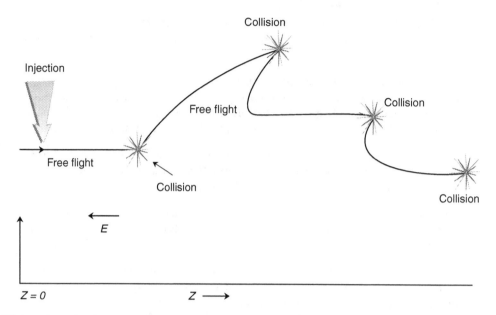

Figure 7.3: A schematic of processes involved in the physical picture used in Monte Carlo methods.

of β_n, in which case once R_n is known, β_n is also known, at least implicitly. In some cases, it is difficult to evaluate the integral on the right-hand side and other approaches are then used to evaluate β.

Let us now examine the various components of the Monte Carlo program as outlined in Figs. 7.3 and 7.4.

7.2.2 Injection of Carriers

The injection conditions are relevant for transport of carriers over short distances or time. The steady state v-$\mathbf{F}$ relations should not be dependent upon the initial conditions under which an electron is injected into the semiconductor. Usually in a device the contact is simply a heavily doped semiconductor region (an ohmic contact) where the electrons simply reside with nearly an equilibrium distribution. For a nondegenerate case the equilibrium distribution is simply a Maxwellian distribution according to which the probability of finding the velocity between v and $v + dv$ is

$$P(\boldsymbol{v}) = A^3 \exp\left[\frac{-m^*}{2k_BT}\left(v_x^2 + v_y^2 + v_z^2\right)\right] \tag{7.12}$$

with

$$A^3 = \left(\frac{m^*}{2\pi k_BT}\right)^{3/2}$$

If the electrons are injected along the z-axis, we are only interested in electrons with positive velocities which provide the current into the device. The x- and y-direction

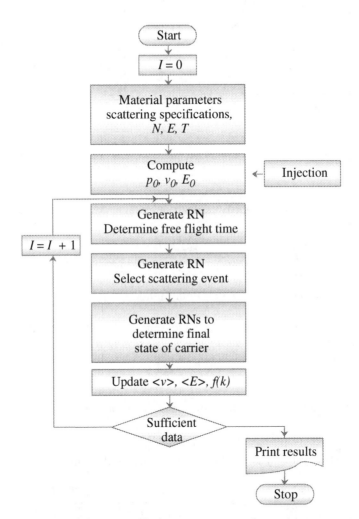

Figure 7.4: A flowchart of the Monte Carlo program to study carrier transport.

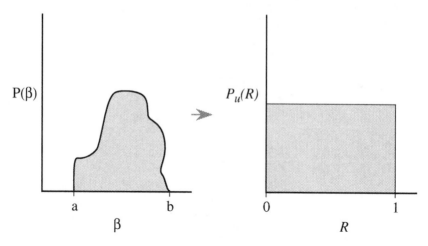

Figure 7.5: Mapping of a random probability function to a randomly generated number from a uniform random number generator.

velocities can be positive or negative since the current flow is along the z-axis. These are then given by

$$\int_0^{R_n} P_u(R)dR = \int_{-\infty}^{v_{xn}} A \exp\left[\frac{-m^*}{2k_BT}v_x^2\right] dv_x$$

with a similar equation for v_y. Two random numbers are therefore used to get v_x and v_y.

To find the velocity along the z-axis one makes the approximation that the current is given by the velocities along the z-direction only, i.e.,

$$\frac{J}{e} = \int v_z\, f(v_z)\, dv_z$$

The flux between the velocity v_z and $v_z + dv_z$ is just

$$P(v_z)\, dv_z = v_z\, f(v_z)\, dv_z$$

Then the z-direction velocity is weighted into the probability function. The Monte Carlo method gives

$$\int_0^{R_n} P_u(R)dR = \frac{\int_0^{v_{zn}} v_z \exp\left[-m^*v_z^2/(2k_BT)\right] dv_z}{\int_0^{\infty} v_z \exp\left[-m^*v_z^2/(2k_BT)\right] dv_z}$$

The denominator is present to ensure that the probability is normalized. The right-hand side can now be integrated to get

$$v_{zn} = \sqrt{\frac{-2k_BT \ln(1 - R_n)}{m^*}} \tag{7.13}$$

The term $-\ln(1 - R_n)$ is positive since $0 \le R_n \le 1$. The z-direction velocity then has only a positive value.

If the contact injects the electrons according to some other distribution, the Monte Carlo method has to simulate that particular distribution.

7.2.3 Free Flight

As shown in Fig. 7.3 the Monte Carlo approach treats the transport problem as a series of free flights followed by scattering events. During the free flight the electron simply moves in the electric field according to the free electron equations of motion

$$\frac{d\boldsymbol{p}}{dt} = e\mathbf{F}$$

If the electric field is along the z-axis, we have the following changes in the various properties of the electron (the time interval for free flight is typically 10^{-13} s so that on can linearize the equations):

Momentum after time t

$$
\begin{aligned}
p_x(t) &= p_x(0) \\
p_y(t) &= p_y(0) \\
p_z(t) &= p_z(0) + eFt
\end{aligned}
\tag{7.14}
$$

Position after time t

$$
\begin{aligned}
x(t) &= x(0) + \frac{p_x(0)}{m^*}t \\
y(t) &= y(0) + \frac{p_y(0)}{m^*}t \\
z(t) &= z(0) + \frac{E(t) - E(0)}{eF}
\end{aligned}
\tag{7.15}
$$

where the energy change is given by

$$E(t) - E(0) = \frac{p^2(t)}{2m^*} - \frac{p^2(0)}{2m^*} \tag{7.16}$$

7.2.4 Scattering Times

An important question to address next is: How long does free flight last? To evaluate this, we note that:

1. The scattering rate for the ith scattering process is given by the Fermi golden rule

$$
\begin{aligned}
W_i(\boldsymbol{k}) &= \frac{2\pi}{\hbar} \sum_{\mathbf{k}'} \left|M_{\mathbf{k},\mathbf{k}'}\right|^2 \delta(E_f - E_i) \\
R_{\text{tot}} &= \sum_{i=1}^{m} W_i(\boldsymbol{k})
\end{aligned}
\tag{7.17}
$$

where R_{tot} represents the sum of all scattering rates due to the m scattering processes. All of the scattering rates are calculated and in some cases tabulated in look up tables.

2. The angular dependence of the scattering process is given by the matrix element

$$W(\boldsymbol{k}, \boldsymbol{k}') = \frac{2\pi}{\hbar} \left| M_{\boldsymbol{k}\boldsymbol{k}'} \right|^2 \delta(E_f - E_i) \tag{7.18}$$

If $\lambda(\boldsymbol{k}, t)$ is the total scattering rate for an electron with momentum $\boldsymbol{k}(t)$, the probability that the electron will drift without scattering for time t and then scatter in time Δt is

$$P(t)\Delta t = \lambda(\boldsymbol{k}, t) \, \Delta t \, [1 - \lambda(\boldsymbol{k}_i, t_i)\Delta t]^n \tag{7.19}$$

with

$$
\begin{aligned}
n\Delta t &= t \\
i\Delta t &= t_i \\
\boldsymbol{k}_i &= \boldsymbol{k}(t_i)
\end{aligned}
$$

The first term in Eqn. 7.19 is the probability of having a collision in time Δt and the second term in the parenthesis is the combined probability of not having a collision in n time intervals of Δt. If Δt is small or n is large, Eqn. 7.19 becomes

$$P(t) = \lambda(\boldsymbol{k}, t) \exp\left[-\int_0^t \lambda(\boldsymbol{k}, t')dt' \right] \tag{7.20}$$

We use the Monte Carlo approach to find the times interval t_f between collisions. We have

$$\int_0^{R_n} P_u(r)dr = \int_0^{t_f} P(t)dt$$

or

$$R_n = \int_0^{t_f} \lambda(t) \exp\left\{ -\int_0^t \lambda(t')dt' \right\} dt$$

Noting that

$$\frac{-d}{dt} \exp\left\{ -\int_0^t \lambda(t)'dt' \right\} = \lambda(t) \exp\left\{ -\int_0^t \lambda(t)'dt' \right\}$$

we get

$$R_n = \left[-\exp\left(-\int_0^t \lambda(t)'dt' \right) \right] \Big|_0^{t_f}$$

or

$$R_n = 1 - \exp\left(-\int_0^t \lambda(t)'dt' \right) \tag{7.21}$$

In principle this equation can be used to generate various values of t_f. However, if $\lambda(t)$ is equal to R_{tot}, it has a fairly complex energetic and, therefore, temporal dependence, which makes it difficult to evaluate the integral. One possibility is to develop elaborate look up tables after having solved the problem numerically. However, a simple and ingenious way is based on the concept of self-scattering as shown in Fig. 7.6. We

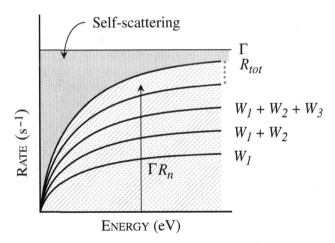

Figure 7.6: A schematic showing how a random number allows one to determine the scattering mechanism responsible for scattering.

define a scattering rate

$$\Gamma \equiv \lambda = R_{\text{tot}} + \lambda_0 \tag{7.22}$$

where λ_0 is such that Γ is a constant scattering rate. The λ_0 part of the scattering rate is just a fictitious scattering rate called the self-scattering rate which does not cause any real scattering. With this choice we get

$$
\begin{aligned}
R_n &= 1 - \exp\left[-\int_0^{t_f} \Gamma dt\right] \\
&= 1 - e^{-\Gamma t_f}
\end{aligned}
$$

or

$$t_f = \frac{-1}{\Gamma} \ln(1 - R_n) \tag{7.23}$$

Eqn. 7.23 allows us to find the free flight time during which the electron simply accelerates in the presence of the electric field. At the end of the free flight time the electron scatters. Depending upon the choice of the parameter Γ, a small fraction of the electrons will "self scatter," i.e., here no real scattering. In this case the electron is simply allowed to continue accelerating in the field.

7.2.5 Nature of the Scattering Event

After the free flight the properties of the electron are updated. This requires us to know which particular mechanism is responsible for scattering. This is determined as shown schematically in Fig. 7.6 and requires a single random number, R, which is used to identify the scattering mechanism ℓ which satisfies the inequality

$$\sum_{i=1}^{\ell-1} W_i(\boldsymbol{k}) < \Gamma R \leq \sum_{i=1}^{\ell} W_i(\boldsymbol{k}) \tag{7.24}$$

The self-scattering process is introduced so that the simple expression of Eqn. 7.23 can be used. As shown in Fig. 7.6 there will be cases where self-scattering (i.e., no scattering) will be chosen. One has to be quite judicious in the choice of Γ. Γ should be larger than R_{tot}, but not by too much, otherwise a larger number of scattering events will be self-scattering events which just consume extra computer time.

7.2.6 Energy and Momentum After Scattering

After the scattering event is identified one has to define the electron's energy and momentum just after the scattering event. Since the scattering time is assumed to be zero, the position of the electron is unaffected by the scattering event. The scattering process immediately tells us how much the energy has changed as a result of scattering. For example, we have the following results:

Ionized impurity:	$\Delta E = 0$
Alloy scattering:	$\Delta E = 0$
Polar optical phonon absorption:	$\Delta E = \hbar\omega_0$
Polar optical phonon emission:	$\Delta E = -\hbar\omega_0$
Acoustic phonon scattering:	$\Delta E \approx 0.$

The updating of the momentum (direction) requires further generation of random numbers. It is convenient to find the scattered angle in the coordinate system where the z-axis is along the direction of the original momentum $\boldsymbol{k}$. This is done since all the angular dependencies calculated in the previous two chapters are in this system. Of course, eventually we will describe the final momentum in the coordinate system where the z-axis is along the electric field. This will require a simple transformation. The two coordinate systems are shown in Fig. 7.7. The angles β and α represent the azimuthal and polar angles of scattering in the initial momentum coordinate system while ϕ and θ are the scattering angles in the $\boldsymbol{F}$-field coordinate system.

The determination of the azimuthal angle β after scattering is given by the probability that $\boldsymbol{k}'$ lies between azimuthal angles β and $\beta + d\beta$

$$P(\beta)d\beta = \frac{d\beta \int_0^\infty \int_0^\pi W(\boldsymbol{k},\boldsymbol{k}') \, \sin\alpha \, d\alpha \, k'^2 \, dk'}{\int_0^{2\pi} d\beta \int_0^\infty \int_0^\pi W(\boldsymbol{k},\boldsymbol{k}') \, \sin\alpha \, d\alpha \, k'^2 \, dk'}$$

Since $W(\boldsymbol{k},\boldsymbol{k}')$ does not depend upon β for any of the isotropic band scattering rates, we simply have

$$P(\beta)d\beta = \frac{d\beta}{2\pi}$$

The generation of a random number R_n determines β by the equation

$$\beta = 2\pi R_n \tag{7.25}$$

The azimuthal angle is then uniformly distributed between 0 and 2π.

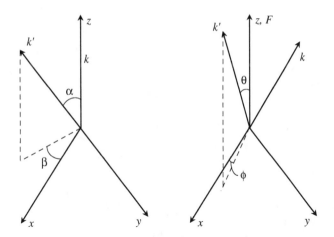

Figure 7.7: The two coordinate systems used in Monte Carlo method. One uses the system (left) where the momentum before scattering is along the z-axis to determine the final state after scattering. This has then to be transformed in the fixed-coordinate system (right) where the **F**-field is along the z-axis.

In general, the scattering rate has a polar angle dependence so that the determination of the polar angle α is somewhat more involved. We have

$$P(\alpha)d\alpha = \frac{\sin\alpha \, d\alpha \int_0^\infty \int_0^{2\pi} W(\mathbf{k},\mathbf{k}') \, d\beta \, k'^2 \, dk'}{\int_0^\pi \int_0^{2\pi} \int_0^\infty \sin\alpha \, d\alpha \, W(\mathbf{k},\mathbf{k}') \, d\beta \, k'^2 \, dk'}$$

Let us first examine isotropic scattering where $W(\mathbf{k},\mathbf{k}')$ has no α dependence. This case occurs for acoustic phonon scattering, alloy scattering (with no clustering), etc. In this case we have

$$P(\alpha)d\alpha = \frac{\sin\alpha \, d\alpha}{2}a$$

In the Monte Carlo approach, a random number is generated and we have

$$R_n = \frac{1}{2}\int_0^\alpha \sin\alpha' \, d\alpha'$$

$$= \frac{1}{2}(1 - \cos\alpha)$$

or

$$\cos\alpha = 1 - 2R_n \tag{7.26}$$

Note that even though the scattering is isotropic, the average value of α is $\pi/2$. This is due to the number of available states between α and $\alpha + d\alpha$ which peak at $\pi/2$ as shown in Fig. 7.8.

If the scattering process is not randomizing, techniques have to be developed to find the scattering angle. We will now examine a few such approaches.

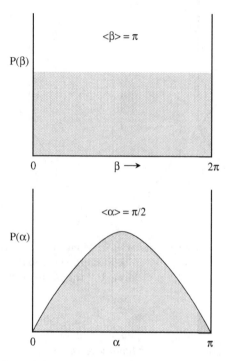

Figure 7.8: The probability of the azimuthal and polar scattering angles in the case of an isotopic scattering involving a randomizing scattering.

Ionized Impurity Scattering

In this case the scattering matrix element has a strong angular dependence as shown in Fig. 7.9. The dependence is given by

$$W(\boldsymbol{k}, \boldsymbol{k}') \propto \frac{1}{\left[4k^2 \sin^2(\alpha/2) + \lambda^2\right]^2}$$

where λ is the inverse screening length. The integration for $P(\alpha)$ can be carried out and the scattering angle is given by

$$\cos \alpha = 1 - \frac{2(1 - R_n)}{1 + 4\frac{E_{ss}}{E_{ss_\beta}} R_n} \tag{7.27}$$

where

$$E_\beta = \frac{\hbar^2 \lambda^2}{2m^*}$$

and

$$E = \frac{\hbar^2 k^2}{2m^*}$$

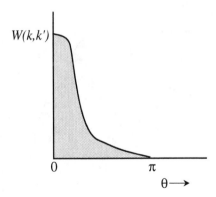

Figure 7.9: Angular dependence of the scattering by ionized impurities. The scattering has a strong forward angle preference.

Polar Optical Phonon Scattering

Another important nonrandomizing scattering is the polar optical phonon scattering. In this case the scattering matrix element has a dependence given by

$$W(\mathbf{k}, \mathbf{k}') \propto \frac{1}{\left|\mathbf{k} - \mathbf{k}'\right|^2}$$

For a parabolic energy momentum relation

$$\frac{\hbar^2 k^2}{2m^*} = E$$

we get

$$P(\alpha)d\alpha = \frac{\left[\frac{\sin\alpha\, d\alpha}{Ess + Ess' - 2(EssEss')^{1/2}\cos\alpha}\right]}{\int_0^\pi \frac{\sin\alpha\, d\alpha}{Ess + Ess' - 2(EssEss')^{1/2}\cos\alpha}}$$

Using the Monte Carlo approach

$$
\begin{aligned}
R_n &= \int_0^\alpha P(\alpha')\, d\alpha' \\
&= \frac{\ln\left[1 + f(1 - \cos\alpha)\right]}{\ln\left[1 + 2f\right]}
\end{aligned}
$$

where

$$f = 2(EE')^{1/2}\left[E^{1/2} - E'^{1/2}\right]^{-2} \tag{7.28}$$

Thus, the polar angle is given by

$$\cos\alpha = \frac{\left[(1 + f) - (1 + 2f)^{R_n}\right]}{f} \tag{7.29}$$

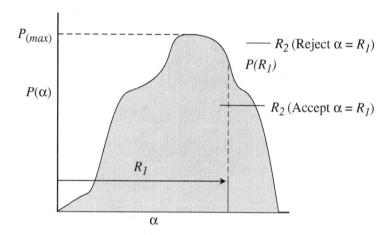

Figure 7.10: The procedure for using the von Neumann method to generate an arbitrary probability function.

There are cases where it is not possible to carry out the integrals needed to obtain a relation like the one given above. This may be due to the bandstructure not being parabolic. In such cases, a technique due to von Neumann is used to obtain α from an arbitrary $P(\alpha)$. In this technique, random numbers are generated in pairs, the first number R_1 representing the value of $\cos \alpha$ in the range -1 to 1, i.e., the limits for $\alpha = 0$ and π. The second number, R_2, is generated between 0 and $P(max)$, where $P(max)$ is the maximum value of the probability. If $P(R_1)$ is less than R_2, the value of R_1 ($= \cos \alpha$) is not accepted and another pair of random numbers is generated until $P(R_1)$ is larger than R_2. In this case the value of $\cos \alpha$ will then mimic $P(\alpha)$. Fig. 7.10 shows schematically how the von Neumann technique works.

Once α and β are known, a simple rotational transformation is made to obtain the angles ϕ and θ in the electric field coordinate system. This is straightforward, since the orientation of the vector $\boldsymbol{k}$ just before scattering and the electric field are known. This updates the momentum of the electron after each scattering event.

The Monte Carlo procedure described here is repeated thousands of times to obtain convergence. Since the trajectory of the electron is followed in real space, as well as energy and momentum space, a tremendous insight is gained. The k-space distribution for the electron ensemble can be mapped out by constructing a k-space grid and determining the relative time spent by the electron around each grid point. Once the distribution function is known, essentially all physical quantities of interest can be determined. However, often it is useful to determine certain physical quantities directly. For example, the drift velocity in the jth valley is determined from the relation

$$
\begin{aligned}
v_j &= \frac{1}{k_j} \sum \int_{k_{zi}}^{k_{zf}} \frac{1}{\hbar} \frac{\partial E(\boldsymbol{k})}{\partial k_z} dk_z \\
&= \frac{1}{\hbar k_j} \sum (E_f - E_i)
\end{aligned}
\tag{7.30}
$$

Here k_{zi} and k_{zf} are the initial and final $\mathbf{k}$-vectors in a free flight, the summation is over all free flights and E_i and E_f are the energies of the electron at the beginning and end of a free flight. The vector k_j is the total length in the direction of the k-space trajectory in the valley and is given by

$$k_j = \frac{T_j e \, |\mathbf{F}|}{\hbar} \tag{7.31}$$

where T_j is the total time spent in the valley and $\mathbf{F}$ is the electric field. The relative population of the electron in different valleys is determined by simply the relative time spent by the electron in different valleys.

The material parameters, including the various deformation potentials, have been collected from various experimental papers and represented by Jacoboni and Raggiani (1983) and Hess (1988). In Table 7.2 we have collected the parameters for GaAs and Si. We will briefly list the total scattering rate $W(\mathbf{k})$ and differential scattering rates $W(\mathbf{k}, \mathbf{k}')$ for electron transport in direct gap semiconductors below. The results for the Γ-valley assume a nonparabolic band.

Ionized Impurity Scattering

$$W(\mathbf{k}) \;=\; 4\pi F \left(\frac{2k}{\lambda}\right)^2 \left[\frac{1}{1 + (\lambda/2k)^2}\right]$$

$$F \;=\; \frac{1}{\hbar}\left[\frac{Ze^2}{\epsilon}\right]^2 \frac{N(E_{\mathbf{k}})}{32k^4} \cdot N_I \tag{7.32}$$

Angular dependence

$$\frac{1}{4k^2 \sin^2(\theta/2) + \lambda^2} \sin\theta \, d\theta \tag{7.33}$$

Alloy Scattering in Alloy $\mathbf{A}_x\mathbf{B}_{1-x}$

$$W(\mathbf{k}) = \frac{3\pi^3}{8\hbar} V_0 \, U_{\text{all}}^2 \, N(E_{\mathbf{k}}) \, x \, (1 - x) \tag{7.34}$$

Angular dependence: randomizing.

Polar Optical Phonon Scattering

In Chapter 6 we discussed the polar optical phonon scattering assuming unity for the central cell overlap integral and a parabolic band. The polar optical phonon is usually the most important scattering mechanism for high field transport. At these high fields the electron energies are quite large and nonparabolic effects quite important. We will, therefore, give the rates for nonparabolic bands (Fawcett, et al., 1970). The angular dependence of the scattering rate is given by

$$W(\mathbf{k}, \mathbf{k}') \;=\; \frac{\pi e^2 \omega_0}{V \left|\mathbf{k} - \mathbf{k}'\right|^2} \left(\frac{1}{\epsilon_\infty} - \frac{1}{\epsilon_s}\right) G(\mathbf{k}, \mathbf{k}')$$

$$\times \begin{cases} n(\omega_0)\, \delta(E(\mathbf{k}') - E(\mathbf{k}) - \hbar\omega_0) & \text{(absorption)} \\ (n(\omega_0) + 1)\, \delta(E(\mathbf{k}') - E(\mathbf{k}) + \hbar\omega_0) & \text{(emission)} \end{cases} \tag{7.35}$$

Parameter	Symbol	Value in Si	Value in GaAs
Mass density (g/cm^3)	ρ	2.329	5.36
Lattice constant (Å)	a_0	5.43	5.642
Low frequency dielectric constant	ϵ_0	11.7	12.90
High frequency dielectric constant	ϵ_∞	—	10.92
Piezoelectric constant (C/m^2)	e_{PZ}	—	0.160
Longitudinal acoustic velocity $(\times 10^5$ cm/s)	v_l	9.04	5.24
Transverse acoustic velocity $(\times 10^5$ cm/s)	v_t	5.34	3.0
Longitudinal optical phonon energy (eV)	$\hbar\omega_0$	0.063	0.03536
Electron effective mass (lowest valley) (m_0)	m^* m_l^*, m_t^*	— 0.916, 0.19 (X)	0.067 (Γ) —
Electron effective mass (upper valley) (m_0)	m^* m^* m_l^*, m_t^*	— — 1.59, 0.12 (L)	0.222 (L) 0.58 (X) —
Nonparabolicity parameter (eV^{-1})	α	0.5 (X)	0.610 (Γ) 0.461 (L) 0.204 (X)
Energy separation between valleys (eV)	$\Delta E_{\Gamma L}$ $\Delta E_{\Gamma X}$	— —	0.29 0.48

Table 7.2: Material parameters for Silicon and Gallium Arsenide, (from Jacoboni and Reggiani, 1983, Hess, 1988).

where $n(\omega_0)$ is the occupation probability for the optical phonons, ϵ_∞, ϵ_s are the high frequency and static dielectric constants, and ω_0 is the optical phonon frequency. If a nonparabolic energy band is assumed with the relation

$$\frac{\hbar^2 k^2}{2m^*} = E(1 + \alpha E)$$

the overlap integral is given by

$$G(\boldsymbol{k}, \boldsymbol{k}') = \left[a_{\boldsymbol{k}} a_{\boldsymbol{k}'} + c_{\boldsymbol{k}} c_{\boldsymbol{k}'} \cos \beta\right]^2 \tag{7.36}$$

Parameter	Symbol	Value in Si	Value in GaAs
Electron acoustic deformation potential (eV)	D_A	9.5	7.0 (Γ) 9.2 (L) 9.0 (X)
Electron optical deformation potential ($\times 10^8$ eV/cm)	D_0	—	3.0 (L)
Optical phonon energy (eV)	ω_0	0.0642	0.0343
Hole acoustic deformation potential (eV)	D_A	5.0	3.5
Hole optical deformation potential (eV/cm)	D_0	6.00	6.48
Intervalley parameters, g-type (X–X) ($\times 10^8$ eV/cm), (eV)	D_y, E_y	0.5, 0.012 (TA) 0.8, 0.019 (LA) 11.0, 0.062 (LO)	—
Intervalley parameters, f-type (X–X) ($\times 10^8$ eV/cm), (eV)	D_y, E_y	0.3, 0.019 (TA) 2.0, 9.947 (LA) 2.0, 0.059 (LO)	—
Intervalley parameters, (X–L) ($\times 10^8$ eV/cm), (eV)	D_y, E_y	2.0, 0.058 2.0, 0.055 2.0, 0.041 2.0, 0.017	—
Intervalley deformation potential ($\times 10^8$ eV/cm)	$D_{\Gamma L}, D_{\Gamma X}$ D_{LL}, D_{LX} D_{XX}	—	10.0, 10.0 10.0, 5.0 7.0
Intervalley phonon energy (eV)	$E_{\Gamma L}, E_{\Gamma X}$ E_{LL}, E_{LX} E_{XX}	– –	0.0278, 0.0299 0.0290, 0.0293 0.0299

Table 7.2: continued.

where β is the angle between $\boldsymbol{k}$ and $\boldsymbol{k}'$ and

$$a_{\mathbf{k}} = \left[\frac{1 + \alpha E(\boldsymbol{k})}{1 + 2\alpha E(\boldsymbol{k})} \right]^{1/2}$$

$$c_{\mathbf{k}} = \left[\frac{\alpha E(\boldsymbol{k})}{1 + 2\alpha E(\boldsymbol{k})} \right]^{1/2}$$

Also, $a_{\mathbf{k}'}$ and $c_{\mathbf{k}'}$ are the same corresponding functions of $E(\boldsymbol{k})$.

If α is 0, the overlap integral is unity. The total scattering rate is

$$W(\boldsymbol{k}) = \frac{e^2 m^{*1/2} \omega_0}{4\pi\sqrt{2}\hbar} \left(\frac{1}{\epsilon_\infty} - \frac{1}{\epsilon_s} \right) \frac{1 + 2\alpha E'}{\gamma^{1/2}(E)} F_0(E, E')$$

$$\times \begin{cases} n(\omega_0) & \text{(absorption)} \\ (n(\omega_0) + 1) & \text{(emission)} \end{cases} \tag{7.37}$$

where

$$\begin{aligned}
E' &= E + \hbar\omega_0 \text{ for absorption} \\
&= E - \hbar\omega_0 \text{ for emission} \\
\gamma(E) &= E(1 + \alpha E) \\
F_0(E, E') &= C^{-1} \left(A \ln \left| \frac{\gamma^{1/2}(E) + \gamma^{1/2}(E')}{\gamma^{1/2}(E) - \gamma^{1/2}(E')} \right| + B \right) \\
A &= \left[2(1 + \alpha E)(1 + \alpha E') + \alpha\{\gamma(E) + \gamma(E')\} \right]^2 \\
B &= -2\alpha\gamma^{1/2}(E)\,\gamma^{1/2}(E') \\
&\quad \times \left[4(1 + \alpha E)(1 + \alpha E') + \alpha\{\gamma(E) + \gamma(E')\} \right] \\
C &= 4(1 + \alpha E)(1 + \alpha E')(1 + 2\alpha E)(1 + 2\alpha E') \tag{7.38}
\end{aligned}$$

Although the results look quite complicated, it is straightforward to use them in the Monte Carlo method. The angular dependence of the momentum after scattering requires the von Neumann technique discussed earlier.

Acoustic Phonon Scattering

Acoustic phonon scattering is an important scattering mechanism at low electric fields especially at low temperatures where the optical phonon occupation number is very small. The total scattering rate is

$$W(\boldsymbol{k}) = \frac{(2m^*)^{3/2} k_B T\, D_{ac}^2}{2\pi\rho v_s^2 \hbar^4} \gamma^{1/2}(E)(1 + 2\alpha E) F_a(E)$$

$$F_a(E) = \frac{(1 + \alpha E)^2 + \frac{1}{3}(\alpha E)^2}{(1 + 2\alpha E)^2} \tag{7.39}$$

Once again if $\alpha = 0$, we recover the results of Chapter 6.

Intervalley Scattering

As noted in Chapter 6, the intervalley scattering is simply treated as a deformation potential scattering described by a deformation potential D_{ij}. For equivalent valleys the total scattering rate is

$$W(\boldsymbol{k}) = \frac{(Z_e - 1)\, m^{*3/2} D_{ij}^2}{\sqrt{2}\pi\rho\, \omega_{ij}\hbar^3} E'^{1/2} \begin{cases} n(\omega_{ij}) & \text{(absorption)} \\ (n(\omega_{ij}) + 1) & \text{(emission)} \end{cases} \tag{7.40}$$

where

$$E' = \begin{cases} E(\boldsymbol{k}) + \hbar w_{ij} & \text{(absorption)} \\ E(\boldsymbol{k}) - \hbar w_{ij} & \text{(emission)} \end{cases}$$

Z_e is the number of equivalent valleys and w_{ij} is the phonon frequency which allows the intervalley scattering. The angular dependence of the scattering rate is randomizing.

For non-equivalent intervalley scattering (e.g., $\Gamma \to L$) the scattering rate has a similar form except the factor $(Z_e - 1)$ is replaced by the number of final valleys available for scattering (e.g., 4 for $\Gamma \to L$ scattering; 1 for $L \to \Gamma$ scattering; 6 for $\Gamma \to X$ scattering, etc.). The scattering rate for the transfer of the electron form i to j valley in this case is

$$\begin{aligned} W(\boldsymbol{k}) &= \frac{Z_j m_j^{*3/2} D_{ij}^2}{\sqrt{2} \pi \rho w_{ij} \hbar^3} \gamma_j^{1/2} (E')(1 + 2\alpha_j E') G_{ij}(E, E') \\ &\times \begin{cases} n(w_{ij}) & \text{(absorption)} \\ (n(w_{ij}) + 1) & \text{(emission)} \end{cases} \end{aligned} \tag{7.41}$$

where

$$G_{ij}(\boldsymbol{k}, \boldsymbol{k}') = \frac{\left[1 + \alpha_i E_i(\boldsymbol{k})\right]\left[1 + \alpha_j E_j(\boldsymbol{k}')\right]}{\left[1 + 2\alpha_i E_i(\boldsymbol{k})\right]\left[1 + 2\alpha_j E_j(\boldsymbol{k}')\right]}$$

and

$$E_j' = \begin{cases} E_i - \Delta_j + \Delta_i + \hbar w_{ij} & \text{(absorption)} \\ E_i - \Delta_j + \Delta_i - \hbar w_{ij} & \text{(emission)} \end{cases}$$

The approach discussed here is extremely versatile, since it can be easily generalized to address any of the following problems:

1. Steady state v–$\mathbf{F}$ relations along with information on carrier distribution in k-space, electron temperature, valley occupation, etc.

2. Noise in electronic transport can also be studied from the carrier distribution information.

3. Transient v–$\mathbf{F}$ relations and its dependence on transit length, scattering processes, injection conditions, etc.

4. Dependence of transit time in a fixed channel on the field distribution which could be calculated in a self-consistent way with the Poisson equation and current continuity equation.

5. Carrier injection and thermalization process.

6. Impact ionization process.

EXAMPLE 7.1 Use the scattering rates given in this section for a Monte Carlo method and obtain some typical scattering rates for GaAs.

In the following we will examine some typical energies. The reader can easily scale for other energies and material parameters. The example includes the non-parabolic band approximations.

In GaAs the bandstructure nonparabolicity parameter α is given as

$$\alpha = \frac{1}{E_g}\left(1 - \frac{m^*}{m_0}\right)^2$$

$$= 0.576 \text{ eV}^{-1}$$

Polar L-O Phonon Scattering

$$W = \frac{e^2 (m^*)^{1/2}\omega_0}{4\pi\sqrt{2}\hbar\epsilon_0}\left(\frac{\epsilon_0}{\epsilon_\infty} - \frac{\epsilon_0}{\epsilon_s}\right)\frac{(1 + 2\alpha E_F)}{\gamma^{1/2}(E_i)}F_0(E_i, E_f)\begin{Bmatrix} N & \text{absorption} \\ N+1 & \text{emission} \end{Bmatrix}$$

$$= 3.69\left(\frac{\omega_0}{\text{rad/s}}\right)\left(\frac{m^*}{m_0}\right)^{1/2}\left(\frac{\epsilon_0}{\epsilon_\infty} - \frac{\epsilon_0}{\epsilon_s}\right)$$

$$\times \frac{(1 + 2\alpha E_f)}{(\gamma(E_i)/\text{eV})^{1/2}}F_0(E_i, E_f)\begin{Bmatrix} N \\ N+1 \end{Bmatrix}\text{s}^{-1}$$

with F_0 given in Eqn. 7.38, $\gamma(E) = E(1 + \alpha E)$ and

$$E_f = E_i \pm \hbar\omega_0 \begin{cases} + \text{ for absorption} \\ - \text{ for emission} \end{cases}$$

$$N = \frac{1}{\exp(\hbar\omega_0/k_B T) - 1}$$

Example: GaAs, $T = 300$ K, Γ valley, L-O phonon absorption, $E_i = 3k_B T/2 = 0.039$ eV.

$$\begin{aligned}
\omega_0 &= 4.5 \times 10^{13} \text{ rad/s} \\
\hbar\omega_0 &= 0.030 \text{ eV} \\
E_f &= E_i + \hbar\omega_0 = 0.069 \text{ eV} \\
m^* &= 0.067 \ m_o \\
\epsilon_s/\epsilon_0 &= 13.2 \\
\epsilon_\infty/\epsilon_0 &= 10.9 \\
N &= \frac{1}{\exp(\hbar\omega_0/k_B T) - 1} = 0.458 \\
\alpha &= \frac{1}{E_g}\left(1 - \frac{m^*}{m_0}\right)^2 = 0.576 \text{ eV}^{-1} \\
\gamma(E_i) &= E_i(1 + \alpha E_i) = 0.040 \text{ eV} \\
\gamma(E_f) &= E_f(1 + \alpha E_f) = 0.072 \text{ eV} \\
1 + \alpha E_i &= 1.022 \\
1 + \alpha E_f &= 1.040 \\
1 + 2\alpha E_i &= 1.045
\end{aligned}$$

$$1 + 2\alpha E_f = 1.079$$
$$2\gamma(E_i) = 0.023$$
$$\alpha\gamma(E_f) = 0.041$$

$$W = 3.69 \left(\frac{\omega_0}{\text{rad/s}}\right) \sqrt{\frac{m^*}{m_0}} \left(\frac{\epsilon_0}{\epsilon_\infty} - \frac{\epsilon_0}{\epsilon_s}\right)$$

$$\times \frac{1 + 2\alpha E_f}{\sqrt{\gamma(E_i)/\text{eV}}} F_0(E_i, E_f) N \text{ s}^{-1}$$

$$= 3.69(4.5 \times 10^{13})\sqrt{0.067} \left(\frac{1}{10.9} - \frac{1}{13.2}\right)$$

$$\times \frac{1.079}{\sqrt{0.040}} F_0(E_i, E_f) \, 0.458 \text{ s}^{-1}$$

$$= 1.70 \times 10^{12} F_0(E_i, E_f) \text{ s}^{-1}$$

$$F_0(E_i, E_f) = \frac{1}{C} \left(A \ln \left|\frac{\sqrt{\gamma(E_i)} + \sqrt{\gamma(E_f)}}{\sqrt{\gamma(E_i)} - \sqrt{\gamma(E_f)}}\right| + B \right)$$

$$= \frac{1}{C} \left(A \ln \left|\frac{\sqrt{0.040} + \sqrt{0.072}}{\sqrt{0.040} - \sqrt{0.072}}\right| + B \right)$$

$$= \frac{1.926A + B}{C}$$

$$A = [2(1 + \alpha E_i)(1 + \alpha E_f) + \alpha(\gamma(E_i) + \gamma(E_f))]^2$$
$$= [2(1.022)(1.040) + 0.023 + 0.041] = 4.795$$

$$B = -2\alpha\sqrt{\gamma(E_i)\gamma(E_f)}$$
$$\times [4(1 + \alpha E_i)(1 + \alpha E_f) + \alpha(\gamma(E_i) + \gamma(E_f))]$$
$$= -2\sqrt{(0.023)(0.041)}$$
$$\times [4(1.022)(1.040) + 0.023 + 0.041] = -0.265$$

$$C = 4(1 + \alpha E_i)(1 + \alpha E_f)(1 + 2\alpha E_i)(1 + 2\alpha E_f)$$
$$= 4(1.022)(1.040)(1.045)(1.079) = 4.794$$

$$F_0(E_i, E_f) = \frac{(1.926)(4.795) + (-0.265)}{4.794} = 1.871$$

$$W = 1.70 \times 10^{12}(1.871) \text{ s}^{-1} = 3.18 \times 10^{12} \text{ s}^{-1}$$

Acoustic Phonon Scattering

$$W = \frac{(2m^*)^{3/2} k_B T \, D_{ac}^2}{2\pi\rho \, v_s^2 \, \hbar^4} \gamma^{1/2}(E)(1 + 2\alpha E) F_a(E)$$

$$= 4.49 \times 10^{21} \frac{\left(\frac{m^*}{m_0}\right)^{3/2} \left(\frac{T}{\text{K}}\right) \left(\frac{D_{ac}}{\text{eV}}\right)^2}{\left(\frac{\rho}{\text{gm/cm}^3}\right) \left(\frac{v_s}{\text{cm/s}}\right)^2} \left(\frac{\gamma(E)}{\text{eV}}\right)^{1/2} (1 + 2\alpha E) F_a(E) \text{ s}^{-1}$$

with $F_a(E)$ as given in Eqn. 7.39.

Example: GaAs, $T = 300$ K, Γ valley, acoustic phonon scattering,

$$E_i = E_f = \frac{3}{2}k_B T = 0.039 \text{ eV} = E$$

$$m^* = 0.067 \, m_0$$

$$D_{ac} = 7.8 \text{ eV}$$

$$\rho = 5.37 \text{ gm/cm}^3$$

$$v_s = 5.22 \times 10^5 \text{ cm/s}$$

$$\alpha = 0.576 \text{ eV}^{-1}$$

$$\gamma(E) = 0.040 \text{ eV}$$

$$\alpha E = 0.022$$

$$1 + \alpha E = 1.022$$

$$1 + 2\alpha E = 1.045$$

$$W = 4.49 \times 10^{21} \frac{\left(\frac{m^*}{m_0}\right)^{3/2} \left(\frac{T}{\text{K}}\right) \left(\frac{D_{ac}}{\text{eV}}\right)^2}{\left(\frac{\rho}{\text{gm/cm}^3}\right) \left(\frac{v_s}{\text{cm/s}}\right)^2} \sqrt{\frac{\gamma(E)}{\text{eV}}} \, (1 + 2\alpha E) \, F_a(E) \text{ s}^{-1}$$

$$= 4.49 \times 10^{21} \frac{(0.067)^{3/2} \, (300) \, (7.8)^2}{(5.37) \, (5.22 \times 10^5)^2} (0.040)^{1/2} \, (1.045) \, F_a(E) \text{ s}^{-1}$$

$$= 2.03 \times 10^{11} F_a(E) \text{ s}^{-1}$$

$$F_a(E) = \frac{(1 + \alpha E)^2 + \frac{1}{3}(\alpha E)^2}{(1 + 2\alpha E)^2} = \frac{(1.022)^2 + \frac{1}{3}(0.022)^2}{(1.045)^2} = 0.957$$

$$W = 2.03 \times 10^{11} \, (0.957) \text{ s}^{-1} = 1.94 \times 10^{11} \text{ s}^{-1}$$

Equivalent Intervalley Scattering

$$W = \frac{(Z_e - 1)(m^*)^{3/2} D_{ij}^2}{\sqrt{2}\pi\rho \, \omega_{ij}\hbar^3} E_f^{1/2} \begin{cases} N & \text{absorption} \\ N+1 & \text{emission} \end{cases}$$

$$= 1.71 \times 10^{10} \frac{(Z_e - 1)\left(\frac{m^*}{m_0}\right)^{3/2} \left(\frac{D_{ij}}{\text{eV cm}^{-1}}\right)^2}{\left(\frac{\rho}{\text{gm/cm}^3}\right)\left(\frac{\omega_{ij}}{\text{rad/s}}\right)} \left(\frac{E_f}{\text{eV}}\right)^{1/2}$$

$$\times \begin{cases} N & \text{absorption} \\ N+1 & \text{emission} \end{cases} \text{s}^{-1}$$

Example: GaAs, $T = 300$ K, L valley, equivalent intervalley scattering by absorption

$$E_i = 0 \text{ eV (at L valley edge; 0.36 eV above } \Gamma \text{ valley edge).}$$

$$Z_e = 4$$

$$m^* = 0.35 m_0$$

$$D_{ij} = 1.0 \times 10^9 \text{ eV cm}^{-1}$$

$$\rho = 5.37 \text{ gm/cm}^{-3}$$

$$\omega_{ij} = 4.56 \times 10^{13} \text{ rad/s}$$

$$E_f = E_i + \hbar\omega_{ij} = 0.030 \text{ eV}$$

$$N = \frac{1}{\exp(\hbar\omega_{ij}/k_B T) - 1} = 0.457$$

$$W = 1.71 \times 10^{10} \frac{(Z_e - 1)\left(\frac{m^*}{m_0}\right)^{3/2}\left(\frac{D_{ij}}{\text{eV cm}^{-1}}\right)^2}{\left(\frac{\rho}{\text{gm/cm}^3}\right)\left(\frac{\omega_{ij}}{\text{rad/s}}\right)}\left(\frac{E_f}{\text{eV}}\right)^{1/2} N \text{ s}^{-1}$$

$$= 1.71 \times 10^{10} \frac{(3)(0.35)(1 \times 10^9)^2}{(5.37)(4.56 \times 10^{13})}(0.030)(0.457) \text{ s}^{-1}$$

$$= 1.01 \times 10^{12} \text{ s}^{-1}$$

Non-Equivalent Intervalley Scattering

$$W = \frac{Z_j(m_j^*)^{3/2}(D_{ij})^2}{\sqrt{2}\pi\rho\,\omega_{ij}\hbar^3}\gamma_j^{1/2}(E_f)\,(1 + 2\alpha_f E_f)\,G_{ij}(E_i, E_f)$$

$$\times \begin{cases} N & \text{absorption} \\ N+1 & \text{emission} \end{cases}$$

$$= 1.71 \times 10^{10} \frac{Z_j\left(\frac{m_j^*}{m_0}\right)^{3/2}\left(\frac{D_{ij}}{\text{eV cm}^{-1}}\right)^2}{\left(\frac{\rho}{\text{gm/cm}^3}\right)\left(\frac{\omega_{ij}}{\text{rad/s}}\right)}\left(\frac{\gamma_j E_f}{\text{eV}}\right)^{1/2}$$

$$\times \; (1 + 2\alpha_f E_f)\,G_{ij}(E_i, E_f) \begin{cases} N & \text{absorption} \\ N+1 & \text{emission} \end{cases} \text{ s}^{-1}$$

with G_{ij} given in Eqn 7.41.

Example: GaAs, $T = 300$ K, Γ valley, non-equivalent intervalley scattering by absorption,

$$\begin{aligned}
E_i &= E_L - E_\Gamma = 0.36 \text{ eV} = 13.9 k_B T \\
Z_j &= 4 \\
m_j^* &= 0.35 m_0 \\
D_{ij} &= 1.0 \times 10^9 \text{ eV cm}^{-1} \\
\rho &= 5.37 \text{ gm/cm}^3 \\
\omega_{ij} &= 4.56 \times 10^{13} \text{ rad/s} \\
N &= 0.457 \\
\alpha_i &= 0.576 \text{ eV}^{-1} \text{ (Γ valley is nonparabolic)} \\
\alpha_f &= 0 \text{ eV}^{-1} \text{ (L valley is parabolic)} \\
E_f &= E_i - (E_L - E_\Gamma) + \hbar\omega_{ij} = \hbar\omega_{ij} = 0.030 \text{ eV} \\
\gamma_i(E_f) &= E_f = 0.030 \text{ eV} \\
1 + \alpha_i E_i &= 1.21 \\
1 + \alpha_f E_f &= 1 \\
1 + 2\alpha_i E_i &= 1.41 \\
1 + 2\alpha_f E_f &= 1
\end{aligned}$$

$$W = 1.71 \times 10^{10} \frac{Z_j\left(\frac{m_j^*}{m_0}\right)^{3/2}\left(\frac{D_{ij}}{\text{eV cm}^{-1}}\right)^2}{\left(\frac{\rho}{\text{gm/cm}^3}\right)\left(\frac{\omega_{ij}}{\text{rad/s}}\right)}\sqrt{\frac{\gamma_j(E_f)}{\text{eV}}}$$

$$\times \quad (1 + 2\alpha_f E_f) \, G_{ij}(E_i, E_f) N \text{ s}^{-1}$$

$$= \quad 1.71 \times 10^{10} \frac{4(0.35)(1 \times 10^9)^2}{(5.37)(4.56 \times 10^{13})} \sqrt{0.030}$$

$$\times \quad (1)(0.457) \, G_{ij}(E_i, E_f) \text{ s}^{-1}$$

$$= \quad 7.74 \times 10^{12} \, G_{ij}(E_i, E_f) \text{ s}^{-1}$$

$$G_{ij}(E_i, E_f) \quad = \quad \frac{(1 + \alpha_i E_i)(1 + \alpha_f E_f)}{(1 + 2\alpha_i E_i)(1 + 2\alpha_f E_f)}$$

$$= \quad \frac{(1.21)(1)}{(1.41)(1)} = 0.858$$

$$W \quad = \quad 7.74 \times 10^{12}(0.858) \text{ s}^{-1} = 6.64 \times 10^{12} \text{ s}^{-1}$$

Ionized Impurity Scattering

$$W \quad = \quad 4\pi F \left(\frac{2k}{\lambda}\right)^2 \left[\frac{1}{1 + \left(\frac{\lambda}{2k}\right)^2}\right]^2$$

$$F \quad = \quad \frac{1}{\hbar} \left[\frac{Ze^2}{\epsilon_s}\right]^2 \frac{N(E_k)}{32k^4} N_I$$

where the density of states $N(E_k)$ is (without spin degeneracy)

$$N(E_k) \quad = \quad \frac{(m^*)^{3/2}}{\sqrt{2}\pi^2\hbar^3} \gamma^{1/2}(E_k)(1 + 2\alpha E_k)$$

$$F \quad = \quad 5.34 \times 10^{25} \frac{Z^2 \left(\frac{N_I}{\text{m}^{-3}}\right) \left(\frac{m^*}{m_0}\right)^{3/2}}{(\epsilon_s/\epsilon_0)^2 \left(\frac{k}{\text{m}^{-1}}\right)^4} \sqrt{\frac{\gamma(E_k)}{\text{eV}}} (1 + 2\alpha E_k) \text{ s}^{-1}$$

Using

$$\left(\frac{k}{\text{m}^{-1}}\right)^4 = 6.89 \times 10^{38} \left(\frac{m^*}{m_0}\right)^2 \left(\frac{\gamma(E_k)}{\text{eV}}\right)^2$$

$$F \quad = \quad 7.62 \times 10^{-14} \frac{Z^2 \left(\frac{N_I}{\text{m}^{-3}}\right)}{(\epsilon_s/\epsilon_0)^2_s \sqrt{\frac{m^*}{m_0}} \left(\frac{\gamma(E_k)}{\text{eV}}\right)^{3/2}} \frac{1 + 2\alpha E_k}{} \text{ s}^{-1}$$

$$\lambda^2 \quad = \quad \frac{(n + p)e^2}{\epsilon_s k_B T}$$

$$= \quad \frac{2.1 \times 10^{-4} \left(\frac{n+p}{\text{m}^{-3}}\right)}{(\epsilon_s/\epsilon_0)\left(\frac{T}{\text{K}}\right)} \text{ m}^{-2}$$

Example: GaAs, $T = 300$ K, Γ valley, ionized impurity scattering

$$E_i \quad = \quad E_f = \frac{3}{2}k_B T = 0.039 \text{ eV} = E_k$$

$$n + p \approx N_I \quad = \quad 1 \times 10^{17} \text{ cm}^{-3} = 10^{23} \text{ m}^{-3}$$

$$
\begin{aligned}
Z &= 1 \,(\text{singly ionized impurity}) \\
m^* &= 0.067 m_0 \\
\epsilon_s/\epsilon_0 &= 13.2 \\
\alpha &= 0.576 \text{ eV}^{-1} \\
\gamma(E_k) &= 0.040 \text{ eV} \\
1 + 2\alpha E_k &= 1.045 \\
\lambda &= 7.28 \times 10^7 \text{ m}^{-1} \\
k &= 5.12 \times 10^9 \left(\frac{m^*}{m_0}\right)^{1/2} \left(\frac{\gamma(E_k)}{\text{eV}}\right)^{1/2} \text{ m}^{-1} \\
&= 2.65 \times 10^8 \text{ m}^{-1} \\
\frac{2k}{\lambda} &= 7.29 \\
W &= 4\pi F \left(\frac{2k}{\lambda}\right)^2 \left[\frac{1}{1 + \left(\frac{\lambda}{2k}\right)^2}\right]^2 = 1.46 \times 10^{13} \text{s}^{-1}
\end{aligned}
$$

Alloy Scattering

$$
W = \frac{3\pi^3}{8\hbar} V_0 \, U_{\text{all}}^2 \, N(E_k) x (1-x)
$$

For zinc-blende and diamond lattices, $V_0 = a_0^3/4$, where a_0 is the lattice constant (e.g., $a_0 = 5.6533 \times 10^{-8}$ cm for GaAs).

$$
\begin{aligned}
N(E_k) &= \frac{(m^*)^{3/2}}{\sqrt{2}\pi^2\hbar^3} \gamma^{1/2}(E_k)(1 + 2\alpha E_k) \\
W &= 1.5 \times 10^{13} \left(\frac{m^*}{m_0}\right)^{3/2} \left(\frac{U_{\text{all}}}{\text{eV}}\right)^2 \left(\frac{a_0}{\text{Å}}\right)^3 x(1-x) \\
&\quad \times \sqrt{\frac{\gamma(E)}{\text{eV}}} (1 + 2\alpha E) \text{ s}^{-1}
\end{aligned}
$$

Example: $Al_{0.1}Ga_{0.9}As$, $T = 300$ K, Γ valley, alloy scattering,

$$
\begin{aligned}
E_i &= E_f = \frac{3}{2}k_B T = 0.039 \text{ eV} \\
m^* &= 0.067 \, m_0 \\
U_{\text{all}} &= 0.2 \text{ eV} \\
a_0 &= 5.65 \text{ Å} \\
x &= 0.1 \\
\gamma(E) &= 0.040 \text{ eV} \\
1 + 2\alpha E &= 1.045 \\
W &= 1.5 \times 10^{13} \left(\frac{m^*}{m_0}\right)^{3/2} \left(\frac{U_{\text{all}}}{\text{eV}}\right)^2 \left(\frac{a_0}{\text{Å}}\right)^3 x(1-x) \\
&\quad \times \sqrt{\frac{\gamma(E)}{\text{eV}}} (1 + 2\alpha E) \text{ s}^{-1}
\end{aligned}
$$

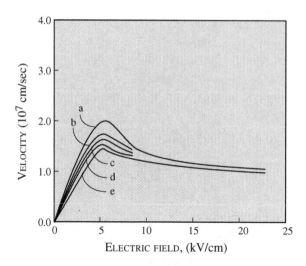

Figure 7.11: Electron velocity as a function of electric field in a GaAs. The impurity density for the different curves is (a) $N_D = 0$; (b) $N_D = 1.0 \times 10^{17}$ cm^{-3}; (c) $N_D = 2.0 \times 10^{17}$ cm^{-3}; (d) $N_D = 4.0 \times 10^{17}$ cm^{-3}; (e) $N_D = 8.0 \times 10^{17}$ cm^{-3}.

$$
\begin{aligned}
&= \quad 1.5 \times 10^{13}(0.067)^{3/2}(0.2)^2(5.65)^3(0.1)(0.9)\sqrt{0.040}(1.045) \text{ s}^{-1} \\
&= \quad 3.5 \times 10^{10} \text{ s}^{-1}
\end{aligned}
$$

It is important to calculate the relative rates of various scattering mechanisms for the parameters used. However, one must note that angular dependence of the scattering event is also very important.

7.3 STEADY STATE AND TRANSIENT TRANSPORT

7.3.1 GaAs, Steady State

As a semiconductor widely used in high speed electronic devices, transport in GaAs has been extensively studied by Monte Carlo methods. At low electric fields the electron moves in the high mobility, low mass ($m^*/m_0 = 0.067$) Γ-valley and has an excellent v– **F** relationship with a room temperature mobility ~ 8000 cm^2/V-s, for pure GaAs. The peak velocity in pure GaAs is $\sim 2 \times 10^7$ cm/s at room temperature, at an electric field of ~ 3.5 kV/cm. Up to this field, most of the electrons are in the Γ-valley as their energy is less than the Γ-L energy separation. However, at higher electric fields there is a transfer of electrons from the Γ to L valley where the electron mass is very large ($m^*/m_0 \sim 0.22$). This causes the negative differential resistance which is a special feature of most direct bandgap semiconductors. The negative differential resistance is exploited to produce charge domains resulting from instabilities and leading to Gunn oscillations.

In Fig. 7.11 we show the doping dependence of the v- **F** relations in GaAs at room temperature. It is interesting to note that while the low field mobility falls as rapidly as the doping is increased, the high field velocity does not show such a

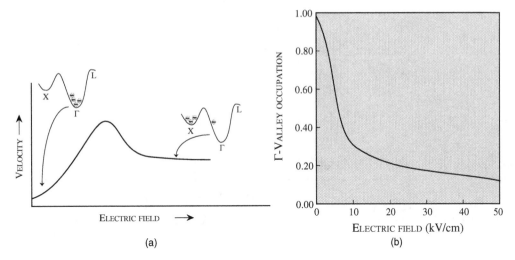

Figure 7.12: (a) A schematic of how electrons transfer from the Γ-valley to the X-valley in GaAs as the field is increased. (b) Occupation of the Γ-valley in GaAs electron transport as a function of electric field.

pronounced dependence. This is because the low field mobility electron energy is small and ionized impurity scattering is quite important. The high field transport, on the other hand, is dominated by polar optical phonon emission and intervalley scattering and since carriers have high energies ionized impurity is not as important.

In Fig. 7.12 we show the Γ valley occupation fraction for the electrons as a function of the electric field. As we can see at high fields the electrons transfer out of the Γ-valley. Also shown are the electron temperatures as a function of the electric field.

While the Monte Carlo calculations are computer intensive, they can be used to develop simple analytical descriptions for the v–$\mathbf{F}$ curves which could be used for certain device simulations. For example, one can use the following function for the field dependence of the mobility

$$\mu(\mathbf{F}) = \frac{\mu^0}{\left[1 + \theta(\mathbf{F} - \mathbf{F}_0)\left\{(\mathbf{F} - \mathbf{F}_0)/\mathbf{F}_{\mathrm{cr}}^2\right\}^2\right]} \tag{7.42}$$

where μ^0 is the low field mobility, $\mathbf{F}_{\mathrm{cr}} = v_{\mathrm{sat}}/\mu^0$ and

$$\mathbf{F}_0 = \frac{1}{2}\left[\mathbf{F}_{\mathrm{th}} + \sqrt{\mathbf{F}_{\mathrm{th}}^2 - 4\mathbf{F}_{\mathrm{cr}}^2}\right]$$

where $\mathbf{F}_{\mathrm{th}}$ is the electric field where the velocity peaks. The function $\theta(\mathbf{F} - \mathbf{F}_0)$ is a step function which is zero for fields below $\mathbf{F}_0$ and 1 otherwise. Similarly, the saturation velocity has a temperature dependence given by (Jacoboni, et al., 1977)

$$v_{\mathrm{sat}} = \frac{2.4 \times 10^7}{1 + 0.8\exp(T_L/600)} \ \mathrm{cm/s} \tag{7.43}$$

where T_L is the lattice temperature in K.

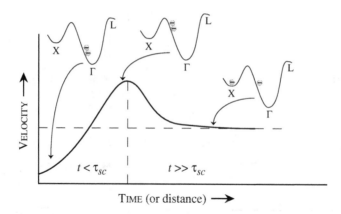

Figure 7.13: A schematic of how velocity overshoot occurs in a direct bandgap material like GaAs.

7.3.2 GaAs, Transient Behavior

The steady state results discussed above are valid when the electrons have sufficient time to suffer several scattering events to reach a steady state distribution. This usually requires a few picoseconds or a transit distance of ≥ 1000 Å. Transient transport is important in all semiconductors during times shorter than the scattering times. For electrons in direct gap semiconductors the electron mass is small in the Γ-valley and large in the upper valleys. In these materials transient behavior can result in very interesting effects which are commonly known as velocity overshoot and velocity undershoot. To observe these effects it is necessary that the electric fields encountered by the injected electrons are quite large.

To understand the overshoot effect consider Fig. 7.13. When electrons are injected into a region of high electric field, for a short time ($t < \tau_{sc}$) the electrons move without scattering or ballistically. During this time the electron velocity is given by

$$v(t) = v(0) + \frac{1}{2}\frac{eF}{m^*}t^2 \qquad (7.44)$$

The electrons stay in the Γ-valley (if they start out there) even though their energy exceeds the intervalley separation. This allows electrons to have velocities well above the steady state value.

Once the scattering processes start (e.g., intervalley scattering) the electron velocity begins to drop and eventually reaches the steady state value.

If electrons are initially injected into an upper valley where the mass is high into a region with low electric field, it is possible for electrons to have velocities smaller than the steady state values for a short time. This leads to velocity undershoot.

In InGaAs and GaAs based devices velocity overshoot effect is very important, especially as device dimensions reach $< 0.1 \ \mu$m.

In Fig. 7.14 we show some typical results for this "velocity overshoot" at various electric fields. It is important to note that the velocity overshoot persists up to a few picoseconds which is the transit time in many modern submicron GaAs devices. The

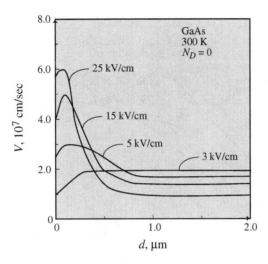

Figure 7.14: A schematic of how velocity overshoot occurs in a direct bandgap material like GaAs.

Monte Carlo method is ideally suited to understanding the nonstationary transport in the transient regime. However, since it is usually very computer intensive, especially for device simulations, key parameters (e.g., average momentum and energy relaxation time, carrier temperature, etc.) are extracted from Monte Carlo methods and then used in simpler numerical techniques. We will discuss some of these aspects later on.

The results described for GaAs are typical of electron transport in most direct bandgap semiconductors. A material system that has become very important for high frequency microwave devices is $In_{0.53}Ga_{0.47}As$ which is lattice matched to InP substrates. This material has a very small carrier mass ($m^*/m_0 = 0.04$) and a large intervalley separation ($\Delta E_{L-X} \sim 0.55$ eV) which gives it a superior v–$\mathbf{F}$ relation than that of GaAs. Also, the velocity overshoot effects persist up to longer transit times and are also much stronger than in GaAs.

7.3.3   High Field Electron Transport in Si

Silicon is by far the most important semiconductor in electronics. Both electron transport and hole transport are important for MOSFET and bipolar junction transistor technology. Unlike GaAs, Si is an indirect gap material and electrons have to sample 6 equivalent valleys as they move in an electric field. The high density of states and the strong g and f scattering causes the velocity to be lower than that in direct gap semiconductors. There is no negative differential resistance region, since there is no low mass–high mass transition in the energy band.

The mobility of electrons in silicon is a factor of 5 smaller than that in GaAs. This is an important reason why GaAs devices are superior. However, as devices become smaller, electrons move under high field conditions. At high fields the velocity of electrons in GaAs and in Si are similar, as can be seen in Fig. 7.15. As a result, Si devices

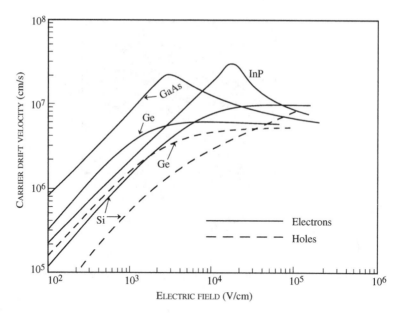

Figure 7.15: Velocity-field curves for several semiconductors.

perform quite well and are not a factor 5 slower than GaAs devices.

Silicon does not show transient transport phenomena, except in very short channels (approaching a few hundred Angstroms).

In Fig. 7.15 we show the velocity field relations for several materials. It is interesting to note that while the low field velocities are quite different for electrons in different materials, at high fields velocities tend to saturate (in all materials) to a value of $\sim 10^7$ cm/s. At very high fields electrons move at high energies where the density of states are quite similar in all materials.

7.4 BALANCE EQUATION APPROACH TO HIGH FIELD TRANSPORT

The Monte Carlo approach provides the most versatile technique for transport studies at high fields not only for bulk semiconductors, but also for heterostructures as will be discussed later in this chapter. However, due to the computer intensive character of the Monte Carlo method, other numerical tools have been developed to study high field transport. These techniques are particularly useful for device simulations where the necessity of self-consistency forces one to study transport iteratively which requires numerous field dependent transport studies. In Chapter 4, we discussed some of the approximate methods to solve the Boltzmann transport equation. In particular we developed the balance equations for carrier density, carrier momentum, and carrier energy. These equations are extremely useful in providing a simple description of the high field transport in semiconductors.

A widely used approach for transport studies is dependent on the momentum and energy balance equations written for the displaced Maxwellian carrier distribution.

The carrier distribution in turn is described by the carrier temperature as discussed in Chapter 4. We will briefly recall some of those results and describe the approach used for high field transport. The displaced Maxwellian distribution function has the general form

$$f \approx e^{-|\mathbf{p}-m^*\mathbf{v}_d|^2/(2m^*k_BT_C)} \tag{7.45}$$

where T_C is the carrier temperature. In Chapter 4, we showed that the energy argument of Eqn. 7.45 can be expanded to give

$$
\begin{aligned}
f &= e^{-(p^2 - 2m^*\mathbf{p}\cdot\mathbf{v}_d + m^{*2}v_d^2)/(k_BT_C)} \\
&\approx \exp\left\{-\left[p^2/(2m^*k_BT_C)\right]\right\}\left[1 + \frac{\boldsymbol{p}\cdot\boldsymbol{v}_d}{k_BT_C}\right]
\end{aligned}
\tag{7.46}
$$

This simplification is valid if the drift part of the energy, $m^*v_d^2/2$, is small compared to the total energy of the electron distribution. This is usually true since $v_d \sim 10^7$ cm s^{-1}, i.e., $m^*v_d^2/2$ is of the order of 20–30 meV while the average energy of the electron is approximately 200–300 meV at high fields. The kinetic energy is given essentially by the carrier temperature

$$W_{zz} = \frac{1}{2}k_BT_C \tag{7.47}$$

The diffusion coefficient also takes on a simple form

$$D = \frac{k_BT_C}{e}\mu \tag{7.48}$$

The approach based on the drifted Maxwellian assumptions are called the electron temperature approach since the unknown is the electron temperature. The balance equations for energy and momentum are used to solve the problem. The velocity under steady state conditions with uniform electric field is

$$v_z = \frac{e}{m^* \ll 1/\tau_m \gg} F_z$$

The energy balance equation then gives for steady state conditions

$$\frac{J_zF_z}{n} = \left\langle\left\langle\frac{1}{\tau_E}\right\rangle\right\rangle\left(\frac{3}{2}k_BT_C - \frac{3}{2}k_BT_L\right) \tag{7.49}$$

which gives us for the carrier temperature

$$\frac{T_C}{T_L} = 1 + \frac{2e^2}{3k_BT_Lm^* \ll 1/\tau_m \gg \ll 1/\tau_E \gg}F_z^2 \tag{7.50}$$

To solve for the drift velocity and the carrier temperature, one has to evaluate the averaged relaxation times which further requires the knowledge of the distribution function. In the displaced Maxwellian approach by ignoring the drift part, the relaxation times can be calculated in terms of the carrier temperature. For example, we showed that if a particular scattering process has a energy dependence given through the scattering time of

$$\tau = \tau_0 \left(\frac{E}{k_B T_C} \right)^{\mathbf{s}}$$

then we have

$$\left\langle\left\langle \frac{1}{\tau_m} \right\rangle\right\rangle = \frac{1}{\tau_0} \frac{\Gamma(5/2)}{\Gamma(s + 5/2)}$$

A similar solution can be obtained for the energy relaxation time. Iterative self-consistent approaches can then be developed to solve the balance equations. Usually one has to compare the results of the balance equation with the Monte Carlo results to make sure that the assumptions made are valid.

Let us examine the dependence of mobility of carriers on carrier temperature and electric field in the electron temperature approach. While this approach is not as accurate as the Monte Carlo approach, it provides good semiquantitative fits. Consider the case where the transport is limited by acoustic phonon scattering. The scattering time is given by

$$
\begin{aligned}
\tau_m &= \frac{\pi \hbar^4 c_\ell}{\sqrt{2}\,(m^*)^{3/2}\,D^2} \frac{E^{-1/2}}{k_B T_L} \\
&= \frac{\pi \hbar^4 c_\ell}{\sqrt{2}\,(m^*)^{3/2}\,D^2} \frac{1}{(k_B T_L)^{3/2}} \left(\frac{T_L}{T_C} \right)^{1/2} \left[\frac{E}{k_B T_C} \right]^{-1/2} \\
&= \tau_0(T_C) \left[\frac{E}{k_B T_C} \right]^{-1/2}
\end{aligned}
\tag{7.51}
$$

The constant $\tau_0(T_C)$ is given by

$$\tau_0(T_C) = \tau_0(T_L)\sqrt{\frac{T_L}{T_C}} \tag{7.52}$$

We then have

$$\left\langle\left\langle \frac{1}{\tau_m} \right\rangle\right\rangle = \left\langle\left\langle \frac{1}{\tau_m^0} \right\rangle\right\rangle \sqrt{\frac{T_C}{T_L}}$$

or

$$\mu(T_C) = \mu_n^0 \sqrt{\frac{T_L}{T_C}} \tag{7.53}$$

The mobility thus decreases as the carriers get hotter.

In the case of mobility limited by ionized impurity scattering, the energy dependence is $E^{3/2}$ for the scattering time. The use of the electron temperature approach then gives

$$\mu_n(T_C) = \mu_n^0 \left[\frac{T_C}{T_L} \right]^{3/2} \tag{7.54}$$

and the mobility increases with carrier temperature. It must be remembered that these expressions are valid in case the density of states does not alter with increased energy

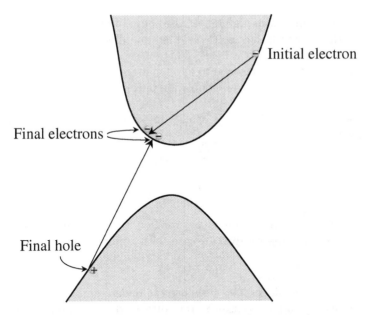

Figure 7.16: The impact ionization process where a high energy electron scatters from a valence band electron producing two conduction band electrons and a hole. Hot holes can undergo a similar process.

in a manner other than $E^{1/2}$. In the case of III–V direct bandgap materials, we know that the presence of the satellite valleys can abruptly alter the density of states so that the equations would be invalid once the electrons transfer to the upper valleys.

7.5   IMPACT IONIZATION IN SEMICONDUTORS

In Chapter 5 we have discussed how "hot" carriers with energies larger than the semi-conductor bandgap can cause impact ionization. Impact ionization becomes important at high electric fields (typically, $F > 10^5$ V/cm) and is usually responsible for break-down in semiconductor devices. As shown in Fig. 7.16, in this process hot electrons (holes) scatter from holes (electrons) via the Coulombic interaction to excite the hole (electron) from the valence band (construction band) to the conduction band (valence band). The number of excess free carriers thus increases (called carrier multiplication) causing a runaway current.

The impact ionization process can be included in a Monte Carlo approach just as another scattering mechanism. While this is simple in principle, it is somewhat tedious in practice. The reason for the difficulty is that the energies required for the impact ionization to start is quite high as shown in Chapter 5. At these high energies the bandstructure is no longer described by simple analytical methods. Also the collision rates are so high at such high carrier energies that the use of the Fermi golden rule becomes questionable. The δ-function in the golden rule arises from the assumption that the perturbation exists for a long time. One has to relax the energy conservation rule for very short duration collisions at high rates. Also the assumption that the collision time is essentially zero during which the electric field has no influence on the electrons, becomes questionable at such high electric fields. This effect (intracollisional field effect) also causes the simple Monte Carlo method to develop inaccuracies. Nevertheless, reasonable results which shed light on the impact ionization phenomenon can be obtained by Monte Carlo methods.

For device applications, one is usually interested in the impact ionization coefficient defined by the equation

$$\frac{\partial I}{\partial t} = \alpha_t I$$

or

$$\frac{\partial I}{\partial Z} = \alpha_z I \tag{7.55}$$

where I is the current and α represents an average rate of ionization per unit time (α_t) or unit distance (α_z). For the steady state case, with uniform time and space dependence

$$\alpha_z = \alpha_\tau / v_d \tag{7.56}$$

Monte Carlo techniques have been used to understand impact ionization breakdown by including the impact ionization rate discussed in Chapter 5. Such techniques have been applied to electrons and holes in several semiconductors and give reasonable agreement with experiments. Impact ionization results from electrons (holes) that are at the very high energy tail of the distribution function. This is because of the threshold energy needed to start the ionization process. In Fig. 7.17 we show a schematic of the carrier distribution function for different electric fields. At no applied field we simply have the Fermi Dirac distribution function $f^0(E)$. As the field is increased the distribution function becomes more like a "displaced Maxwellian" and eventually at very high fields there is a small density of carriers that have energies larger than the threshold energy needed for impact ionization. Monte Carlo methods used to describe such high energy carriers need to accurately account for the bandstructure. Simple effective mass approaches don't give very accurate results.

Several numerical or semi-analytical approaches exist to understand impact ionization. These use the energy balance approach to estimate the carriers that can cause impact ionization. The interested reader can examine the publications listed at the end of this chapter to understand these approaches.

In Fig. 7.18 we show the impact ionization coefficients for electrons (α_{imp}) and holes (β_{imp}) for some semiconductors. Note that, as expected, the smaller the bandgap, the larger the impact ionization coefficient. A commonly used parameter to represent impact ionization is via the "critical field." At the critical field α (or β) reaches a value of $\sim 10^4$ cm^{-1} so that over a 1 μm semiconductor film the probability of breakdown is near unity. In Table 7.3 we show critical fields for some semiconductors.

7.6 TRANSPORT IN QUANTUM WELLS

In Chapter 3 we have discussed how semiconductors can be combined to make quantum wells. Most high performance semiconductor electronic devices are based on heterostructures and electrons (holes) essentially move in 2-dimensional space. Two important

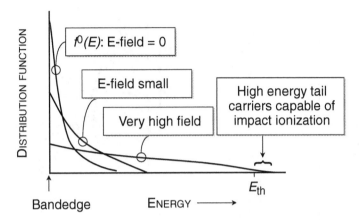

Figure 7.17: A schematic of how the carrier distribution function change as field is increased. The high energy tail responsible for impact ionization is shown.

MATERIAL	BANDGAP (eV)	BREAKDOWN ELECTRIC FIELD (V/cm)
GaAs	1.43	4×10^5
Ge	0.664	10^5
InP	1.34	
Si	1.1	3×10^5
$In_{0.53}Ga_{0.47}As$	0.8	2×10^5
C	5.5	10^7
SiC	2.9	$2\text{-}3 \times 10^6$
SiO_2	9	10^7
Si_3N_4	5	10^7

Table 7.3: Breakdown electric fields in some materials.

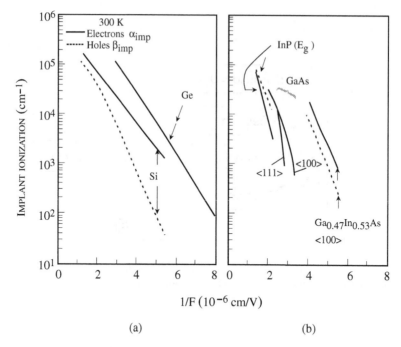

Figure 7.18: Ionization rates at 300 K versus reciprocal electric field for some semiconductors. (Adapted from S.M. Sze, *Physics of Semiconductor Devices*, Wiley, New York (1981).)

classes of devices are the metal-oxide-semiconductor field effect transistor (MOSFET) and modulation doped field effect transistor (MODFET). In Figs. 7.19 and 7.20 we show schematics of the cross-section and band profile of the two devices. The MOSFET is based on silicon technology (the only semiconductor on which a high quality insulator, SiO_2, can be grown), while the MODFET is based on compound semiconductors such as AlGaAs/GaAs, InP/InGaAs, AlGaN/GaN, etc.

In the MOSFET, the most important electronic device, a gate bias, is used to "invert" the silicon bands and create a triangular quantum well where electrons reside. In the MODFET also a triangular quantum well is formed as shown. By modulation doping and using an undoped spacer layer between the electrons and the dopants, ionized impurity scattering is effectively eliminated.

Transport in 2-dimensional systems can be studied by the Monte Carlo method in a manner similar to that discussed for the 3-dimensional system. The key differences are:

- The wavefunctions have the form

$$\psi_k(\rho, z) = u_k g(z) e^{ik \cdot \rho} \tag{7.57}$$

where u_k is the central cell part, $g(z)$ is the envelope function in the 2-dimensional potential defining the quantum well, ρ is a vector in the 2-dimensional plane of transport. Thus in calculating the matrix elements for scattering and scattering rates one has to use the proper envelope functions. One has to evaluate the integral

$$\int g_i(z)g_j(z)dz$$

when one scatters from a state i to state j. This is usually done numerically.

• The density of states describing the final states into which scattering occurs is altered. One has to be careful to include intrasubband scattering and intersubband scattering. For intersubband scattering it is important to note that

$$\int g_i(z)g_j(z) = 0$$

However, if the scattering potential has z-dependence, the scattering matrix element

$$\int g_i(z)V(z)g_j(z)$$

need not be zero.

• Scattering mechanisms absent in 3-dimensional systems can become important in 2-dimensional systems. For example, interface roughness scattering can play an important role in 2-dimensional transport. For example, in bulk pure silicon room temperature electron mobility is ~ 1100 cm^2/V·s. However, in n-type MOSFET the mobility is ~ 600 cm^2/V·s due to the importance of interface roughness scattering.

Once the scattering rates are calculated transport can be studied by the Monte Carlo method. The method proceeds in a manner similar to what has been described for the 3-dimensional problem. Because the envelope functions and potential profile have to be often determined numerically, the scattering rates can usually not be expressed analytically.

The acoustic phonon scattering in a two dimensional system from state i to j is

$$W_{ij} = \frac{m^* k_B T D^2}{\hbar^3 \rho s_l^2} \left| \int g_i(z)g_j(z)dz \right|^2 \tag{7.58}$$

where ρ is the density of the material, s_l is the longitudinal sound velocity and D is the acoustic phonon deformation potential. The envelope function is $g(z)$.

The polar optical phonon scattering is given by

$$W_{ij} = \frac{eE_0}{2\hbar} \left[N(\omega_0) + \frac{1}{2} \pm \frac{1}{2} \right] \int_0^{2\pi} \frac{H_{ij}(Q_\pm)}{Q_\pm} d\theta \tag{7.59}$$

with

$$E_0 = \frac{m^* e \omega_0}{\hbar} \left[\frac{1}{\epsilon_\infty} - \frac{1}{\epsilon_s} \right]$$

$$N(\omega_0) = \frac{1}{\exp\left(\frac{\hbar\omega_0}{k_B T}\right) - 1}$$

$$H_{ij} = \int_{-\infty}^{\infty} dz \int_{-\infty}^{\infty} dz' \rho_{ij}(z)\rho_{ij}^*(z') \exp\left(-Q|z - z'|\right)$$

$$\rho_{ij}(z) = g_i^* g_j$$

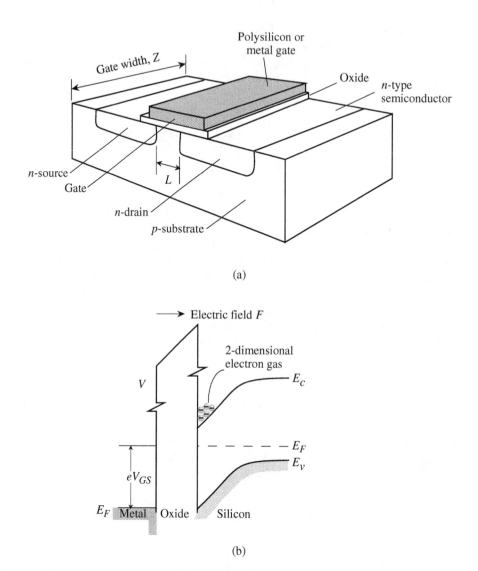

(a)

(b)

Figure 7.19: (a) A schematic of a MOSFET with gate length L. (b) By applying a gate bias the semiconductor bands can be "invented," as shown, inducing electrons in the traingular quantum well.

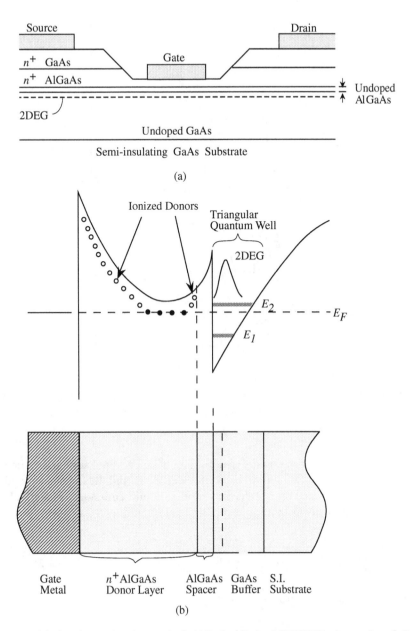

(a)

(b)

Figure 7.20: (a) A schematic of a typical AlGaAs/GaAs MEDFET. An undoped "spacer" layer ($\sim 30 - 50$ Å) is used to minimize ionized impurity scattering. (b) The conduction band profile showing a triangular quantum well in which a 2-dimensional electrons gas resides. Under applied bias the electrons move from the source to the drain.

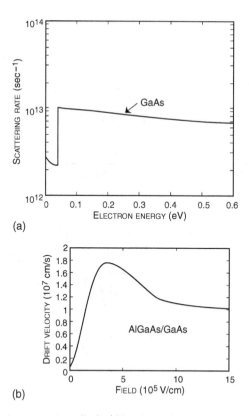

(a)

(b)

Figure 7.21: (a) Scattering rates in a GaAs/Al$_{0.3}$Ga$_{0.7}$As channel as a function of energy. (b) Velocity-field curves calculated in the channel. Electron density is 2×10^{12} cm^{-2}.

Here m^* is the effective mass, i and j denote the initial and final subband indices, ϵ_s and ϵ_∞ are the low and high-frequency dielectric constants, and $\hbar\omega_0$ is the optical phonon energy. The $+$ and $-$ signs on Q refer to absorption and emission, respectively, for the scattered wave vector given by

$$
Q = |\mathbf{k} - \mathbf{k}'| = \left[2k^2 \pm \frac{2\omega_{ij}m^*}{\hbar} - 2k \left(k^2 \pm \frac{2\omega_{ij}m^*}{\hbar} \right)^{1/2} \cos(\theta) \right]^{1/2}
$$

$$
\hbar\omega_{ij} = \hbar\omega_0 \pm (E_i - E_j) \tag{7.60}
$$

Form factor H_{ij} is evaluated numerically using the envelope functions g.

As an example of Monte Carlo results for transport in a 2D system in Fig. 7.21 we show results for an AlGaAs/GaAs quantum well with an electron density of 2×10^{12} cm^{-2}. It is interesting to note that the velocity-field results are not very different from those in pure GaAs, especially at high electric fields.

7.7 TRANSPORT IN QUANTUM WIRES AND DOTS

In the previous section we have discussed how ionized impurity scattering, important at low field and low temperatures, can be essentially eliminated in 2-dimensional systems. There are some other differences between 3-dimensional transport and 2-dimensional transport which are due to the differences in the density of states. However, there are no major qualitative differences, since in 3D and 2D the density of states are continuous and increase as one moves away from the banedges into the allowed band. In sub-2-dimensional systems the density of states shape is qualitatively different from 2- or 3-dimensional systems. It is reasonable to expect then that scattering rates and transport will be dramatically different in quantum wires. Also scattering rates in quantum dots will be quite different due to the unique density of states (a series of δ-functions).

There has been considerable interest in pursuing research in quasi-1-D structures due to the nature of the eigenstates, eigenenergies, and density of states. In a quasi-1-D wire, the electron function is confined in two dimensions and is "free" only in one direction. For a simple rectangular wire of dimensions L_x and L_y, the wavefunction is of the form

$$\psi_{\mathbf{k}} = \frac{1}{\sqrt{L_z}} \, g_n(x) \, g_\ell(y) \, e^{ik_z z} \tag{7.61}$$

where the envelope functions for an infinite barrier case have the usual form

$$g_n(x) = \sqrt{\frac{2}{L_x}} \, \cos\left(\frac{n\pi x}{L_x}\right), \text{for } n \text{ odd}$$

$$g_n(x) = \sqrt{\frac{2}{L_x}} \, \sin\left(\frac{n\pi x}{L_x}\right), \text{for } n \text{ even} \tag{7.62}$$

where the well extends from $x = -L_x/2$ to $x = +L_x/2$. Similar eigenfunctions pertain for $g_\ell(y)$.

The electron energies are

$$E_{n,\ell} = \frac{\hbar^2}{2m^*} \left\{ \left(\frac{n\pi}{L_x}\right)^2 + \left(\frac{\ell\pi}{L_y}\right)^2 \right\} + \frac{\hbar^2 k_z^2}{2m^*} \tag{7.63}$$

Such a dispersion relation produces a density of states of the form

$$N(E) = \sum_n \frac{\sqrt{2}m^{*1/2}}{\pi\hbar} (E - E_{n\ell})^{-1/2} \tag{7.64}$$

where the $E_{n\ell}$ are the various subband levels. An important effect of this quantization is that once L_x and L_y approach 100 Å or less, there is enough inter-subband separation (≥ 50 meV) that there is no inter-subband scattering. The scattering in the same subband is severely restricted for the 1D system as can be seen from Fig. 7.22 where elastic scattering is considered. The only possible final state for elastic scattering is k or $-k$ unlike for the 2D or 3D system. Scattering to the state k has no influence on transport while scattering to $-k$ state requires a very short ranged potential to conserve momentum. Overall this results in a severe suppression of scattering and mobilities as high as 10^7 cm^2 V^{-1}s^{-1} have been predicted.

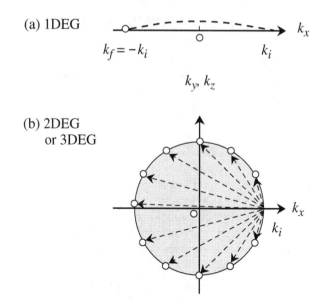

Figure 7.22: Equal energy surface of a 1-DEG (a) in comparison with that of 2-DEG or 3-DEG (b) in k-space. Note that scattering of a 1-DEG takes place only between k and $-k$, whereas for the 2D or 3D system, the electron can be scattered into various states.

The experimental picture of superior transport in 1D systems has not been very rosy so far. This appears to be linked to the growth and fabrication difficulties. Unlike quantum wells, the fabrication of quantum wires usually involves complex etching/regrowth steps which probably introduce serious defects which affect the transport properties.

In quasi-0-dimensional systems (quantum dots) we cannot talk about transport in the usual sense of electrons moving in space. However, we can talk about an electron entering the quantum dot potential well and then trickling down to the ground state. This process of carrier "relaxation" is quite important for optoelectronic devices such as lasers and detectors based on quantum dots.

As shown in Fig. 7.23 in a quantum dot system the density of states are finite only at certain energies and go to zero in between these confined energies. In other higher dimensional systems the density of states is always nonzero once the band starts. Since there are energy regions where no electronic states are allowed in the quantum dot, the electron cannot scatter into these energy regions. This creates several interesting issues in carrier scattering times that can be exploited for device design. A most interesting situation arises when the separation between the ground state E_1 and the next excited state, E_2, is larger than the phonon energy, i.e.,

$$E_2 - E_1 = E_{21} > \hbar\omega_0 \tag{7.65}$$

where ω_0 is the frequency of the optical phonons. In this case energy conservation forbids an electron in state E_2 to go to state E_1 by emitting an optical phonon. Since optical

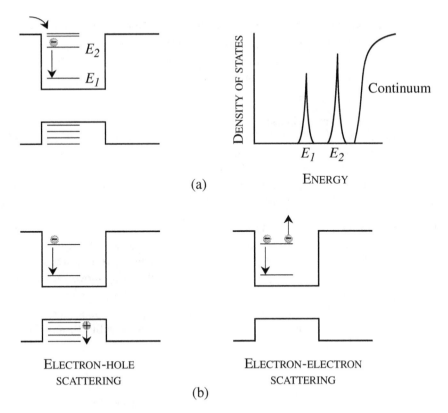

(a)

ELECTRON-HOLE
SCATTERING

ELECTRON-ELECTRON
SCATTERING

(b)

Figure 7.23: (a) Electronic levels and density of states in a quantum dot. If $E_2 - E_1$ is larger than the optical phonon energy an electron in level E_2 cannot go to level E_1 by first order phonon scattering. (b) Processes that can allow electrons to by-pass the phonon bottleneck.

phonon emission is the strongest scattering rate in higher dimensional systems, the words "phonon bottleneck" are used to describe the carrier relaxation process in dots.

Experimental studies on carrier relaxation show that electrons can indeed relax to the ground state quite fast. Theoretical studies have shown that processes represented in Fig. 7.23b are responsible for bypassing the phonon bottleneck. Thus an electron in the excited state can scatter from holes or other electrons and lose energy to relax into the ground state. However, even though the phonon bottleneck is bypassed, the relaxation times have been shown to be longer than those observed in quantum wells where the density of states is continuous.

7.8 TECHNOLOGY ISSUES

The velocity-field relationship is a critical ingredient of device design. As devices become smaller and more complicated and expensive to build, accurate simulations can save enormous amounts of money. Results used from Monte Carlo programs are widely used to model devices, especially as devices become so small that velocity overshoot effects become important. It is important to note that in semiconductor electronic devices both low field and high field transport properties are important.

In Table 7.4 we show a schematic of the electric field profiles in a field effect transistor and a bipolar junction transistor. These two classes of devices are the basis of most of modern electronics.

Field Effect Transistors (FETs)
As shown in Table 7.4, the field effect device is a "lateral" device in which either electrons or holes move from the source to the drain. The electric field is small between the source and the gate and increases rapidly at the drain side as shown. The low field mobility is very important to ensure a small source-gate access resistance. High field velocity is crucial in determining the device high frequency behavior.

Bipolar Junction Transistors (BJTs)
As with FETs, bipolar devices also have important low field and high field regions as shown in Table 7.4. In BJTs both electron and hole transport are important in an *npn* BJT the holes move under low field conditions, while electrons move under both low and high field conditions.

Both FETs and BJTs show improved high speed performance as device dimensions shrink. In addition, by using different materials it is possible to enhance device performance as outlined in Table 7.4.

7.9   PROBLEMS

7.1 Calculate and plot the average speed of electrons in GaAs and Si between 10 K and 400 K. How do these speeds compare to the saturation drift velocities?

7.2 Calculate the low field mobility in undoped and doped ($N_D = 10^{17}$ cm^{-3}) Si between 77 K and 300 K. Assume that all the donors are ionized.

7.3 Calculate the low field mobility in pure GaAs and In$_{0.53}$Ga$_{0.47}$As at 10 K and 300 K. Note that the room temperature mobility in In$_{0.53}$Ga$_{0.47}$As is higher than in GaAs, but at 10 K, this is not the case. What is the key reason for this turnaround? Assume an alloy scattering potential of 1.0 eV.

7.4 Calculate and plot the energy dependence of the various scattering rates listed in Section 13.3 for electrons in GaAs from 0 to 400 meV. Examine carefully the relative strengths of the various scattering processes.

7.5 Write a Monte Carlo computer program based on the flowchart of Fig. 7.4 and using the scattering rates and their angular dependence for GaAs based upon the Γ-L ordering. Compare the output of your results with the results presented in the paper by Fawcett, et al., (1970). Note that in their paper they had used a Γ-X ordering of the valleys.

7.6 Develop a Monte Carlo computer program to study electron transport in Si using the six equivalent X-valleys model and without including any other valleys. Examine the differences in the electron transport if you use an isotropic band with the density of states mass for the E vs k relation or if you use the longitudinal and transverse mass

DEVICE	IMPORTANT CONSIDERATIONS

FIELD EFFECT TRANSISTORS 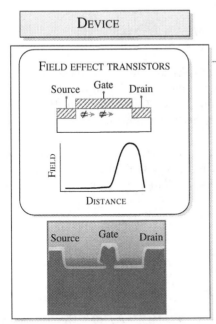	• Superior low field moblity
	    ≫ improved source-gate resistance
	 • Si ≫ GaAs ≫ InGaAs
	 • Superior high field velocity
	    ≫ high frequency/high speed operation
	 • Low impact ionization coefficient
	    ≫ high power devices
	       Si ≫ GaAs ≫ SiC ≫ GaN ≫ C(?)

BIPOLAR JUNCTION TRANSISTORS 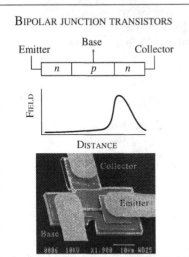	• Superior hole mobility ≫ low base resistance • Superior electron low field mobility (diffusion coefficient) ≫ small base transit time • Superior high field electron velocity ≫ small base-collector transit time, high frequency/high field response • Power devices ≫ high bandgap collector region.

Table 7.4: An overview of two important classes of electronic devices and influences of carrier transport on their performance.

relation, i.e.,

$$E(\mathbf{k}) = \frac{\hbar^2 k^2}{2m_{\text{dos}}^*}$$

vs.

$$E(\mathbf{k}) = \frac{\hbar^2 k_l^2}{2m_l^*} + \frac{\hbar^2 k_t^2}{2m_t^*}$$

7.7 Use your Monte Carlo program developed for GaAs to study the transient electron transport. Compare your results to the results in the paper by Maloney and Frey.

7.8 Study the transient transport in Si using your Monte Carlo program. How small would Si devices have to be before transient effects become important. Compare this to the case for GaAs devices.

7.9 When devices operate at high frequency, the conductivity of the semiconductor material changes. The propagation of the signal is then described by the description used in Chapter xx for electromagnetic waves. Using the equation of motion for the drift velocity,

$$\frac{d\langle v \rangle}{dt} = -\frac{\langle v \rangle}{\tau} + \frac{\mathbf{F}}{m}$$

show that the conductivity has the frequency dependent form

$$\sigma(\omega) = \frac{\sigma_0}{1 + i\omega\tau}$$

where σ_0 is the dc conductivity. Using $\mu = 8000$ cm^2/V-s for GaAs, plot the frequency dependence of the conductivity up to a frequency of 200 GHz.

7.10 Using a Monte Carlo method, calculate the electron temperature as a function of electric field for GaAs. Study the temperature up to a field of 10^5 V/cm.

7.10 REFERENCES

- **Low Field Transport**

    - "Transport: The Boltzmann Equation," by E. M. Conwell, in *Handbook on Semiconductors* North Holland, Amsterdam (1982), vol. 1.

    - C. Jacoboni, C. Canali, G. Ottoviani and A. Alberigi- Quaranta, *Sol. St. Electron.*, **20**, 77 (1977).

    - Lundstrom, M., *Fundamentals of Carrier Transport* (Modular Series on Solid State Devices, edited by G. W. Neudeck and R. F. Pierret, Addison-Wesley, Reading, 1990), vol. X.

    - "Low Field Electron Transport," by D. L. Rode, in *Semiconductors and Semimetals* edited by R. K. Willardson and A. C. Beer, Academic Press, New York (1975), vol. 10.

    - G. E. Stillman, C. M. Wolfe and J. O. Dimmock, *J. Phys. Chem. Solids*, **31**, 1199 (1970).

- **Monte Carlo Method**

  - W. Fawcett, A. D. Boardman and S. Swain, *J. Phys. Chem. Solids*, **31**, 1963 (1970).

  - D. K. Ferry, *Semiconductors* Macmillan, New York (1991).

  - K. Hess, *Advanced Theory of Semiconductor Devices* Prentice-Hall, Englewood Cliffs (1988).

  - R. W. Hockney and J. W. Eastwood, *Computer Simulation Using Particles* McGraw-Hill, New York (1981).

  - C. Jacoboni and P. Lugli, *The Monte Carlo Method for Semiconductor Device Simulation* Springer-Verlag, New York (1989).

  - C. Jacoboni and L. Reggiani, *Rev. Mod. Phys.*, **55**, 645 (1983).

  - M. Lundstrom, *Fundamentals of Carrier Transport* (Modular Series on Solid State Devices, edited by G. W. Neudeck and R. F. Pierret, Addison-Wesley, Reading, 1990), vol. X.

- **General High Field Transport**

  - C. S. Chang and H. R. Felterman, *Sol. St. Electron.*, **29**, 1295 (1986).

  - C. Jacoboni, C. Canali, G. Ottoviani and A. Alberigi- Quaranta, *Sol. St. Electron.*, **20**, 77 (1977).

  - C. Jacoboni and L. Reggiani, *Adv. in Phys.*, **28**, 493 (1979).

  - B. R. Nag, *Electron Transport in Compound Semiconductors* Springer-Verlag, New York (1980).

  - M. A. Omar and L. Reggiani, *Sol. St. Electron.*, **30**, 693 (1987).

  - J. G. Ruch and G. S. Kino, *Appl. Phys. Lett.*, **10**, 40 (1967).

  - J. G. Ruch and G. S. Kino, *Phys. Rev.*, **174**, 921 (1968).

  - K. Seeger, *Semiconductor Physics* Springer-Verlag, New York (1985).

  - P. Smith, M. Inoue and J. Frey, *Appl. Phys. Lett.*, **37**, 797 (1980).

- **Transient Transport Issues**

  - A good review is provided in IEEE Spectrum, February 1986. Excellent articles by J. M. Poate and R. C. Dynes; by L. Eastman; and by M. I. Nathan and M. Heiblum cover important issues in ballistic transport and its use in devices.

  - T. H. Glisson, C. K. Williams, J. R. Hauser and M. A. Littlejohn, in *VLSI Electronics: Microstructure Science* Academic Press, New York (1982), vol. 4.

  - J. R. Hays, A. F. J. Levy and W. Wiegmann, *Electron. Lett.*, **20**, 851 (1984).

  - M. Heiblum, M. L. Nathan, D. C. Thomas and C. M. Knoedler, *Phys. Rev. Letts.*, **55**, 2200 (1985).

- T. J. Maloney and J. Frey, *J. Appl. Phys.*, **48**, 781 (1977).

- "The HEMT: A Superfast Transistor," by Morkoc, H. and P. M. Solomon, *IEEE Spectrum*, **21**, 28 (February 1984).

• **Transport of Holes**

- A A. lberigi-Quaranta, M. Martini, G. Ottoviani, G. Radgelli and G. Zangarini, *Sol. St. Electron.*, **11**, 685 (1968).

- J. M. Hinckley and J. Singh, *Phys. Rev. B.*, **41**, 2912 (1990).

- J. M. Hinckley, *The Effect of Strain on Pseudomorphic p- $Si_x Ge_{1-x}$: Physics and Modeling of the Valence Bandstructure and Hole Transport*, Ph.D. Thesis, (The University of Michigan, Ann Arbor, 1990).

- F. L. Madarasz and F. Szmulowicz, *Phys. Rev. B*, **24**, 4611 (1981).

- E. J. Ryder, *Phys. Rev.*, **90**, 766 (1953).

- "Mobility of Holes in III-V Compounds," by J. D. Wiley, in *Semiconductors and Semimetals* (edited by R. K. Willardson and A. C. Beer, Academic Press, New York, 1975), vol. 10, chap. 2.

• **Balance of Approach**

- G. Baccarani and M. R. Wordeman, *Sol. St. Electron.*, **28**, 407 (1985).

- K.Blotekjaar, *IEEE Trans. Electron. Dev.*, **ED-17**, 38 (1970).

- "High Field Transport in Semiconductors," by E. M. Conwell, in *Solid State Physics* (edited by F. Seitz, D. Turnbull and H. Ehrenreich, Academic Press, New York, 1967), vol. Suppl. 9.

- R. L. Liboff, *Kinetic Equations* Gordon and Breech, New York (1971).

- M. Lundstrom, *Fundamentals of Carrier Transport* (Modular Series on Solid State Devices, edited by G. W. Neudeck and R. F. Pierret, Addison-Wesley, Reading, 1990), vol. X.

• **Impact Ionization Coefficients**

- G. A. Baraff, *Phys. Rev.*, **128**, 2507 (1962).

- G. A. Baraff, *Phys. Rev.*, **133**, A26 (1964).

- D. J. Bartelink, J. L. Moll and N. Meyer, *Phys. Rev.*, **130**, 972 (1963).

- D. K. Ferry, *Semiconductors* Macmillan, New York (1991).

- W. N. Grant, *Sol. St. Electron.*, **16**, 1189 (1973).

- K. Hess, *Advanced Theory of Semiconductor Devices* Prentice-Hall, Englewood Cliffs (1988).

- R. A. Logan and S. M. Sze, Proc. Int. Conf. Phys. Semicond., Kyoto, *J. Phys. Soc. Japan*, **Suppl. 21**, 434 (1966).

- J. L. Moll and R. van Overstraeten, *Sol. St. Electron.*, **6**, 147 (1963).

– T. P. Pearsall, F. Capasso, R. E. Nahory, M. A. Pollack and J. R. Che-likowsky, *Sol. St. Electron.*, **21**, 297 (1978).

– T. P. Pearsall, *Appl. Phys. Lett.*, **36**, 218 (1980).

– B. K. Ridley, *J. Phys.* C, **16**, 3373 (1983).

– B. K. Ridley, *J. Phys.* C, **16**, 4733 (1983).

– H. Shichijo and K. Hess, *Phys. Rev.* B, **23**, 4197 (1981).

– Shockley, W., *Sol. St. Electron.*, **2**, 35 (1961).

– G. E. Stillman and C. M. Wolfe, in *Semiconductors and Semimetals* (edited by R. K. Willardson and A. C. Beer, Academic Press, New York, 1977), vol 12.

- **Transport and Scattering in Low Dimensional Structures**

  – T. Ando, *Journal of Physical Society of Japan*, **51**, 3900 (1982).

  – J.A. Brum and G. Bastard, *Physical Review*, B, **33**, 1420 (1986).

  – H. Jiang and J. Singh, *IEEE Journal of Quantum Electronics*, **34**, 1188 (1998).

  – P.J. Prince, *Surface Science*, **143**, 145 (1984).

  – H. Sakaki, *Japanese Journal of Applied Physics*, **19**, L735 (1980).

  – I. Vurgaftman, Y. Lam, and J. Singh, *Physical Review*, B, **50**, 309 (1994).

Chapter

8

COHERENCE, DISORDER, AND MESOSCOPIC SYSTEMS

8.1 INTRODUCTION

In the previous chapters on transport we have applied Born approximation or the Fermi golden rule to describe the scattering processes in semiconductors. While the approach described in these chapters and the outcome is most relevant to modern microelectronic devices there are a number of important issues that are not described by this approach. As semiconductor devices and technology evolve, these issues are becoming increasingly important. In this chapter we will discuss some transport issues that are not described by the formalisms of the previous three chapters.

In Fig. 8.1 we show several types of structural properties of materials. In Fig. 8.1a we show a perfect crystal where there are no sources of scattering. Of course, in a real material we have phonon related fluctuations even in a perfect material. However, for short times or at very low temperatures it is possible to consider a material with no scattering. There are several types of transport that are of interest when there is no scattering: i) ballistic transport, where electrons move according to the modified Newton's equation. This kind of transport has been discussed in Section 7.3.2; and ii) Bloch oscillations, where electrons oscillate in k-space as they reach the Brillouin zone edge, as will be discussed in Section 8.2. In addition we can have tunneling type transport as well as quantum interference effects. These are discussed in Sections 8.3 and 8.4.

In Fig. 8.1b we show the case where there is a small degree of disorder. This is the situation where Born approximation can be used and transport under these conditions

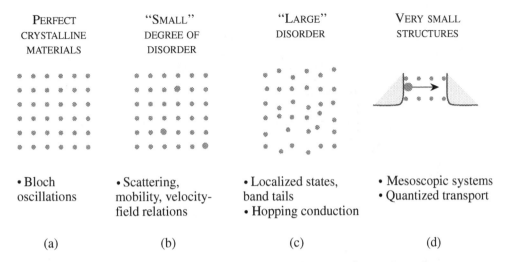

| PERFECT CRYSTALLINE MATERIALS | "SMALL" DEGREE OF DISORDER | "LARGE" DISORDER | VERY SMALL STRUCTURES |

- Bloch oscillations

- Scattering, mobility, velocity-field relations

- Localized states, band tails
- Hopping conduction

- Mesoscopic systems
- Quantized transport

(a)   (b)   (c)   (d)

Figure 8.1: A schematic of how levels of structural disorder and size impact electronic properties of a material. The various cases have been described in the text.

has been discussed in the previous chapters. In Fig. 8.1c we show the case where the structural disorder is large. This happens in amorphous materials and leads to localized states (band tails) and transport that is described by "hopping" behavior. We will discuss these issues in Sections 8.5 and 8.6.

Finally in Fig. 8.1d we show the case for devices that are very small (several tens of atoms across). Such structures are called *mesoscopic structures* and are increasingly becoming important as fabrication technology improves. Mesoscopic structures have a number of very interesting and potentially important transport properties. We will discuss these in Section 8.6.

8.2   ZENER-BLOCH OSCILLATIONS

In the discussion so far on transport, we discussed the effects of scattering on the response of electrons to an electric field. It is interesting to look at the electron response in absence of any scattering. We briefly mentioned this in the discussion of ballistic transport. The equation of motion is simply

$$\hbar\frac{d\boldsymbol{k}}{dt} = e\mathbf{F} \tag{8.1}$$

In the absence of any collisions the electron will simply start from the bottom of the band (Fig. 8.2) and go along the E vs k curve until it reaches the Brillouin zone edge. As discussed in Chapter 2, the bandstructure E vs $\boldsymbol{k}$ is represented within the first Brillouin zone since it is periodic in k-space. The electron at the zone edge is thus "reflected" as shown in Fig. 8.2 and now starts to lose energy in its motion in the field. The k-direction of the electron changes sign as the electron passes through the zone edge representing oscillations in k-space and consequently in the real space. These oscillations are called the Zener-Bloch oscillations.

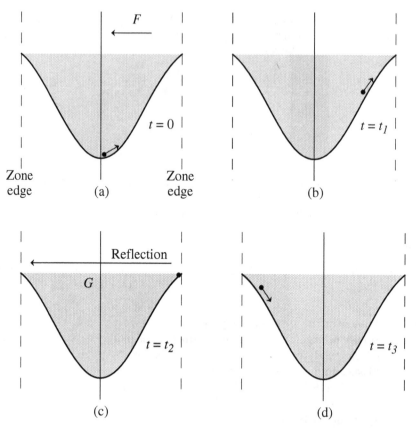

Figure 8.2: A schematic showing how an electron starting at $t = 0$ at the bottom of the Γ-valley travels up the E vs k diagram and gets reflected at the zone edge.

If we have a spatial periodicity defined by distance a the bandstructure is periodic in the reciprocal vector $\Gamma = 2\pi/a$. As a result the frequency of this oscillation is

$$\omega = \frac{e\mathbf{F}a}{\hbar} \tag{8.2}$$

The oscillation frequency is quite high and can easily be in the several terrahertz regime. Due to this possibility of high frequency oscillations, there is a tremendous interest in this phenomenon. However, since the scattering mechanisms are usually strong enough to cause the electron to scatter before it can go through a complete oscillation, it has not been possible to observe these oscillations. Also, at high electric fields, the tunneling between bands is quite strong, reducing the possibility of oscillations.

The Bloch oscillations are localized in space and the electron function is centered on a particular unit cell. The simple plane wave description is no longer adequate since the presence of the electric field causes a perturbation in the bands making k in the direction of the field no longer a good quantum number. It is easy to see the nature of the state in presence of the field by using perturbation theory. The equation satisfied by the electron is

$$(H_0 - e\mathbf{F} \cdot \mathbf{r})\psi_b = E_b\psi_b \tag{8.3}$$

where H_0 is the Hamiltonian that leads to the zero-field bandstructure and E_b are the energy levels (say, measured from the bandedge). The general wavefunction is

$$\psi_b = \frac{1}{\sqrt{V}} \sum_{\mathbf{k}} C_{\mathbf{k}} \, \phi_{\mathbf{k}}(\mathbf{r}) \tag{8.4}$$

where $\phi_{\mathbf{k}}(\mathbf{r})$ are the unperturbed Bloch functions for the crystal. Substituting ψ_b in Eqn. 8.3 and taking a dot product with $\phi_{\mathbf{k}'}$ we have

$$C_{\mathbf{k}'} = \sum_{\mathbf{k}} \frac{C_{\mathbf{k}} \langle \mathbf{k}' | (-e\mathbf{F} \cdot \mathbf{r}) | \mathbf{k} \rangle}{E_b - E_{\mathbf{k}'}} \tag{8.5}$$

The matrix element to be determined is

$$M_{\mathbf{k}\mathbf{k}'} = -\frac{e\mathbf{F}}{(2\pi)^3} \cdot V \int d^3k \, e^{-i\mathbf{k}' \cdot \mathbf{r}} \, C_{\mathbf{k}} \, \mathbf{r} \, e^{i\mathbf{k} \cdot \mathbf{r}} \tag{8.6}$$

We use the identity

$$\mathbf{r} e^{i\mathbf{k} \cdot \mathbf{r}} = -i\nabla_{\mathbf{k}} e^{i\mathbf{k} \cdot \mathbf{r}} \tag{8.7}$$

and integrate by parts using the orthogonality of the plane waves to get

$$\sum_{\mathbf{k}} C_{\mathbf{k}} \langle \mathbf{k}' | (-e\mathbf{F} \cdot \mathbf{r}) | \mathbf{k} \rangle = -e\mathbf{F} \, \nabla_{\mathbf{k}} C_{\mathbf{k}} \, \delta_{\mathbf{k}\mathbf{k}'}$$

This gives upon substitution in Eqn. 8.5

$$\frac{dC_{\mathbf{k}}}{d\mathbf{k}} = \frac{i(E_b - E_{\mathbf{k}})C_{\mathbf{k}}}{e\mathbf{F}}$$

or

$$C_{\mathbf{k}} = C_0 \exp\left\{ \int \frac{i(E_b - E_{\mathbf{k}})}{e\mathbf{F}} d\mathbf{k} \right\} \tag{8.8}$$

We recall that $E_{\mathbf{k}}$ is periodic in k-space. The localized function ψ_b must be periodic in unit cells, i.e., for the same value of E_b measured from the bandedge there have to be series of degenerate energy levels which are simply centered around different unit cells. To retain this periodicity, the function $C_{\mathbf{k}}$ must be periodic.

Choosing

$$E_{\mathbf{k}} = E_0 + E(\mathbf{k}) \tag{8.9}$$

where E_0 is the bandedge, we get a periodic value for C_0 if

$$E_b - E_0 = -e\mathbf{F}na \tag{8.10}$$

where n is an integer and a is the unit cell dimension. Thus, the wavefunction which has an energy E_b is centered around the unit cell at the point na ($= r_0$). This gives us

$$C_{\mathbf{k}} = C_0 \exp \left\{ -(\mathbf{k} \cdot \mathbf{r}_0) + \int \frac{E(\mathbf{k})}{e\mathbf{F}} d\mathbf{k} \right\} \tag{8.11}$$

and the wavefunction is

$$\psi_b = \frac{1}{\sqrt{V}} C_0 \sum_{\mathbf{k}} u_{\mathbf{k}}(\mathbf{r}) \exp \left\{ i \int \frac{E(\mathbf{k})}{e\mathbf{F}} d\mathbf{k} - i\mathbf{k} \cdot (\mathbf{r}_0 - \mathbf{r}) \right\} \tag{8.12}$$

where $u_{\mathbf{k}}(\mathbf{r})$ is the central cell part of the states.

As shown schematically in Fig. 8.2 the electron oscillates in k-space as it goes from the bottom of a band to the zone edge and then suffers a reflection and reverses its k-space trajectory. It also oscillates in real space. Such charge oscillations could be exploited as a source of radiation. The frequency of the Bloch-Zener oscillations is

$$\omega_b = \frac{e\mathbf{F}}{2\pi\hbar\mathbf{G}} \tag{8.13}$$

where $\mathbf{G}$ is the reciprocal lattice vector. This frequency can be adjusted to be in the range of 10^2 Hz. This frequency range is quite important in technology, but lower frequencies (~ 100 Ghz) can be reached by high speed electronic devices, while higher frequencies can be reached by optical devices.

Due to the potential applications of the Bloch oscillations, a considerable amount of work has focused on their realization. However, these attempts have not been very successful. The reason for this is the scattering processes and interband transitions which do not allow the coherent movement of the electrons that is necessary for the oscillations to occur. Oscillations can occur if we have the condition

$$\omega_b \tau_{SC} \geq 1 \tag{8.14}$$

where τ_{SC} is the scattering time. Use of superlattice concepts can, in principle, make it easier for the condition to be satisfied.

In Fig. 8.3 we show a schematic of the effect of enlarging the periodic distance (by making superlattices) on an energy band. On the top panel of the figure we show the energy band schematic of a crystal with a unit cell periodicity represented by the distance a. The zone edge in k-space is at $2\pi/a$. Now if a superlattice with a period na is made as shown in the lower panel the zone edge occurs at $2\pi/na$. Assuming that the scattering time is not changed much due to superlattice formation, it can be expected that an electron will be able to reach the superlattice zone edge without scattering, thus Bloch-Zener oscillations could occur. In reality, however, it has proven to be difficult to observe such oscillations in superlattices.

8.3 RESONANT TUNNELING

Resonant tunneling is a very interesting phenomenon in which an electron passes through two or more classically forbidden regions sandwiching a classically allowed region. A particularly interesting outcome of resonant tunneling is "negative differential resistance."

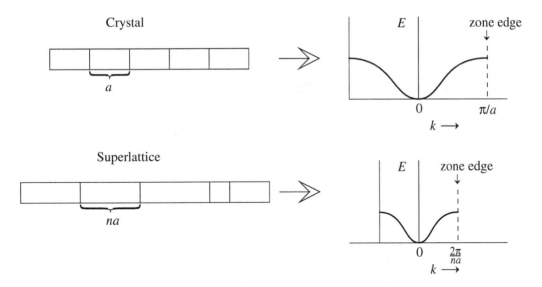

Figure 8.3: A schematic of how the use of a superlattice can reduce the k-space an electron has to traverse before it reaches the zone edge. The reduced zone edge may allow the possibility of Bloch-Zener oscillations.

In Fig. 8.4a we show a typical potential profile for a resonant tunneling structure. Such a structure will be discussed in detail in this section. As shown the double barrier structure of Fig. 8.4 has a quasi-bound state at energy E_0 as shown. If electrons injected from the left side region have energies that correspond to E_0 they are reflected back. As a result when a bias is applied across the structure interesting effects are observed. The operation of a resonant tunneling structure is understood conceptually by examining Fig. 8.4. At zero bias, point A, no current flows through the structure. At point B, when the Fermi energy lines up with the quasibound state, a maximum amount of current flows through the structure. Further increasing the bias results in the structure of point C, where the current through the structure has decreased with increasing bias (negative resistance). Applying a larger bias results in a strong thermionic emission current and thus the current increases substantially as shown at point D.

To understand the tunneling behavior we will discuss an approach known as the transfer matrix approach in which the potential profile (say, the conduction band lineup) is divided into regions of constant potential. The Schrödinger equation is solved in each region and the corresponding wavefunction in each region is matched at the boundaries with the wavefunctions in the adjacent regions. For simplicity the electronic wavefunction is expanded in terms of a single band on either side of a barrier and the Schrödinger equation is separated into a parallel and perpendicular part. The one-dimensional Schrödinger's equation for the direction perpendicular to a barrier interface can be written as

$$\left[-\frac{\hbar^2}{2m}\frac{d^2}{dz^2} + V(z) - \boldsymbol{E} \right] \psi_E(z) = 0 \tag{8.15}$$

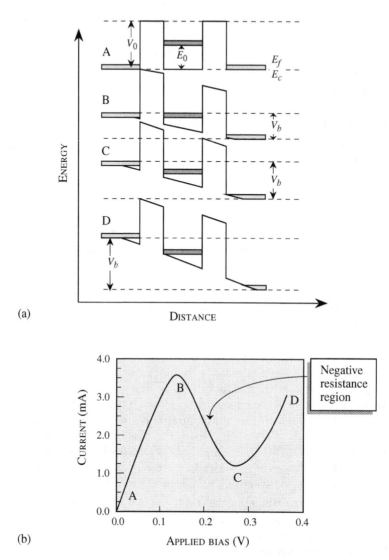

(a)

(b)

Figure 8.4: (a) Schematic explanation of the operation of resonant tunneling devices showing the energy band diagram for different bias voltages. (b) Typical current–voltage characteristic for the resonant tunneling diode.

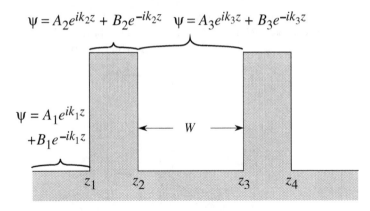

Figure 8.5: Typical resonant tunneling structure with two barriers. The wavefunction in each region has a general form shown.

When the potential $V(z)$ is constant in a given region, the general solution of Eqn. 8.15 has the form (see Fig. 8.5)

$$\psi(z) = A \exp\left[ikz\right] + B \exp\left[-ikz\right] \tag{8.16}$$

with

$$\frac{\hbar^2 k^2}{2m} = E - V \tag{8.17}$$

When $E - V > 0$, k is real, and the wave functions are plane waves. When $E - V < 0$, k is imaginary and the wave functions are growing and decaying waves. Thus, the overall wavefunction for a single barrier profile is an exponentially decaying wave for the barrier region and a plane-wave everywhere else.

At the boundaries between two materials one applies the boundary conditions

$$\psi(z^-) = \psi(z^+) \tag{8.18}$$

and

$$\left.\frac{1}{m_1}\frac{d\psi}{dz}\right|_{z^-} = \left.\frac{1}{m_2}\frac{d\psi}{dz}\right|_{z^+} \tag{8.19}$$

where m_1 and m_2 are the masses in the two regions. The second boundary condition ensures current continuity across the interface. The boundary conditions at the interface then determine the coefficients A and B (the subscripts denote the different regions) and can be described by a 2×2 matrix, M, such that

$$\begin{bmatrix} A_1 \\ B_1 \end{bmatrix} = M \begin{bmatrix} A_2 \\ B_2 \end{bmatrix} \tag{8.20}$$

where M is known as the transfer matrix. It can be written as

$$M = \frac{1}{2k_1 m_2} \begin{bmatrix} C \exp\left[i(k_2 - k_1)z_1\right] & D \exp\left[-i(k_2 + k_1)z_1\right] \\ D \exp\left[i(k_2 + k_1)z_1\right] & C \exp\left[-i(k_2 - k_1)z_1\right] \end{bmatrix} \tag{8.21}$$

where $C = (k_1 m_2 + k_2 m_1)$ and $D = (k_1 m_2 - k_2 m_1)$.

In general, if the potential profile consists of n , characterized by the potential values V_x and the masses m_x separated by $n-1$ interfaces at positions z_x then

$$\begin{bmatrix} A_1 \\ B_1 \end{bmatrix} = [M_1 \cdots M_{n-1}] \begin{bmatrix} A_n \\ B_n \end{bmatrix} \tag{8.22}$$

The elements of M_x are

$$M_x(1,1) = \left(\frac{1}{2} + \frac{k_{x+1}}{2k_x}\frac{m_x}{m_{x+1}}\right) \exp\left[i(k_{x+1} - k_x)z_x\right]$$

$$M_x(1,2) = \left(\frac{1}{2} - \frac{k_{x+1}}{2k_x}\frac{m_x}{m_{x+1}}\right) \exp\left[-i(k_{x+1} + k_x)z_x\right]$$

$$M_x(2,1) = \left(\frac{1}{2} - \frac{k_{x+1}}{2k_x}\frac{m_x}{m_{x+1}}\right) \exp\left[i(k_{x+1} + k_x)z_x\right]$$

$$M_x(2,2) = \left(\frac{1}{2} + \frac{k_{x+1}}{2k_x}\frac{m_x}{m_{x+1}}\right) \exp\left[-i(k_{x+1} - k_x)z_x\right]$$

If an electron is incident from the left only a transmitted wave will appear in region n, and therefore $B_n = 0$.

A simple application of this formalism is the tunneling of electrons through a single barrier of height V_0 and width a. The tunneling probability is given by

$$\begin{aligned} T_{1B}(E) &= \left|\frac{A_3}{A_1}\right|^2 \\ &= \frac{4E(V_0 - E)}{V_0^2 \sinh^2(\gamma a) + 4E(V_0 - E)} \end{aligned} \tag{8.23}$$

with

$$\gamma = \frac{1}{\hbar}\sqrt{2m(V_0 - E)} \tag{8.24}$$

This tunneling probability does not have any useful features. On the other hand, the double barrier structure has very interesting behavior in its tunneling probability. If we have two barriers as shown in Fig. 8.4, the tunneling through the double barrier is given by

$$T_{2B} = \left[1 + \frac{4R_{1B}}{T_{1B}^2}\sin^2(k_1 W - \theta)\right]^{-1} \tag{8.25}$$

where R_{1B} is the reflection probability from a single barrier

$$R_{1B} = \frac{V_0^2 \sinh^2 \gamma a}{V_0^2 \sinh^2 \gamma a + 4E(V_0 - E)} \tag{8.26}$$

and θ is given by

$$\tan\theta = \frac{2k_1\gamma \cosh \gamma a}{(k_1^2 - \gamma^2)\sinh \gamma a} \tag{8.27}$$

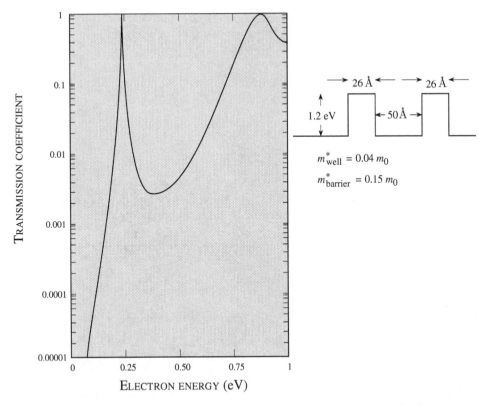

Figure 8.6: Transmission coefficient vs. electron longitudinal energy for a double barrier structure at zero bias (solid curve) and with an applied voltage (dashed curve).

The wavevector k_1 is given by

$$\frac{\hbar^2 k_1^2}{2m^*} = E$$

The tunneling through a double barrier structure has interesting resonances as can be seen from the expression for T_{2B}. The calculated transmission probability as a function of longitudinal electron energy for a typical RTD is shown in Fig. 8.6. For this calculation the barriers are assumed to be 1.2 eV and 26 Å wide. The well is 50 Å wide. The mass of the particle is taken to be $0.15m_0$ in the barriers and $0.042m_0$ outside of the barriers. The sharp peaks in the transmission probability correspond to resonant tunneling through the quasi-bound states in the quantum well formed between the two barriers. Note the relative widths of the ground and first resonant state. Also note that the transmission probability reaches unity. The tunneling probability reaches unity at energies corresponding to the quasi-bound states in the quantum well, of width W, formed by the barriers.

Although the transfer-matrix method was developed for rectangular barriers, it can be generalized to profiles of arbitrary shape, by dividing the barrier into steps of infinitesimal width. This enables one to calculate the probability of transmission through the double barrier structure in the presence of an applied field.

The current in the system is given by

$$
\begin{aligned}
\boldsymbol{J} &= ne\boldsymbol{v} \\
&= \frac{e}{4\pi^3 \hbar} \int_0^\infty dk_\ell \int_0^\infty d^2 k_t \left[f(\boldsymbol{E}) - f(\boldsymbol{E}') \right] T(\boldsymbol{E}_\ell) \frac{\partial \boldsymbol{E}}{\partial k_\ell}
\end{aligned}
\tag{8.28}
$$

where the longitudinal velocity is

$$
v = \frac{1}{\hbar} \frac{\partial \boldsymbol{E}}{\partial k_\ell}
$$

and the net current is due to the electrons going from the left-hand side with energy $\boldsymbol{E}$ and from the right-hand side with energy $\boldsymbol{E}' = \boldsymbol{E} + eFl = \boldsymbol{E} + eV$ where F is the electric field and l is the distance between the contacts on the two sides.

$$
\begin{aligned}
J &= \frac{e}{4\pi^3 \hbar} \int dk_\ell T(\boldsymbol{E}_\ell) \frac{\partial \boldsymbol{E}}{\partial k_\ell} \int d^2 k_t \left[\frac{1}{\exp\left[(\boldsymbol{E}_t + \boldsymbol{E}_\ell - \boldsymbol{E}_F)/k_B T \right] + 1} \right. \\
&\quad \left. - \frac{1}{\exp\left[(\boldsymbol{E}_t + \boldsymbol{E}_\ell + eV - \boldsymbol{E}_F)/k_B T \right] + 1} \right]
\end{aligned}
$$

The transverse momentum integral can be simplified by noting that

$$
\begin{aligned}
d^2 k_t &= k_t \, dk_t \, d\phi \\
&= \frac{m^* \, d\boldsymbol{E}_t \, d\phi}{\hbar^2}
\end{aligned}
$$

The current then becomes

$$
\begin{aligned}
J &= \frac{em^*}{2\pi^2 \hbar^3} \int d\boldsymbol{E}_\ell T(\boldsymbol{E}_\ell) \int_0^\infty d\boldsymbol{E}_t \left[\frac{1}{\exp\left[(\boldsymbol{E}_t + \boldsymbol{E}_\ell - \boldsymbol{E}_F)/k_B T \right] + 1} \right. \\
&\quad \left. - \frac{1}{\exp\left[(\boldsymbol{E}_t + \boldsymbol{E}_\ell + eV - \boldsymbol{E}_F)/k_B T \right] + 1} \right] \\
&= \frac{em^*}{2\pi^2 \hbar^3} \int_0^\infty T(\boldsymbol{E}_\ell) \, \ln \left[\frac{1 + \exp\left[(\boldsymbol{E}_F - \boldsymbol{E}_\ell)/k_B T \right]}{1 + \exp\left[(\boldsymbol{E}_F - \boldsymbol{E}_\ell - eV)/k_B T \right]} \right] d\boldsymbol{E}_\ell
\end{aligned}
\tag{8.29}
$$

The current shows peaks because of the peaks in the tunneling probability. The reader is advised to be careful to distinguish between $T(\boldsymbol{E}_\ell)$, the tunneling probability and T, the temperature.

In Fig. 8.7 we show typical current-voltage characteristics measured in resonant double barrier structures. The results show are for a InGaAs/AlAs structure with parameters shown. As can be seen a large peak to valley current ration can be obtained at room temperature. While the formalism described above does provide the qualitative features observed in the I–V characteristics of the resonant tunneling diodes, it does not provide good quantitative agreement. Approaches which include scattering effects have been used to improve the agreement between experiments and theory. The techniques are too complex to be discussed here. Nevertheless, the need for such complexity shows how electronic devices have approached the regime where simpler quantum mechanics concepts are not adequate.

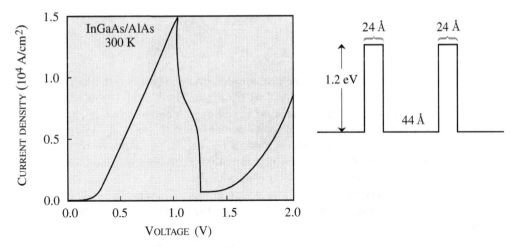

Figure 8.7: Static current–voltage characteristics of an InGaAs/AlAs resonant tunneling diode at room temperature.

8.4 QUANTUM INTERFERENCE EFFECTS

In Chapter 2 we have seen that in a perfectly periodic potential the electron wavefunction has the form

$$\psi_k(r) = u_k(r)e^{ik \cdot r}$$

In this perfect structure the electron maintains its phase coherence as it propagates in the structure. However, in a real material electrons scatter from a variety of sources. As a result, the particle wave's coherence is lost after a distance of a mean free path. If v is the average speed of the electron and τ the time interval between scattering, the particle loses its phase coherence in a distance

$$\lambda \sim v\tau$$

In the absence of phase coherence, effects such as interference are not observable unless the dimensions of the material (or device) are smaller than the mean free path.

In high-quality semiconductors (the material of choice for most information-processing devices) the mean free path is ~ 100 Å at room temperature and ~ 1000 Å at liquid helium. It is possible to see quantum interference effects at very low temperatures in semiconductor devices. These effects can be exploited to design digital devices and switches operating at very low power levels. The general principle of operation is shown in Fig. 8.9. Electron waves travel from a source to a drain via two paths. At the output the intensity of the electron wave is

$$I(d) = \mid \psi_1(d) + \psi_2(d)^2 \tag{8.30}$$

If the waves are described by

$$
\begin{aligned}
\psi_1(x) &= Ae^{ik_1 x} \\
\psi_2(x) &= Ae^{ik_2 x}
\end{aligned}
\tag{8.31}
$$

we have

$$I(d) = 2A^2[1 - \cos(k_1 - k_2)d] \qquad (8.32)$$

If we can now somehow alter the wavelengths of the electron (i.e., the value of $(k_1 - k_2)$) we can modulate the signal at the drain. This modulation can be done by using an electric bias to alter the kinetic energy of the electrons in one arm. In Fig. 8.8b we show a schematic of a split-gate device in which electrons propagate from the source to the drain either under one gate or the other. The ungated region is such that it provides a potential barrier for electron transport as shown by the band profile. Interference effects are then caused by altering the gate bias.

In quantum interference transistors, a gate bias is alters the potential energy seen by the electrons. The electron k-vector at the Fermi energy is given by (E_c is the bandedge)

$$E = E_c + \frac{\hbar^2 k^2}{2m^*} \qquad (8.33)$$

By changing the position of E_F, one can alter the k-value. Thus one can develop quantum interference transistors. Unfortunately, these effects are only observable at very low temperatures.

8.5   DISORDERED SEMICONDUCTORS

The first seven chapters of this text have dealt with the properties of the nearly perfect crystalline structure and defects or "disorder" has been assumed to be small. However, there are cases where "disorder" is very large and the basic nature of the electronic states is altered. In this section we will deal with situations where there is some "defect" at almost every site of the crystal.

A most important example is the Si/SiO_2 interface where, because of the lattice mismatch between Si and SiO_2, the interface region has an amorphous nature with small but significant distortions in the bond angles and bond lengths of the interfacial atoms. Another important example are the random alloy semiconductors, in which, even though there is a perfect underlying lattice, the atoms on each site are not well-defined leading to random potential fluctuations. Similarly the interfaces of even high quality heterostructures have interfacial disorder on the scale of a few monolayers.

The amorphous semiconductors also form an important class of materials. These semiconductors are usually grown under highly non-equilibrium growth conditions so that the growing structure is unable to reach the equilibrium crystalline structure. The advantage, of course, is that the structure is grown at a very low cost and, perhaps, can be grown over large areas. Amorphous silicon is a prime example of such semiconductors and offers the possibility of a cheap material for solar energy conversion. In such amorphous semiconductors, the nearest neighbor coordination is maintained, but long-range order is lost. Since often such semiconductors have broken bonds, certain additional defects are introduced in the structure to passivate the defects. For example, in amorphous silicon, hydrogen serves this purpose.

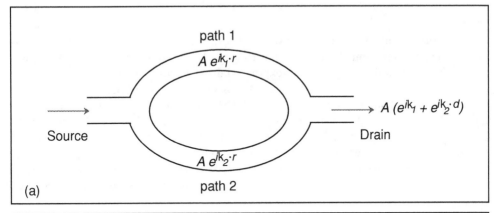

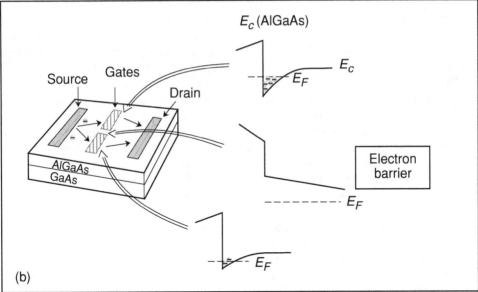

Figure 8.8: (a) A schematic of a coherent electron beam traveling along two paths and interfering. (b) A schematic of a split-gate transistor to exploit quantum interference effects. Electrons can propagate from the source to the drain under the two gates in the 2-dimensional channel of AlGaAs/GaAs as shown.

8.5.1   Extended and Localized States

In the discussion of levels produced by dopants, we saw that the wavefunction associated with the point defect is not an extended or Bloch state of the form

$$\psi_{\text{ex}} = \frac{1}{\sqrt{V}} u_{\mathbf{k}}(r) e^{i\mathbf{k}\cdot\mathbf{r}} \tag{8.34}$$

but a localized state with a finite extent in space. The defect states have a general form $\psi_{\text{loc}}(r, r_0)$ representing the fact that they are localized around a point r_0 in space. Typical localized states may have an exponentially decaying behavior.

In amorphous semiconductors electrons see a random background potential in contrast to the periodic potential of a crystal. In the random potential electrons can find local potential wells where they can be trapped or localized. At low energies in a random potential (in an infinite structure) we get a continuum of localized states. As one goes up to higher electronic energies the electron wavefunction spreads over a larger volume until eventually it becomes extended to the entire volume. Of course, the extended state is not the Bloch state of the crystal—it simply spreads over the entire sample.

The qualitative differences between the extended and localized states can theoretically be brought out by an approach suggested by Thouless. Consider a volume V_1 of the disordered sample. If one solves for the electronic states of the sample with boundary conditions that the wavefunctions go to zero at the boundary, then both "extended" and localized states will be confined to this volume. Now if one increases the volume, the amplitude of the "extended" states will decrease as $\sim 1/\sqrt{V_1}$ while that of the localized states will essentially remain constant. The energy points which separate the extended and localized states are called the mobility edge, the term arising from the consensus that the dc conductivity of the localized states goes to zero (at low temperatures). The effect of disorder on the nature of electronic states was first studied by Anderson in his classic paper. It was shown that as the disorder is increased, the extent of the localized states increases as shown schematically in Fig. 8.9. As can be seen, near the bandedges one gets localized states forming "bandtails." The width of these bandtails is related to the level of disorder in the system.

There have been many different approaches to tackle the problem of disordered systems. We will briefly discuss a statistical model to examine the bandtail states. Such models are useful to address not only the "band tail states" (i.e., states in the bandgap away from the mobility edges), but also their success in addressing problems of broadening of excitonic transitions and general effects of fluctuations.

We will discuss the issue of how bandtails are produced by examining a random alloy in which two materials are randomly mixed. Due to the random nature of the potential bandtail states are formed. Lifshitz was the first to introduce a statistical theory to attack the problem of a random A-B alloy. Fig. 8.10 shows individual eigenspectra of any A and B components forming the alloy. The problem that Lifshitz addressed was to find the density of states near the extremities $E_{\min}^A$ and $E_{\max}^B$. The region of interest for statistical-variational theories is between $E_{\min}^A$, $E_{\min}^B$ and $E_{\max}^A$, $E_{\max}^B$. The mean concentrations of A and B atoms are C_A^0 and C_B^0, and it is assumed that no correlations exist between the atoms. Lifshitz asserted that states near $E_{\min}^A$ (in the alloy) will arise

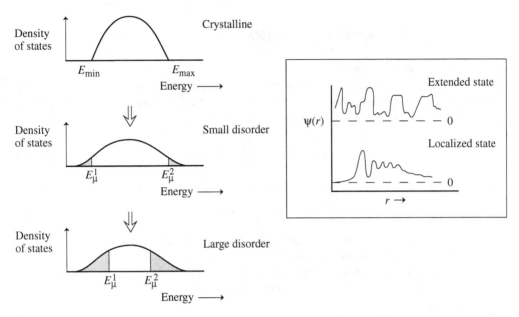

Figure 8.9: Density of states and the influence of disorder. The shaded region represents the region where the electronic states are localized in space. The mobility edges E_μ separate the region of localized and extended states. The inset shows a schematic of an extended and a localized state.

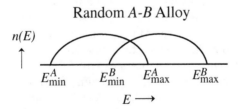

Figure 8.10: The individual eigenspectra of the A and B components forming a random alloy.

from regions in space which are purely A-like, i.e., regions which represent extreme potential fluctuations. The probability of a volume V_0 having a concentration C_A when the mean concentration is C_A^0 is given by the usual statistical techniques

$$P(V_0) = \exp\left[-\frac{V_0}{v_0}\{C_A \ln(C_A/C_A^0) + C_B \ln(C_B/C_B^0)\}\right] \tag{8.35}$$

where v_0 is the atomic volume.

In addition Lifshitz argued that it is the ground state of such a fluctuation which is important in determining the density of states near $E_{\min}^A$. Using these prescriptions, then the probability in Eqn. 8.35 becomes (for $C_A = 1$)

$$P(V_0) = \exp\left[-\frac{V_0}{v_0} \ln(1/C_A^0)\right] \tag{8.36}$$

and the energy of a localized wavefunction describing ground state may be written as

$$E = E_{\min}^A + B/R_0^2 \tag{8.37}$$

where B is a constant ($\sim h^2/m_A^*$) and $R_0^3 \sim V_0$ for 3D. The first term in Eqn. 8.37 represents the lowest "potential energy" that an electron could have in the pure A environment, while the second term represents the "kinetic energy" or the energy of confining the electron to a volume V. From Eqn. 8.37, one then derives the relation between V_0 and E, viz.

$$V_0 = \left[\left(\frac{E - E_{\min}^A}{B}\right)\right]^{-3/2} \tag{8.38}$$

and substituting in Eqn. 8.36, one gets

$$
\begin{aligned}
n(E) &\propto P(E) \\
&= \exp\left[-\frac{1}{v_0} \ln(1/C_A^0)\left(\frac{E - E_{\min}^A}{B}\right)^{-3/2}\right]
\end{aligned}
\tag{8.39}
$$

The density of band tail states thus depends upon the smallest volume of disorder (v_0 which is the unit cell volume for a perfectly random system or could be larger for a clustered alloy). The Lifshitz results are applicable only at the extreme end of the band tail states, since it is assumed that these states arise from regions of pure A material. In general one has to relax this condition along with the condition that the localized states result from the ground state function.

With the availability of computers capable of handling large matrix solutions, it is now possible to study the electronic spectrum of a disordered system numerically. These approaches then allow one to examine localized and extended states. A typical plot of band tail states is shown in Fig. 8.11.

8.5.2 Transport in Disordered Semiconductors

There are some important qualitative differences between transport in a crystalline material with a small degree of imperfections and transport in disordered materials. As

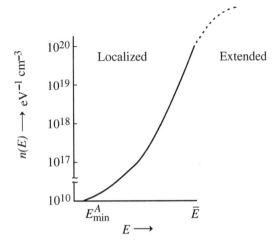

Figure 8.11: Behavior of the density of states for system with small disorder. Band tails which can be described by exponentially decaying terms are produced.

we have discussed in Chapters 4 through 7, in crystalline materials electrons scatter between one state and another due to the imperfections. Transport is then described by the Boltzmann transport equation or by Monte Carlo methods. In disordered systems the electrons are trapped in localized states. Thus below the mobility edge transport occurs via i) "hopping" of electrons from one localized state to another; ii) "hopping" from a localized state to an extended state; and iii) transport in extended states. The approaches used to understand transport in disordered systems are listed in Fig. 8.12.

In the case of the disordered system we calculate the electronic spectra first in presence of the disorder. This is important, as discussed earlier, because of the qualitative differences between a crystalline material with defects and a disordered material. Transport in the localized states is by hopping conduction, details of which we will discuss shortly. The extended states are described by a phase coherence distance and we

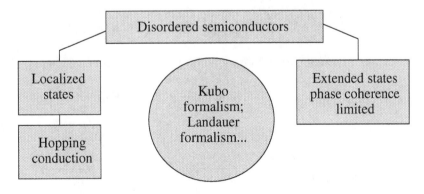

Figure 8.12: The different approaches used to address defects in disordered systems.

need to develop a formalism which will connect this to the conductivity of the material. This connection is provided by the Kubo formalism which is a general approach to transport and is particularly useful for cases of strong scattering. We will present the Kubo formalism for transport now.

We will calculate the conductivity $\sigma(\omega)$ of a disordered system at frequency ω and then obtain the d_c conductivity by letting ω go to zero. Let us consider an electric field, $F \cos \omega t$, acting on a system of volume V. The probability, P, per unit time that an electron makes a transition from a state E to any of the degenerate states $E + \hbar\omega$ is

$$P = \frac{1}{4} e^2 \, F^2 \, \left(\frac{2\pi}{\hbar} \right) |\langle E + \hbar\omega | x | E \rangle|^2 \, V N \, (E + \hbar\omega) \tag{8.40}$$

The matrix element for the transition is

$$\langle E' | x | E \rangle = \int \psi^*_{Ess'} \, x \, \psi_{Ess} \, d^3 x \tag{8.41}$$

where ψ_{Ess} is the electronic state with energy E and is normalized to volume V. The states need not be a Bloch state. It is useful to rewrite the matrix element in the form.

$$\langle E + \hbar\omega | x | E \rangle = \frac{\hbar}{m\omega} \int \psi^*_{Ess+\hbar\omega} \, \frac{\partial}{\partial x} \psi_{Ess} \, d^3 x \tag{8.42}$$

This form could be obtained if we use the perturbation potential as eA/mc instead of eFx, and use the relation $A = cF/\omega$. The transition rate is now

$$P = \frac{\pi e^2 \hbar V}{2m^2 \omega^2} \, F^2 \, |D|^2 N(E + \hbar\omega) \tag{8.43}$$

where

$$D = \int \psi^*_{Ess+\hbar\omega} \, \frac{\partial}{\partial x} \psi_{Ess} \, d^3 x$$

We now define the conductivity of the system via the relationship between the conductivity and the rate of loss of energy per unit volume. This quantity is just $\sigma(\omega) F^2 / 2$. The rate of energy gain is given by

$$R_{\text{gain}} = 2 \, \hbar\omega \int P \, f(E) \, [1 - f(E + \hbar\omega)] \, N(E) \, dE \tag{8.44}$$

The factor of 2 is to count both spin directions. The rate of loss of the energy is

$$R_{\text{loss}} = 2 \, \hbar\omega \int P \, f(E + \hbar\omega) \, [1 - f(E)] \, N(E + \hbar\omega) \, dE \tag{8.45}$$

Thus, the conductivity is given by

$$\sigma(\omega) = \frac{2\pi e^2 \hbar^3 V}{m^2} \int \frac{[f(E) - f(E + \hbar\omega)] \, |D|^2_{\text{av}} N(E) N(E + \hbar\omega) \, dE}{\hbar\omega} \tag{8.46}$$

The term $|D|^2_{\text{av}}$ is $|D|^2$ averaged over all states ψ_{Ess} with energy E.

At zero temperature, this becomes

$$\sigma(\omega) = \frac{2\pi e^2 \hbar^3 V}{m^2} \int \frac{|D|^2_{\text{av}} N(E) N(E + \hbar\omega)}{\hbar\omega} \, dE \tag{8.47}$$

The *dc* conductivity is given by taking the limit $\omega \to 0$

$$\sigma_{Ess} = \frac{2\pi e^2 \hbar^2 \Omega}{m^2} |D_{Ess}|^2_{\text{av}} \{N(E)\}^2 \tag{8.48}$$

where

$$D_{Ess} = \int \psi^*_{Ess'} \frac{\partial}{\partial x} \psi_{Ess} \, d^3x, \text{ at } E' = E \tag{8.49}$$

The Kubo formalism reduces to the Boltzmann formalism

$$\sigma = \frac{ne^2 \langle \tau \rangle}{m^*} \tag{8.50}$$

when the mean free path is long. To evaluate the matrix element D, it is useful to consider the volume ℓ^3 over which the wavefunction in the localized state maintains phase coherence. Thus, even though the extended state is not a Bloch state, we argue that in the volume ℓ^3 it can be written as a Bloch state with wavevector k. The coherence is then lost outside the volume ℓ^3. We can then divide the spatial integral over V for D, into volume regions ℓ^3, where the integral is nonzero.

$$D = \left(\frac{V}{\ell^3}\right)^{1/2} \delta \tag{8.51}$$

where

$$\delta = k \int_{\ell^3} \frac{\exp[i(k' - k) \cdot r]}{V} \, d^3r \tag{8.52}$$

We may write $|k' - k| = 2k\sin(\theta/2) \approx k\theta$ where θ is the angle between k and k'. The integral is approximately evaluated as

$$\begin{aligned} \delta &= \frac{k\ell^3}{V}, \text{ if } k\ell\theta < 1 \\ &= 0 \quad \text{otherwise} \end{aligned} \tag{8.53}$$

To obtain $|D_{Ess}|^2$, we have to do an averaging over the angle θ

$$\begin{aligned} \langle |D|^2 \rangle &= \frac{V}{\ell^3} \frac{k^2\ell^6}{V^2} \frac{1}{4\pi} \int_0^\pi \frac{\sin(k\ell\theta/2)}{(k\ell\theta/2)} 2\pi \, \sin\theta \, d\theta \\ &\approx \frac{\pi\ell}{3V} \end{aligned} \tag{8.54}$$

We note that ℓ is the mean free path used in the Boltzmann approach, i.e.,

$$\begin{aligned} \ell &= v\tau \\ &= \frac{\hbar k}{m^*}\tau \end{aligned} \tag{8.55}$$

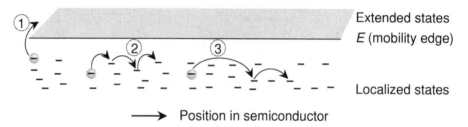

Extended states
E (mobility edge)

Localized states

⟶ Position in semiconductor

Figure 8.13: Mechanisms for transport in a disordered system. In case 1, the electron is thermal activated to the states above the mobility edge. In case 2, the electron hops to the nearest localized state while in the case 3, the electron hops to the "optimum" site as explained in the text.

The Kubo formalism provides new results for conductivity only when the mean free path becomes very small. When ℓ approaches a, the atomic spacing, the conductivity that results is called the minimum conductivity.

The effects of phonon scattering which is not reflected in the electronic spectra directly have to be treated separately. For the extended states the phonon scattering reduces the phase coherence length and reduce the conductivity. However, for conductivity in the localized states, phonons actually increase the mobility of the carriers.

The localized states are not connected in real space so that the electron needs to hop from one state to another. The important conduction mechanisms are shown schematically in Fig. 8.13. In the first mode of conduction the electrons are excited to the mobility edge by phonons, and the conduction behavior is described by the thermally activated behavior

$$\sigma = \sigma_0 \exp\left[\frac{-(E_c - E_F)}{k_B T}\right] \tag{8.56}$$

where σ_0 is the conductivity at the mobility edge. The conductivity at the mobility edge has been a subject of great interest and is still being examined. Mott has shown that it has a form

$$\sigma_0 = \frac{C e^2}{\hbar a} \tag{8.57}$$

where $C \sim 0.03$ and a is the minimum distance over which phase coherence could occur. This quantity is also called the minimum metallic conductivity. For $a = 3$ Å, the value is $2 \times 10^2 \ \Omega^{-1} \ \text{cm}^{-1}$.

The second process indicated in Fig. 8.13 involves thermal activation from one localized state to the nearest state in space, above the Fermi level. This process has been used to explain the impurity conduction in doped semiconductors. The electron is always assumed to move to the nearest empty localized state. To estimate this conductivity, we assume that the wavefunctions are described by (assuming a center at the origin)

$$\psi = e^{-\alpha r} \tag{8.58}$$

where α^{-1} is the localization length. The current density is proportional to the overlap between the wavefunctions, the density of states at the Fermi level $N(E_F)$, the width of the Fermi distribution $k_B T$, the effective velocity of transport which is chosen as ν_{ph}, the attempt frequency ($\approx$ phonon frequency) times the average spacing between the states R. If ΔE is the average separation between the energies of the two states, and F the applied field, the current density is

$$
\begin{aligned}
J \quad \sim \quad & ek_B T \; N(E_F) \; R \; \nu_{\mathrm{ph}} \; \exp(-2\alpha R) \\
\times \quad & \left[\exp\left(\frac{-\Delta E + eFR}{k_B T} \right) - \exp\left(\frac{-\Delta E - eFR}{k_B T} \right) \right] \\
= \quad & 2ek_B T \; N(E_F) \; R \; \nu_{\mathrm{ph}} \; \exp\left(-2\alpha R - \frac{\Delta E}{k_B T} \right) \sinh\left(\frac{eFR}{k_B T} \right) \qquad (8.59)
\end{aligned}
$$

For small electric fields, the sinh function can be expanded and the conductivity becomes

$$
\begin{aligned}
\sigma \quad = \quad & \frac{J}{F} \\
= \quad & 2e^2 R^2 \; \nu_{\mathrm{ph}} \; N(E_F) \; \exp\left[-2\alpha R - \frac{\Delta E}{k_B T} \right] \qquad (8.60)
\end{aligned}
$$

An estimate of the energy spacing of the levels is simply obtained from the definition of the density of states, i.e.,

$$
\Delta E \approx \frac{1}{N(E_F) R_0^3} \qquad (8.61)
$$

where R_0 is the average separation between nearest neighbor states. This kind of nearest neighbor hopping is dominant if nearly all states are strongly localized, e.g., as in impurity states due to dopants.

In most disordered semiconductors, where the disorder is not too strong, one has another important transport process indicated by the third process in Fig. 8.13. This process, known as variable range hopping, was introduced by Mott and is a dominant transport mode at low temperatures. At low temperatures, the hop would not occur to the nearest spatial state, but the electron may prefer to go a potentially farther state but one which is closer in energy so that a lower phonon energy is needed.

In a range R of a given localized state, the density of states per unit energy range near the Fermi level are

$$
\left(\frac{4\pi}{3} \right) R^3 N(E_F)
$$

Thus, for the hopping process involving a distance within R, the average separation of level energies will be

$$
\Delta E = \frac{3}{4\pi R^3 \; N(E_F)}
$$

As can be seen, the farther the electron hops, the smaller the activation barrier that it needs to overcome. On the other hand, a hop of a distance R will involve an overlap function which falls as $\exp(-2\alpha R)$. Thus, there will be an optimum distance for which the term

$$
\exp(-2\alpha R) \exp\left(\frac{-\Delta E}{k_B T} \right)
$$

is a maximum. This can easily be seen to occur when

$$\frac{d}{dR}\left[2\alpha R + \frac{3}{4\pi R^3 \, N(E) \, k_B T}\right] = 0$$

or

$$R_m = \left[\frac{1}{8\pi N(E) \, \alpha k_B T}\right]^{1/4}$$

Using this value of R_m, we get for the conductivity behavior

$$\sigma = A\exp\left(-\frac{B}{T^{1/4}}\right) \tag{8.62}$$

where

$$B = 2\left(\frac{3}{2\pi}\right)^{1/4}\left(\frac{\alpha^3}{k_B N(E_F)}\right)^{1/4}$$

This variable range hopping temperature behavior has been observed in numerous disordered systems. It is straightforward to show that for 2-dimensional systems the temperature dependence is

$$\sigma = A\exp\left(-\frac{B'}{T^{1/3}}\right) \tag{8.63}$$

The temperature dependence of conductivity given above has been observed in 2-dimensional electron systems.

The formalisms given above for transport in amorphous semiconductors provide a mere glimpse into the complexities of transport in disordered materials. We have not discussed high field transport in amorphous materials. It is expected that at high fields where electron energies are large, carriers will primarily reside in extended states above the mobility edge. Here the transport will be similar to transport in crystalline materials, except the scattering rates would be higher.

We have discussed transport in essentially infinite systems. When the sample length starts to approach the localization length (coherence length) interesting new effects arise. These are discussed in the next section.

8.6   MESOSCOPIC STRUCTURES

In our studies of transport we have assumed so far that the sample is quite large compared to the coherence length (or distance between successive scattering events). New physical phenomena emerge when sample dimensions become comparable to or smaller than the coherence length. An area where continuous development has been occurring in technology is the fabrication of small structures. Advances in electron beam and X-ray lithography combined with low damage etching techniques has allowed fabrication of structures as small as ~ 100 Å. This has enabled the fabrication of semiconductor devices that are small enough that electrons move from one contact to the other maintaining their phase. This is unlike tunneling structures discussed in Section 8.3 in which phase coherence is present only in the tunneling region. Such structures known generally as mesoscopic structures exhibit two interesting effects.

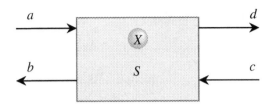

Figure 8.14: A schematic showing the effect of the scattering center S on electron waves a and c incident from the left and right respectively. The waves b and d emerge as a result of reflection and transmission.

- The transport and conductance may be described by coherent movement of the electron wave.

- It is also possible that the capacitance of the structure is so small that addition of a single electron can alter the potential energy of the system. This can cause interesting effects generally described by the term Coulomb blockade. Both of these effects offer the potential of new devices that can operate at very small power levels. The structures also offer the study of few-electron physics.

8.6.1 Conductance Fluctuations and Coherent Transport

As noted above, in very small structures electron waves can flow from one contact to another maintaining phase coherence. In structures that are $\sim$ 100–500 Å this occurs at low temperatures, since at high temperatures the random scattering due to phonons removes the coherence in the transport process. Since phase coherence is maintained in transport, the macroscopic averaging procedures we have been using in defining mobility or conductivity do not hold anymore. A dramatic manifestation of the phase coherence is the fluctuation seen in conductivity of mesoscopic structures as a function of magnetic field, electron concentration, etc.

The origin of the fluctuations can be understood on the basis of Landauer formalism which allows one to study transport in terms of the scattering processes directly. For simplicity consider a one-dimensional system with scattering centers. Each of these scatterers is characterized in terms of a transfer matrix which describes what fraction of the incident electron is "reflected" after scattering and what fraction is transmitted. The scatterer is described by the reflection and transmission coefficients shown in Fig. 8.14. The reflection and transmission coefficients are R and T for an incident wave from the left or right. To calculate the current flow one needs to identify the relative change in the carrier density on the left and right side of the scatterer. If there is an applied bias δV, the excess carrier density (at low temperatures) is

$$\delta n \approx \frac{dn}{dE}(e\,\delta V) \tag{8.64}$$

The excess carriers on the left over the right side can also be evaluated as the magnitude of the particle currents on the left divided by the velocity minus the magnitude of the particle currents on the right.

$$\begin{aligned}\delta n &= \frac{j_a + j_b}{v} - \frac{j_c + j_d}{v} \\ &= \frac{(j_a - j_c) + (j_b - j_d)}{v}\end{aligned} \tag{8.65}$$

Now it can be easily seen that

$$(j_b - j_d) = (R - T)(j_a - j_c) \tag{8.66}$$

Thus, since $R + T = 1$

$$\delta n = \frac{2R\,(j_a - j_c)}{v} \tag{8.67}$$

The electrical current on the left side is

$$\begin{aligned}I_\ell &= eT\,j_a - eT\,j_c \\ &= eT\,(j_a - j_c)\end{aligned} \tag{8.68}$$

These equations then give for the conductance

$$\begin{aligned}G &= \frac{I_\ell}{\delta V} \\ &= \frac{T}{2R}\,e^2\,\frac{dn}{dE}\,\frac{1}{\hbar}\,\frac{\partial E}{\partial k}\end{aligned} \tag{8.69}$$

Now for a 1-dimensional case

$$\frac{dn}{dk} = \frac{1}{\pi} \tag{8.70}$$

so that (including the spin degeneracy factor of 2)

$$G = \frac{2e^2}{h}\,\frac{T}{R} \tag{8.71}$$

The expression shows that the fundamental unit of conductance is $2e^2/h$ which will be modified by the values of T/R. In general in the Landauer formalism one has to sum the electron contributions from all different paths the electron could take in going from one contact to another. It may appear that when such an averaging is carried out the conductance fluctuations arising from different paths would average out, especially as the sample size is increased. However, it turns out that there is a remarkable universality in the magnitude of the fluctuations independent of the sample size, dimensionality and extent of disorder, provided the disorder is weak and the temperature is low (a few Kelvin). Such universal conductance fluctuations have been measured in a vast range of experiments involving magnetic field and Fermi level position (voltage).

In Fig. 8.15 we show experimental results of Wees, et al., carried out on a GaAs/AlGaAs MODFET at low temperatures. As shown, a pair of contacts are used to create a short channel of the high mobility region, and conductance is measured. The gates form a 1-dimensional channel in which the Fermi level and thus the electron

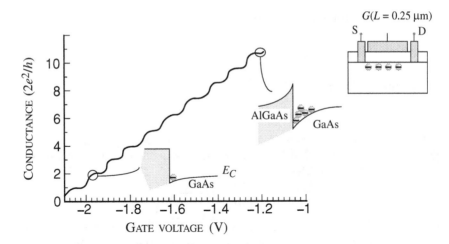

Figure 8.15: Experimental studies on conductance fluctuations arising in a GaAs/AlGaAs channel constricted by the structure shown. The results are for the channel conductance in units of $e^2/\pi\hbar$ ($= 2e^2/h$). (From the paper by B. J. van Wees, et al., Phys. Rev. Lett., **60**, 848 (1988).)

wavefunctions can be altered. As can be seen from the Fig. 8.15, there are oscillations in the conductance as suggested by the Landauer formalism.

There is a great deal of ongoing work in mesoscopic systems and improvements in fabrication technology have allowed demonstrations of mesoscopic effects at temperatures larger than liquid nitrogen temperatures. As traditional semiconductor devices shrink further their physics will increasingly be described by mesoscopic effects.

8.6.2 Coulomb Blockade Effects

In addition to coherent transport effects that can occur in very small systems, electron charging energy effects can become significant in small structures. An interesting and potentially important phenomenon manifests itself when structures have very small capacitance. This is called the Coulomb blockade effect. We are familiar with the parallel plate capacitor with capacitance C and the relation between a charge increment ΔQ and the potential variation ΔV

$$C = \frac{\Delta Q}{\Delta V} \text{ or } \Delta V = \frac{\Delta Q}{C} \tag{8.72}$$

The capacitance is given by the spacing of the plates (d) and the area (A)

$$C = \frac{\epsilon A}{d} \tag{8.73}$$

Now consider a case where the capacitance becomes smaller and smaller until a single electron on the capacitor causes a significant change in the voltage. The charging energy to place a single electron on a capacitor is

$$\Delta E = \frac{e^2}{2C} \tag{8.74}$$

and the voltage needed is

$$\frac{e}{2C} = \frac{80 \ mV}{C(aF)} \tag{8.75}$$

where the capacitance is in units of 10^{-18} F(aF). If we write the charging energy as a thermal energy, $k_B T_0$, the temperature associated with the charging energy is

$$T_0 = \frac{e^2}{2k_B C} = \frac{928.5 \ K}{C(aF)} \tag{8.76}$$

Coulomb blockade effects will manifest themselves if the sample temperature T is smaller than this effective charging temperature T_0. Here are the implications of the simple equations described above:

• Once a capacitor reaches values approaching $\sim 10^{-18}$ F, each electron causes a shift in voltage of several 10s of millivolts.

• The charging energy of the capacitor, i.e., the energy needed to place a single extra electron becomes comparable to or larger than $k_B T$ with T reaching 10 K or even 100 K if the capacitance becomes comparable to 10^{-18} F.

These observations which arise from "Coulomb blockade" can be exploited to design very interesting devices that have the potential for low power/high density electronic devices. To get the small capacitors needed to generate Coulomb blockade effects at reasonable temperatures one has to use areal dimensions of $\overset{<}{\sim} 1000$ Å$\times$ 1000 Å with spacing between the contacts reaching ~ 50–100 Å. With such dimensions (using a relative dielectric constant of ~ 10) we get capacitors with capacitances of the order of $\sim 10^{-16}$ F. The charging voltages are then ~ 1 mV and $T_0 \sim 10$ K. If the area of the capacitor is reduced further these values increase. It is possible to fabricate small capacitors with capacitance approaching 10^{-18} F.

In Fig. 8.16 we show the band profile of a typical tunnel junction capacitor which consists of two metal contacts separated by a thin tunneling barrier. In the absence of any Coulomb blockade (i.e., for a large capacitor) as the bias is applied electrons tunnel across the barrier and we simply get an ohmic behavior as shown in Fig. 8.16a.

In Fig. 8.16b we show the behavior for a structures where the charging energy is large enough to have measurable effects. At zero bias there is no net flow of electrons as usual. However at small biases smaller than the charging energy, an electron cannot move from the left to the right because that would raise the energy of the right side by $e^2/2C$ as shown. Once the voltage level (times electron charge) exceeds the charging energy, electrons can flow across the junction and we have ohmic behavior. The current–voltage relation shows a highly non-linear behavior as shown in Fig. 8.16b and has been exploited to demonstrate interesting devices with applications in logic, storage, etc.

The effects sketched in Fig. 8.16b have a strong temperature dependence. As the temperature rises, the distribution of carriers in the contact is smeared by $\sim k_B T$. As a result the temperature dependence of the current–voltage curves is shown schematically

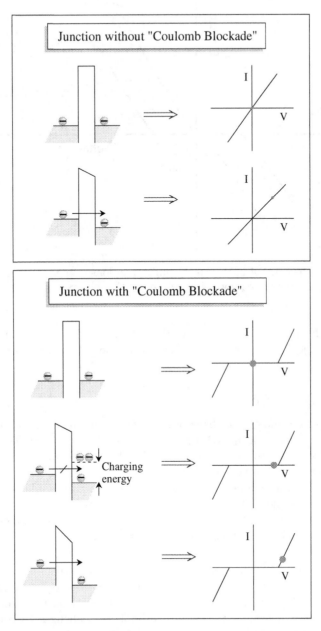

Figure 8.16: (a) A normal tunnel junction with large capacitance shows ohmic I–V characteristics. (b) In very small capacitance tunnel junctions the presence of a Coulomb blockade ensures no current flows until the voltage reaches a certain point.

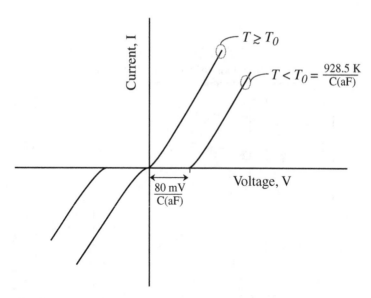

Figure 8.17: A schematic of how current voltage relations change as temperature is raised. Above T_0, defined in the figure, normal ohmic conduction occurs.

in Fig. 8.17. The temperature, T_0 defined above gives the upper temperature up to which Coulomb blockade effect can be observed in the I–V characteristics.

8.7   TECHNOLOGY ISSUES

Semiconductor electronic devices have been driven by miniturization and (to a smaller extent) new materials. Silicon MOSFET is, by far, the dominant device and is used for memory cells, transistor, and capacitors. The saga of semiconductor technology is captured by two laws observed by Gordon Moore, the co-founder of Intel. According to Moore's fist law, the number of active elements on a chip double every 18 months! The second law says the cost of a fabrication facility grows on a semi-log scale! With lab facilities already approaching the cost of well over a billion dollars, technologists are desperate to violate the second law! And the first law is also getting close to hitting a wall.

In Table 8.1 we show how semiconductor technology is expected to develop according to the roadmap prepared by Semiconductor Industry Associates (SIA). We can see from the SIA roadmap that gate length of MOSFETs will approach ~ 500 Å by ~ 2010. Of course, numerous material and technology challenges (two important ones are listed in Table 8.1) will have to be overcome for this to happen.

At channel lengths below 50 nm (500 Å) electrons will move from the source to drain without scattering. Also Coulomb blockade effects will become increasingly important. Thus physics considerations discussed in this chapter will become increasingly important.

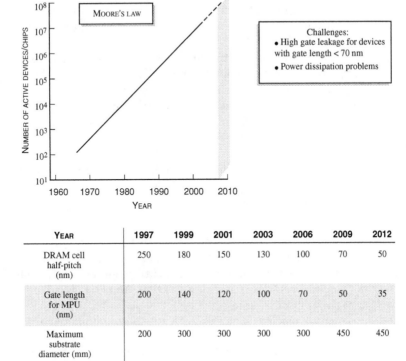

YEAR	1997	1999	2001	2003	2006	2009	2012
DRAM cell half-pitch (nm)	250	180	150	130	100	70	50
Gate length for MPU (nm)	200	140	120	100	70	50	35
Maximum substrate diameter (mm)	200	300	300	300	300	450	450
Acceptable defect density at 60% yield for DRAM	2080	1455	1310	1040	735	520	370
Defect density for MPU	1940	1710	1510	1355	1120	940	775
Power supply (V)	1.8-2.5	1.5-1.8	1.2-1.5	1.2-1.5	0.9-1.2	0.6-0.9	0.5-0.6
Power dissipation with heat sink (W)	70	90	110	130	160	170	175
Power dissipation without heat sink (for portable electronics) (W)	1.2	1.4	1.7	2.0	2.4	2.8	3.2
Cost per function DRAM (μcents/function)	120	60	30	15	5.3	1.9	0.66
Cost per function MPU (μcents/function)	3000	1735	1000	580	255	110	49

Table 8.1: An overview of semiconductor device roadmap prepared by the Semiconductor Industry Associates.

8.8   PROBLEMS

Section 8.2

8.1   Consider a GaAs sample in which a field of 10 kV/cm is applied. Discuss the conditions under which Bloch oscillations can occur. Also calculate the frequency of oscillations.

8.2 Design a GaAs/AlAs superlattice structure in which Bloch oscillations could occur when the scattering rate is 10^{13} s^{-1} and the applied field is 10 kV/cm. Discuss possible effects that could prevent the observation of the oscillations.

8.3 Consider a Si crystal in which a field of 10^4 V/cm is applied. Calculate the Bloch oscillation period if the field is applied along the i) [100]; ii) [110], and iii) [111] directions.

Section 8.3

8.4 In the resonant tunnel structure the transmission probability vs. energy plot has resonances with a linewidth ΔE_n. Show that if E_n is the energy of the nth resonance,

$$\Delta E_n \sim \frac{E_n T_{1B}}{\pi n}$$

where T_{1B} is the transmission through a single barrier.

8.5 Estimate the time an electron will take to tunnel through a resonant tunnel double barrier structure. You can use the Heisenberg relation $\Delta t \Delta E \sim \hbar$, where ΔE is the energy linewidth of the transmission resonance.

8.6 Consider a resonant tunneling structure with the following parameters:

$$
\begin{aligned}
\text{Barrier height}, V_0 &= 0.3 \text{ eV} \\
\text{Well size}, W &= 60 \text{ Å} \\
\text{Barrier width}, a &= 25 \text{ Å} \\
\text{Effective mass}, m^* &= 0.07 \, m_0
\end{aligned}
$$

Calculate and plot the tunneling probability of electrons as a function of energy for $0 < E < V_0$.

Section 8.4

8.7 Consider a 0.1 μm AlGaAs/GaAs device in which a 2-dimensional gas is formed with a density of $n_{2D} = 10^{12}$ cm^{-2}. A split gate device is made from the structure. Calculate the gate voltage needed (assume this is $\Delta E_F/e$) to switch a quantum interference transistor.

8.8 In normal transistors the ON and OFF states of the device are produced by injecting and removing electrons in the device. Consider a Si device with an area of 2.0 μm×0.1 μm in which a 1 V gate bias changes the electron density in the channel from 10^{12} cm^{-2} to 10^8 cm^{-2}, thus switching the device from ON to OFF. What is the switching energy?

   Estimate the switching energy if quantum interference effects were used in the same device.

8.9 Discuss what kinds of scattering processes will destroy quantum interference effects in devices and what kinds will not. Consider the various scattering mechanisms considered in Chapters 5 and 6.

Section 8.5

8.10 In a perfect crystalline material indirect gap materials have very weak interaction with photons. Explain why. Discuss why amorphous Si has a strong interaction with photons.

8.11 The room temperature mobility in amorphous silicon is in the range of 0.1 to 1.0 cm^2/V·s. What is the phase coherence distance for electrons?

Section 8.6

8.13 Consider a 2-dimensional electron channel in a AlGaAs/GaAs device. The gate length is 0.1 μm and gate width is 2.0 μm. The device is biased so that the electron density in the channel is 10^{12} cm^{-2}. How much will the electron number in the channel change if $\Delta\sigma = e^2/h$? Use a semi-classical model with mobility 10^5 cm^2/V·s.

8.14 Consider a metal-oxide-silicon capacitor. At what areal dimensions will it display Coulomb blockade effects at 300 K? The relative dielectric constant of SiO$_2$ is 3.9 and the oxide thickness is 50 Å.

8.15 Consider a MOSFET in which the capacitance is 10^{-18} F. The gate capacitor state is altered by a single electron (at very low temperatures). Calculate the change in the device channel current if the device transconductance is

$$g_m = \frac{\delta I}{\delta V_G} = 1.0 \text{ S}$$

Such devices are called *single electron transistors*.

8.9 REFERENCES

- **Disordered Semiconductors**

    - Anderson, P. W., Phys. Rev., **109**, 1492 (1958).

    - Cohen, M. H., H. Fritzsche and S. R. Ovshinsky, Phys. Rev. Lett., **22**, 1065 (1977).

    - Halperin, B. I. and M. Lax, Phys. Rev., **153**, 802 (1967).

    - Lifshitz, I. M., Adv. Phys., **13**, 483 (1969).

    - Lloyd, P. and J. Best, J. Phys. C, **8**, 3752 (1975).

    - Morigaki, Kazuo, *Physics of Amorphous Semiconductors* (World Scientific Publishing Company, 1999).

    - Mott, N. F. and E. A. Davis, *Electronic Processes in Non-Crystalline Materials* (Clarendon Press, Oxford, 1971).

    - Singh, J. and A. Madhukar, *Band-Tailing in Disordered Systems in Excitations in Disordered Systems* (edited by M. F. Thorpe, Plenum Press, New York, 1982). See other articles in this book, also.

  – Thouless, D. J., Phys. Rept., **13C**, 93 (1974).

- **Mesoscopic Structures**

  – Articles in *Nanostructure Physics and Fabrication* (edited by M. A. Reed and W. P. Kirk, Academic Press, New York, 1989).

  – Datta, Supriyo, *Electronic Transport in Mesoscopic Systems* (Cambridge University Press, 1995).

  – Ferry, D. K., *Semiconductors* (Macmillan, New York, 1991).

  – Gradert, Hermann and H. Michel, Editors, *Single Charge Tunneling: Coulomb Blockade Phenomenon in Nanostructures* (NATO ASI Series, **B**, Physics, vol. 294), Plenum Publishing Corporation, 1992.

  – Janssen, Martin, *Fluctuations and Localization in Mesoscopic Electron Systems* (World Scientific Publishing Company, 1991).

  – Landauer, R., Philos. Mag., **21**, 863 (1970).

  – Murayama, Yoshinasa, *Mesoscopic Systems*, (John Wiley and Sons, 2001).

  – Physics Today, (Dec. 1988). Covers the important aspects of physics in mesoscopic structures.

  – Van Wees, B. J., H. Van Houten, C. W. J. Beenakker, J. L. Williamson, L. P. Kauwenhoven, D. van der Marel, and C. T. Foxon, Phys. Rev. Lett., **60**, 848 (1988).

Chapter 9

OPTICAL PROPERTIES OF SEMICONDUCTORS

9.1 INTRODUCTION

Interactions of electrons and photons in semiconductors form the basis of technologies such as optical communications, display, and optical memories. In this and the next chapter we will discuss how electrons in a semiconductor interact with light. To describe this interaction, light has to be treated as particles (i.e., photons). The problem is mathematically quite similar to the electron-phonon (lattice vibration) scattering problem discussed in Chapter 6. Electron-photon interactions are described via scattering theory through an absorption or emission of a photon. Both intraband and interband processes can occur as shown in Fig. 9.1. Intraband scattering in semiconductors is an important source of loss in lasers and can usually be described by a Drude-like model where a sinusoidal electric field interacts with electrons or holes. Monte Carlo methods or other transport models can account for it quite adequately. The interband scattering involving valence and conduction band states is, of course, most important for optical devices such as lasers and detectors. In addition to the band-to-band transitions, increasing interest has recently focussed on excitonic states especially in quantum well structures. The exciton-photon interaction in semiconductor structures contains important physics and is also of great technical interest for high speed modulation devices and optical switches. Excitonic effects will be discussed in the next chapter.

We will briefly review some important concepts in electromagnetic theory and then discuss the interactions between electrons and photons. We will focus on the special aspects of this interaction for semiconductor electrons, especially those relating to selection rules. Both 3-dimensional and lower-dimensional systems will be covered. We will also focus on the selection rules and "gain" in semiconductor structures, considerations which are extremely important in solid state lasers.

While the single electron picture which gives us the E vs. k relation provides a

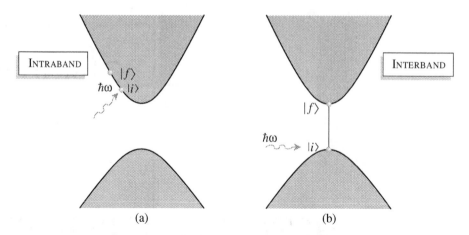

Figure 9.1: Intraband and interband scattering of an electron from an initial state k_i to a final state k_f.

good description of optical processes for photon energies above the bandgap, just below the bandgap there are extremely important optical interactions involving the electron-hole system. These processes, known as excitonic processes, are becoming increasingly important for technological applications such as high speed switches and modulators. We will discuss the exciton related effects in Chapter 10.

9.2   MAXWELL EQUATIONS AND VECTOR POTENTIAL

The properties of electromagnetic fields in a medium are described by the four Maxwell equations. Apart from the electric ($\boldsymbol{F}$) and magnetic ($\boldsymbol{B}$) fields and velocity of light, the effects of the material are represented by the dielectric constant, permeability (we will assume that the permeability $\mu = \mu_o$), electrical conductivity, etc., We start with the four Maxwell equations

$$
\begin{aligned}
\nabla \times \boldsymbol{F} + \frac{\partial \boldsymbol{B}}{\partial t} &= 0 \\
\nabla \times \boldsymbol{H} - \frac{\partial \boldsymbol{D}}{\partial t} &= \boldsymbol{J} \\
\nabla \cdot \boldsymbol{D} &= \rho \\
\nabla \cdot \boldsymbol{B} &= 0
\end{aligned}
\tag{9.1}
$$

where $\boldsymbol{F}$ and $\boldsymbol{H}$ are the electric and magnetic fields, $\boldsymbol{D} = \epsilon \boldsymbol{F}$, $\boldsymbol{B} = \mu \boldsymbol{H}$, $\boldsymbol{J}$, and ρ are the current and charge densities. In dealing with the electron-photon interactions, it is convenient to work with the vector and scalar potentials $\boldsymbol{A}$ and ϕ respectively, which are defined through the equations

$$
\begin{aligned}
\boldsymbol{F} &= -\frac{\partial \boldsymbol{A}}{\partial t} - \nabla \phi \\
\boldsymbol{B} &= \nabla \times \boldsymbol{A}
\end{aligned}
\tag{9.2}
$$

The first and fourth Maxwell equations are automatically satisfied by these definitions. The potentials $\boldsymbol{A}$ and ϕ are not unique, but can be replaced by a new set of potentials $\boldsymbol{A}'$ and ϕ' given by

$$\begin{aligned} \boldsymbol{A}' &= \boldsymbol{A} + \nabla\chi \\ \phi' &= \phi - \frac{\partial\chi}{\partial t} \end{aligned} \tag{9.3}$$

The new choice of potentials does not have any effect on the physical fields $\boldsymbol{F}$ and $\boldsymbol{B}$. To at least partially remove the arbitrariness of the potentials $\boldsymbol{A}$ and ϕ, we define certain transformations called the gauge transformations which restrict them. Before considering these transformations let us rewrite the Maxwell equations in terms of $\boldsymbol{A}$ and ϕ.

The second and third Maxwell equations become

$$\begin{aligned} \frac{1}{\mu_o}\nabla \times \nabla \times \boldsymbol{A} + \epsilon\frac{\partial^2 \boldsymbol{A}}{\partial t^2} + \epsilon\nabla\frac{\partial\phi}{\partial t} &= \boldsymbol{J} \\ \frac{\partial}{\partial t}\nabla \cdot \boldsymbol{A} + \nabla^2\phi &= \frac{\rho}{\epsilon} \end{aligned}$$

Now

$$\nabla \times \nabla \times \boldsymbol{A} = \nabla(\nabla \cdot \boldsymbol{A}) - \nabla^2\boldsymbol{A}$$

giving us

$$\nabla\left(\frac{1}{\mu_o}\nabla \cdot \boldsymbol{A} + \epsilon\frac{\partial\phi}{\partial t}\right) - \frac{1}{\mu_o}\nabla^2\boldsymbol{A} + \epsilon\frac{\partial^2 \boldsymbol{A}}{\partial t^2} = \boldsymbol{J}$$

$$\frac{\partial}{\partial t}(\nabla \cdot \boldsymbol{A}) + \nabla^2\phi = -\frac{\rho}{\epsilon}$$

We will now impose certain restrictions on $\boldsymbol{A}$ and ϕ to further simplify these equations. Note that these restrictions have no implications on $\boldsymbol{F}$ and $\boldsymbol{B}$ fields as can be easily verified from their descriptions in terms of $\boldsymbol{A}$ and ϕ. The purpose of working in different gauges is mathematical elegance and simplicity.

Since the vector potential $\boldsymbol{A}$ is defined in terms of its curl: $\boldsymbol{B} = \nabla \times \boldsymbol{A}$, its divergence $(\nabla \cdot \boldsymbol{A})$ is arbitrary. Choice of a particular gauge is equivalent to the choice of the the value of $\nabla \cdot \boldsymbol{A}$.

In a gauge known as the Lorentz gauge, widely used in relativistic electrodynamics, we choose

$$\nabla \cdot \boldsymbol{A}' + \frac{\partial\phi'}{\partial t} = 0$$

This is equivalent to imposing the following restriction on the arbitrary quantity χ

$$\nabla^2\chi - \frac{\partial^2\chi}{\partial t^2} = -\left(\nabla \cdot \boldsymbol{A} + \frac{\partial\phi}{\partial t}\right)$$

With this choice of χ we get for the Maxwell equations

$$\frac{1}{\mu_o}\nabla^2 \boldsymbol{A} - \epsilon \frac{\partial^2 \boldsymbol{A}}{\partial t^2} = \boldsymbol{J}$$

$$\nabla^2 \phi - \frac{\partial^2 \phi}{\partial t^2} = -\frac{\rho}{\epsilon} \qquad (9.4)$$

This form of Maxwell's equations is extremely useful for generalizing to relativistic electrodynamics. In dealing with electron-photon interactions, a most useful gauge is the radiation or Coulomb gauge. If $\boldsymbol{J} = 0$ and $\rho = 0$, we can choose the constant background potential $\phi' = 0$. In addition we can choose $\nabla \cdot \boldsymbol{A}' = 0$. In this case the solutions for the vector potential are represented by plane wave transverse electromagnetic waves. We will be working in this gauge. It is useful to establish the relation between the vector potential $\boldsymbol{A}$ and the photon density which represents the optical power. The time dependent solution for the vector potential solution of Eqn. 9.10 with $\boldsymbol{J} = 0$ is

$$A(\boldsymbol{r}, t) = \boldsymbol{A}_0 \left\{ \exp\left[i(\boldsymbol{k} \cdot \boldsymbol{r} - \omega t)\right] + \text{ c.c.} \right\} \qquad (9.5)$$

with

$$k^2 = \epsilon \mu_o \omega^2$$

Note that in the MKS units $(\epsilon_o \mu_o)^{1/2}$ is the velocity of light c (3×10^8 ms^{-1}). The electric and magnetic fields are

$$\boldsymbol{F} = \frac{\partial \boldsymbol{A}}{\partial t}$$

$$= -2\omega \boldsymbol{A}_0 \sin(\boldsymbol{k} \cdot \boldsymbol{r} - \omega t)$$

$$\boldsymbol{B} = \nabla \times \boldsymbol{A}$$

$$= -2\boldsymbol{k} \times \boldsymbol{A}_0 \sin(\boldsymbol{k} \cdot \boldsymbol{r} - \omega t) \qquad (9.6)$$

The Poynting vector $\boldsymbol{S}$ representing the optical power is

$$\boldsymbol{S} = (\boldsymbol{F} \times \boldsymbol{H})$$

$$= \frac{4}{\mu_o} v k^2 |\boldsymbol{A}_0|^2 \sin^2(\boldsymbol{k} \cdot \boldsymbol{r} - \omega t)\hat{k} \qquad (9.7)$$

where v is the velocity of light in the medium ($= c/\sqrt{\tilde{\epsilon}}$) and $\hat{k}$ is a unit vector in the direction of $\boldsymbol{k}$. Here $\tilde{\epsilon}$ is the relative dielectric constant. The time averaged value of the power is

$$<\boldsymbol{S}>_{\text{time}} = \hat{k} \frac{2vk^2 |\boldsymbol{A}_0|^2}{\mu_o}$$

$$= 2v\epsilon\mu_o\omega^2 |\boldsymbol{A}_0|^2 \hat{k} \qquad (9.8)$$

since

$$|\boldsymbol{k}| = \omega/v \qquad (9.9)$$

The energy density is then

$$\left|\frac{\boldsymbol{S}}{v}\right| = \frac{2\epsilon\,\omega^2\,|\boldsymbol{A}_0|^2}{c^2} \tag{9.10}$$

Also, if the photon mode occupation is n_{ph}, the energy density is (for a volume V)

$$\frac{n_{ph}\hbar\omega}{V} \tag{9.11}$$

Equating these two, we get

$$|A_0|^2 = \frac{n_{ph}\hbar}{2\epsilon\omega V} \tag{9.12}$$

Having established these basic equations for the electromagnetic field, we will now develop a macroscopic picture of the photon-material interactions. Going back to the Maxwell's equations and writing $\boldsymbol{J} = \sigma\boldsymbol{F}$, we get the wave equation for the electric field (after eliminating the $\boldsymbol{B}$-field)

$$\nabla^2\boldsymbol{F} = \epsilon\mu_o\frac{\partial^2\boldsymbol{F}}{\partial t^2} + \sigma\mu_o\frac{\partial\boldsymbol{F}}{\partial t} \tag{9.13}$$

This represents a wave propagating with dissipation. The general solution can be chosen to be of the form

$$\boldsymbol{F} = \boldsymbol{F}_0 \exp\left\{i(\boldsymbol{k}\cdot\boldsymbol{r} - \omega t)\right\} \tag{9.14}$$

so that k is given by

$$-k^2 = -\epsilon\mu_o\omega^2 - \sigma\mu_o i\omega$$

or $\left(c = (\epsilon_o\mu_o)^{-1/2}\right)$

$$k = \frac{\omega}{c}\left(\tilde{\epsilon} + \frac{\sigma\mu_o i}{\omega}\right)^{1/2} \tag{9.15}$$

In general, k is a complex number. In free space where $\sigma = 0$ we simply have $(\tilde{\epsilon} = 1)$

$$k = \omega/c$$

In a medium, the phase velocity is modified by dividing c by a complex refractive index given by

$$n_r = \left(\tilde{\epsilon} + \frac{\sigma\mu_o i}{\omega}\right)^{1/2} \tag{9.16}$$

We can write the complex refractive index in terms of its real and imaginary parts

$$n_r = n'_r + in''_r \tag{9.17}$$

so that

$$k = \frac{n'_r\omega}{c} + in''_r\frac{\omega}{c}$$

The electric field wave Eqn. 9.14 now becomes (for propagations in the +z direction)

$$\boldsymbol{F} = \boldsymbol{F}_0 \exp\left\{ i\omega \left(\frac{n_r' z}{c} - t \right) \right\} \exp\left(\frac{-n_r'' \omega z}{c} \right) \tag{9.18}$$

The velocity of the wave is reduced by n_r' to c/n_r' and its amplitude is damped exponentially by a fraction $\exp\left(-2\pi n_r''/n_r'\right)$ per wavelength. The damping of the wave is associated with the absorption of the electromagnetic energy. The absorption coefficient α is described by the absorption of the intensity (i.e., square of Eqn. 9.18)

$$\alpha = \frac{2n_r'' \omega}{c} \tag{9.19}$$

Note that in the absence of absorption, $n_r' = n_r$, and the refraction index will simply be denoted by n_r. The absorption coefficient can be measured for any material system and it provides information on n_r''.

In the above formalism we have seen that the response of the medium to the outside world is represented by a complex dielectric constant or a complex refractive index. The real and imaginary parts of the relative dielectric constants are related to the refractive index components by the following relations (from Eqns. 9.16 and 9.17)

$$\begin{aligned} \tilde{\epsilon}_1 &= n_r'^2 - n_r''^2 \\ \tilde{\epsilon}_2 &= 2n_r' n_r'' \end{aligned} \tag{9.20}$$

where $\tilde{\epsilon}_1$ and $\tilde{\epsilon}_2$ are the real and imaginary parts of the relative dielectric constant. An important relation is satisfied by the real and imaginary parts of all response functions provided the causality principle is obeyed (effect follows the cause in time). The relation is given by an expression known as the Kramers Kronig relation which when applied to the refractive index and absorption coefficient gives the relation

$$n(\omega_o) - 1 = \frac{c}{\pi} P \int_o^\infty \frac{\alpha(\omega)d\omega}{\omega^2 - \omega_o^2} \tag{9.21}$$

This relation is extremely useful since it allows one to calculate the refractive index if the absorption coefficient is known. In particular, if the absorption spectra is modulated by some external means (say, an electric field), the effect on the refractive index can be calculated.

EXAMPLE 9.1 An electromagnetic radiation with a power density of 1 $\mu W/m^2$ impinges upon a receiver. Calculate the electric field amplitude of this radiation if the photon energy is 0.8 eV.

The power density is extremely small, but low noise detectors available in today's technology can detect such levels.

The photon density is related to the power density by

$$\text{Power density} = n_{ph} \hbar w c$$

Thus,

$$\frac{n_{ph}}{V} = \frac{(10^{-6} \ \text{Wm}^{-2})}{(0.8 \times 1.6 \times 10^{-19} \ \text{J})(3 \times 10^8 \ \text{m/s})}$$

$$= 2.6 \times 10^4 \text{ m}^{-3}$$

From Eqn. 9.12 we have

$$|A_o|^2 = \frac{(2.6 \times 10^4 \text{ m}^{-3})(1.05 \times 10^{-34} \text{ Js})}{2 \times (8.84 \times 10^{-12} \text{ } F/m)(1.22 \times 10^{15} \text{ rad/s})}$$

The electric field amplitude is

$$|F_o| = 2\omega A_o = 2.745 \times 10^{-2} \text{ V/m}$$
$$= 2.745 \times 10^{-4} \text{ V/cm}$$

9.3 ELECTRONS IN AN ELECTROMAGNETIC FIELD

The electron carries a negative charge which interacts with the electric and magnetic fields of the electromagnetic radiation. The force on the electron due to the electric field is simply the charge times the field and that due to the magnetic field is the Lorentz force determined by the cross product of the electron velocity and the field. The energy associated with this interaction determines the interaction Hamiltonian that is responsible for electron scattering. The Hamiltonian describing the interactions between a charge, e, and the electromagnetic field is

$$
\begin{aligned}
H &= \frac{1}{2m_0} \left(\boldsymbol{p} - e\boldsymbol{A} \right)^2 + e\phi + V(\boldsymbol{r}) \\
&= \frac{p^2}{2m_0} - \frac{e}{2m_0}(\boldsymbol{p} \cdot \boldsymbol{A} + \boldsymbol{A} \cdot \boldsymbol{p}) + \frac{e^2}{2m_0}A^2 + e\phi + V(\boldsymbol{r}) \quad (9.22)
\end{aligned}
$$

Here $\boldsymbol{A}$ is the vector potential and $V(\boldsymbol{r})$ is any background crystal potential.

In quantum mechanics the momentum $\mathbf{p}$ is a differential operator so we have

$$\frac{e}{2m_0}\boldsymbol{p} \cdot \boldsymbol{A} = \frac{e\boldsymbol{A} \cdot \boldsymbol{p}}{2m_0} - \frac{ie\hbar}{2m_0}\nabla \cdot \boldsymbol{A}$$

The Hamiltonian becomes

$$H = \frac{p^2}{2m_0} - \frac{e}{m_0}\boldsymbol{A} \cdot \boldsymbol{p} + \frac{ie\hbar}{2m_0}\nabla \cdot \boldsymbol{A} + \frac{e^2}{2m_0}A^2 + e\phi + V(\boldsymbol{r})$$

We will now use the perturbation theory to study the effect of the electromagnetic radiation on the electron. In the quantum theory of radiation, the electromagnetic field is written in terms of creation and destruction operators, in analog with the harmonic oscillator problem. In Chapter 6 we have applied the same approach to the electron-phonon interaction.

The Schrödinger equation to be solved is

$$
\begin{aligned}
i\hbar \frac{\partial \psi}{\partial t} &= \left[-\frac{\hbar^2}{2m_0}\nabla^2 + \frac{ie\hbar}{m_0 c}\boldsymbol{A} \cdot \nabla + \frac{ie\hbar}{2m_0}(\nabla \cdot \boldsymbol{A}) \right. \\
&\quad + \left. \frac{e^2}{2m_0}A^2 + e\phi + V(\boldsymbol{r}) \right] \psi \quad (9.23)
\end{aligned}
$$

We will work in the radiation gauge ($\nabla \cdot \boldsymbol{A} = \phi = 0$), and use the time dependent theory which results in scattering rates given by the Fermi golden rule. We assume that the optical power and consequently A is small, so that

$$\left| \frac{ie\hbar}{m_0} \boldsymbol{A} \cdot \nabla \right| : \left| \frac{\hbar^2}{2m_0} \nabla^2 \right| \approx \left| \frac{e^2 A^2}{2m_0^2} \right| : \left| \frac{ie\hbar}{m} \boldsymbol{A} \cdot \nabla \right|$$

$$\approx \frac{eA}{p}$$

For an optical power of 1 W/cm^2 the photon density of a 1 eV energy beam is $\sim 10^9$ cm^{-3}. Using an electron velocity of 10^6 cm/s, one finds that

$$\frac{eA}{p} \sim 10^{-5}$$

Thus, even for an optical beam carrying 1 MW/cm^2, the value of eA/p is small enough that perturbation theory can be used. We will, therefore, only retain the first order term in $\boldsymbol{A}$.

The Schrödinger equation is now written as

$$i\hbar \frac{\partial \psi}{\partial t} = (H_0 + H^{'})\psi \tag{9.24}$$

where

$$H_0 = -\frac{\hbar}{2m_0} \nabla^2 + V(\boldsymbol{r}) \tag{9.25}$$

and

$$H^{'} = \frac{ie\hbar}{m_0} \boldsymbol{A} \cdot \nabla \tag{9.26}$$

The scattering problem is schematically represented in Fig. 9.2 where the perturbation due to the electromagnetic field causes a scattering of the electron. The vector potential A represents the electromagnetic field which in the "second quantization" is represented by an operator. This operator can be written in terms of creation and destruction operators $b^\dagger$ and b as discussed for the harmonic oscillator in Chapter 6.

From the Fermi golden rule the scattering rate from the initial electron state $|i\rangle$ to the final state $|f\rangle$ is

$$W(i) = \frac{2\pi}{\hbar} \sum_f \left| \langle f|H^{'}|i\rangle \right|^2 \delta\left(\boldsymbol{E}_f - \boldsymbol{E}_i \mp \hbar\omega\right)$$

where the upper sign is for photon absorption and the lower one is for emission.

Note that the initial and final electron and photon states can be represented by the momentum states of the electron, along with the photon densities as shown in Fig. 9.3. The initial and final electron-photon states are represented by the following: Absorption:

$$|i\rangle = |k_i, n_{ph}\rangle$$
$$|f\rangle = |k_f, n_{ph} - 1\rangle$$

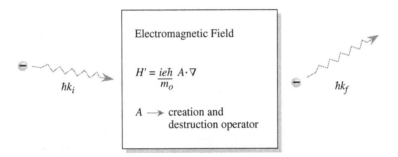

Figure 9.2: A schematic of the scattering of an electron by the electromagnetic field.

Emission:

$$|i\rangle = |k_i, n_{ph}\rangle$$
$$|f\rangle = |k_f, n_{ph} + 1\rangle$$

where k_i and k_f are the electron's initial and final wave vectors and n_{ph} is the photon density in the initial state. The vector potential is written as

$$A_o = \sqrt{\frac{\hbar}{2\omega\epsilon V}}\left(b^\dagger + b\right) \tag{9.27}$$

Here $b^\dagger$ and b are the photon creation and destruction operators. Note that this choice gives the value for $|A_o|^2$ given in Eqn. 9.12.

The creation and destruction operations have the matrix elements (focusing only on the photon term of the matrix element; a is the polarization unit vector).
Absorption:

$$\langle i \mid A \cdot \nabla \mid f \rangle \Rightarrow \sqrt{\frac{\hbar}{2\omega\epsilon}}\left\langle k_f, n_{ph} - 1 \left| (b^\dagger + b) \cdot a \cdot \nabla \right| k_i, n_{ph} \right\rangle$$

$$= \sqrt{\frac{\hbar}{2\omega\epsilon}}(n_{ph})^{1/2} \langle k_f \mid a \cdot \nabla \mid k_i \rangle$$

Emission:

$$\langle i \mid A \cdot \nabla \mid f \rangle \Rightarrow \sqrt{\frac{\hbar}{2\omega\epsilon}}\left\langle k_f, n_{ph} + 1 \left| (b^\dagger + b) a \cdot \nabla \right| k_i, n_{ph} \right\rangle$$

$$= \sqrt{\frac{\hbar}{2\omega\epsilon}}(n_{ph} + 1)^{1/2} \langle k_f \mid a \cdot \nabla \mid k_i \rangle$$

Notice the prefactors $(n_{ph})^{1/2}$ for the absorption process and $(n_{ph} + 1)^{1/2}$ for the emission process. This difference is extremely important, as will be discussed later.

Let us consider the photon absorption process where a photon with momentum $\hbar k_{ph}$ and energy $\hbar\omega$ is absorbed by an electron system. To calculate this rate we need

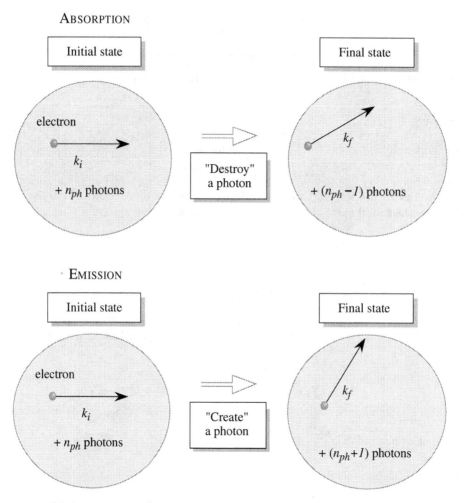

Figure 9.3: (a) A schematic of an absorption process where a photon is absorbed (destroyed) and the energy and momentum of the electron is altered; (b) the emission of a photon where a photon is created.

to sum over all possible electron states which can allow such a process to occur. The photon energy $\hbar\omega$ is transferred to the electron. To find the total scattering rate we integrate over the final electron density of states

$$W_{abs} = \frac{2\pi}{\hbar} \frac{e^2}{m_0^2} \left(\frac{\hbar n_{ph}}{2\omega\epsilon}\right) \sum_{\text{final states}} \left| \int \psi_f^*(a \cdot p)e^{ik_{ph}\cdot r}\psi_i d^3r \right|^2 \cdot \delta(E_i - E_f + \hbar\omega) \quad (9.28)$$

In the above expression, if we are considering the rate at which a photon is absorbed, this involves summing over any possible electronic levels, as long as the energy conservation is obeyed. One may, instead, be interested in the problem where a single electron with momentum k_i is scattered into a final state with momentum k_f. In this case the final states sum would be over all photon states that can cause such transitions and would involve the photon density of states. These two approaches are illustrated in Fig. 9.4 These issues will become apparent when we calculate the absorption and emission rates.

The rate of an electron $\hbar k_i$ to emit any photon and reach a state with momentum $\hbar k_f$ is similarly

$$W_{em} = \frac{2\pi}{\hbar} \frac{e^2}{m_o^2} \left(\frac{\hbar(n_{ph}+1)}{2\omega\epsilon}\right) \sum_{\text{final states}} \left| \int \psi_f^*(a \cdot p)e^{-ik_{ph}\cdot r}\psi_i d^3r \right|^2 \cdot \delta(E_i - E_f - \hbar\omega)$$

$$(9.29)$$

we notice that the emission term can be rewritten as stimulated and spontaneous emission terms

$$W_{em} = W_{st} + W_{spon}$$

where

$$W_{st} = \frac{2\pi}{\hbar} \frac{e^2}{m_0^2} \frac{\hbar n_{ph}}{2\omega\epsilon} \sum_{\text{final states}} \left| \int \psi_f^* e^{-ik_{ph}\cdot r}(a \cdot p)\psi_i d^3r \right|^2$$

$$\cdot \quad \delta(E_i - E_f - \hbar\omega) \quad (9.30)$$

$$W_{spon} = \frac{2\pi}{\hbar} \frac{e^2}{m_0^2} \frac{\hbar}{2\omega\epsilon} \sum_{\text{final states}} \left| \int \psi_f^* e^{-ik_{ph}\cdot r}\psi_i d^3r \right|^2 \quad (9.31)$$

$$\cdot \quad \delta(E_i - E_f - \hbar\omega)$$

The stimulated emission is due to the initial photons present in the system and the emitted photons maintain phase coherence with the initial photons. The spontaneous emission comes from the perturbations due to the vacuum state (i.e., $n_{ph} = 0$) energy fluctuations and the emitted photons are incoherent with no phase relationship. The difference between these two kinds of processes is the key to understanding the differences between a light emitting diode and the laser diode (Section 9.10).

Let us now focus on the semiconductor electronic states involved in the absorption or emission process. The photon momentum $\hbar k_{ph}$ for most energies of interest in solid state devices ($\hbar\omega \sim 0.1$ - 2.0 eV) is extremely small compared to the electron momentum so that momentum conservation requires that

$$k_i = k_f$$

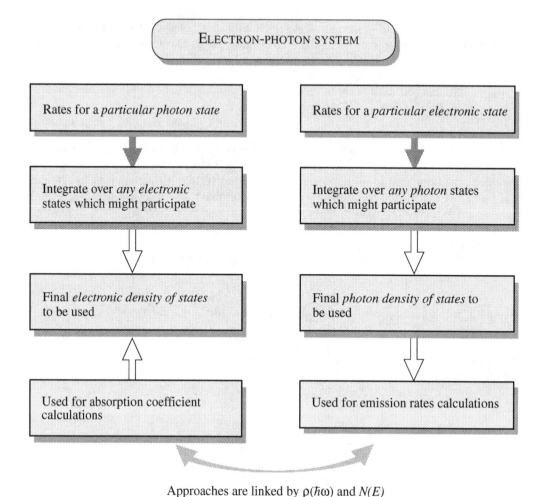

Approaches are linked by $\rho(\hbar\omega)$ and $N(E)$

Figure 9.4: Final states used in scattering rates in the Fermi golden rule. Depending upon the process being considered one may sum over final photon density of states or final electron density of states.

Thus, in first order perturbation theory, the electronic transitions due to photons are "vertical" in the $E\text{-}k$ description. An example for interband transitions is shown in Fig. 9.5. The approximation of neglecting k_{ph} is called the dipole approximation. In the dipole approximation the momentum matrix element, (the integral within the vertical bars in Eqns. 9.28 and 9.29) which we denote by p_{if} becomes quite simple.

Let us consider the initial and final states which have the Bloch function form. The momentum matrix element is in the dipole approximation

$$\boldsymbol{p}_{\mathrm{if}} = -i\hbar \int \psi_{\mathbf{k}_f \ell'}^* \, \nabla \psi_{\mathbf{k}_i \ell} \, d^3 r$$

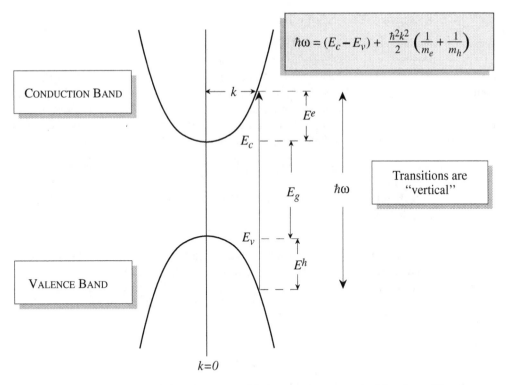

$$\hbar\omega = (E_c - E_v) + \frac{\hbar^2 k^2}{2}\left(\frac{1}{m_e} + \frac{1}{m_h}\right)$$

CONDUCTION BAND

Transitions are "vertical"

VALENCE BAND

$k=0$

Figure 9.5: The positions of the electron and hole energies at vertical k-values. The electron and hole energies are determined by the photon energy and the carrier masses. Since the photon momentum is negligible the transitions are vertical.

where we choose

$$
\begin{aligned}
|i\rangle &= \psi_{\mathbf{k}_i \ell} \\
&= e^{i\mathbf{k}_i \cdot \mathbf{r}} u_{\mathbf{k}_i \ell} \\
|f\rangle &= \psi_{\mathbf{k}_f \ell'} \\
&= e^{i\mathbf{k}_f \cdot \mathbf{r}} u_{\mathbf{k}_f \ell'}
\end{aligned}
$$

where $u_{\mathbf{k}\ell}$ is the cell periodic part of the Bloch state and ℓ, ℓ' are the band indices. Carrying out the differentiation we can write

$$\mathbf{p}_{\mathrm{if}} = \hbar\mathbf{k}_i \int \psi^*_{\mathbf{k}_f \ell'}\,\psi_{\mathbf{k}_i \ell}\,d^3 r - i\hbar \int u^*_{\mathbf{k}_f \ell'}\,(\nabla u_{\mathbf{k}_i \ell})\,e^{i(\mathbf{k}_i - \mathbf{k}_f)\cdot\mathbf{r}}\,d^3 r \qquad (9.32)$$

In the next section, we will examine the selection rules explicitly, for semiconductor systems of interest.

9.4   INTERBAND TRANSITIONS

9.4.1   Interband Transitions in Bulk Semiconductors

Let us consider the selection rules for *band-to-band* transitions in direct gap semiconductors as shown in Fig. 9.5. We will focus on direct gap materials based on the zinc blende structures (GaAs, InAs, InP, etc.) and will now use our understanding of the nature of conduction and valence band state central cell functions. The first term on the right-hand side of the momentum matrix element (Eqn. 9.32) is zero because of orthogonality of Bloch states. The second term requires ($u_{\mathbf{k}'\ell'}^* \nabla u_{\mathbf{k}\ell}$ is periodic)

$$\mathbf{k}_i - \mathbf{k}_f = 0$$

so that the interband transitions are "vertical" transitions. The interband matrix element is, therefore ($k_i = k_f = k$)

$$\langle u_{c\mathbf{k}} | \mathbf{p}_a | u_{v\mathbf{k}} \rangle$$

where $u_{c\mathbf{k}}$ and $u_{v\mathbf{k}}$ represent the conduction band and valence band central cell states.

For near bandedge transitions we will assume that $u_{c\mathbf{k}}$ and $u_{v\mathbf{k}}$ are given by their zone center values. We remind ourselves that in this case the central cell states are (See Chapter 2 for a discussion on the central cell states in the conduction and valence band states),

- Conduction band:

$$u_{c0} = |s\rangle \tag{9.33}$$

  where $|s\rangle$ is a spherically symmetric state.

- Valence band:

$$
\begin{aligned}
\text{Heavy hole states:} \quad & |3/2, 3/2\rangle = \frac{-1}{\sqrt{2}} \left(|p_x\rangle + i|p_y\rangle \right) \uparrow \\
& |3/2, -3/2\rangle = \frac{1}{\sqrt{2}} \left(|p_x\rangle - i|p_y\rangle \right) \downarrow \\
\text{Light hole states:} \quad & |3/2, 1/2\rangle = \frac{-1}{\sqrt{6}} \left[(|p_x\rangle + i|p_y\rangle) \downarrow -2|p_z\rangle \uparrow \right] \\
& |3/2, -1/2\rangle = \frac{1}{\sqrt{6}} \left[(|p_x\rangle - i|p_y\rangle) \uparrow +2|p_z\rangle >\downarrow \right]
\end{aligned}
\tag{9.34}
$$

From symmetry we see that *only* the matrix elements of the form

$$\langle p_x|p_x|s\rangle = \langle p_y|p_y|s\rangle = \langle p_z|p_z|s\rangle = p_{cv}$$

are nonzero. Thus, for band-to-band transition, the only allowed transitions have the following matrix elements

$$\langle \text{HH}|p_x|s\rangle = \langle \text{HH}|p_y|s\rangle = \frac{1}{\sqrt{2}} \langle p_x|p_x|s\rangle$$

$$\langle \text{LH}|p_x|s\rangle = \langle \text{LH}|p_y|s\rangle = \frac{1}{\sqrt{6}} \langle p_x|p_x|s\rangle$$

and

$$\langle \mathrm{LH}|p_z|s\rangle = \frac{2}{\sqrt{6}}\langle p_x|p_x|s\rangle = \frac{2}{\sqrt{6}}p_{cv}$$

It is important to note that

$$\langle \mathrm{HH}|p_z|s\rangle = 0$$

It is very useful to examine the matrix element square for light polarized along various orientations. This polarization dependence is accentuated in quantum wells because the HH and LH states are no longer degenerate in that case.

$$
\begin{aligned}
z\text{-polarized light:} \quad & \mathrm{HH} \rightarrow \text{c-band: No coupling} \\
& \mathrm{LH} \rightarrow \text{c-band:} \quad |\boldsymbol{p}_{\mathrm{if}}|^2 = \tfrac{2}{3}\,|\langle p_x|p_x|s\rangle|^2 \\
x\text{-polarized light:} \quad & \mathrm{HH} \rightarrow \text{c-band:} \quad |\boldsymbol{p}_{\mathrm{if}}|^2 = \tfrac{1}{2}\,|\langle p_x|p_x|s\rangle|^2 \\
& \mathrm{LH} \rightarrow \text{c-band:} \quad |\boldsymbol{p}_{\mathrm{if}}|^2 = \tfrac{1}{6}\,|\langle p_x|p_x|s\rangle|^2 \\
y\text{-polarized light:} \quad & \mathrm{HH} \rightarrow \text{c-band:} \quad |\boldsymbol{p}_{\mathrm{if}}|^2 = \tfrac{1}{2}\,|\langle p_x|p_x|s\rangle|^2 \\
& \mathrm{LH} \rightarrow \text{c-band:} \quad |\boldsymbol{p}_{\mathrm{if}}|^2 = \tfrac{1}{6}\,|\langle p_x|p_x|s\rangle|^2
\end{aligned}
\tag{9.35}
$$

We see that the z-polarized light has no coupling to the HH states. Of course, the states have the pure form only at $\boldsymbol{k} = 0$. Away from $\boldsymbol{k} = 0$, the HH state and LH states have some mixture. For x-y polarized light the light couples three times as strongly to the HH states as to the LH states. In quantum well structures where the HH, LH degeneracy is lifted, the selection holes have important consequences for lasers and detectors and their polarization dependent properties.

It is convenient to define a quantity

$$E_p = \frac{2}{m_0}\,|\langle p_x|p_x|s\rangle|^2 \tag{9.36}$$

The values of $\boldsymbol{E}_p$ for several semiconductors are given in Table 9.1.

As a result of the vertical transitions we have, as shown in Fig. 9.5 the equality (using the parabolic band approximation)

$$
\begin{aligned}
\hbar\omega - E_g &= \frac{\hbar^2 k^2}{2}\left(\frac{1}{m_e^*} + \frac{1}{m_h^*}\right) \\
&= \frac{\hbar^2 k^2}{2m_r^*}
\end{aligned}
\tag{9.37}
$$

where m_r^* is the reduced mass of the e-h system. The final state density of states in the summation are thus the reduced density of states given by

$$N_{cv}(\hbar\omega) = \sqrt{2}\frac{(m_r^*)^{3/2}(\hbar\omega - E_g)^{1/2}}{\pi^2\hbar^3} \tag{9.38}$$

and we have from Eqn. 9.28

$$W_{abs} = \frac{\pi e^2 \hbar n_{ph}}{\epsilon m_o^2 \hbar\omega}\,|\,(a \cdot p)_{cv}\,|^2\,N_{cv}(\hbar\omega) \tag{9.39}$$

Semiconductor	E_p (eV)
GaAs	25.7
InP	20.9
InAs	22.2
CdTe	20.7

Table 9.1: Values of E_p for different semiconductors. (After P. Lawaetz, *Physical Review*, B, **4**, 3460 (1971).)

In bulk semiconductors the expression for unpolarized light becomes (from Eqns. 9.35 and 9.39)

$$W_{abs} = \frac{\pi e^2 \hbar n_{ph}}{2\epsilon m_0 \hbar \omega} \left(\frac{2p_{cv}^2}{m_0} \right) \frac{2}{3} N_{cv}(\hbar \omega) \tag{9.40}$$

Before summarizing and comparing the results for the optical absorption and electron-hole recombination in bulk and quantum well structures, we will briefly relate the absorption coefficient to the absorption rate. It is useful to talk about absorption coefficient rather than the rate at which a photon is absorbed. If we consider a beam of photons traveling along the x-axis, we can write the continuity equation for the photon density

$$\frac{dn_{ph}}{dt} = \left. \frac{\partial n_{ph}}{\partial t} \right|_{vol} + \frac{\partial(vn_{ph})}{\partial x} \tag{9.41}$$

where the first term represents the absorption rate of photons and the second term represents the photons leaving due to the photon current. Here v is the velocity of light. In steady state we have, in general

$$n_{ph}(x) = n_0 \exp(-\alpha x)$$

which defines the absorption coefficient. Also

$$\frac{\partial n_{ph}}{\partial t} = W_{abs}$$

and in steady state we have

$$W_{abs} = \alpha v n_{ph}$$

or

$$\alpha = \frac{W_{abs}}{v n_{ph}} \tag{9.42}$$

The absorption coefficient is now given from Eqns. 9.40 and 9.42. In Fig. 9.6 we show absorption coefficients for several semiconductors.

We now come to e-h recombination time. As discussed earlier, if we are interested in calculating the recombination rate of an electron with a hole at the same k

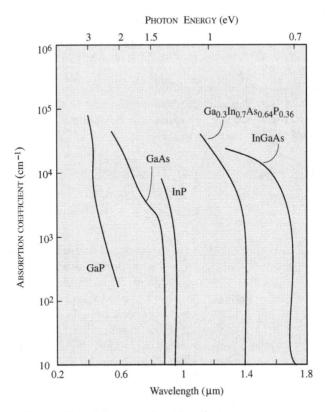

Figure 9.6: Absorption coefficient for several semiconductors.

value, we integrate over all possible photon states into which emission could occur. For total emission, we have

$$W_{\text{em}} = \frac{\pi e^2 \hbar}{m_o^2 \hbar \omega \epsilon} \left(n_{ph} + 1 \right) \left| \boldsymbol{a} \cdot \boldsymbol{p}_{\text{if}} \right|^2 \rho_a(\hbar\omega) \tag{9.43}$$

where ρ_a is the photon density of states for the polarization $\boldsymbol{a}$. The total photon density of states is given by (there are 2 transverse modes for each $\boldsymbol{k}$ value)

$$\rho(\hbar\omega) = \frac{2\omega^2}{2\pi^2 \hbar v^3} \tag{9.44}$$

for photons emitted in the 3-dimensional space. For $n_{ph} = 0$, the emission rate is called the spontaneous emission rate W_{spon} and it's inverse is the e-h recombination time τ_o. The time τ_o represents the time taken by an electron in a state k to recombine with an available hole in the state k.

9.4.2 Interband Transitions in Quantum Wells

The formalism developed so far can be extended in a straightforward manner to the case of the quantum well structures. The central cell functions in the quantum wells are relatively unaffected by the presence of the confining potential. The two changes that occur are the nature of the wavefunctions, which for the low lying states are confined to the well region, and the density of states which have the usual step-like form for parabolic 2-dimensional bands.

The absorption rates are calculated only for the well region since the barrier material has a higher bandgap and does not participate in the optical process till the photon energies are much higher. The quantum well conduction and valence band states are given in the envelope function approximation by

$$
\psi_c^n = \frac{1}{\sqrt{AW}} \, e^{i\mathbf{k}_e \cdot \rho} \, g_c^n(z) \, u_{c\mathbf{k}_e}^n
$$

$$
\psi_v^m = \frac{1}{\sqrt{AW}} \, e^{i\mathbf{k}_h \cdot \rho} \sum_\nu g_v^{\nu m}(z) \, u_{v\mathbf{k}_h}^{\nu m} \tag{9.45}
$$

Here W is the well size and A is the area considered and we have used (as discussed in Chapter 2) the scalar description of the conduction band states and a multiband (indexed by ν) description of the valence band. The envelope functions g_c^n and $g_v^{\nu m}$ correspond to the n and m subband levels in the conduction and valence bands respectively. If we ignore the band mixing effects between the HH and LH states, i.e., ignore the off-diagonal terms in the Kohn-Luttinger Hamiltonian, we obtain simple analytic results for the absorption and emission rates. Note that the momentum matrix element now undergoes the following change when we go from our 3-dimensional calculation to a quasi-2-dimensional one

$$
p_{\text{if}}^{3D} = \frac{1}{V} \int e^{i(\mathbf{k}_e - \mathbf{k}_h) \cdot \mathbf{r}} \langle u_v^\nu | p_a | u_c \rangle \, d^3 r
$$

$$
\rightarrow p_{\text{if}}^{2D} = \frac{1}{AW} \sum_\nu \langle g_v^{\nu m} | g_c^n \rangle \int e^{i(\mathbf{k}_e - \mathbf{k}_h) \cdot \rho} \langle u_v^{\nu m} | p_a | u_c \rangle \, d^2 \rho \tag{9.46}
$$

where $\langle g_v^{\nu m} | g_c^n \rangle$ denotes the overlap between the z-dependent envelope functions of the conduction and valence bands. For symmetrical potentials one has the approximate condition that

$$
\sum_\nu \langle g_v^{\nu m} | g_c^n \rangle \approx \delta_{nm} \tag{9.47}
$$

This condition is not exact and can be changed if there is any asymmetry present (e.g., if there is a transverse electric field present).

The final state reduced density of states (Eqn. 9.38) has to be replaced by the 2-dimensional reduced density of states.

$$
\frac{N_{cv}^{2D}(\hbar\omega)}{W} = \frac{m_r^*}{\pi \hbar^2 W} \sum_{nm} \langle g_v^m | g_c^n \rangle \, \theta(\mathbf{E}_{\text{nm}} - \hbar\omega) \tag{9.48}
$$

and

$$
\mathbf{E}_{\text{nm}} = E_{\text{gap}} + \mathbf{E}_c^n + \mathbf{E}_v^m
$$

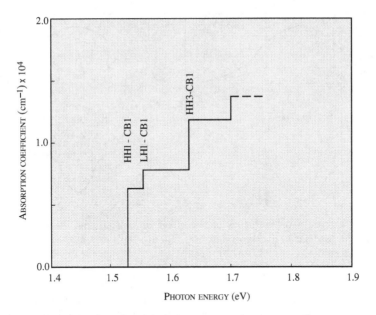

Figure 9.7: Calculated absorption coefficient in a 100 Å GaAs/Al$_{0.3}$Ga$_{0.7}$As quantum well structure for in-plane polarized light. The HH transition is about three times stronger than the LH transition in this polarization. In a real material excitonic transition dominate near the bandedges as disucssed in the next chapter.

Here m_r is the reduced electron hole mass. The θ-function is the Heaviside step function.

$$\alpha(\hbar\omega) = \frac{\pi e^2 \hbar}{m_o^2 c n_r \epsilon_o} \frac{1}{(\hbar\omega)} |\boldsymbol{a} \cdot \boldsymbol{p}_{\text{if}}|^2 \frac{N_{2D}(\hbar\omega)}{W} \sum_{n,m} f_{nm}\, \theta(\boldsymbol{E}_{nm} - \hbar\omega) \tag{9.49}$$

where the overlap integral f_{nm} is

$$f_{nm} = \left| \sum_\nu \langle g_\nu^{\nu m} | g_c^n \rangle \right|^2 \tag{9.50}$$

It is useful to examine some numerical values of the absorption coefficient and the recombination time for, say, a common system like GaAs. In Fig. 9.7 we show the absorption coefficient for a 100 Å GaAs /Al$_{0.3}$Ga$_{0.7}$As quantum well structure. In the bulk semiconductor the absorption coefficient starts at $\hbar\omega = E_g$, with a zero value and initially increases as $(\hbar\omega - \boldsymbol{E}_g)^{1/2}$. It also has a $1/\hbar\omega$ behavior which only influences the absorption coefficients at high energies where the density of states is not parabolic anymore. Because of the degeneracy of the HH and LH states, there is no polarization dependence of the absorption coefficient near the bandgap region. The absorption coefficient in the quantum well structure is quite distinct from the bulk case mainly because of the density of states function. Another difference arises because of the lifting of the HH, LH degeneracy which makes the absorption coefficient strongly polarization dependent as discussed earlier.

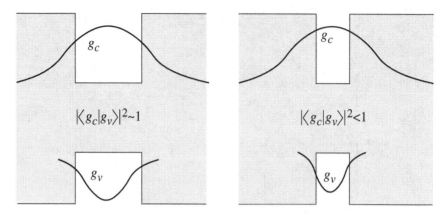

Figure 9.8: Schematic representation of the effect of well size on electron and hole wavefunctions and the associated overlap integrals. The overlap integral decreases at very narrow quantum well sizes.

In quantum wells, the $1/W$ dependence of the absorption coefficient is quite interesting and somewhat misleading. We note that this dependence came from our assumption that the wavefunction is localized a distance equal to the well size. The $1/W$ dependence suggests that the absorption coefficient can be increased indefinitely by decreasing W, the well size. This is, however, not true. As shown schematically in Fig. 9.8, as the well size is narrowed, the wavefunctions of the electron and hole no longer are confined to the well size. The region of interest for the electrons starts increasing beyond W. Also, because of the different masses of the electron and hole, the electron function starts spreading beyond the well at a larger well size making the overlap from less than unity. For most semiconductor systems the optimum well size is ~ 80 to 100 Å.

9.5   INDIRECT INTERBAND TRANSITIONS

In materials such as Si, Ge, AlAs, i.e., indirect semiconductors, near-bandedge optical transitions require photons and phonons to satisfy k-concentration. As a result, optical absorption is observed in Si and Ge although the absorption rate is far weaker than for GaAs.

Typical processes in the interband transition are shown in Fig. 9.9. The process is second order in which the electron is first scattered by a photon to the direct band conserving momentum and then scattering to the indirect band by a phonon. While momentum is conserved in the intermediate process, *the energy is not conserved since this process is virtual and the time–energy uncertainty ensures that there is no energy conservation requirement.* The overall process, however, does conserve energy. The scattering rate is again given by the Fermi golden rule except that the matrix element is second order

$$W_{if} = \frac{2\pi}{\hbar} \int \left| \frac{\sum \langle f|H_{\mathrm{per}}|n\rangle \langle n|H_{\mathrm{per}}|i\rangle}{E_i - E_n} \right|^2 \delta(E_f - E_i) \frac{d^3k}{(2\pi)^3} \qquad (9.51)$$

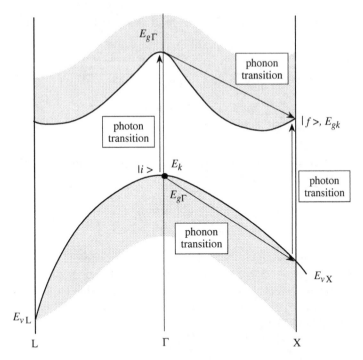

Figure 9.9: Two processes showing how a photon and a phonon can take an electron from state $|i\rangle$ to state $|f\rangle$. The photon energy need not be equal to the vertical energy, since the intermediate transitions are "virtual," i.e., the electron does not reside there for any length of time.

where $|n\rangle$ is an intermediate state, and the perturbation is

$$H_{\mathrm{per}} = H_{\mathrm{ph}} + H_{\mathrm{ep}}$$

Here H_{ph} is the electron-photon interaction discussed so far and H_{ep} is the electron-phonon interaction discussed in Chapter 6. In general, the two processes shown in Fig. 9.9 can contribute to the scattering process. However, the process which first involves a photon interaction are stronger since the denominator is smaller (denominator is $\sim$ direct bandgap), than for the processes involving the phonon first (denominator is $\sim$ $|\boldsymbol{E}_{v\Gamma} - \boldsymbol{E}_{vX}|$ or $|\boldsymbol{E}_{v\Gamma} - \boldsymbol{E}_{vL}|$ which are larger than the direct bandgap). The scattering rate is then

$$W_{if}(\boldsymbol{k}) = \frac{2\pi}{\hbar} \int_f \left\{ |M_{\mathrm{em}}|^2 + |M_{\mathrm{abs}}|^2 \right\} \delta(\boldsymbol{E}_f - \boldsymbol{E}_i) \frac{d^3k}{(2\pi)^3}$$

The matrix elements M_{em} and M_{abs} correspond to the cases where first a photon is absorbed and then a phonon is either emitted or absorbed. Note that the photon energy $\hbar\omega$ is smaller than the direct bandgap, but the intermediate transition can occur since energy need not be conserved. The form of the matrix elements is

$$M_{\text{abs}} = \frac{\left|\langle c, \boldsymbol{k} + \boldsymbol{q}|H_{\text{ep}}^{\text{abs}}|c, \boldsymbol{k}\rangle\right|^2 \left|\langle c, \boldsymbol{k}|H_{\text{ph}}^{\text{abs}}|v, \boldsymbol{k}\rangle\right|^2}{(E_{g\Gamma} - \hbar\omega)^2}$$

$$M_{\text{em}} = \frac{\left|\langle c, \boldsymbol{k} - \boldsymbol{q}|H_{\text{ep}}^{\text{em}}|c, \boldsymbol{k}\rangle\right|^2 \left|\langle c, \boldsymbol{k}|H_{\text{ph}}^{\text{em}}|v, \boldsymbol{k}\rangle\right|^2}{(E_{g\Gamma} - \hbar\omega)^2} \tag{9.52}$$

The phonon scattering is due to the optical phonon intervalley scattering with a matrix element (as discussed in Chapter 6)

$$M_q^2 = \frac{\hbar D_{ij}^2}{2\rho V \omega_{ij}} \left\{ \begin{array}{c} n(\omega_{ij}) \\ n(\omega_{ij}) + 1 \end{array} \right\}$$

for the absorption and emission processes respectively. Here D_{ij} is the deformation potential, ρ is the mass density, and ω_{ij} is the phonon frequency which connects the Γ valley to the zone edge valley. It is useful to point out that due to the *indirect* nature of the transition, the rates calculated earlier for direct gap semiconductors are essentially *lowered* by a factor equal to

$$\frac{M_q^2}{(E_{g\Gamma} - \hbar\omega)^2} \tag{9.53}$$

This factor is typically 10^{-2} to 10^{-3} and has a temperature dependence due to the temperature dependence of the phonon occupation $n(\omega_{ij})$. In case of indirect transitions, for a given initial states $|v, \boldsymbol{k}\rangle$, there is a spread in the final states due to the phonon scattering. The scattering rate sums over this spread, giving

$$\begin{aligned} W_{ij}(k) &= \frac{2\pi}{\hbar} \frac{M_{\text{ph}}^2}{(E_{g\Gamma} - \hbar\omega)^2} \frac{\hbar D_{ij}^2}{2\rho\omega_{ij}} J_v \\ &\times \ [n(\omega_{ij})\, N_c(E_1 + \hbar\omega_{ij}) \\ &+ \ \{n(\omega_{ij}) + 1\}\, N_c(E_1 - \hbar\omega_{ij})] \end{aligned}$$

where J_v is the number of equivalent valleys, N_c is the density of states for a given spin in a valley, and

$$E_1 = \hbar\omega - E_{g\mathbf{k}'} - E_{\mathbf{k}}$$

where $E_{g\mathbf{k}'}$ is the *indirect gap* and the $E_{\mathbf{k}}$ is the energy of the initial electron measured from the top of the valence band.

To find the absorption coefficient we need to sum the above rate over all possible starting states which could absorb a photon with energy $\hbar\omega$. This means we must sum over all possible initial states from $E_{\mathbf{k}} = 0$ to E_{kmax} where

$$\begin{aligned} E_{kmax} &= \hbar\omega - E_{g\mathbf{k}'} + \hbar\omega_{ij} \text{ (phonon absorption)} \\ E_{kmax} &= \hbar\omega - E_{g\mathbf{k}'} - \hbar\omega_{ij} \text{ (phonon emission)} \end{aligned}$$

We multiply $W(\boldsymbol{k})$ by $2VN_v(\boldsymbol{E_k})d\boldsymbol{E_k}$ where N_v is the single spin density of states in the valence band and integrate from 0 to $\boldsymbol{E}_{kmax}$. For parabolic bands the integral is simple and we get

$$
\begin{aligned}
W_{\text{abs}}(\hbar\omega) &= \frac{M_{\text{ph}}^2 \, D_{ij}^2 \, J_v \, (m_c m_v)^{3/2}}{8\pi^2(\boldsymbol{E}_{g\Gamma} - \hbar\omega)^2 \, \hbar^6 \, \rho \, \omega_{ij}} \\
&\times \left[n(\omega_{ij}) \left(\hbar\omega - \boldsymbol{E}_{g\boldsymbol{k}'} + \hbar\omega_{ij} \right)^2 \right. \\
&+ \left. \{ n(\omega_{ij}) + 1 \} \left(\hbar\omega - \boldsymbol{E}_{g\boldsymbol{k}'} - \hbar\omega_{ij} \right)^2 \right]
\end{aligned}
\tag{9.54}
$$

with the photon related matrix element as before,

$$
M_{\text{ph}}^2 = \frac{e^2 \hbar n_{\text{ph}} \, |a \cdot \boldsymbol{p}_{\text{if}}|^2}{2m_o^2 \epsilon \omega}
\tag{9.55}
$$

The absorption coefficient is then given by $W_{\text{abs}}/(n_{\text{ph}} v_{\text{ph}})$. Once the threshold photon energy is reached, the absorption coefficient increases as $(\hbar\omega - \boldsymbol{E}_{\text{th}})^2$, in contrast to the direct gap case where the energy dependence was $(\hbar\omega - \boldsymbol{E}_g)^{1/2}$.

In Fig. 9.10 we show typical absorption measurements for Si and Ge. We note the low absorption coefficient at near the bandgap in Si when compared to the results for a direct gap material like GaAs. Once the photon energies reach the direct gap region, the absorption coefficient increases rapidly, since direct transitions are possible. Notice that here we have only considered phonons as a scattering process. Other scattering processes such as alloy scattering, impurity scattering, etc., can also cause optical absorption in indirect semiconductors. As a result, "poor" quality indirect semiconductors have a better absorption coefficient than pure indirect materials. An example of absorption in amorphous silicon (a-Si) is shown in Fig. 9.10. In an amorphous material the k-vector is not a "good" quantum number (there is no periodic lattice) and effectively k-conservation is not required. As a result, the absorption coefficient is quite strong.

EXAMPLE 9.2 A 1.6 eV photon is absorbed by a valence band electron in GaAs. If the bandgap of GaAs is 1.41 eV, calculate the energy of the electron and heavy hole produced by the photon absorption.

The electron, heavy-hole, and reduced mass of GaAs are 0.067 m_0, 0.45 m_0, and 0.058 m_0, respectively. The electron and the hole generated by photon absorption have the same momentum. The energy of the electron is

$$
\begin{aligned}
E^e &= E_c + \frac{m_r^*}{m_e^*}(\hbar\omega - E_g) \\
E^e - E_c &= \frac{0.058}{0.067}(1.6 - 1.41) = 0.164 \text{ eV}
\end{aligned}
$$

The hole energy is

$$
\begin{aligned}
E^h - E_v &= -\frac{m_r^*}{m_h^*}(\hbar\omega - E_g) = -\frac{0.058}{0.45}(1.6 - 1.41) \\
&= -0.025 \text{ eV}
\end{aligned}
$$

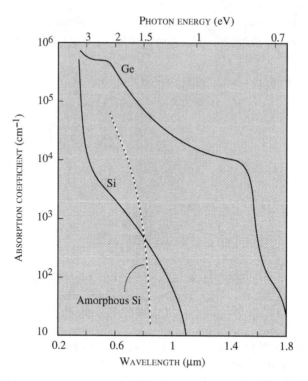

Figure 9.10: Absorption coefficient of Si and Ge. Also shown is absorption coefficient for amorphous silicon which is almost like a direct gap semiconductor, since k-selection is not applicable.

The electron by virtue of its lower mass is created with a much greater energy than the hole.

EXAMPLE 9.3 In silicon, an electron from the top of the valence band is taken to the bottom of the conduction band by photon absorption. Calculate the change in the electron momentum. Can this momentum difference be provided by a photon?

The conduction band minima for silicon are at a k-value of $\frac{2\pi}{a}$ (0.85, 0, 0). There are five other similar bandedges. The top of the valence band has a k-value of 0. The change in the momentum is thus

$$\hbar\Delta k = \hbar\frac{2\pi}{a}(0.85) = (1.05 \times 10^{-34})\left(\frac{2\pi}{5.43 \times 10^{-10}}\right)(0.85)$$
$$= 1.03 \times 10^{-24} \text{ kg m s}^{-1}$$

A photon which has an energy equal to the silicon bandgap can only provide a momentum of

$$\hbar k_{ph} = \hbar \cdot \frac{2\pi}{\lambda}$$

The λ for silicon bandgap is 1.06 μm and thus the photon momentum is about a factor of 600 too small to balance the momentum needed for the momentum conservation. The lattice vibrations produced by thermal vibration are needed for the process.

EXAMPLE 9.4 The absorption coefficient near the bandedges of GaAs and Si are $\sim 10^4$ cm^{-1} and 10^3 cm^{-1} respectively. What is the minimum thickness of a sample in each case which can absorb 90% of the incident light?

The light absorbed in a sample of length L is

$$\frac{I_{abs}}{I_{inc}} = 1 - \exp\left(-\alpha L\right)$$

or
$$L = = \frac{1}{\alpha} \ln\left(1 - \frac{I_{abs}}{I_{inc}}\right)$$

Using $\frac{I_{abs}}{I_{inc}}$ equal to 0.9, we get

$$L(\text{GaAs}) = -\frac{1}{10^4} \ln(0.1) = 2.3 \times 10^{-4} \text{ cm}$$
$$= 2.3 \ \mu\text{m}$$
$$L(\text{Si}) = -\frac{1}{10^3} \ln(0.1) = 23 \ \mu\text{m}$$

Thus an Si detector requires a very thick active absorption layer to function.

EXAMPLE 9.5 Calculate the absorption coefficient of GaAs as a function of photon frequency.

The joint density of states for GaAs is (using a reduced mass of $0.065 m_0$)

$$N_{cv}(E) = \frac{\sqrt{2}(m_r^*)^{3/2}(E - E_g)^{1/2}}{\pi^2 \hbar^3}$$
$$= \frac{1.414 \times (0.065 \times 0.91 \times 10^{-30} \text{ kg})^{3/2}(E - E_g)^{1/2}}{9.87 \times (1.05 \times 10^{-34})^3}$$
$$= 1.78 \times 10^{54}(E - \hbar\omega)^{1/2} \text{ J}^{-1} \text{ m}^{-3}$$

The absorption coefficient is for unpolarized light

$$\alpha(\hbar\omega) = \frac{\pi e^2 \hbar}{2 n_r c \epsilon_o m_0} \left(\frac{2 p_{cv}^2}{m_0}\right) \frac{N_{cv}(\hbar\omega)}{\hbar\omega} \cdot \frac{2}{3}$$

The term $\frac{2 p_{cv}^2}{m_0}$ is ~ 23.0 eV for GaAs. This gives

$$\alpha(\hbar\omega) = \frac{3.1416 \times (1.6 \times 10^{-19} \text{ C})^2 (1.05 \times 10^{-34} \text{ Js})}{2 \times 3.4 \times (3 \times 10^8 \text{ m/s})(8.84 \times 10^{-12} \text{ (F/m)}^2)}$$
$$\cdot \frac{(23.0 \times 1.6 \times 10^{-19} \text{ J})}{(0.91 \times 10^{-30} \text{ kg})} \frac{(\hbar\omega - E_g)^{1/2}}{\hbar\omega} \times 1.78 \times 10^{54} \times \frac{2}{3}$$
$$\alpha(\hbar\omega) = 2.25 \times 10^{-3} \frac{(\hbar\omega - E_g)^{1/2}}{\hbar\omega} \text{ m}^{-1}$$

Here the energy and $\hbar\omega$ are in units of Joules. It is usual to express the energy in eV, and the absorption coefficient in cm^{-1}. This is obtained by multiplying the result by

$$\left[\frac{1}{(1.6 \times 10^{-19})^{1/2}} \times \frac{1}{100}\right]$$

$$\alpha(\hbar\omega) = 5.6 \times 10^4 \frac{(\hbar\omega - E_g)^{1/2}}{\hbar\omega} \text{ cm}^{-1}$$

For GaAs the bandgap is 1.5 eV at low temperatures and 1.43 eV at room temperatures. From the value of α, we can see that a few microns of GaAs are adequate to absorb a significant fraction of light above the bandgap.

EXAMPLE 9.6 Calculate the electron-hole recombination time in GaAs.

The recombination rate is given by

$$W_{em} = \frac{e^2 n_r}{6\pi\epsilon_o m_0 c^3 \hbar^2} \left(\frac{2p_{cv}^2}{m_0} \right) \hbar\omega$$

with $\frac{2p_{cv}^2}{m_0}$ being 23 eV for GaAs.

$$
\begin{aligned}
W_{em} &= \frac{(1.6 \times 10^{-19}\ \text{C})^2 \times 3.4 \times (23 \times 1.6 \times 10^{-19}\ \text{J})\hbar\omega}{6 \times 3.1416 \times (8.84 \times 10^{-12}\ \text{F/m}) \times (0.91 \times 10^{-30}\ \text{kg})} \\
&\quad \cdot \frac{1}{(3 \times 10^8\ \text{m/s})^3 \times (1.05 \times 10^{-34}\ \text{Js})^2} \\
&= 7.1 \times 10^{27} \hbar\omega\ \text{s}^{-1}
\end{aligned}
$$

If we require the value of $\hbar\omega$ in eV instead of Joules we get

$$
\begin{aligned}
W_{em} &= 7.1 \times 10^{27} \times (1.6 \times 10^{-19})\hbar\omega\ \text{s}^{-1} \\
&= 1.14 \times 10^9\ \hbar\omega\ \text{s}^{-1}
\end{aligned}
$$

For GaAs, $\hbar\omega \sim 1.5$ eV so that

$$W_{em} = 1.71 \times 10^9\ \text{s}^{-1}$$

The corresponding recombination time is

$$\tau_o = \frac{1}{W_{em}} = 0.58\ \text{ns}$$

Remember that this is the recombination time when an electron can find a hole to recombine with. This happens when there is a high concentration of electrons and holes, i.e., at high injection of electrons and holes or when a minority carrier is injected into a heavily doped majority carrier region.

9.6 INTRABAND TRANSITIONS

When we consider intraband transitions (i.e., either within the conduction band or valence band), it is not possible to have first order "vertical" transitions in k-space. Thus intraband transitions cannot be first order processes for bulk semiconductors. However, in quantum wells we have seen that one forms subbands due to quantum confinement. It is possible to have inter-subband transitions. Such intraband (or inter-subband) transitions make quantum well structures quite exciting for long wavelength optical devices.

9.6.1 Intraband Transitions in Bulk Semiconductors

As noted above, in bulk semiconductors, the intraband transitions must involve a phonon or some other scattering mechanism (ionized impurity, defects, etc.,) to ensure momentum conservation. The second order process is essentially similar to the one we dealt with for indirect processes. The intraband transitions, also known as free carrier absorption, is quite important particularly for lasers, since it is responsible for losses in the cladding layers of the laser.

Free carrier absorption is described quite well by the Drude model of transport. The optical radiation is represented by a sinusoidal electric field which causes the electron to oscillate in the energy band. If there is no scattering the electron will simply gain and lose energy, and there will be no net absorption of energy. However, because of scattering mechanisms present, the electron absorbs net energy from the field and then emits phonons so that optical absorption occurs. The absorption coefficient has the dependence

$$\propto (\hbar\omega) \quad \propto \quad \frac{1}{\omega^2}$$
$$\propto \quad \frac{1}{\mu} \tag{9.56}$$

i.e., the absorption coefficient falls as $1/\omega^2$ and is inversely proportional to the carrier mobility. If the mobility is large (i.e., scattering is weak) the absorption coefficient becomes very small.

9.6.2 Intraband Transitions in Quantum Wells

We have seen in Chapter 3 that quantum well structures can produce remarkable changes in the electronic properties of semiconductor structures. In quantum well structures, the electronic states are no longer of the plane wave form in the growth direction making it possible to have intraband transitions for certain polarizations of light. Since the intraband (or inter-subband) transition energy can be easily varied by changing the well size, these transitions have great importance for far infrared detectors.

Let us consider the intraband (inter-subband) transitions for a quantum well grown along the z-direction. Due to the confinement in the z-direction, the subband functions can be written as (say, the first two functions)

$$\begin{aligned}
\psi^1(\mathbf{k}, z) &= g^1(z)\, e^{i\mathbf{k}\cdot\rho}\, u^1_{n\mathbf{k}}(\mathbf{r}) \\
\psi^2(\mathbf{k}, z) &= g^2(z)\, e^{i\mathbf{k}\cdot\rho}\, u^2_{n\mathbf{k}}(\mathbf{r})
\end{aligned} \tag{9.57}$$

In first order we have to examine vertical transitions. The functions g^1 and g^2 are orthogonal and to a good approximation the central cell functions are same for the different subbands (this especially true for the conduction band). Thus, the momentum matrix element is given by

$$\mathbf{p}_{\mathrm{if}} = -\frac{i\hbar}{W} \int g^{2*}(z)\, e^{-i\mathbf{k}\cdot\rho}\, \mathbf{a}\cdot\nabla g^1(z)\, e^{i\mathbf{k}\cdot\rho}\, d^2\rho\, dz$$

where W is the well width.

As in the case of the 3-dimensional system, the momentum matrix element is zero if the polarization vector (or the ∇ function) is in the ρ-plane. Thus, for the x-y polarized light, the *transitions rate is still zero*. (Note that if there is strong mixing of the central cell functions as in the valence bands, this condition can be relaxed.) However, if the light is z-polarized we get

$$\boldsymbol{p}_{\text{if}} = \frac{-i\hbar}{W} \int g^{2*}(z)\, \hat{z}\, \frac{\partial}{\partial z} g^1(z)\, dz \tag{9.58}$$

Since $g^1(z)$ and $g^2(z)$ have even and odd parities respectively, as shown in Fig. 9.11, $g^2(z)$ and $\partial g^1(z)/\partial z$ both have odd parity. The momentum matrix element for z-polarization is then approximately

$$|\boldsymbol{p}_{\text{if}}| \approx \frac{\hbar}{W} \tag{9.59}$$

This result is reasonably accurate if both the ground and excited states are confined to the well size. Often the excited state may have less confinement than the ground state in which case one has to explicitly evaluate the integral.

In the simple parabolic approximation, we see that the dispersion relations of the two subbands are parallel and shifted by the subband energy levels difference $\boldsymbol{E}_2 - \boldsymbol{E}_1$ as shown in Fig. 9.11b. Therefore the joint density of states is a δ-function with infinite density of states at the transition energy. However, we must include the occupation function and broadening into the problem. In the inter-subband transitions, the electrons have to be introduced by doping the material. Ideally, for maximum absorption, the electrons are present in the first subband and, hopefully, not present in the second subband. Introducing the Fermi factors, we get for the absorption process

$$W_{\text{abs}} = \frac{\pi e^2 n_{ph}}{m_o^2 \omega \epsilon} \frac{1}{W} \sum_f |p_{\text{if}}|^2 \, \delta\left(\boldsymbol{E}_2 - \boldsymbol{E}_1 - \hbar\omega\right)\, f(\boldsymbol{E}_1)\left[1 - f(\boldsymbol{E}_2)\right] \tag{9.60}$$

If we assume that the second subband is empty, the sum over the final states is zero except at resonance where we have

$$\sum_{2nd\ subband} \delta\left(\boldsymbol{E}_2 - \boldsymbol{E}_1 - \hbar\omega\right) f(\boldsymbol{E}_1) = N_1 \tag{9.61}$$

where N_1 is the electron concentration in the first subband. The density of states at resonance is infinite for the simple parabolic band giving an infinite absorption rate. However, in reality the nonparabolicity and scattering mechanisms introduce some broadening in the density of states. According to the uncertainty principle levels 1 and 2 have a broadening in their energy position due to the finite time an electron can spend before being scattered. If we assume a Gaussian broadening, the two-dimensional density of states becomes

$$N(E) = \frac{N_1 \exp\left(-\frac{(E-E_{12})^2}{1.44\sigma^2}\right)}{\sqrt{1.44\pi}\sigma}$$

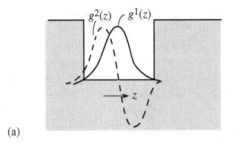

(a)

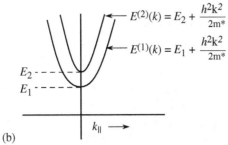

(b)

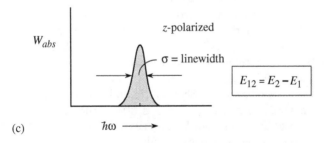

(c)

Figure 9.11: A schematic presentation of the (a) envelope functions for two levels in a quantum well, (b) subband structure in the well, and (c) absorption rate for z-polarized light in a quantum well.

where σ is the linewidth of the transition. The linewidth may have contributions from both homogeneous (phonon related) and inhomogeneous (structural imperfections) effects.

The absorption coefficient for the z-polarized light now becomes

$$\alpha(\hbar\omega) = \frac{\pi e^2 \hbar}{m_o^2 c n_r \epsilon_o \, \hbar\omega} \frac{|\boldsymbol{p}_{\mathrm{if}}|^2}{W} \frac{N_1 \exp\left(-\frac{(\hbar\omega - E_{12})^2}{1.44\sigma^2}\right)}{\sqrt{1.44\pi}\sigma} \tag{9.62}$$

The absorption coefficient increases rapidly with decreasing well size although it is important to note again that as the well size becomes very small, the electronic states are no longer confined to the well size as assumed by us. For a transition which is only 1–2 meV wide, the absorption coefficient could reach 10^4 cm^{-1}, making such transitions useful for detectors or modulators.

We note that z-polarized light for a z-confinement quantum well implies that light is traveling in the plane of the substrate. Thus there is no absorption for vertical incident light. As noted above, for valence band quantum wells there is strong band mixing which allows vertical incident absorption also. However, for conduction band quantum wells this is a drawback for applications in detectors. In imaging arrays usually mirrors are etched into detector chips so that vertical incident light can be reflected to make it incident along the x-y plane.

9.6.3   Intraband Transitions in Quantum Dots

In Chapter 1, Section 1.4, we have discussed how high strain epitaxy causes growth to occur by island growth mode, which can then be exploited to make self-assembled quantum dots. Such structures can be used in optoelectronic devices such as lasers and detectors. One area which has drawn considerable attention is intersubband detectors. As in quantum well intersubband photo-detectors (QWIPs) intersubband separations can be controlled for applications in long wavelength detection. Thus applications such as night vision, thermal imaging for medical diagnosis, etc., can benefit from such devices. There is an important advantage to the use of quantum dot intersubband photo-detectors (QDIPs) as opposed to QWIPs. Due to 3D confinement of the electrons vertical incident transitions are very strong in QDIPs. As noted above, in QWIPs these transitions are forbidden for the conduction band and are only allowed in the valence band where hole mixing effects make them possible. Thus it is possible to use QDIPs for imaging applications without the use of mirrors to convert vertical incidence light into in-plane incidence light.

As discussed in Chapter 1, Section 1.4, self-assembled quantum dots have a very large built-in strain. As a result of this strain the electronic properties of the dot is influenced more by strain than by quantization. For example in the InAs/GaAs self-assembled dot the bandgap of InAs is only $\sim$0.4 eV, while the dot effective gap is $\sim$1.0 eV. As a result of the large strain the problem of the self-assembled dot cannot be solved by a simple effective mass description where the conduction and valence bands are decoupled. Use has been made of the 8 band $k \cdot p$ method as well as the psuedopotential method to examine the electronic properties of dots.

In the previous subsection we considered the intersubband absorption for quantum wells. In the conduction band the strength of the transition is due to the overlap integral between the envelope functions, since the character of the central cell function is assumed to be the same (s-type) for the ground and excited states. In the case of the self-assembled dots this is not the case, since there is a strong p-type mixture in the conduction band states. If an 8 band model is used we can write the wavefunction for the electronic levels in the dot as

$$\psi_n(\mathrm{x}) = \sum_{j=1}^{8} \phi_{nj}(\boldsymbol{r}) u_j(\mathrm{x}) \tag{9.63}$$

where ϕ_{nj} is the envelope function part and u_j is the central cell part.

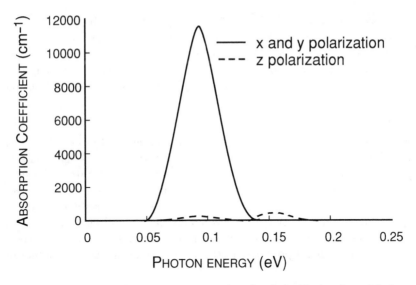

Figure 9.12: Intersubband absorption spectra for the InAs/GaAs dot with base width 124 Å and height 62 Å. The linewidth is taken to be 30 meV. It is assumed there is only one electron in the ground states of the dot.

The transition matrix element between initial state $|\psi_i\rangle$ and final state $|\psi_f\rangle$ has the form

$$\mathbf{p}_{fi} = \sum_{jj'} \left(\langle \phi_{fj'}|\mathbf{p}|\phi_{ij}\rangle \delta_{j'j} + \langle u_{j'}|\mathbf{p}|u_j\rangle \langle \phi_{fj'}|\phi_{ij}\rangle \right) \qquad (9.64)$$

An important point to note is that the very strong transition is due to the bandedge Bloch part of the state and not due to the envelope part. As noted above this is unlike the wide quantum well case, where in the conduction band the intersubband transition is due to the envelope function part.

The intersubband absorption coefficient is

$$\alpha(\hbar\omega) = \frac{\pi e^2 \hbar}{\epsilon_0 n_0 c m_0^2 V_{av}} \sum_f \frac{1}{\hbar\omega} |\epsilon \cdot \mathbf{p}_{fi}|^2 \frac{1}{\sqrt{2\pi}\sigma} \exp\left((E_{fi,av} - \hbar\omega)^2 / 2\sigma^2 \right) \qquad (9.65)$$

The optical absorption between the ground and excited levels is found to have a value

$$\alpha \sim \frac{3.5 \times 10^5}{\sigma} \ \text{cm}^{-1} \qquad (9.66)$$

where σ is the linewidth of the transition in meV. It must be noted that while this is a very large value, the absorbing region is quite small (a few hundred Angstroms in length). As a result, the uniformity in dot sizes is quite critical. In Fig. 9.12 we show the intersubband absorption calculated for the InAs/GaAs QD. The transition between the first excited states and the ground states is dominantly in-plane polarized. This makes possible to detect the light impinging vertically on the surface.

9.7   CHARGE INJECTION AND RADIATIVE RECOMBINATION

Under equilibrium conditions, electron occupation in the valence band is close to unity while the occupation in the conduction band is close to zero. In actual experimental situations this may, of course, not be true. In this section we will discuss charge injection which plays a key role in the performance of light emitting diodes and laser diodes.

9.7.1   Spontaneous Emission Rate

When electrons and holes are injected into the conduction and valence bands of a semiconductor, they recombine with each other as we have discussed earlier. In then absence of any photon density in the cavity (i.e., $n_{ph} = 0$), the emission rate is the spontaneous emission rate which has a value of $\sim 1/(0.5\ \text{ns})$, provided *an electron is present in the state k and a hole is present in the same state k in the valence band*. In reality, however, the rate depends upon the occupation probabilities of the electron and hole with the same k-value. Therefore, we have to include the distribution functions for electrons and holes and integrate over all possible electronic states. Thus, the recombination rate is (units are $\text{cm}^{-3}\text{s}^{-1}$)

$$
\begin{aligned}
R_{\text{spon}} \;=\; & \frac{2}{3}\int d(\hbar\omega)\frac{2e^2 n_r \hbar\omega}{m_0^2 c^3 \hbar^2}\left[\int \frac{1}{(2\pi)^3}d^3k\,|p_{\text{if}}|^2\right. \\
\times\; & \delta(E^e(\boldsymbol{k}) - E^h(\boldsymbol{k}) - \hbar\omega) \\
\times\; & \left. f_e(E^e(\boldsymbol{k}))f_h(E^h(\boldsymbol{k}))\right]
\end{aligned}
\tag{9.67}
$$

The integral over $d(\hbar\omega)$ is to find the rate for all photons emitted and the integration over d^3k is to get the rate for all the occupied electron and hole states. The prefactor $2/3$ comes about since we are considering emission into any photon polarization so that we average the matrix element square $|\boldsymbol{a}\cdot\boldsymbol{p}_{\text{if}}|^2$.

The extension to quantum well structures is obtained by converting the 3D density of states to the 2D density of states (units are $\text{cm}^{-2}\text{s}^{-1}$)

$$
\begin{aligned}
R_{\text{spon}} \;=\; & \frac{2}{3}\int d(\hbar\omega)\frac{2e^2 n_r \hbar\omega}{m_0^2 c^3 \hbar^2}\sum_{nm}\left[\int \frac{d^2k}{(2\pi)^2}|p_{\text{if}}|^2\right. \\
\times\; & \delta(E_n^e(\boldsymbol{k}) - E_m^h(\boldsymbol{k}) - \hbar\omega) \\
\times\; & \left. f_e(E_n^e(\boldsymbol{k}))f_h(E_m^h(\boldsymbol{k}))\right]
\end{aligned}
\tag{9.68}
$$

Using the definition of the time τ_o we have ($\tau_o = \frac{1}{W_{spon}}$)

$$
R_{spon} = \frac{1}{\tau_o}\int d(\hbar\omega)N_{cv}\{f^e(E^e)\}\{f^h(E^h)\}
\tag{9.69}
$$

The spontaneous recombination rate is quite important for both electronic and optoelectronic devices. It is important to examine the rate for several important cases. We will give results for the electron hole recombination for the following cases:

i) **Minority carrier injection:** If $n \gg p$ and the sample is heavily doped, we can assume that $f^e(E^e)$ is close to unity. We then have for the rate at which holes will recombine with electrons,

$$
\begin{aligned}
R_{spon} &\cong \frac{1}{\tau_o} \int d(\hbar\omega) N_{cv} f^h(E^h) \cong \frac{1}{\tau_o} \int d(\hbar\omega) N_h f^h(E^h) \left(\frac{m_r^*}{m_h^*}\right)^{3/2} \\
&\cong \frac{1}{\tau_o} \left(\frac{m_r^*}{m_h^*}\right)^{3/2} p
\end{aligned}
\tag{9.70}
$$

Thus the recombination rate is proportional to the minority carrier density (holes in this case).

ii) **Strong injection:** This case is important when a high density of both electrons and holes is injected and we can assume that both f^e and f^h are step functions with values 1 or zero. We get for this case

$$
R_{spon} = \frac{n}{\tau_o} = \frac{p}{\tau_o}
\tag{9.71}
$$

iii) **Weak injection:** In this case we can use the Boltzmann distribution to describe the Fermi functions. We have

$$
f^e \cdot f^h \cong \exp\left\{-\frac{(E_c - E_{Fn})}{k_B T}\right\} \exp\left\{-\frac{(E_{Fp} - E_v)}{k_B T}\right\} \cdot \exp\left\{-\frac{(\hbar\omega - E_g)}{k_B T}\right\}
$$

The spontaneous emission rate now becomes

$$
R_{spon} = \frac{1}{2\tau_o} \left(\frac{2\pi\hbar^2 m_r^*}{k_B T m_e^* m_h^*}\right)^{3/2} np
\tag{9.72}
$$

If we write the total charge as equilibrium charge plus excess charge,

$$
n = n_o + \Delta n; \, p = p_o + \Delta n
\tag{9.73}
$$

we have for the excess carrier recombination (note that at equilibrium the rates of recombination and generation are equal)

$$
R_{spon} \cong \frac{1}{2\tau_o} \left(\frac{2\pi\hbar^2 m_r^*}{k_B T m_e^* m_h^*}\right)^{3/2} (\Delta n p_o + \Delta p n_o)
\tag{9.74}
$$

If $\Delta n = \Delta p$, we can define the rate of a single excess carrier recombination as

$$
\frac{1}{\tau_r} = \frac{R_{spon}}{\Delta n} = \frac{1}{2\tau_o} \left(\frac{2\pi\hbar^2 m_r^*}{k_B T m_e^* m_h^*}\right) (n_o + p_o)
\tag{9.75}
$$

At low injection τ_r is much larger than τ_o, since at low injection, electrons have a low probability to find a hole with which to recombine.

iv) **Inversion condition:** Another useful approximation occurs when the electron and hole densities are such that $f^e + f^h = 1$. *This is the condition for inversion when the emission and absorption coefficients become equal.* If we assume in this case $f^e \sim f^h = 1/2$, we get the approximate relation

$$R_{spon} \cong \frac{n}{4\tau_o} \cong \frac{p}{4\tau_o} \tag{9.76}$$

The recombination lifetime is approximately $4\tau_o$ in this case. This is a useful result to estimate the threshold current of semiconductor lasers.

The gain and recombination processes discussed here are extremely important in both electronic and optoelectronic devices that will be discussed later. We point out from the above discussion that the recombination time for a single excess carrier can be written in many situations in the form

$$\tau_r = \frac{\Delta n}{R_{spon}} \tag{9.77}$$

For minority carrier injection or strong injection $\tau_r \cong \tau_o$. In general, R_{spon} has a strong carrier density dependence as does τ_r. A typical curve showing the dependence of τ_r on the carrier density is shown in Fig. 9.13 for GaAs. Note that the radiative lifetime in GaAs can range from microseconds to nanoseconds, depending upon the injection density.

9.7.2   Gain in a Semiconductor

If electrons are injected into the conduction band and holes into the valence band (as happens for light emitting devices discussed later) *the electron-hole pairs could recombine and emit more photons than could be absorbed.* Thus one must talk about the emission coefficient minus the absorption coefficient. This term is called the gain of the material. If the gain is positive, an optical beam will grow as it moves through the material instead of decaying. In the simple parabolic bands we have the gain $g(\hbar\omega)$ given by the generalization of our result for the absorption coefficient (gain = emission coefficient − absorption coefficient)

$$g(\hbar\omega) = \frac{\pi e^2 \hbar}{n_r c m_0^2 \epsilon_0 (\hbar\omega)} \mid a \cdot p_{if} \mid^2 N_{cv}(\hbar\omega)[f^e(E^e) - (1 - f^h(E^h))] \tag{9.78}$$

The term in the square brackets arises since the emission of photons is proportional to $f^e \cdot f^h$, while the absorption process is proportional to $(1 - f^e) \cdot (1 - f^h)$. The difference of these terms appears in the equation above.

The energies E^e and E^h are as shown in Fig. 9.14

$$\hbar\omega - E_g = \frac{\hbar^2 k^2}{2m_r^*}$$

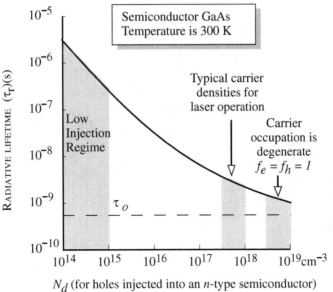

N_d (for holes injected into an *n*-type semiconductor)

$n = p$ (for excess electron hole pairs injected into a region)

Figure 9.13: The dependence of the radiative lifetime in GaAs as a function of carrier injection ($n = p$) or minority carrier injection into a doped region with doping density as shown.

$$E^e = E_c + \frac{\hbar^2 k^2}{2m_e^*} = E_c + \frac{m_r^*}{m_e^*}(\hbar\omega - E_g)$$

$$E^h = E_v - \frac{\hbar^2 k^2}{2m_h^*} = E_v - \frac{m_r^*}{m_h^*}(\hbar\omega - E_g)$$

The occupation of electrons and holes in equilibrium is given by the Fermi level E_F. As excess electrons and holes are injected into the conduction and valence bands, respectively, occupation is given by quasi-Fermi levels, E_{Fn} and E_{Fp}. As shown in Fig. 9.14, at high injection the quasi-Fermi levels start to penetrate the conduction band and the valence band. If $f^e(E^e) = 0$ and $f^h(E^h) = 0$, i.e., if there are no electrons in the conduction band and no holes in the valence band, we see that the gain is simply $-\alpha(\hbar\omega)$ which we had discussed earlier. A positive value of gain occurs for a particular energy when

$$f^e(E^e) > 1 - f^h(E^h) \tag{9.79}$$

a condition that is called inversion. In this case the light wave passing in the material has the spatial dependence

$$I(z) = I_o \exp{(gz)} \tag{9.80}$$

which grows with distance instead of diminishing as it usually does if $g(\hbar\omega)$ is negative. The gain in the optical intensity is the basis for the semiconductor laser.

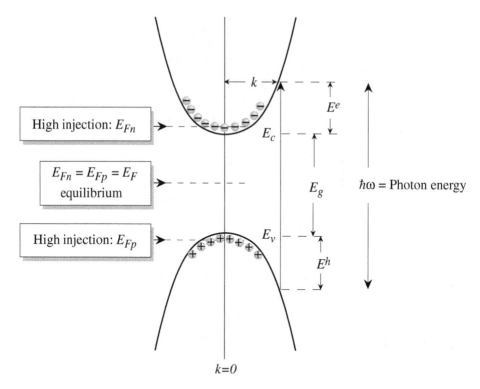

Figure 9.14: The positions of the quasi-Fermi levels, the electron and hole energies at vertical k-values. The electron and hole energies are determined by the photon energy and the carrier masses. At high injection the quasi-Fermi levels start to penetrate the conduction and valence bands.

EXAMPLE 9.7 According to the Joyce-Dixon approximation, the relation between the Fermi level and carrier concentration is given by

$$E_F - E_c = k_B T \left[\ell n \frac{n}{N_c} + \frac{1}{\sqrt{8}} \frac{n}{N_c} \right]$$

where N_c is the effective density of states for the band. Calculate the carrier density needed for the transparency condition in GaAs at 300 K and 77 K. The transparency condition is defined at the situation where the maximum gain is zero (i.e., the optical beam propagates without loss or gain).

At room temperature the valence and conduction band effective density of states is

$$N_v = 7 \times 10^{18} \text{ cm}^{-3}$$
$$N_c = 4.7 \times 10^{17} \text{ cm}^{-3}$$

The values at 77 K are

$$N_v = 0.91 \times 10^{18} \text{ cm}^{-3}$$
$$N_c = 0.61 \times 10^{17} \text{ cm}^{-3}$$

In the semiconductor laser, an equal number of electrons and holes are injected into the active region. We will look for the transparency conditions for photons with energy equal to the bandgap. The approach is very simple: i) choose a value of n or p; ii) calculate E_F from the Joyce-Dixon approximation; iii) calculate $f^e + f^h - 1$ and check if it is positive at the bandedge. The same approach can be used to find the gain as a function of $\hbar\omega$.

For 300 K we find that the material is transparent when $n \sim 1.1 \times 10^{18}$ cm^{-3} at 300 K and $n \sim 2.5 \times 10^{17}$ cm^{-3} at 77 K. Thus a significant decrease in the injected charge occurs as temperature is decreased.

9.8 NONRADIATIVE RECOMBINATION

In this chapter we have focused on optical interactions in semiconductors. We have discussed how electrons and holes recombine to produce photons. Such radiative processes compete with nonradiative processes in which electron-hole recombination does not create photons. Nonradiative processes produce phonons (or heat) instead of light and can occur through defect levels or through the Auger processes (discussed in Chapter 5).

9.8.1 Charge Injection: Nonradiative Effects

According to Bloch's theorem, in perfect semiconductors, there are no electronic states in the bandgap region. However, in real semiconductors there are always intentional or unintentional impurities which produce electronic levels which are in the bandgap. These impurity levels can arise from chemical impurities or from native defects such as a vacancies. The bandgap levels are states in which the electron is "localized" in a finite space near the defect unlike the usual Bloch states which represent the valence and conduction band states, and which are extended in space. As the "free" electrons move in the allowed bands, they can be trapped by the defects as shown in Fig. 9.15. The defects can also allow the recombination of an electron and hole without emitting a photon as was the case in the previous section. This nonradiative recombination competes with radiative recombination. We will briefly discuss the non-radiative processes involving a midgap level with density N_t.

An empty state can be assigned a capture cross-section σ so that physically if an electron comes within this area around a trap, it will be captured. If v_{th} is the velocity of the electron we can define a nonradiative time $\tau_{nr}(e)$ and $\tau_{nr}(p)$ through the following relations

$$\tau_{nr}(e) = \frac{1}{N_t v_{th} \sigma_n} \quad \text{and} \quad \tau_{nr}(p) = \frac{1}{N_t v_{th} \sigma_p} \tag{9.81}$$

where σ_n and σ_p are the cross-sections for electrons and holes. Based on this time constant a theory for nonradiative recombination has been given by Shockley, Read, and Hall. This treatment is greatly simplified if we make the following assumptions: i) $\tau_{nr} = \tau_{nr}(e) = \tau_{nr}(p)$; ii) $E_t = E_{Fi}$, i.e., the trap levels are essentially at midgap level; iii) $np \gg n_i^2$ under the injection conditions. This gives, for the nonradiative rate,

$$R_t = \frac{np}{\tau_{nr}(n + p)} \tag{9.82}$$

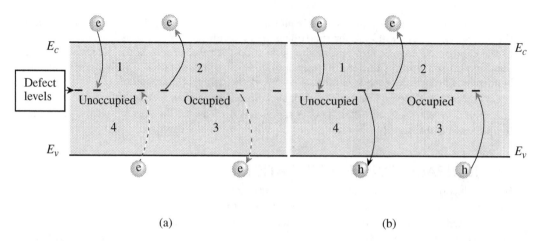

Figure 9.15: Various processes that lead to trapping and recombination via deep levels in the bandgap region (dashed line). The processes 1 and 2 in (a) represent trapping and emission of electrons while 3 and 4 represent the same for holes. The *e-h* recombination is shown in part (b).

The time constant τ_{nr} depends upon the impurity density, the cross-section associated with the defect and the electron thermal velocity. Typically the cross-sections are in the range 10^{-13} to 10^{-15} cm^2.

EXAMPLE 9.8 A silicon sample has an impurity level of 10^{15} cm^{-3}. These impurities create a midgap level with a cross-section of 10^{-14} cm^{-2}. Calculate the electron trapping time at 300 K and 77 K.

We can obtain the thermal velocities of the electrons by using the relation

$$\frac{1}{2}m^*v_{th}^2 = \frac{3}{2}k_BT$$

This gives

$$v_{th}(300\ K) = 2 \times 10^7\ \text{cm/s}$$
$$v_{th}(77\ K) = 1 \times 10^7\ \text{cm/s}$$

The electron trapping time is then

$$\tau_{nr}(300\ \text{K}) = \frac{1}{10^{15} \times 2 \times 10^7 \times 10^{-14}} = 5 \times 10^{-9}\ \text{s}$$
$$\tau_{nr}(77\ \text{K}) = 10^{15} \times 1 \times 10^7 \times 10^{-14} = 10^{-8}\ \text{s}$$

In silicon the *e-h* recombination by emission of photon is of the range of 1 ms to 1 μs so that the non-radiative (trap related) lifetime is much shorter.

9.8.2 Non-Radiative Recombination: Auger Processes

In Chapter 5 we have discussed the Auger process in which an electron and hole recombine and the excess energy is transferred to either an electron or a hole as shown in Fig.

9.16. The important point to note is that no photons are produced in this process so that an electron-hole pair is lost without any photon output.

In Fig. 9.16a, we show a process called conduction-conduction-heavy hole-conduction or CCHC, the individual letters representing the state of the carriers involved in the Auger process. After the scattering, an electron-hole pair is lost and one is left with a hot electron. The hot electron subsequently loses its excess energy by emitting phonons as discussed in Chapter 6. In Fig. 9.16b we show a different process called CHHS (standing for conduction-heavy hole-heavy hole-split off) in which two holes and an electron interact to produce a hot hole in the split-off band. Depending upon the nature of the semiconductor bandstructure, one or both of these processes can dominate the Auger process.

In Chapter 5 we have discussed many of the details of the Auger process. The process involves 3 particles (2 electrons + 1 hole, or 2 holes + 1 electron) and is proportional to $n^2 p$ or $p^2 n$. If the electron and hole injection is equal, the Auger rate is often written in the form

$$
\begin{aligned}
W &= R_{\text{Auger}} \\
&= Fn^3
\end{aligned}
\tag{9.83}
$$

where F is the Auger coefficient.

The Auger rates increase exponentially as the bandgap is decreased. They also increase exponentially as the temperature increases. These are direct results of the energy and momentum conservation constraints and the carrier statistics. Auger processes are more or less unimportant in semiconductors with bandgaps larger than approximately 1.5 eV (e.g., GaAs, AlGaAs, InP). However, they become quite important in narrow bandgap materials such as $In_{0.53}Ga_{0.47}As$ ($E = 0.8$ eV) and HgCdTe ($E < 0.5$ eV), and are thus a serious hindrance for the development of long wavelength lasers.

EXAMPLE 9.9 Consider an Auger process that involves the scattering shown in Fig. 9.16b. This process denoted by CHHS is a dominant process for long distance communication lasers. The masses of the conduction, heavy hole and split off band are m_c, m_h, m_s, respectively, and the bands are parabolic. Calculate the threshold energy for the initial electron and hole states for such a transition using momentum and energy conservation.

Let k_1 and k_1' be the initial k-vectors for the electron and hole, respectively, and k_2 and k_2' be the vectors corresponding to the split-off hole and the heavy hole, respectively (as shown in Fig. 9.16b). Momentum conservation gives us

$$
k_2 = k_1 + k_1' + k_2'
$$

Energy conservation give us

$$
E_{k_1} + E_g - E_{k_2} - \Delta = -E_{k_1'} - E_{k_2'}
$$

We choose $k_1' = k_2'$, and define a variational parameter α, where

$$
k_1' = \alpha k_1
$$

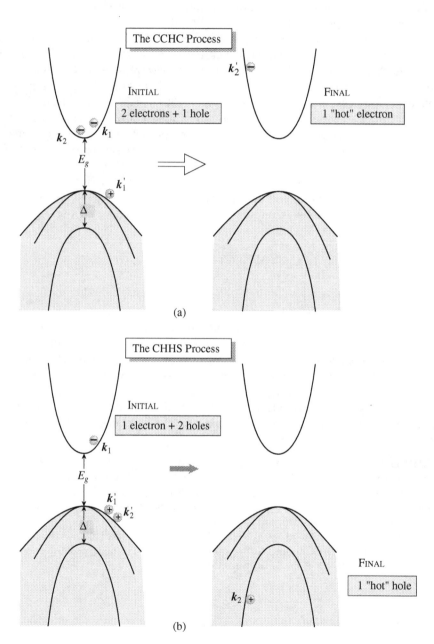

Figure 9.16: (a) The Auger process involving 2 electrons and 1 hole in the initial state and 1 hot electron after the scattering. (b) The process where two holes and an electron suffer an Auger process and give a hot hole in the split-off band.

We have

$$k_2 = k_1(2\alpha + 1)$$

Also, from energy conservation, we get

$$\frac{\hbar^2 k_1^2}{2m_c} + \frac{\alpha^2 \hbar^2 k_1^2}{m_h} + (E_g - \Delta) = \frac{\hbar^2 k_2^2}{2m_s} = \frac{\hbar^2 k_1^2}{2m_s}(1 + 4\alpha^2 + 4\alpha)$$

Solving for k_1^2, we get

$$k_1^2 = \frac{(E_g - \Delta)(2m_s/\hbar^2)}{1 + 4\alpha^2 + 4\alpha - \frac{m_s}{m_c} - \frac{2\alpha^2 m_s}{m_h}}$$

To minimize E'_{k_2} we must minimize

$$\frac{k_1^2}{2m_c} + \frac{k_2'^2}{m_h} = \frac{k_1^2}{2m_h}\left(2\alpha^2 + \frac{m_h}{m_c}\right)$$

Substituting for k_1^2, we find that the function to be maximized is

$$f(\alpha) = \frac{\left(2\alpha^2 + \frac{m_h}{m_c}\right)}{1 + 4\alpha^2 + 4\alpha - \frac{m_s}{m_c} - \frac{2\alpha^2 m_s}{m_h}}$$

Equating $\frac{\partial f}{\partial \alpha} = 0$, we get

$$\left(1 + 4\alpha^2 + 4\alpha - \frac{m_s}{m_c} - 2\alpha^2 \frac{m_s}{m_h}\right) 4\alpha = \left(2\alpha^2 + \frac{m_h}{m_c}\right)\left(8\alpha + 4 - 4\alpha \frac{m_s}{m_h}\right)$$

so that

$$2\alpha^2 + \alpha \left(1 - \frac{2m_h}{m_c}\right) - \frac{m_h}{m_c} = 0$$

or,

$$\alpha = \frac{m_h}{m_c} = \mu$$

We can now calculate the energy of the initial electron (i.e., the threshold energy)

$$\frac{\hbar^2 k_1^2}{2m_c} = \frac{(E_g - \Delta)m_s/m_c}{1 + 4\mu^2 + 4\mu - \frac{m_s}{m_c} - 2\mu^2 \frac{m_s}{m_h}}$$

An important outcome of this result is that as E_g approaches Δ, the threshold for the Auger process goes to zero. This causes an extremely high Auger recombination in materials where the bandgap and the split-off energy are comparable.

9.9   SEMICONDUCTOR LIGHT EMITTERS

Radiative-recombination is the basis of light emitters, like light emitting diodes (LEDs) and laser diodes (LDs). The LEDs and LDs are both very important technology devices with application in displays and communications. The LED operates on the basis of spontaneous emission while the LD operates on the basis of stimulated emission. As discussed in Section 9.3 stimulated emission depends upon the "photon number" present in the material. In spontaneous emission the photons that energy from e-h recombination are incoherent, i.e., each one has a random phase. However, in stimulated emission the photons that energy have the same phase as the ones already present. This difference highlighted in Fig. 9.17 is responsible for coherent light emission in laser diodes. In this section we will review the two important devices.

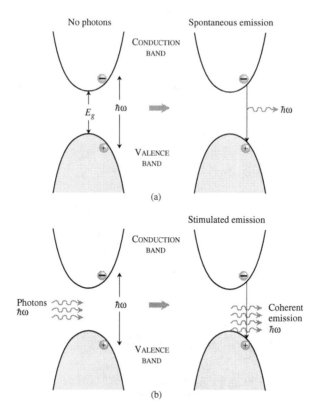

Figure 9.17: (a) In the absence of any photons, electron-hole recombination produces photons with no phase coherence. (b) In the presence of photons, electron-hole recombination produces photons which are coherent with the previously existing photons.

9.9.1   Light Emitting Diode

The LED is essentially a forward biased p-n diode as shown in Fig. 9.18. Electrons and holes are injected as minority carriers across the diode junction and they recombine either by radiative recombination or non-radiative recombination. The diode must be designed so that the radiative recombination can be made as strong as possible.

In the forward bias conditions the electrons are injected from the n-side to the p-side while holes are injected from the p-side to the n-side as shown in Fig. 9.18. The LED is designed so that the photons are emitted close to the top layer and not in the buried layer as shown in Fig. 9.18. The reason for this choice is that photons emitted deep in the device have a high probability of being reabsorbed. Thus one prefers to have only one kind of carrier injection for the diode current. Usually the top layer of the LED is p-type, and for photons to be emitted in this layer one must require the diode current to be dominated by the electron current. From diode theory this requires asymmetric doping on the n and p sides.

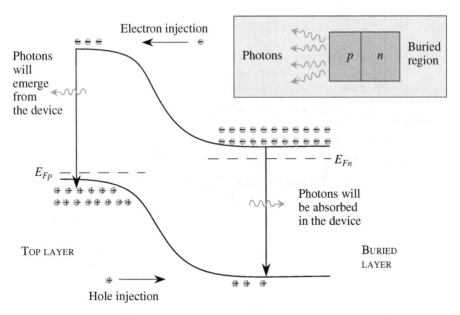

Figure 9.18: In a forward biased *p-n* junction, electrons and holes are injected as shown. In the figure, the holes injected into the buried n region will generate photons which will not emerge from the surface of the LED. The electrons injected will generate photons which are near the surface and have a high probability to emerge.

9.9.2 Laser Diode

The laser diode, like the LED, uses a forward biased *p-n* junction to inject electrons and holes to generate light. However, the laser structure is designed to create an "optical cavity" which can "guide" the photons generated. The optical cavity is essentially a resonant cavity in which the photons have multiple reflections. Thus, when photons are emitted, only a small fraction is allowed to leave the cavity. As a result, the photon density starts to build up in the cavity. For semiconductor lasers, the most widely used cavity is the Fabry-Perot cavity shown in Fig. 9.19a. The important ingredient of the cavity is a polished mirror surface which assures that resonant modes are produced in the cavity as shown in Fig. 9.19b. These resonant modes are those for which the wavelengths of the photon satisfy the relation

$$L = q\lambda/2 \tag{9.84}$$

where q is an integer, L is the cavity length, and λ is the light wavelength in the material

$$\lambda = \frac{\lambda_o}{n_r} \tag{9.85}$$

where n_r is the refractive index of the cavity. As can be seen from Fig. 9.19a, the Fabry-Perot cavity has mirrored surfaces on two sides. The other sides are roughened so that photons emitted through these sides are not reflected back.

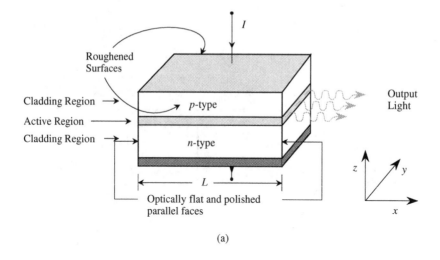

(a)

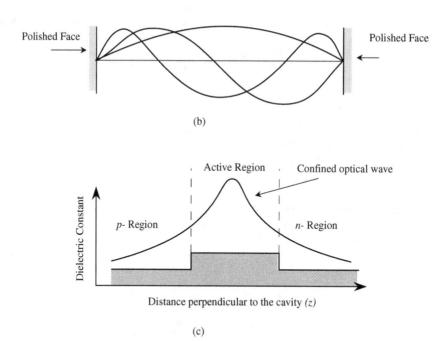

(b)

(c)

Figure 9.19: (a) A typical laser structure showing the cavity and the mirrors used to confine photons. The active region can be quite simple as in the case of double heterostructure lasers or quite complicated as in the case of quantum well lasers. (b) The stationary states of the cavity. The mirrors are responsible for these resonant states. (c) The variation in dielectric constant is responsible for the optical confinement.

If a planar heterostructure of the form shown in Fig. 9.19c is used to confine the optical wave in the z-direction, the optical equation has the form

$$\frac{d^2 F_k(z)}{d^2 z} + \left(\frac{\epsilon(z)\omega^2}{c^2} - k^2\right) F_k(z) = 0 \qquad (9.86)$$

where F is the electric field representing the optical wave. The dielectric constant $\epsilon(z)$ is chosen to have a z-direction variation so that the optical wave is confined in the z-direction as shown in Fig. 9.19c. This requires the cladding layers to be made from a large bandgap material.

An important parameter of the laser cavity is the optical confinement factor Γ, which gives the fraction of the optical wave in the active region (i.e., where e-h recombination occurs)

$$\Gamma = \frac{\int_{\text{active region}} |F(z)|^2 dz}{\int |F(z)|^2 dz} \qquad (9.87)$$

This confinement factor is almost unity for "bulk" double heterostructure lasers where the active region is $\overset{>}{\sim} 1.0$ μm, while it is as small as 1% for advanced quantum well lasers. However, in spite of the small value of Γ, quantum well lasers have superior performance because of their superior electronic properties owing to their 2- dimensional density of states.

Optical Absorption, Loss and Gain

As excess electrons and holes are injected into the active region of a laser (either through forward biasing the p-n diode or by optically pumping) the gain in the laser turns from negative to positive over some region of energy. Typical gain versus injection curves are shown in Fig. 9.20.

The gain comes only from the active region where the recombination is occurring. Often this active region is of very small dimensions. In this case, one needs to define the cavity gain which is given by

$$\text{Cavity gain} = g(\hbar\omega)\Gamma \qquad (9.88)$$

where Γ is the fraction of the optical intensity overlapping with the gain medium. The value of Γ is almost unity for double heterostructure (or bulk) lasers and $\sim$0.01 for quantum well lasers. In quantum well lasers, the overall cavity gain can still be very high, since the gain in the quantum well is very large for a fixed injection density when compared to bulk semiconductors.

In order for the laser oscillations to start, it is essential that when photons are emitted in the laser cavity, the gain associated with the cavity is able to surmount the loss suffered by the photons. The photon loss consists of two parts: i) loss because of absorption of the photons in the cladding regions and contacts of the laser; ii) loss due to the photons emerging from the cavity.

The cavity loss α_{loss} is primarily due to free carrier absorption of the light. This is a second order process and in high quality materials this loss can be as low as 10 cm^{-1}. It must be noted that the loss is dependent upon doping and defects in the

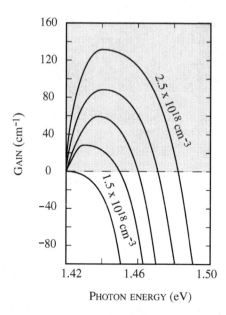

Figure 9.20: Gain vs. photon energy curves for a variety of carrier injections for GaAs at 300 K. The electron and hole injections are the same. The injected carrier densities are increased in steps of 0.25×10^{18} cm^{-3} from the lowest value shown.

material and, therefore, the material quality should be very good, especially in regions where the optical wave is confined.

An additional loss of photons is due to the escape from the cavity. Thus loss is given by

$$\alpha_{cavity} = \frac{-1}{L} \ell n R \qquad (9.89)$$

where R is the reflectivity of the mirror. For a semiconductor-air interface, the value of the reflection coefficient is

$$R = \frac{(n_r - 1)^2}{(n_r + 1)^2} \qquad (9.90)$$

where n_r is the refractive index of the semiconductor.

Laser Below and Above Threshold

In Fig. 9.21 we show the light output as a function of injected current density in a laser diode. This is the light output in the lasing mode. If we compare this with the output from an LED we notice an important difference. The light output from a laser diode in the lasing modes displays a rather abrupt change in behavior below the "threshold" condition and above this condition. The threshold condition is defined as the condition where the cavity gain overcomes the cavity loss for any photon energy, i.e., when

$$\Gamma g(\hbar\omega) = \alpha_{\text{loss}} - \frac{\ell n R}{L} \qquad (9.91)$$

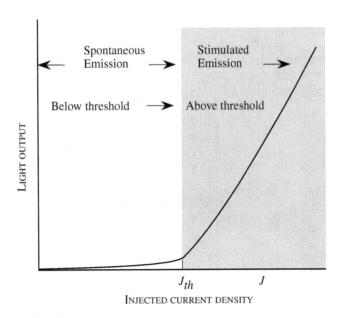

Figure 9.21: The light output in the lasing modes as a function of current injection in a semiconductor laser. Above threshold, the presence of a high photon density causes stimulated emission to dominate.

In high quality lasers $\alpha_{\text{loss}} \sim 10$ cm^{-1} and the reflection loss may contribute a similar amount. Another useful definition in the laser is the condition of transparency when the light suffers no absorption or gain, i.e.,

$$\Gamma g(\hbar\omega) = 0 \tag{9.92}$$

When the laser diode is forward biased, electrons and holes are injected into the active region of the laser. These electrons and holes recombine to emit photons. It is important to identify two distinct regions of operation of the laser. Referring to Fig. 9.22, when the forward bias current is small, the number of electrons and holes injected are small. As a result, the gain in the device is too small to overcome the cavity loss. The photons that are emitted are either absorbed in the cavity or lost to the outside. As the forward bias increases, more carriers are injected into the device until eventually the threshold condition is satisfied for some photon energy. As a result, the photon number starts to build up in the cavity. As the device is further biased beyond threshold, stimulated emission starts to occur and dominates the spontaneous emission. The light output in the photon mode for which the threshold condition is satisfied becomes very strong.

If n_{th} is the carrier density at which the gain reaches the threshold value the radiative part of the threshold current density is

$$J_r(th) = \frac{en_{th}d_{las}}{\tau_r} = \frac{en_{th}(2D)}{\tau_r} \tag{9.93}$$

where d_{las} is the thickness of the active region and $n_{th}(2D)$ is the areal carrier density

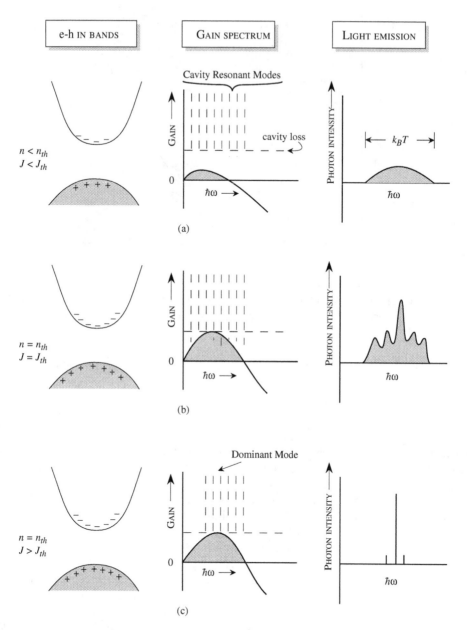

Figure 9.22: (a) The laser below threshold. The gain is less than the cavity loss and the light emission is broad as in an LED. (b) The laser at threshold. A few modes start to dominate the emission spectrum. (c) The laser above threshold. The gain spectrum does not change but, due to the stimulated emission, a dominant mode takes over the light emission.

(appropriate in quantum wells). The radiative lifetime is τ_r which is given by (see Eqn. 9.76)

$$\tau_r \sim \frac{\tau_0}{4} \tag{9.94}$$

under lasing conditions.

In addition to the radiative current we also have nonradiative current J_{nr}. For Auger processes we have

$$J_{nr} = eFn^3 d_{las}$$

The total threshold current density is then

$$J_{th} = \frac{en_{th}(2D)}{\tau_r} + eFn_{th}^3 d_{las} \tag{9.95}$$

EXAMPLE 9.10 In two n^+p GaAs LEDs, $n^+ \gg p$ so that the electron injection dominates the diode current. If the non-radiative recombination time is 10^{-7} s, calculate the 300 K internal radiative efficiency for the diodes when the doping in the p-region for the two diodes is 10^{16} cm^{-3} and 5×10^{17} cm^{-3}.

When the p-type doping is 10^{16} cm^{-3}, the hole density is low and the e-h recombination time for the injected electrons is given by (see Section 9.7)

$$\frac{1}{\tau_r} = \frac{1}{2\tau_o} \left(\frac{2\pi\hbar^2 m_r^*}{k_B T m_e^* m_h^*} \right)^{3/2} p$$

We see that for p equal to 10^{16} cm^{-3}, we have

$$\tau_r = 5.7 \times 10^{-7} \text{ s}$$

In the case where the p doping is high, the recombination time is given by the high density limit (see Eqn. 9.70) as

$$\frac{1}{\tau_r} = \frac{R_{spon}}{n} = \frac{1}{\tau_o} \left(\frac{m_r^*}{m_h^*} \right)^{3/2}$$

$$\tau_r = \frac{\tau_o}{0.05} \sim 20\tau_o \sim 12 \text{ ns}$$

For the low doping case, the internal quantum efficiency for the diode is

$$\eta_{Qr} = \frac{1}{1 + \frac{\tau_r}{\tau_{nr}}} = \frac{1}{1 + (5.7)} = 0.15$$

For the heavier doped p-region diode, we have

$$\eta_{Qr} = \frac{1}{1 + \frac{10^{-7}}{20 \times 10^{-9}}} = 0.83$$

Thus, there is an increase in the internal efficiency as the p doping is increased.

EXAMPLE 9.11 Calculate the carrier density needed for the transparency condition in GaAs at 300 K and 77 K. The transparency condition is defined at the situation where the maximum gain is zero (i.e., the optical beam propagates without loss or gain). Calculate the transparency condition at $\hbar\omega = E_g$.

At room temperature the valence and conduction band effective density of states is

$$N_v = 7 \times 10^{18} \text{ cm}^{-3}$$
$$N_c = 4.7 \times 10^{17} \text{ cm}^{-3}$$

The values at 77 K are

$$N_v = 0.91 \times 10^{18} \text{ cm}^{-3}$$
$$N_c = 0.61 \times 10^{17} \text{ cm}^{-3}$$

In the semiconductor laser, an equal number of electrons and holes are injected into the active region. We will look for the transparency conditions for photons with energy equal to the bandgap. The approach is very simple: i) choose a value of n or p; ii) calculate E_{Fn} and E_{Fp} from the Joyce-Dixon approximation; iii) calculate $f^e + f^h - 1$ and check if it is positive at the bandedge. The same approach can be used to find the gain as a function of $\hbar\omega$.

For 300 K we find that the material is transparent when $n \sim 1.1 \times 10^{18}$ cm^{-3} at 300 K and $n \sim 2.5 \times 10^{17}$ cm^{-3} at 77 K. Thus a significant decrease in the injected charge occurs as temperature is decreased.

EXAMPLE 9.12 Consider a GaAs double heterostructure laser at 300 K. The optical confinement factor is unity. Calculate the threshold carrier density assuming that it is 20% larger than the density for transparency. If the active layer thickness is 2.0 μm, calculate the threshold current density.

From the previous example we see that at transparency

$$n = 1.1 \times 10^{18} \text{ cm}^{-3}$$

The threshold density is then

$$n_{th} = 1.32 \times 10^{18} \text{ cm}^{-3}$$

The radiative recombination time is approximately four times τ_o, i.e., $\sim$2.4 ns. The current density then becomes

$$J_{th} = \frac{e \cdot n_{th} \cdot d_{las}}{\tau_r} = \frac{(1.6 \times 10^{-19} \text{ C})(1.32 \times 10^{18} \text{ cm}^{-3})(2 \times 10^{-4} \text{ cm})}{2.4 \times 10^{-9} \text{ s}}$$
$$= 1.76 \times 10^4 \text{ A/cm}^2$$

EXAMPLE 9.13 Consider two double heterostructure GaAs/AlGaAs lasers at 300 K. One laser has an undoped active region while the other one is doped p-type at 8×10^{17} cm^{-3}. Calculate the threshold current densities for the two lasers if the cavity loss is 50 cm^{-1} and the radiative lifetime at lasing is 2.4 ns for both lasers. The active region width is 0.1 μm.

This example is chosen to demonstrate the advantages of p-type doping in threshold current reduction. Since holes are already present in the active region, one does not have to inject as much charge to create gain. However, it must be noted that too much p-doping can cause an increase in cavity loss and even non-radiative traps (some dopants can be incorporated on unintended sites in the crystal).

To solve this problem, a computer program should be written. This program should calculate the quasi-Fermi levels for the electrons and holes and then evaluate the gain.

We have

$$
E_{Fn} = E_c + k_B T \left[\ell n \frac{n}{N_c} + \frac{1}{\sqrt{8}} \frac{n}{N_c} \right]
$$

$$
E_{Fp} = E_v - k_B T \left[\ell n \frac{p_{tot}}{N_v} + \frac{1}{\sqrt{8}} \frac{p_{tot}}{N_V} \right]
$$

where n is the electron (and hole) density injected and

$$
p_{tot} = p + p_A
$$

where p_A is the acceptor density. For the undoped laser, one finds that at approximately 1.1×10^{18} cm^{-3} the laser reaches the threshold condition. For the doped laser we get a value of $n = p = 8.5 \times 10^{17}$ cm^{-3}. The threshold current densities in the two cases are

$$
J(\text{undoped}) = \frac{(1.1 \times 10^{18} \text{ cm}^{-3})(0.1 \times 10^{-4} \text{ cm})(1.6 \times 10^{-19} \text{ C})}{(2.4 \times 10^{-9} \text{ s})}
$$

$$
= 733 \text{ A/cm}^2
$$

$$
J(\text{doped}) = \frac{(8.5 \times 10^{17} \text{ cm}^{-3})(0.1 \times 10^{-4} \text{ cm})(1.6 \times 10^{-19} \text{ C})}{(2.4 \times 10^{-9} \text{ s})}
$$

$$
= 566 \text{ A/cm}^2
$$

9.10   CHARGE INJECTION AND BANDGAP RENORMALIZATION

In our discussions of the semiconductor bandstructure we have not discussed whether the E vs. k relation will change if one introduces extra electrons in the conduction band or holes in the valence band. In fact, without explicitly stating this, the bandstructure we have discussed so far is for the case where all the electrons are in the valence band and the conduction band is empty. What happens when the situation is changed, either by doping or by *e-h* injection? An *e-h* pair will have a Coulombic interaction with each other. A manifestation of this interaction is the exciton, which will be discussed in the next chapter. However, the presence of the excess electrons in the conduction band or holes in the valence band also shifts the bandgap energy, decreasing it slightly. This phenomenon is often called *bandgap renormalization*.

The bandgap renormalization effect has to be treated by the many body theory, since the energy seen by a carrier depends upon the presence of other electrons and holes. Extensive experimental and theoretical work has been done to study how the bandgap changes with carrier density for both bulk and quantum well systems. We will not discuss the many body treatment here, but simply provide an approximate result for the bandgap shrinkage. In bulk materials, the bandgap shrinkage is given by

$$
\Delta E_g = -K(n^{1/3} + p^{1/3}) \tag{9.96}
$$

For GaAs, the constant K is such that we have (expressing n and p in cm^{-3})

$$
\Delta E_g = -1.6 \times 10^{-8} (n^{1/3} + p^{1/3}) \text{ eV} \tag{9.97}
$$

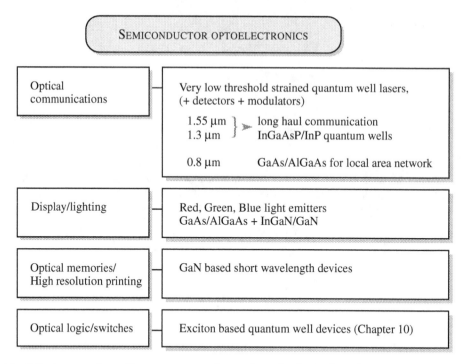

Table 9.2: An overview of important optoelectronic technologies based on semiconductors.

From this equation we see that if $n = p = 10^{18}$ cm^{-3}, the change in the bandgap is ~ -32 meV. While this is a small quantity, it can have an important effect on devices such as laser diodes, since the light emission will shift due to the bandgap change.

9.11 TECHNOLOGY ISSUES

Semiconductor optoelectronics is now an important enabling technology in the information processing age. Communication, storage, printing, and display are dependent on high performance semiconductor devices as shown in Table 9.2. Advances in these areas are coming from low-dimensional systems (quantum wells and quantum dots) and new materials (GaN, InN, for example).

Semiconductor optoelectronics devices have been used to demonstrate all of the important logic functions needed for computation. These devices have not produced "optical computers" as yet and it is unlikely they will do so in near future. However, these devices have resulted in high speed modulators and switches, as will be discussed in the next chapter.

9.12 PROBLEMS

Sections 9.4-9.5

9.1 Identify the various semiconductors (including alloys) that can be used for light emission at 1.55 μm. Remember that light emission occurs at an energy near the band-gap.

9.2 Calculate and plot the optical absorption coefficients in GaAs, InAs, and InP as a function of photon energy assuming a parabolic density of states.

9.3 Consider a 100 Å GaAs/Al$_{0.3}$Ga$_{0.7}$As quantum well structure. Assuming that the problem can be treated as that with an infinite barrier, calculate the absorption spectra for in-plane and out-of-plane polarized light for $E \leq 100$ meV from the effective band-edge. Assume the simple uncoupled model for the HH and LH states.

9.4 Calculate and plot the overlap of an electron and heavy hole ground state envelope function in a GaAs/Al$_{0.3}$Ga$_{0.7}$As quantum well as a function of well size from 20 Å to 200 Å. Assume a 60:40 value for ΔE_c:ΔE_v. At what well size does the overlap significantly differ from unity?

9.5 Calculate the gain in a GaAs region as a function of injected carrier density at room temperature. Plot your results in the form of gain vs. energy.

9.6 Estimate the strength of the intraband transitions in a quantum well structure of 100 Å GaAs/Al$_{0.3}$Ga$_{0.7}$As at an electron carrier concentration of 10^{12} cm^{-2}. Assume an infinite barrier model and a linewidth (full width at half maximum) of 1 meV for the transition.

9.7 Calculate the *e-h* radiative recombination time τ_0 (i.e., for $f^e = 1 = f^h$) for carriers in Hg$_{1-x}$Cd$_x$Te (E_g(eV)$= -0.3 + 1.9x$) for x between 0.5 and 1.0. The momentum matrix element is given by

$$\frac{2p_{cv}^2}{m_0} = 22 \text{ eV}$$

Assume that the refractive index is 3.7 and is independent of composition.

Section 9.7

9.8 In a GaAs sample at 300 K, equal concentrations of electrons and holes are injected. If the carrier density is $n = p = 10^{17}$ cm^{-3}, calculate the electron and hole Fermi levels using the Boltzmann and Joyce-Dixon approximations.

9.9 In a *p*-type GaAs doped at $N_a = 10^{18}$ cm^{-3}, electrons are injected to produce a minority carrier concentration of 10^{15} cm^{-3}. What is the rate of photon emission assuming that all *e-h* recombination is due to photon emission ? What is the optical output power? The photon energy is $\hbar\omega = 1.41$ eV.

9.10 Calculate the electron carrier density needed to push the electron Fermi level to the conduction bandedge in GaAs. Also calculate the hole density needed to push the hole Fermi level to the valence bandedge. Calculate the results for 300 K and 77 K.

9.11 Calculate and plot the 300 K recombination rate in GaAs as a function of electron (hole) carrier density. Cover the electron (hole) density range from 10^{14} cm^{-3} to 10^{18} cm^{-3} in your calculations. Estimate the carrier density dependence of the recombination rate in the low carrier density and high carrier density regime.

9.12 Consider an In$_{0.53}$Ga$_{0.47}$As sample at 300 K in which excess electrons and holes are injected. The excess density is 10^{15} cm^{-3}. Calculate the rate at which photons are generated in the system. The bandgap is 0.8 eV and carrier masses are $m_e^* = 0.042 \ m_0; m_h^* = m_0$. Also calculate the photon emission rate if the same density is injected at 77 K. (Use the low injection approximation.) Assume that the refractive index and the momentum matrix element is the same as in GaAs (given in the text).

Section 9.8

9.13 The radiative lifetime of a GaAs sample is 1.0 ns. The sample has a defect at the midgap with a capture cross-section of 10^{-15} cm^2. At what defect concentration does the non-radiative lifetime become equal to the radiative lifetime at i) 77 K and ii) 300 K?

9.14 Consider a semiconductor with a bandgap of 0.75 eV and Auger coefficient of 10^{-28} cm^{-6}s^{-1} at 300 K. The material is used for 0.1 μm active region laser where $n_{2D} = p_{2D} = 10^{12}$ cm^{-2}. Calculate the non-radiative current density.

Section 9.9

9.15 Consider the semiconductor alloy InGaAsP with a bandgap of 0.8 eV. The electron and hole masses are 0.04 m_0, and 0.35 m_0, respectively. Calculate the injected electron and hole densities needed at 300 K to cause inversion for the electrons and holes at the bandedge energies. How does the injected density change if the temperature is 77 K? Use the Joyce-Dixon approximation.

9.16 Consider a GaAs based laser at 300 K. Calculate the injection density required at which the inversion condition is satisfied at i) the bandedges; and ii) at an energy of $\hbar\omega = E_g + k_B T$. Use the Joyce-Dixon approximation.

9.17 Consider a GaAs based laser at 300 K. A gain of 30 cm^{-1} is needed to overcome cavity losses at an energy of $\hbar\omega = E_g + 0.026$ eV. Calculate the injection density required. Also, calculate the injection density if the laser is to operate at 400 K.

9.18 Two GaAs/AlGaAs double heterostructure lasers are fabricated with active region thicknesses of 2.0 μm and 0.5 μm. The optical confinement factors are 1.0 and 0.8, respectively. The carrier injection density needed to cause lasing is 1.0×10^{18} cm^{-3} in the first laser and 1.1×10^{18} cm^{-3} in the second one. The radiative recombination times are 1.5 ns. Calculate the threshold current densities for the two lasers.

9.19 Consider a 1.55 μm laser operating at 77 K and 300 K. A cavity loss of 40 cm^{-1} exists in this laser. The other laser parameters are:

$$\Gamma = 0.1$$
$$m_e^* = 0.04\ m_0$$
$$m_h^* = 0.4\ m_0$$

Auger coefficient

$$F = 10^{-28}\ \text{cm}^6\ \text{s}^{-1}\ \text{at 300 K}$$
$$F = 10^{-30}\ \text{cm}^6\ \text{s}^{-1}\ \text{at 77 K}$$
$$d_{las} = 0.1\ \mu\text{m}$$

Calculate the threshold carrier density needed at 77 K and 300 K. If the radiative lifetime at threshold is 2ns calculate the radiative and non-radiative current densities at threshold.

9.20 Calculate the carrier density needed to reach a peak TE gain of 1000 cm^{-1} in a

100 Å GaAs/AlGaAs quantum well at 300 K. Assume the following parameters:
($m_e^* = 0.067\ m_0; m_{HH}^* = 0.45\ m_0; m_{LH}^* = 0.45\ m_0$):

$$E_{e1} - E_c(\text{GaAs}) = 33\ \text{meV}$$
$$E_{e2} - E_c(\text{GaAs}) = 120\ \text{meV}$$
$$E_v(\text{GaAs}) - E_{HH1} = 8\ \text{meV}$$
$$E_v(\text{GaAs}) - E_{LHI} = 20\ \text{meV}$$

Use a Gaussian broadening function of halfwidth 1.3 meV. What is the peak TM gain
at the same injection?

9.21 Consider a 80 Å GaAs/Al$_{0.3}$Ga$_{0.7}$As quantum well laser with $\Gamma = 0.02$. Calculate
the threshold carrier density and threshold current for the following device parameters.

$$E_{e1} - E_c(\text{GaAs}) = 40\ \text{meV}$$
$$E_v(\text{GaAs}) - E_{HH1} = 12\ \text{meV}$$
$$E_v(\text{GaAs}) - E_{LH} = 20\ \text{meV}$$
$$m_e^* = 0.067\ m_0$$
$$m_{HH}^* = 0.45\ m_0$$
$$m_{LH}^* = 0.4\ m_0$$
$$\alpha_c = 50\ \text{cm}^{-1}$$

The radiative lifetime is 2.5 ns. Assume a halfwidth of 1.0 meV for a Gaussian broadening for the gain.

9.22 Consider a 100 Å GaAs/Al$_{0.3}$Ga$_{0.7}$As quantum well laser and a 100 Å
In$_{0.3}$Ga$_{0.7}$As/Al$_{0.3}$Ga$_{0.7}$As laser. Compare the gain curves for the two lasers near threshold. The devices are defined by the following parameters.

GaAs well (cavity loss for both lasers is 40 cm^{-1} and confinement factor is 0.002):

$$m_e^* = 0.067\ m_0$$
$$m_{HH}^* = 0.5\ m_0$$
$$m_{LH}^* = 0.5\ m_0$$
$$E_e^1 - E_c(\text{GaAs}) = 33\ \text{meV}$$
$$E_v(\text{GaAs}) - E_{HH}^1 = 8\ \text{meV}$$
$$E_v(\text{GaAs}) - E_{LH}^1 = 20\ \text{meV}$$

InGaAs well:

$$m_e^* = 0.067\ m_0$$
$$m_{HH}^* = 0.2\ m_0$$
$$E_e^1 - E_c(\text{InGaAs}) = 33\ \text{meV}$$
$$E_v(\text{InGaAs}) - E_{HH}^1 = 8\ \text{meV}$$
$$E_v(\text{InGaAs}) - E_{LH}^1 = 120\ \text{meV}$$

Assume that the gain is broadened by a Gaussian function with halfwidth of 1.5 meV. Temperature is 300 K.

$$E_{gap}(\text{GaAs}) = 1.43 \text{ eV}$$
$$E_{gap}(\text{In}_{0.3}\text{Ga}_{0.7}\text{As}) = 1.25 \text{ eV}$$

9.13 REFERENCES

- **Maxwell's Equations; Gauge Transformation**

  – J. D. Jackson, *Classical Electrodynamics*, John Wiley and Sons, New York (1975).

- **Macroscopic Theory of Optical Processes**

  – J. D. Jackson, *Classical Electrodynamics*, John Wiley and Sons, New York (1975).

  – W. Jones and N. H. March, *Theoretical Solid State Physics*, Dover, New York, vol. 2 (1973).

  – E. Kreyszig, *Advanced Engineering Mathematics*, John Wiley and Sons, New York (1962).

  – F. Stern, "Elementary Theory of the Optical Properties of Solids," *Solid State Physics*, Academic Press, New York, vol. 15 (1963).

  – B.K. Ridley, *Quantum Processes in Semiconductors*, Oxford Press, New York (1982).

  – J. M. Ziman, *Principles of the Theory of Solids*, Cambridge (1972).

- **Microscopic Theory of Optical Processes in Semiconductors**

  – F. Bassani and G. P. Parravicini, *Electronic States and Optical Transitions in Solids*, Pergamon Press, New York (1975).

  – W. Jones and N. H. March, *Theoretical Solid State Physics*, Dover, New York, vol. 2 (1973).

- **Optical Processes in Indirect Semiconductors**

  – F. Bassani and G. P. Parravicini, *Electronic States and Optical Transitions in Solids*, Pergamon, New York (1975).

  – W. C. Dash and R. Newman, *Phys. Rev.*, **99**, 1151 (1955).

  – H. R. Phillip and E. A. Taft, *Phys. Rev.*, **113**, 1002 (1959).

  – H. R. Phillip and E. A. Taft, *Phys. Rev. Lett.*, **8**, 13 (1962).

- **Intraband Transitions in Quantum Wells**

  – D. D. Coon and R. P. G. Karunasiri, *Appl. Phys. Lett.*, **45**, 649 (1984).

  – A. Harwitt and J. S. Harris, *Appl. Phys. Lett.*, **50**, 685 (1987).

- B. F.Lavine , K. K. Choi, C. G. Bathea, J. Walker and R. J. Malik, *Appl. Phys. Lett.*, **50**, 1092 (1987).

- **Auger Processes**

  - A.R. Beattie, *J. Phys. Chem. Solids*, **49**, 589 (1988).
  - A.R. Beattie and P.T. Landsberg, *Proceedings of the Royal Society*, A 249, 16 (1959).
  - A. Haug, *J. Phys. Chem. Solids*, 49, 599 (1988).
  - Y. Jiang, M.C. Teich, and W.I. Wang, *Appl. Phys. Lett.*, **57**, 2922 (1990).
  - J.P. Loehr and J. Singh, *IEEE J. Quant. Electron.*, **QE-29**, 2583 (1993).
  - A. Sugimura, *IEEE J. Quant. Electron.*, **QE-18**, 352 (1982).

- **Bandgap Renormalization**

  - H.C. Casey and F. Stern, *J. Appl. Phys.*, **47**, 631 (1976).
  - S Schmitt-Rink, C. Ell, and H. Haug, *Phys. Rev.*, B, **38**, 3342 (1988).
  - D.A. Kleinmann and R.C. Miller, *Phys. Rev.*, B, **32**, 2266 (1985).

Chapter

10

EXCITONIC EFFECTS
AND MODULATION OF
OPTICAL PROPERTIES

10.1 INTRODUCTION

The bandstructure and optical properties of semiconductors we have discussed so far are based on the assumption that the valence band is filled with electrons and the conduciton band is empty. The effect of electrons in the conduction band and holes in the valence band is only manifested through the occupation probabilities without altering the bandstructure. In reality, of course, there is a Coulombic interaction between an electron and another electron or hole. Some very important properties are modified by such interactions. The full theory of the electron-electron interaction depends upon many body theory, which is beyond the scope of this text. However, there is one important problem, that of excitonic effects in semiconductors, that can be addressed by simpler theoretical techniques.

In Fig. 10.1 we show how exciton effects arise. On the left-hand side, we show the bandstructure of a semiconductor with a full valence band and an empty conduction band. There are no allowed states in the bandgap. Now consider the case where there is one electron in the conduction band and one hole in the valence band. In this new configuration, the Hamiltonian describing the electronic system has changed. We now have an additional Coulombic interaction between the electron and the hole. The electronic bandstructure should thus be modified to reflect this change. The electron-hole system, coupled through the Coulombic interaction, is called the exciton and will be the subject of this chapter.

We will discuss the exciton problem in bulk semiconductors and in heterostructures. The study of excitonic transitions in quantum wells has become an extremely important area, both from the point of view of new physics and of new technology. We

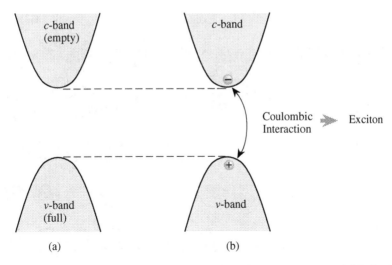

Figure 10.1: (a) The bandstructure in the independent electron picture and (b) the Coulombic interaction between the electron and hole which would modify the band picture.

will see in this chapter that due to quantum confinement, the exciton binding energy is greatly increased. This, and improved optical transition strength, allows one to observe extremely sharp resonances in quantum well optical spectra. Moreover, the energy, or strength, of these resonances can be controlled easily by simple electronics or optics. This ability allows one to use the excitonic transitions for high speed modulation of optical signals, as well as for optoelectronic switches, which could serve important functions in future information processing systems.

The modulation of optical properties is an essential ingredient for advanced optoelectronic systems. While a number of different stimulii can modulate the optical properties of a material (electric field, magnetic field, strain field) only electric field or electromagnetic field-induced modulation can produce high speed operation. In the bulk form, most semiconductors do not have very good modulation properties. However, these properties can be greatly enhanced in quantum wells. In this chapter we will examine the physics behind various optical modulation approaches.

10.2 EXCITONIC STATES IN SEMICONDUCTORS

The electron-hole pair shown in Fig. 10.1 forms a bound state which is described by an envelope function. The form of the envelope function describing the bound state can be calculated using the Coulombic interaction as a perturbation. Two important classes of excitons exist depending upon the extent of the periodic envelope function as shown in Fig. 10.2. When the envelope function is confined to just a few unit cells, the excitons are classified as Frenkel excitons. Due to their restricted spatial extent, the Heisenberg uncertainty principle indicates that their treatment necessitates dealing with the full bandstructure of the semiconductors. On the other hand, if the envelope function extends over several hundred Angstroms, near bandedge electron and hole

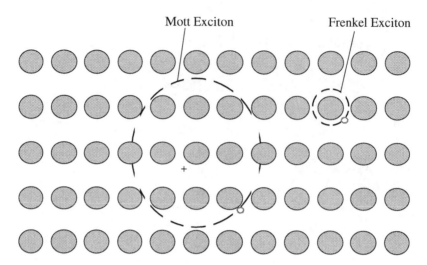

Figure 10.2: A conceptual picture of the periodic envelope function extent of the Frenkel and Mott excitons. The Frenkel exciton periodic function is of the extent of a few unit cells while the Mott exciton function extends over many unit cells.

states can be used to describe them. Such excitons are called Mott excitons and are responsible for the excitonic physics in semiconductors. The effective mass theory can be used to describe these excitons and accordingly the problem is represented by the following Schrödinger equation

$$\left[-\frac{\hbar^2}{2m_e^*}\nabla_e^2 - \frac{\hbar^2}{2m_h^*}\nabla_h^2 - \frac{e^2}{4\pi\epsilon \left| r_e - r_h \right|} \right] \psi_{\text{ex}} = E\psi_{\text{ex}} \qquad (10.1)$$

Here m_e^* and m_h^* are the electron and hole effective masses and $\left| r_e - r_h \right|$ is the difference in coordinates defining the Coulombic interaction between the electron and the hole. We will shortly discuss in more detail the makeup of the exciton wavefunction ψ_{ex}. The problem is now the standard two-body problem, which can be written as a one-body problem, by using the following transformation

$$
\begin{aligned}
\boldsymbol{r} &= \boldsymbol{r}_e - \boldsymbol{r}_h \\
\boldsymbol{k} &= \frac{m_e^*\boldsymbol{k}_e + m_h^*\boldsymbol{k}_h}{m_e^* + m_h^*} \\
\boldsymbol{R} &= \frac{m_e^*\boldsymbol{r}_e + m_h^*\boldsymbol{r}_h}{m_e^* + m_h^*} \\
\boldsymbol{K} &= \boldsymbol{k}_e - \boldsymbol{k}_h
\end{aligned}
\qquad (10.2)
$$

The Hamiltonian then becomes

$$H = \frac{\hbar^2 K^2}{2(m_e^* + m_h^*)} + \left\{ \frac{\hbar^2 k^2}{2m_r^*} - \frac{e^2}{4\pi\epsilon \left| r \right|} \right\} \qquad (10.3)$$

where m_r^* is the reduced mass of the electron-hole system. The Hamiltonian consists of two parts, the first term giving the description for the motion of center of mass of the electron-hole system, while the second term describing the relative motion of the electron-hole system. The first term gives a plane wave solution

$$\psi_{cm} = e^{i\mathbf{K}\cdot\mathbf{R}} \tag{10.4}$$

while the solution to the second term satisfies

$$\left(\frac{\hbar^2 k^2}{2m_r^*} - \frac{e^2}{4\pi\epsilon \, |\mathbf{r}|} \right) F(\mathbf{r}) = E F(\mathbf{r}) \tag{10.5}$$

This is the usual hydrogen atom problem and $F(\mathbf{r})$ can be obtained from the mathematics of that problem. The general exciton solution is now (writing $\mathbf{K}_{ex} = \mathbf{K}$)

$$\psi_{n\mathbf{K}_{ex}} = e^{i\mathbf{K}_{ex}\cdot\mathbf{R}} \, F_n(\mathbf{r}) \, \phi_c(\mathbf{r}_e) \, \phi_v(\mathbf{r}_h) \tag{10.6}$$

where ϕ_c and ϕ_v represent the central cell nature of the electron and hole bandedge states used in the effective mass theory. The excitonic energy levels are then

$$E_{n\mathbf{K}_{ex}} = E_n + \frac{\hbar^2}{2(m_e^* + m_h^*)} K_{ex}^2 \tag{10.7}$$

with E_n being the eigenvalues of the hydrogen atom-like problem

$$E_n = -\frac{m_r^* e^4}{2(4\pi\epsilon)^2 \hbar^2} \frac{1}{n^2} \tag{10.8}$$

and the second term in Eqn. 10.7 represents the kinetic energy of the center of mass of the electron-hole pair.

The energy of the excitonic state is measured with respect to the energy of the state without the Coulomb interaction, i.e., the bandgap. Thus, excitonic levels appear slightly below the bandgap since typical values for E_1 are ~2-6 meV for most semiconductors. The dispersion then looks as shown in Fig. 10.3.

This dispersion relation looks quite different from the usual E vs k relation we are used to. This is because we are describing the system not in terms of the electron crystal momentum, but the electron-hole crystal momentum $\mathbf{K}_{ex}$. This is obviously the appropriate quantum number to describe the problem once the electron-hole Coulombic interaction is turned on. If the Coulombic interaction is turned off, the parabolas below the bandgap (bond states) all disappear and we simply have the free electron-hole dispersion which is the same as the bandstructure discussed in earlier chapters except that we plot the dispersion in a $\mathbf{k}_e - \mathbf{k}_h$ description.

We notice now that unlike the cases discussed in Chapter 9 where the electron-hole joint density of states started at the bandedge position, we now have a density of states below the bandedge energies. However, not all these states will couple to the photon because of momentum conservation.

In order to examine the absorption spectra of excitonic transitions in semi-conductors, it is useful to examine the problem in a little greater detail. As discussed

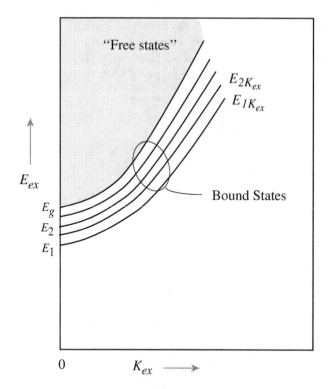

Figure 10.3: Dispersion curves for the electron-hole system in the exciton framework.

earlier, the independent electron picture provides us the conduction and valence band states. The Coulombic interaction of the electron-hole pairs will now be treated as a perturbation, and the new wavefunction can be expressed in terms of the independent electron wavefunction basis. The general form of the excitonic problem is given by the Hamiltonian

$$H_e = H_0 + \frac{1}{2} \sum_{i \neq j} \frac{e^2}{4\pi\epsilon \left| \boldsymbol{r}_i - \boldsymbol{r}_j \right|} \tag{10.9}$$

where H_0 is the independent electron Hamiltonian giving rise to the usual bandstructure. The indices i and j represent the different electron pairs, with the factor $1/2$ to prevent double counting.

Since the Hamiltonian has the symmetry of the crystal, the Bloch theorem applies to the wavefunction, which must satisfy the condition

$$\psi_{\text{ex}}(\boldsymbol{r}_1 + \boldsymbol{R}, \boldsymbol{r}_2 + \boldsymbol{R}, \boldsymbol{r}_3 + \boldsymbol{R}, \ldots) = e^{i\mathbf{K}_{\text{ex}} \cdot \mathbf{R}} \, \psi_{\text{ex}}(\boldsymbol{r}_1, \boldsymbol{r}_2, \boldsymbol{r}_3, \ldots) \tag{10.10}$$

where $\boldsymbol{R}$ is a lattice vector of the crystal.

The exciton state can be written in terms of a basis function $\Phi_{c,\mathbf{k}_e,S_e;v,\mathbf{k}_h,S_h}$ which represents a state where an electron, with momentum $\boldsymbol{k}_e$ and spin S_e, is in the conduction band and a hole, with momentum and spin $\boldsymbol{k}_h$ and S_h, is in the valence band,

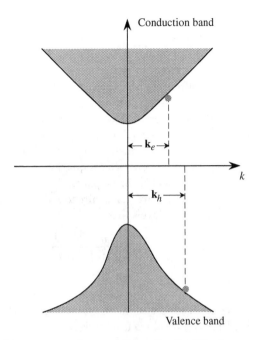

Figure 10.4: Schematic picture of an exciton in the Bloch representation. The state $\Phi_{c,\mathbf{k}_e,S_e;v,\mathbf{k}_h,S_h}$ represents an extra electron of wave vector k_e and spin S_e in the conduction band and a hole of wave vector k_h and spin S_h in the valence band.

as shown in Fig 10.3. The difference $\mathbf{k}_e - \mathbf{k}_h$ represents the momentum of the exciton state. The exciton state is made up of a proper expansion of the Φ states. However, because of the Bloch theorem, the combination $\mathbf{k}_e - \mathbf{k}_h$ in the expansion must be constant for any given excitonic state. This greatly simplifies our exciton wavefunction, which can now be written as

$$\psi_{\text{ex}}^{n\ell m} = \sum_{\mathbf{k}} A_{n\ell m}(\mathbf{k})\, \Phi_{c,\mathbf{k}+\mathbf{K}_{\text{ex}}/2,S_e;v,\mathbf{k}-\mathbf{K}_{\text{ex}}/2,S_h}^{n\ell m} \tag{10.11}$$

Here n is the energy eigenvalue index, ℓ and m are angular momentum indices representing the multiplicity of the excitonic state and $A_{n\ell m}(\mathbf{k})$ are the expansion coefficients. The exciton solution is given with the determination of $A_{n\ell m}(\mathbf{k})$. Since we are dealing with large envelope functions (order of $\sim$100 Å), the coefficients $A_{n\ell m}(\mathbf{k})$ are expected to be localized sharply in k-space. We can define the Fourier transform of $A_{n\ell m}(\mathbf{k})$ as

$$F_{n\ell m}(\mathbf{r}) = \sum_{\mathbf{k}} A_{n\ell m}(\mathbf{k})\, e^{i\mathbf{k}\cdot\mathbf{r}} \tag{10.12}$$

This real space envelope function $F_{n\ell m}(\mathbf{r})$ is the same as we introduced in our simple derivation earlier and which obeys the hydrogen atom-like equation (ignoring exchange interactions):

$$\left[E_{cv}(-i\nabla, \boldsymbol{K}_{ex}) - \frac{e^2}{4\pi\epsilon r} \right] F_{n\ell m}(\boldsymbol{r}) = E_{ex} F_{n\ell m}(\boldsymbol{r}) \qquad (10.13)$$

Here $E_{cv}(-i\nabla, \boldsymbol{K}_{ex})$ represents the operator obtained by expanding $E_c(\boldsymbol{k} + \boldsymbol{K}_{ex}/2) - E_v(\boldsymbol{k} - \boldsymbol{K}_{ex}/2)$ in powers of $\boldsymbol{k}$ and replacing $\boldsymbol{k}$ by $-i\nabla$. The exchange term is usually very small and will be ignored. The dielectric constant, in general, can be quite complicated, especially if the free carrier density is large. At low carrier densities ($n < 10^{14}$ cm^{-3}), the static dielectric constant is a good approximation to ϵ. It is important to note that this effective mass like equation is valid only if $F_{n\ell m}(\boldsymbol{r})$ is extended in space, i.e., $A_{n\ell m}(\boldsymbol{k})$ is peaked in k-space.

For a simple parabolic band, we have already discussed the solution of the exciton problem. The exciton energy levels are

$$\begin{aligned} E_n^{ex} &= E_g - \frac{m_r^* e^4}{2\hbar^2 (4\pi\epsilon)^2} \frac{1}{n^2} \\ &= E_g - \frac{R_{ex}}{n^2} \end{aligned} \qquad (10.14)$$

where R_{ex} denotes the exciton Rydberg. The kinetic energy of the electron-hole pair is to be added to Eqn. 10.14 for the total exciton energy.

The exciton envelope functions are the hydrogen atom-like functions, e.g., the ground state is

$$F_{100}(\boldsymbol{r}) = \frac{1}{\sqrt{\pi a_{ex}^3}} e^{-r/a_{ex}} \qquad (10.15)$$

with $a_{ex} = (\epsilon m_0/\epsilon_0 m_r^*) a_B$ (a_B = Bohr radius = 0.529 Å). The exciton radius a_{ex} is $\sim$100 Å for most semiconductors. Thus, the exciton is spread over a large number of unit cells, and the use of the effective mass equation is justified.

10.3   OPTICAL PROPERTIES WITH INCLUSION OF EXCITONIC EFFECTS

Excitonic effects have very dramatic consequences for the optical properties of semiconductors, especially near the bandedges. Below the bandedge, there is a strong and sharp excitonic absorption/emission transition. Also just above the bandgap, there is a strong enhancement of the absorption process especially in 3D systems.

As discussed in Chapter 9, in the absence of excitonic effects, the absorption coefficient can be written as

$$\alpha(\hbar\omega) = \frac{\pi e^2}{m_0^2 c n_r \epsilon_0} \frac{\hbar}{\hbar\omega} \int \frac{2\, d^3 k}{(8\pi^3)} |\boldsymbol{a} \cdot \boldsymbol{p}_{if}(\boldsymbol{k})|^2 \, \delta(E_c(\boldsymbol{k}) - E_v(\boldsymbol{k}) - \hbar\omega) \qquad (10.16)$$

For allowed transitions we can assume that $\boldsymbol{p}_{if}$ is independent of $\boldsymbol{k}$ giving us the absorption coefficient

$$\begin{aligned} \alpha(\hbar\omega) &= 0 & \text{if } \hbar\omega < E_g \\ &= \frac{\pi e^2}{m^2 c n_r \epsilon_0} \frac{\hbar}{\hbar\omega} |\boldsymbol{a} \cdot \boldsymbol{p}_{if}|^2 \cdot N_{cv}(\hbar\omega) & \text{if } \hbar\omega \geq E_g \end{aligned} \qquad (10.17)$$

where $N_{cv}(\hbar\omega)$ is the joint density of states.

If the excitonic effects are accounted for, these expressions are modified. We will again work in the dipole approximation and consider a transition from the ground state (all electrons are in the valence band) to the excited exciton state. This transition rate is, according to the Fermi golden rule

$$W(\psi_0 \rightarrow \psi_{\mathbf{K}_{ex}}) = \frac{2\pi}{\hbar}\left(\frac{eA}{m_0}\right)^2 \delta_{\mathbf{K}_{ex}} \left|\sum_k A(\mathbf{k})\, \mathbf{a} \cdot \mathbf{p}_{cv}(\mathbf{k})\right|^2 \delta(E_{ex} - E_0 - \hbar\omega) \quad (10.18)$$

where E_0 is the energy corresponding to the ground state.

Once again, if we assume $\mathbf{p}_{cv}(\mathbf{k})$ is independent of $\mathbf{k}$

$$W(\psi_0 \rightarrow \psi_{\mathbf{K}_{ex}}) = \frac{2\pi}{\hbar}\left(\frac{eA}{m_0}\right)^2 \delta_{\mathbf{K}_{ex}} |\mathbf{a} \cdot \mathbf{p}_{if}(0)|^2 \left|\sum_k A(\mathbf{k})\right|^2 \delta(E_{ex} - E_0 - \hbar\omega) \quad (10.19)$$

From the definition of the Fourier transform $F_{n\ell m}$, we see that from Eqn. 10.12

$$F_{n\ell m}(0) = \sum_k A_{n\ell m}(\mathbf{k}) \quad (10.20)$$

We also know from the theory of the hydrogen atom problem that $F_{n\ell m}(0)$ is nonzero only for s-type states, and, in general

$$F_{n\ell m}(0) = \frac{1}{\sqrt{\pi a_{ex}^3 n^3}} \delta_{\ell,0}\delta_{m,0} \quad (10.21)$$

Thus, the absorption rate is given by

$$W(\psi_0 \rightarrow \psi_{\mathbf{K}_{ex}}) = \frac{2\pi}{\hbar}\left(\frac{eA_0}{m_0}\right)^2 \delta_{\mathbf{K}_{ex}} |\mathbf{a} \cdot \mathbf{p}_{if}(0)|^2 \frac{\delta(E_{ex}^n - E_0 - \hbar\omega)}{\pi a_{ex}^3 n^3} \quad (10.22)$$

Comparing this result with the case for free band-to-band transitions, we note that the density of states in the free case is replaced by the term

$$N_{cv}(\hbar\omega) \rightarrow \frac{\delta(E_{ex}^n - E_0 - \hbar\omega)}{\pi a_{ex}^3 n^3} \quad (10.23)$$

with the δ-function eventually being replaced by a broadening function. If, for example, we assume a Gaussian broadening, we have (σ is the half width)

$$\delta(\hbar\omega - E) \rightarrow \frac{1}{\sqrt{1.44\pi}\,\sigma}\exp\left(\frac{-(\hbar\omega - E)^2}{1.44\,\sigma^2}\right)$$

For the ground state $n = 1$ exciton, the absorption coefficient becomes, after using the Gaussian form for the δ-function,

$$\alpha(\hbar\omega) = \frac{\pi e^2\hbar}{2n_r\epsilon_0 cm_0(\hbar\omega)}\frac{2|p_{cv}|^2}{m_0}a_p\left(\frac{1}{\sqrt{1.44\pi}}\frac{1}{\sigma}\frac{1}{\pi a_{ex}^3}\exp\left(\frac{-(\hbar\omega - E_{ex})^2}{1.44\sigma^2}\right)\right) \quad (10.24)$$

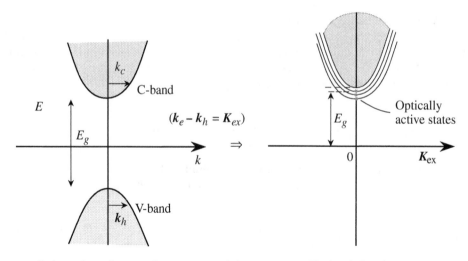

Independent electron picture   +   e-h interact ⇒ Excitonic bandstructure

Figure 10.5: The effect of the electron-hole Coulombic interaction is to create exciton bands as shown. Only $K_{ex} = 0$ states are optically active.

This result suggests that the excitonic transitions occur *as if* each exciton has a spatial extent of $1/(\pi a_{ex}^3 n^3)$. However, we note that this is not really the correct picture, since the excitons are extended states. We also note that since only the $\boldsymbol{K}_{ex} = 0$ state of the exciton is optically active, the transitions are *discrete*, even though the exciton density of states is *continuous*. The strength of the successive transitions decreases as $1/n^3$, so that the $n = 2$ resonance has one-eighth the strength of the $n = 1$ transition.

It is useful to examine again the independent electron picture and the exciton picture. This is done in Fig. 10.5. On the left-hand side we show the usual bandstructure with the valence and conduction band. The presence of the Coulomb interaction causes us to use $\boldsymbol{k}_e - \boldsymbol{k}_h = \boldsymbol{K}_{ex}$ as an appropriate quantum number for the description. As discussed earlier, this leads to the exciton bands below the bandgap and free states above the bandgap. Due to the momentum conservation, only the $\boldsymbol{K}_{ex} = \boldsymbol{k}_e - \boldsymbol{k}_h = 0$ states are optically active. Above the bandgap, these states are just the ones we considered earlier, in the band-to-band transitions. However, below the bandgap, these are the discrete excitonic resonances. Strong changes in the optical properties occur near the bandedge.

As one approaches the bandedge, the exciton lines become closer and closer and (see Eqn. 10.8), even though each transition becomes weaker, the absorption over an infinitesimal energy range reaches a finite value. In fact, the concept of the density of states of $K_{ex} = 0$ states becomes a meaningful concept. This density of states is from Eqn. 10.8

$$D_{ex}(E) \;=\; 2\frac{\partial n}{\partial E}$$

$$= \frac{n^3}{R_{\text{ex}}} \tag{10.25}$$

Extending the expression for the transition rates (Eqn. 10.22), by including a final state density of excitonic states we get

$$
\begin{aligned}
W(\psi_0 \to \psi_{\text{ex}}) &= \frac{2\pi}{\hbar} \left(\frac{eA_0}{m_0} \right)^2 \delta_{\mathbf{K}_{\text{ex}}} |\boldsymbol{a} \cdot \boldsymbol{p}_{\text{if}}(0)|^2 \\
&\quad \times \sum_n \left| \sum_{\boldsymbol{k}} A(\boldsymbol{k}) \right|^2 \delta \left(E_{\text{ex}}^n - E_0 - \hbar\omega \right) \\
&= \frac{2\pi}{\hbar} \left(\frac{eA_0}{m_0} \right)^2 |\boldsymbol{a} \cdot \boldsymbol{p}_{\text{if}}(0)|^2 \frac{1}{\pi a_{\text{ex}}^3} \frac{1}{R_{\text{ex}}} \tag{10.26}
\end{aligned}
$$

This expression is valid near the bandedge. If we compare this expression with the free electron-hole absorption rate near the bandedge, we see that the difference is that the density of states has been replaced by $1/(\pi a_{\text{ex}}^3 R_{\text{ex}})$ or, near the bandedge, the absorption coefficient is

$$\alpha_{\text{ex}}(\hbar\omega \approx E_g) = \alpha_F \cdot \frac{2\pi \, R_{\text{ex}}^{1/2}}{(\hbar\omega - E_g)^{1/2}} \tag{10.27}$$

where α_F is the absorption without excitonic effects. Thus, instead of α going to zero at the bandedge, it *becomes a constant*. By examining the nature of the "free" hydrogen atom-like states for the exciton *above* the bandedge, it can be shown that the absorption coefficient is given by the relation

$$\alpha_{\text{ex}}(\hbar\omega > E_g) = \alpha_F \cdot \frac{\pi x \, e^{\pi x}}{\sinh \pi x} \tag{10.28}$$

where

$$x = \frac{R_{\text{ex}}}{(\hbar\omega - E_g)^{1/2}}$$

When $E_g - \hbar\omega \gg R_{\text{ex}}$, the results reduce to the band-to-band transitions calculated in Chapter 9. These effects are shown in Fig. 10.6.

From these discussions it is clear that the excitonic transitions greatly modify the independent electron absorption spectra, especially near the bandedges. In Fig. 10.7, we show a low temperature measurement of the excitonic and band-to-band absorption in GaAs. The excitonic peak is clearly resolved here. The binding energy of the exciton in GaAs is ~ 4 meV. Since the exciton line is broadened by background impurity potential fluctuations, as well as phonons, it is possible to see such transitions only in high purity semiconductors, at low temperatures. In fact, the observation of excitons in bulk semiconductors is a good indication of the quality of the sample.

In Fig. 10.7 we show how the excitonic structure in high quality GaAs is altered as temperature is increased. For narrow bandgap semiconductors like $In_{0.53}Ga_{0.47}As$, the excitons are difficult to observe, because of the very small binding energy. In such

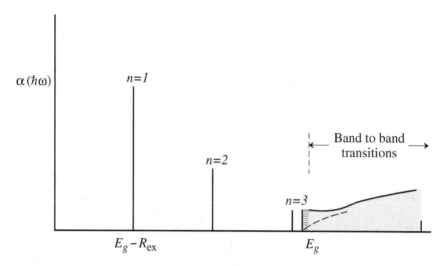

Figure 10.6: A schematic picture of the absorption spectra with (solid line) and without (dashed line) excitonic effects.

materials, the exciton-related transitions are seen better in low temperature lumines-cence experiments. In Appendix B we discuss these experiments.

EXAMPLE 10.1 Consider a ground state exciton in GaAs having a halfwidth of $\sigma = 1.0$ meV. Calculate the peak absorption coefficient for the excitonic resonance.

The optical absorption coefficient is given for excitonic transitions by

$$\alpha(\hbar\omega) = \frac{\pi e^2 \hbar}{2n_r \epsilon_0 c m_0 (\hbar\omega)} \frac{2\,|p_{cv}|^2}{m_0} \; a_p \left(\frac{1}{\sqrt{1.44\pi}} \frac{1}{\sigma} \frac{1}{\pi a_{ex}^3} \exp\left(\frac{-(\hbar\omega - E_{ex})^2}{1.44\sigma^2} \right) \right)$$

We have for GaAs (we will express σ in units of meV for convenience)

$$\frac{2p_{cv}^2}{m_0} \cong 25 \text{ eV}$$

Using $a_p = \frac{2}{3}$ for unpolarized light, we get ($a_{ex} = 120$ Å)

$$
\begin{aligned}
\alpha(\hbar\omega) &= \frac{\pi(1.6 \times 10^{-19} \text{ C})^2 (1.05 \times 10^{-34} \text{ Js})}{2 \times 3.4 (8.85 \times 10^{-12} \text{ F/m})(3 \times 10^8 \text{ m/s})(0.91 \times 10^{-30} \text{ kg})} \left(\frac{25}{1.5} \right) \left(\frac{2}{3} \right) \\
&\quad \cdot \frac{1}{\sqrt{1.44\pi}} \frac{1}{\sigma(meV)(10^{-3} \times 1.6 \times 10^{-19} \text{ J})} \frac{1}{(\pi \times 120 \times 10^{-10} \text{ m})^3} \\
&\quad \exp\left(\frac{-(\hbar\omega - E_{ex})^2}{1.44\sigma^2} \right) \\
&= \frac{1.45 \times 10^6}{\sigma(meV)} \exp\left(\frac{-(\hbar\omega - E_{ex})^2}{1.44\sigma^2} \right) \text{ m}^{-1}
\end{aligned}
$$

For our case, where $\sigma = 1$ meV, we get, for the peak absorption coefficient

$$
\begin{aligned}
\alpha(\text{peak}) &= 1.45 \times 10^6 \text{ m}^{-1} \\
&= 1.45 \times 10^4 \text{ cm}^{-1}
\end{aligned}
$$

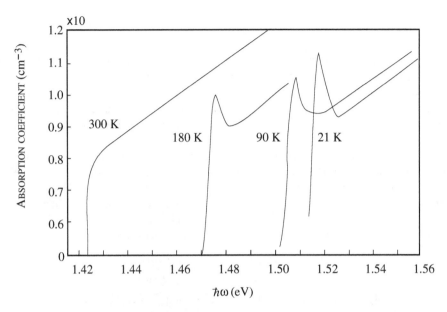

Figure 10.7: Typical optical transitions GaAs. As can be seen, the excitonic peak essentially merges with the band-to-band absorption onset at room temperature. (see M.D. Sturge, *Physical Review*, **127**, 768 (1962).)

The strong dependence upon σ is quite evident and significant, since it is very difficult to control σ.

10.4 EXCITONIC STATES IN QUANTUM WELLS

The ability to make heterostructures has made a tremendous impact in the area of excitons in quantum wells. The ability to fabricate quantum well structures, where the electrons and holes can be strongly confined in the growth direction, has allowed excitonic resonances to assume an important technological aspect. The highly controllable nature of the excitonic resonances lends itself to many versatile devices. The main reason for the interest in excitonic resonances in quantum well structures is the enhanced binding energy of the confined electron-hole system. Simple variational calculations show that the binding energy of a 2-dimensional electron-hole system with Coulombic interaction is four times that of the 3-dimensional system. In reality, of course, a quantum well system is not a 2-dimensional system, but is a quasi-2-dimensional system. The actual binding energy is, therefore, somewhat smaller than the 4 R_{ex} value. Nevertheless, the increased binding energy allows the excitonic transitions to persist up to high temperatures.

In a quantum well system grown along the z-axis the exciton problem can be written as

$$H = \frac{-\hbar^2}{2m_r^*}\left(\frac{1}{\rho}\frac{\partial}{\partial p}\rho\frac{\partial}{\partial p} + \frac{1}{\rho^2}\frac{\partial^2}{\partial\phi^2}\right) - \frac{\hbar^2}{2m_e}\frac{\partial^2}{\partial z_e^2} - \frac{\hbar^2}{2m_h}\frac{\partial^2}{\partial z_h^2}$$

$$- \frac{e^2}{4\pi\epsilon |\boldsymbol{r}_e - \boldsymbol{r}_h|} + V_{ew}(z_e) + V_{hw}(z_h) \tag{10.29}$$

Here V_{ew}, V_{hw} are the confining potentials, m_e and m_h are the in-plane effective masses of the electron and the hole, and μ is the reduced mass. The first term is the kinetic energy operator in the plane of the quantum well. The dielectric constant ϵ is the static dielectric constant, if there are no free carriers present. Otherwise, it is the screened dielectric constant, of the form used in the ionized impurity scattering problem. We will return to the screening problem later.

The Hamiltonian for the finite quantum well exciton problem has no simple analytical solution, even if one assumes that the electron and hole states have a parabolic bandstructure. The function $F(\rho)$, describing the relative motion envelope, cannot be directly obtained. This function is usually obtained by assuming its form to be an exponential, or Gaussian or some combination of Gaussian functions, etc., with some variational constants which are then adjusted to minimize the energy

$$E = \frac{\int \psi_{\text{ex}}^* \, H \, \psi_{\text{ex}} \, dz_e \, dz_n \, \rho \, d\rho \, d\phi}{\int \psi_{\text{ex}}^* \, \psi_{\text{ex}} \, dz_e \, dz_n \, \rho \, d\rho \, d\phi} \tag{10.30}$$

This approach gives quite reliable results, with the effects of choosing different forms of variational functions being no more than $\sim 10\%$ of the exciton binding energies. In Fig. 10.8 we show the effect of well size on the ground state excitonic states in GaAs/Al$_{0.3}$Ga$_{0.7}$As quantum well structures. As can be seen, the exciton binding energies can increase by up to a factor of ~ 2.5 in optimally designed quantum well structures.

10.5　EXCITONIC ABSORPTION IN QUANTUM WELLS

We can now generalize our absorption coefficient results for bulk to a quantum well of size W. The absorption coefficient is given by

$$\alpha_{nm}^{\text{ex}}(\hbar\omega) = \frac{\pi e^2 \hbar}{n_r m_0^2 cW \hbar\omega} \left| \sum_{\mathbf{k},n,m} G_{nm}(\boldsymbol{k}) \, \boldsymbol{a} \cdot \boldsymbol{p}_{nm}(\boldsymbol{k}) \right|^2 \delta\left(\hbar\omega - E_{nm}^{\text{ex}}\right) \tag{10.31}$$

The matrix elements $\boldsymbol{p}_{nm}(\boldsymbol{k})$ are given by the central cell part used for the 3-dimensional problem, as well as the overlap of the envelope function as used in the Chapter 9 for band-to-band absorption

$$\boldsymbol{p}_{nm}(\boldsymbol{k}) = \sum_{\nu,\mu} \int d^2 r \, U_0^\nu(\boldsymbol{r}) \, \boldsymbol{p} \, U_0^\mu(\boldsymbol{r}) \int dz \, f_n^\mu(z) \, g_m^\nu(k, z)$$

The selection rules and polarization dependencies discussed in Chapter 9 for band to band transitions, still hold for the excitonic transitions.

The absorption coefficient is often also written in terms of oscillator strengths f_{nm} as

$$\alpha_{nm}(\hbar\omega) = \sum_{nm} \frac{\pi e^2 \hbar}{n_r \epsilon_0 m_0 cW} \, f_{nm} \, \delta\left(E_{nm}^{\text{ex}} - \hbar\omega\right) \tag{10.32}$$

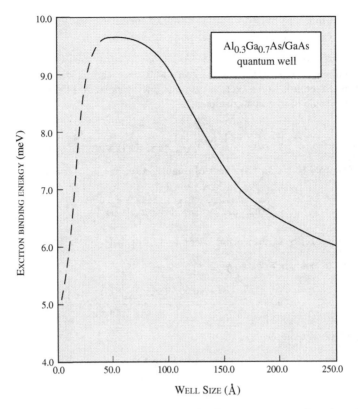

Figure 10.8: Variation of the heavy-hole exciton binding energy as a function of well size in GaAs/Al$_{0.3}$Ga$_{0.7}$As wells. The binding energy of the infinite barrier well should approach four times the 3D exciton binding energy as the well size goes to zero.

where the oscillator strength per unit area is defined by ($\hbar\omega = E^{\text{ex}}_{nm}$)

$$f_{nm} = \frac{2}{m_0 E^{\text{ex}}_{nm}(2\pi)^2} \left| \int d^2k \, G_{nm}(\boldsymbol{k}) \, \boldsymbol{a} \cdot \boldsymbol{p}_{nm}(\boldsymbol{k}) \right|^2 \tag{10.33}$$

The oscillator strength is a better measure of the excitonic absorption because it does not involve the δ-function. The δ-function will eventually be replaced by a broadening function whose value will be sample and temperature dependent as we will see below. Since quantum confinement decreases the spatial extent of the exciton function, the oscillator strength increases as the exciton is confined.

If the 2D exciton function can be represented by an exciton radius a_{ex}, then as in the 3D case, the absorption coefficient becomes

$$\alpha(\hbar\omega) = \frac{\pi e^2 \hbar}{2n_r \epsilon_0 c m_0 \hbar\omega} \left(\frac{2|p_{cv}|^2}{m_0} \right) a_p \left(\frac{1}{\sqrt{1.44\pi}} \frac{1}{\sigma} \frac{1}{W\pi a^2_{\text{ex}}} \exp\left(\frac{-(\hbar\omega - E_{\text{ex}})^2}{1.44\sigma^2} \right) \right) \tag{10.34}$$

In a quantum well as the exciton binding energy increases, the exciton radius decreases (by up to a factor of 2) and the exciton peak becomes stronger.

EXAMPLE 10.2 Consider a 100 Å GaAs/Al$_{0.3}$Ga$_{0.7}$As quantum well structure grown along (001) with a halfwidth of 1 meV. The exciton radius is $\frac{2}{3}$ the bulk value in GaAs. Calculate the peak of the HH ground state exciton resonance for x-polarized light.

The excitonic absorption coefficient is

$$\alpha(\hbar\omega) = \frac{\pi e^2 \hbar}{2n_r \epsilon_0 cm_0 \hbar\omega} \left(\frac{2|p_{cv}|^2}{m_0} \right) a_p \left(\frac{1}{\sqrt{1.44\pi}} \frac{1}{\sigma} \frac{1}{W\pi a_{ex}^2} \exp\left(\frac{-(\hbar\omega - E_{ex})^2}{1.44\sigma^2} \right) \right)$$

Notice from Example 10.1 that the effect of quantization has been essentially the replacement

$$\frac{1}{\pi a_{ex}(3D)^3} \longrightarrow \frac{1}{W\pi a_{ex}(QW)^3}$$

Thus, the excitonic absorption becomes $\left(\text{using} a_{ex}(QW) \cong \frac{2}{3} a_{ex}(3D) \right)$ for light polarized along the x-axis where $a_p = \frac{1}{2}$

$$\alpha(\hbar\omega) = \frac{2.9 \times 10^6}{\sigma(meV)} \exp\left(\frac{-(\hbar\omega - E_{ex})^2}{1.44\sigma^2} \right)$$

For $\sigma = 1$ meV, we get the peak value of

$$\alpha(\text{peak}) = 2.9 \times 10^6 \text{ m}^{-1}$$
$$= 2.9 \times 10^4 \text{ cm}^{-1}$$

10.6   EXCITON BROADENING EFFECTS

In an ideal system the exciton resonance is represented by a δ-function due to momentum conservation. However, in a real system inhomogeneous (structural effects) and homogeneous (finite lifetime effects) sources cause the δ-function to broaden. This line broadening or linewidth is extremely important from both a physics and a technological point of view. In general, the peak of the exciton absorption is inversely proportional to the linewidth of the resonance. This puts a tremendous premium on reducing the exciton linewidth (in a reliable manner). Unfortunately, small variations in growth of quantum wells can change the exciton linewidth by as much as 100%!

As noted above excitonic transitions are broadened by two important kinds of fluctuations: inhomogeneous and homogeneous. The main inhomogeneous broadening sources are:

1. Interface roughness.

2. Alloy potential fluctuations.

3. Well to well fluctuations in multiquantum well spectra.

4. Background impurity broadening.

In high purity materials the impurity broadening is usually negligible (if background density $< 10^{15} \text{cm}^{-3}$). The homogeneous broadening mechanisms are:

1. Acoustic phonon scattering.

2. Optical phonon scattering.

3. Other mechanisms which may affect the exciton lifetime such as tunneling, recombination, etc.

The treatment of inhomogeneous broadening usually follows the approach developed to study electronic states in disordered materials. Line broadening arises due to spatially localized fluctuations which are capable of shifting the excitonic emission energy. In general, one may describe these fluctuations by concentration fluctuations in the mean compositions C_A^0 and C_B^0 of the structure. For example C_A^0 and C_B^0 may represent the fraction of the islands and the valleys representing interface roughness in a quantum well, or the mean composition of the alloy in the treatment of alloy broadening. The width of the concentration fluctuation occurring over a region β (which is an area for interface roughness treatment and a volume for the treatment of alloy broadening) is given by (Lifshitz, 1969):

$$\delta P = 2\sqrt{\frac{1.4\, C_A^0\, C_B^0\, \alpha}{\beta}} \tag{10.35}$$

where α is the smallest extent over which the fluctuation can take place. Again for the treatment of interface roughness, α is the average area of the two-dimensional islands representing the interface roughness and for alloy broadening, α is the volume of the smallest cluster in the alloy. For a perfectly random alloy, this is the volume per cation. The shift in the excitonic energy due to this fluctuation is then

$$2\sigma(\beta) = \delta P\, |\Psi|_\beta^2\, \frac{\partial E_{\text{ex}}}{\partial C} \tag{10.36}$$

where $|\Psi|_\beta^2$ is the fraction of the exciton sensing the region β, and $\partial E_{\text{ex}}/\partial C$ represents the rate of change in the exciton energy with change in the concentration. To obtain the linewidth of the absorption, one needs to identify the volume (or area) β. For interface roughness broadening, this area is $\sim 3r_{\text{ex}}^2$ and the linewidth is given by

$$2\sigma_{\text{IR}} = 2\sqrt{\frac{1.4\, C_A^0\, C_B^0\, a_{\text{2D}}}{3r_{\text{ex}}^2 \pi}} \cdot \left.\frac{\partial E_{\text{ex}}}{\partial W}\right|_{W_0} \cdot \delta_0 \tag{10.37}$$

where a_{2D} is the real extent of the two-dimensional islands representing the interface roughness, δ_0 their height, and r_{ex} is the exciton Bohr radius in the lateral direction parallel to the interface. $\partial E_{\text{ex}}/\partial W$ is the change of the exciton energy as a function of well size.

Apart from interface roughness fluctuations in the same well, often one has to contend with intra-well fluctuations. Most optical devices using excitonic transitions use more than one well, i.e., use multiquantum wells that are produced by opening

and shutting off fluxes of one or other chemical species. There is invariably a monolayer or so of variation in the well sizes across a stack of quantum wells. For a 100 Å
GaAs/Al$_{0.3}$Ga$_{0.7}$As quantum well, a monolayer fluctuation can produce ~ 1.5 meV
change in the exciton resonance energy. This change arises almost entirely from the
changes in subband levels since the *exciton binding energy* does not vary significantly
with a monolayer change in the well size. In high quality GaAs/AlGaAs multiquantum
well samples, an absorption linewidth of 3–4 meV is achievable.

In the GaAs based system, the well material is a binary material with no alloy
scattering. In cases like In$_{0.53}$Ga$_{0.47}$As/InP or In$_{0.53}$Ga$_{0.47}$As/In$_{0.52}$Al$_{0.48}$As, one additionally has alloy broadening effects. These effects can also be of the order of 3–5 meV
and are given by similar treatments. If C_A^0 and C_B^0 are the mean concentrations of the
two alloy components and V_C is the average alloy cluster size (= unit cell for random
alloy), the linewidth is given approximately by

$$2\sigma_{\text{alloy}}^{\text{internal}} = 2\sqrt{\frac{1.4\,C_A^0\,C_B^0\,V_C}{3\pi r_{\text{ex}}^2 W_0}} \cdot \left. \frac{\partial E_{\text{ex}}}{\partial C_A} \right|_{C_A^0} \tag{10.38}$$

where $\partial E_{\text{ex}}/\partial C_A$ represents the change in the exciton resonance with changes in composition. If the barrier material is also an alloy, there is a contribution to the linewidth
from the barrier as well

$$2\sigma_{\text{alloy}}^{\text{external}} = 2\sqrt{\frac{1.4\,C_A^0\,C_B^0\,V_C}{3\pi r_{\text{ex}}^2 L_{\text{eff}}}} \cdot \left. \frac{\partial E_{\text{ex}}}{\partial C_A} \right|_{C_A^0} \tag{10.39}$$

where L_{eff} is an effective length to which the exciton wavefunction penetrates in the
barrier and has to be calculated numerically. L_{eff} is approximately equal to twice the
distance over which the exciton wavefunction falls to $1/e$ of its initial value. Typically
this is on the order of a few monolayers. The well size dependence of the exciton linewidth
is quite apparent from the results of Fig. 10.9 and are due to the increase in $\partial E_{\text{ex}}/\partial W$
with decreasing well size.

Finally, to examine the homogeneous broadening note that the exciton state is
optically created at the $\boldsymbol{K}_{\text{ex}} = 0$ state. The acoustic and optical phonons interact with
the exciton causing either a transition to a $\boldsymbol{K}_{\text{ex}} \neq 0$ state in the same quantum level or
even ionization of the exciton. This homogeneous broadening effect can be calculated in
a manner similar to the one used for electron-phonon scattering. The linewidth is given
through the particle lifetime ΔT by the relation

$$2\sigma \sim \frac{\hbar}{\Delta T} = \hbar \sum W_i \tag{10.40}$$

where W_i is the scattering rate due to processes such as acoustic phonon scattering and
optical phonon scattering. The linewidth can be shown to have the form

$$2\sigma = \alpha T + \frac{\beta}{\exp\left(\hbar\omega_0/k_B T\right) - 1} \tag{10.41}$$

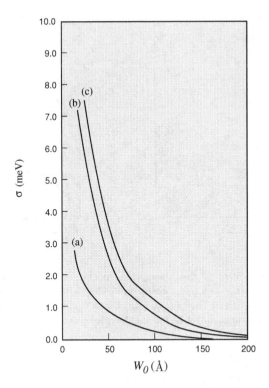

Figure 10.9: Variation of exciton linewidth as a function of well size for a one monolayer interface roughness. The half width σ is given for island size extents of (a) 20 Å; (b) 80 Å; (c) 100 Å for the GaAs/Al$_{0.3}$Ga$_{0.7}$As system.

where the first term is due to acoustic phonon scattering and the second is due to optical phonons. At room temperature, the homogeneous linewidth due to phonons is $\sim$3–4 meV, implying that the exciton lifetime is $\sim$ 0.2 ps.

In Fig. 10.10 we show the comparison of typical excitonic spectra in a 100 Å GaAs/Al$_{0.3}$Ga$_{0.7}$As and 100 Å In$_{0.53}$Ga$_{0.47}$As/In$_{0.52}$Al$_{0.48}$As quantum well structure. The large reduction in the InGaAs spectra is primarily due to the alloy broadening of the exciton peak.

EXAMPLE 10.3 A multiquantum well (MQW) stack is used in an exciton absorption measurement. The nominal thickness of the GaAs/Al$_{0.3}$Ga$_{0.7}$As wells is 100 Å. Consider two cases: i) There is no structural disorder and the exciton halfwidth is given by a homogeneous width $\sigma = 1$ meV; ii) In addition to the above broadening, there is a structural disorder in the MQW stack described by a one monolayer rms fluctuation in the well size. Calculate the peak absorption coefficients in the two cases.

A one monolayer charge in the well size at 100 Å produces a shift in the HH exciton energy of $\sim$ 1.5 meV, due to the shift in the subband energies (there is essentially no change in the binding energies). For simplicity, we assume that the linewidths from the homogeneous and inhomogeneous contributions simply add (in reality, the homogeneous broadening may be better described by a Lorentzian function, rather than a Gaussian function; also one has to use

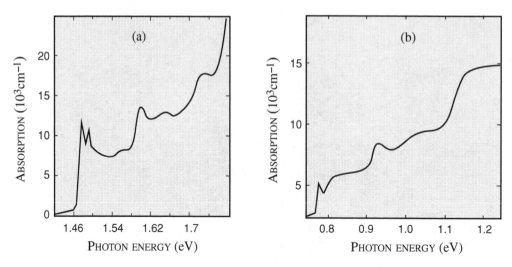

Figure 10.10: A comparison of the absorption spectra in (a) 100 Å GaAs /Al$_{0.3}$Ga$_{0.7}$As and (b) In$_{0.53}$Ga$_{0.47}$As /In$_{0.52}$Al$_{0.48}$As quantum wells. The excitons in InGaAs suffer alloy broadening which reduces their sharpness. (After D.S. Chemla, *Nonlinear Optics: Materials and Devices,* eds. C. Flytzanis and J.L. Oudar, Springer-Verlag, New York (1986).)

a convoluted linewidth).

From Example 10.2 we get, for the two cases,

i) α(peak) $= 2.9 \times 10^4$ cm^{-1}

i) α(peak) $= 1.2 \times 10^4$ cm^{-1}

This example shows the extreme sensitivity of excitonic resonances on minute structural imperfections.

10.7   MODULATION OF OPTICAL PROPERTIES

We have seen that excitonic effects in semiconductors produce sharp resonances. These resonances can be exploited for information processing applications if they can be controlled (or modulated) by electrical or optical signals. The modulation of the optical properties of a semiconductor can be exploited for a number of "intelligent" optoelectronic devices such as switches, logic gates, memory elements, etc.

When the optical properties of a material are modified, the effect on a light beam propagating in the material can be classified into two categories, depending upon the photon energy. As shown in Fig. 10.11, if the photon energy is in a region where the absorption coefficient is zero (beam with frequency ω_1), the effect of the modification of the refractive index is to alter the velocity of propagation of light. On the other hand, if the photon energy is in a region where the absorption coefficient is altered (beam with frequency ω_2), the intensity of light emerging from the sample will be altered. These two approaches for the modification of the optical properties by a applied electric field are called the electro-optic and the electro-absorption approaches, respectively.

In the electro-optic effect, an applied electric field is used to alter the phase

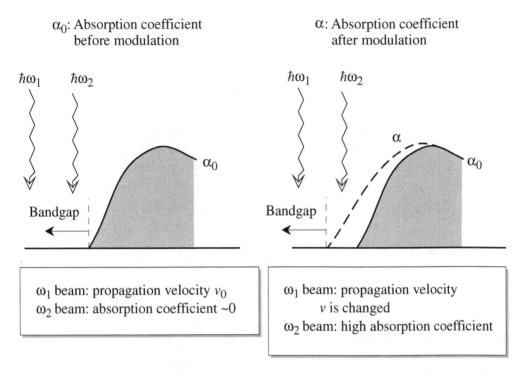

Figure 10.11: A schematic of the effect of a change in optical properties of a material on an optical beam. For energy $\hbar\omega_1$, the main effect of the change in the optical properties is a change in propagation velocity. For $\hbar\omega_2$, the effect is a change in intensity.

velocity of a propagating signal and this effect can be exploited in an interference scheme to alter the polarization or intensity of the light. We will first discuss this approach.

10.7.1 Electro-Optic Effect

The electro-optic effect depends upon the modification of the refractive index of a material by an applied electric field. Before discussing the electro-optic effect, let us review some important optical properties of crystalline materials.

We know from electromagnetic theory that Maxwell's equations and the material properties (ϵ, μ) determine the propagation of light in a material. In general, the medium can be anisotropic and the material properties are described by the relations between displacement field and electric field

$$
\begin{aligned}
D_x &= \epsilon_{xx}F_x + \epsilon_{xy}F_y + \epsilon_{xz}F_z \\
D_y &= \epsilon_{yx}F_x + \epsilon_{yy}F_y + \epsilon_{yz}F_z \\
D_z &= \epsilon_{zx}F_x + \epsilon_{zy}F_y + \epsilon_{zz}F_z
\end{aligned}
\tag{10.42}
$$

or in the short form

$$D_i = \sum_j \epsilon_{ij} F_j \qquad (10.43)$$

If one examines the energy density of the optical wave and imposes the conditions that the energy must be independent under the transformation $i \to -i$, we get

$$\epsilon_{ij} = \epsilon_{ji} \qquad (10.44)$$

Further reduction in the number of the independent ϵ_{ij} can be achieved, depending upon the symmetry of the crystal structure. A most useful way to describe the propagation of light in the medium is via the index ellipsoid. The equation for the index ellipsoid is

$$\frac{x^2}{\epsilon_x} + \frac{y^2}{\epsilon_y} + \frac{z^2}{\epsilon_z} = 1 \qquad (10.45)$$

or in terms of the refractive indices,

$$\sum_{i=1}^{3} \frac{x_i^2}{n_i^2} = 1 \qquad (10.46)$$

Here ϵ_x, ϵ_y *and* ϵ_z are the principal dielectric constants, expressed along the principle axes of the ellipsoid.

Now consider a situation where an electric field is applied to the crystal. The applied electric field modifies the bandstructure of the semiconductor through a number of interactions. These interactions may involve:

i) Strain: In a piezoelectric material where there is no inversion symmetry (e.g., GaAs, CdTe, etc.) the two atoms in the basis of the crystal have different charges. The electric field may cause a distortion in the lattice and, as a result, the bandstructure may change. This may cause a change in the refractive index.

ii) Distortion of the excitonic features: In the previous sections we discussed the optical properties of the exciton. The presence of an electric field can modify the excitonic spectra and thus alter the electronic spectra and, hence, the optical spectra of the material.

In general, the change in the refractive index may be written as

$$n_{ij}(F) - n_{ij}(0) = \Delta n_{ij} = r_{ijk} F_k + s_{ijk\ell} F_k F_\ell \qquad (10.47)$$

where F_i is the applied electric field component along the direction i and r_{ijk} and $s_{ijk\ell}$ are the components of the electro-optic tensor. In materials like GaAs where the inversion symmetry is missing, r_{ijk} is non-zero and one has a linear term in the electro-optic effect. The linear effect is called the Pockel effect. In materials like Si where one has inversion symmetry $r_{ijk} = 0$ and the lowest order effect is due to the quadratic effect (known as the Kerr effect).

In general, r_{ijk} has 27 elements, but since the tensor is invariant under the exchange of i and j, there are only 18 independent terms. It is common to the use of contracted notation $r_{\ell m}$ where $\ell = 1, \ldots 6$ and m = 1, 2, 3. The standard contraction

MATERIAL	$\lambda(\mu m)$	$r_{ij}(10^{-12}$ m/V)
GaAs	0.8 –10	$r_{41} = 1.2$
Quartz	0.6	$r_{11} = -0.47$ $r_{41} = -0.2$
LiNbO$_3$	0.5	$r_{13} = 9.0; r_{22} = 6.6$ $r_{42} = 30$
KDP	0.5	$r_{41} = 8.6; r_{63} = 9.5$

Table 10.1: Electro-optic coefficients for some materials. (After P.K Cheo, *Fiber Optics, Devices and Systems,* Prentice-Hall, New Jersey (1985).)

arises from the identification of $i, j = 1,1; 2,2; 3,3; 2,3; 3,1;1,2$ by $\ell = 1, 2, 3, 4, 5, 6$, respectively. The 18 coefficients are further reduced by the symmetry of the crystals. In semiconductors such as GaAs, it turns out that the only non-zero coefficients are

$$r_{41}$$
$$r_{52} = r_{41}$$
$$r_{63} = r_{41} \tag{10.48}$$

Thus, a single parameter describes the linear electro-optic effect. The value of GaAs is $r_{41} = 1.2 \times 10^{-12}$ m/V. An important class of electro-optic materials are ferroelectric perovskites such as LiNbO$_3$ and LiTaO$_3$ which have trigonal symmetry and materials such as KDP (potassium dihydrogen phosphate) which have tetragonal symmetry). For the trigonal materials the non-zero tensor components are

$$r_{22} = -r_{12} = -r_{61}$$
$$r_{51} = r_{42} = r_{33}$$
$$r_{13} = r_{23} \tag{10.49}$$

For KDP the non-zero elements are

$$r_{41} = r_{52}$$
$$r_{63} \tag{10.50}$$

In Table 10.1, we give the values of the electro-optic coefficients for some materials. The second order electro-optic coefficients $s_{ijk\ell}$ are usually not important for materials unless the optical energy $\hbar\omega$ is very close to the bandgap. In materials like GaAs, the second order coefficients that are non-zero from symmetry considerations are in the contracted form $s_{pq}, p = 1\ldots6, q = 1\ldots6$,

$$s_{11} = s_{22} = s_{33}$$

$$s_{12} = s_{13}$$

$$s_{44} = s_{55} = s_{66} \tag{10.51}$$

The electro-optic effect is used to create a modulation in the frequency, intensity or polarization of an optical beam. To see how this occurs, consider a material like GaAs.

Let us first consider the linear electro-optic effect where a field F is applied to the crystal. The index ellipsoid is

$$\left(\frac{1}{n_x^2} + r_{1k} F_k \right) x^2 \ + \ \left(\frac{1}{n_y^2} + r_{2k} F_k \right) y^2 + \left(\frac{1}{n_z^2} + r_{3k} F_k \right) z^2$$

$$+ \ 2yz r_{4k} F_k + 2zx r_{5k} F_k + 2xy r_{6k} F_k = 1 \tag{10.52}$$

where F_k (k = 1,2,3) is the component of the electric field in the x, y, and z directions. Using the elements of the electro-optic tensor for GaAs, we get

$$\frac{x^2}{n_x^2} + \frac{y^2}{n_y^2} + \frac{z^2}{n_z^2} + 2yz r_{41} F_x + 2zx r_{41} F_y + 2xy r_{41} F_z = 1 \tag{10.53}$$

Let us now simplify the problem by assuming that the electric field is along the $< 001 >$ direction

$$F_x = F_y = 0, \qquad F_z = F \tag{10.54}$$

We now rotate the axes by $45°$ so that the new principal axes are

$$x' = \frac{x}{\sqrt{2}} - \frac{y}{\sqrt{2}}$$

$$y' = \frac{x}{\sqrt{2}} + \frac{y}{\sqrt{2}}$$

$$z' = z \tag{10.55}$$

In terms of this new set of axes, the index of ellipsoid is written as

$$\frac{x'^2}{n_x'^2} + \frac{y'^2}{n_y'^2} + \frac{z'^2}{n_z'^2} = 1 \tag{10.56}$$

where the new indices are

$$n_x' = n_o + \frac{1}{2} n_o^3 r_{41} F$$

$$n_y' = n_o - \frac{1}{2} n_o^3 r_{41} F$$

$$n_z' = n_o \tag{10.57}$$

where n_o is the index in absence of the field ($= n_x = n_y = n_z$). As a result of this change in the indices along the x' and y' axes, for light along $< 01\bar{1} > (x')$ and $< 011 > (y')$

directions, a phase retardation occurs due to the field. The phase retardation for a wave that travels a distance L is, $(n_z' = n_o)$

$$\Delta\phi(x') = \frac{\omega}{c}\left(n_z' - n_x'\right)L\xi_1 = -\frac{\pi}{\lambda}n_o^3 r_{41}FL\xi_1$$

$$\Delta\phi(y') = \frac{\omega}{c}\left(n_z' - n_y'\right)L\xi_1 = \frac{\pi}{\lambda}n_o^3 r_{41}FL\xi_1 \qquad (10.58)$$

The quantity ξ_1 represents the overlap of the optical wave with the region where the electric field is present:

$$\xi_1 = \frac{1}{F}\int\int F\,|\,F_{\text{photon}}\,|^2\,dA \qquad (10.59)$$

where F_{photon} is the photon field. For bulk devices $\xi_1 \sim 1$.

Let us now extend our study to the second order term in the elctro-optic effect.

Using the contracted notiation, for quadratic electro-optic coefficients, the index ellipsoid can be written as

$$\left(\frac{1}{n_x^2} + s_{11}F_x^2 + s_{12}F_y^2 + s_{12}F_z^2\right)x^2$$

$$+ \left(\frac{1}{n_y^2} + s_{12}F_x^2 + s_{11}F_y^2 + s_{12}F_z^2\right)y^2$$

$$+ \left(\frac{1}{n_z^2} + s_{12}F_x^2 + s_{12}F_y^2 + s_{11}F_z^2\right)z^2$$

$$+2yz(2s_{44}F_yF_z) \quad + \quad 2zx(2s_{44}F_xF_z) + 2xy(2s_{44}F_yF_x) = 1 \qquad (10.60)$$

In the presence of an electric field, F, in the z direction, Eqn. 10.60 can be rewritten as

$$\left(\frac{1}{n_x^2} + s_{12}F^2\right)x^2 \quad + \quad \left(\frac{1}{n_y^2} + s_{12}F^2\right)y^2$$

$$+ \quad \left(\frac{1}{n_z^2} + s_{11}F^2\right)z^2 = 1 \qquad (10.61)$$

This index ellipsoid can be rewritten as

$$\frac{x^2 + y^2}{n_o^2} + \frac{z^2}{n_e^2} = 1 \qquad (10.62)$$

with

$$n_o = n - \frac{1}{2}n^3 s_{12}F^2 \qquad (10.63)$$

and

$$n_e = n - \frac{1}{2}n^3 s_{11}F^2 \qquad (10.64)$$

The phase retardation due to the applied field is thus given by

$$\Delta\Phi = \frac{\omega}{c}(n_e - n_o)L\xi_2 = \frac{\pi}{\lambda}n^3(s_{12} - s_{11})F^2L\xi_2$$

where ξ_2 is an overlap of the square of the electric field and the optical field given by

$$\xi_2 = \frac{1}{F^2} \int \int F^2 \mid F_{\text{photon}} \mid^2 dA \tag{10.65}$$

The total phase change between waves travelling along x' and y' then becomes after adding the effects of the linear and quadratic electro-optic effects

$$\Delta\phi(x') = -\frac{\pi L}{\lambda} n_o^3 \left[r_{41}F\xi_1 + (s_{12} - s_{11})F^2\xi_2 \right]$$

$$\Delta\phi(y') = \frac{\pi L}{\lambda} n_o^3 \left[r_{41}F\xi_1 + (s_{12} - s_{11})F^2\xi_2 \right] \tag{10.66}$$

The phase changes produced by the electric field can be exploited for a number of important switching or modulation devices. From Table 10.1 we see that the electro-optic coefficients in materials like LiNbO$_3$ are much larger than those in traditional semiconductors. As a result LiNbO$_3$ is widely used in optical directional couplers and switches.

EXAMPLE 10.4 A bulk GaAs device is used as an electro-optic modulator. The device dimension is 1 mm and a phase change of 90° is obtained between light polarized along $< 01\bar{1} >$ and $< 011 >$. The wavelength of the light is 1.5 μm. Calculate the electric field needed if $\xi_1 = 1$.

The phase change produced is $(\xi = 1)$

$$\Delta\phi = \frac{2\pi}{\lambda} n_o^3 r_{41} FL = \frac{\pi}{4}$$

$$F = \frac{\lambda}{8 n_o^3 r_{41} L}$$

$$= \frac{(1.5 \times 10^{-6} \text{ m})}{8(3.3)^3 (1.2 \times 10^{-12} \text{ m/V})(10^{-3} \text{ m})}$$

$$= 4.35 \times 10^6 \text{ V/m}$$

If the field is across a 10 μm thickness, the voltage needed is 4.35 V.

10.7.2   Modulation of Excitonic Transitions: Quantum Confined Stark Effect

Quantum confined stark effect (QCSE) refers to the changes that occur in the electronic and optical spectra of a quantum well when an electric field is applied. In Fig. 10.12 we show schematically a quantum well without and with an electric field in the confinement direction. There are several effects that occur in the presence of the transverse electric field:

1. The intersubband separations change. The field pushes the electron and hole functions to opposite sides making the ground state intersubband separation smaller. This effect is the dominant term in changing the exciton resonance energy. This effect is shown in Fig. 10.13.

2. Due to the separation of the electron and hole wavefunction, the binding energy of the exciton decreases. This effect is shown in Fig. 10.14.

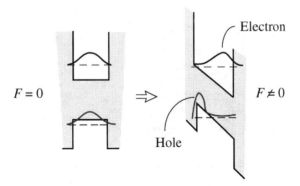

Figure 10.12: A schematic showing how an electric field alters the quantum well shape and the electron and hole wavefunctions.

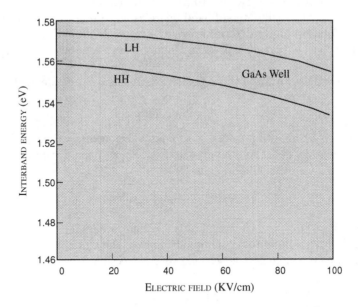

Figure 10.13: Calculated variation of the ground state HH and LH (to conduction band ground state) intersubband transition energies as a function of electric field. The well is a 100 Å GaAs/Al$_{0.3}$Ga$_{0.7}$As.

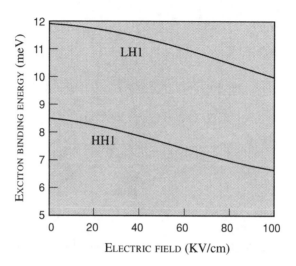

Figure 10.14: Calculated variation in the exciton binding energy in a 100 Å GaAs/Al$_{0.3}$Ga$_{0.7}$As quantum well as a function of electric field. The ground state HH and LH exciton results are shown.

As can be seen from Figs. 10.13 and 10.14, the change in the exciton binding energy is only ∼2–3 meV while the intersubband energies are altered by up to 20 meV. The QCSE is, therefore, primarily determined by the intersubband effect.

While the exact calculation of the intersubband separation requires numerical techniques, one can estimate these changes by using perturbation theory. This approach gives reasonable results for low electric fields. The problem can be defined by the Hamiltonian

$$H = H_0 + eFz \tag{10.67}$$

where H_0 is the usual quantum well Hamiltonian. The eigenfunctions of H_0 are ψ_n and in the square quantum well, the ground state function ψ_1 has an even parity so that the first order correction to the subband energy is

$$\Delta E^{(1)} = \langle \psi_1 | eFz | \psi_1 \rangle \tag{10.68}$$

which is zero. Thus, one has to calculate the second order perturbation by using the usual approach. For the infinite barrier quantum well, the states ψ_n are known and it is, therefore, possible to calculate the second order correction. If the field is small enough such that

$$|eFW| \ll \frac{\hbar^2 \pi^2}{2m^* W^2} \tag{10.69}$$

i.e., the perturbation is small compared to the ground state energy then it can be shown that the ground state energy changes by

$$\Delta E_1^{(2)} = \frac{1}{24\pi^2} \left(\frac{15}{\pi^2} - 1 \right) \frac{m^* e^2 F^2 W^4}{\hbar^2} \tag{10.70}$$

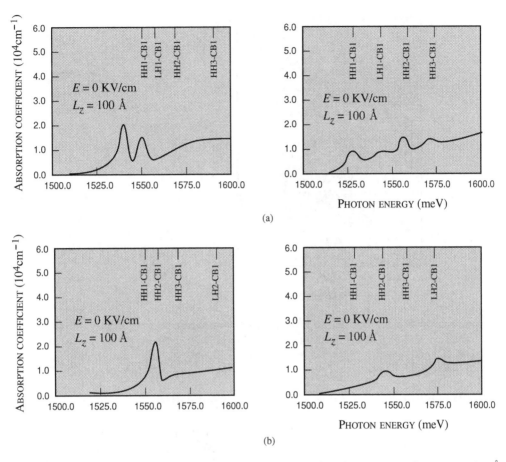

Figure 10.15: Calculated absorption coefficients for TE and TM modes in a 100 Å GaAs/AlGaAs quantum well structure at (a) $F = 0$ and (b) $F = 70$ kV/cm.

One sees that the second order effect increases with m^* and has a strong well size dependence. This would suggest that for best modulation one should use a wide well. However, in wide wells the exciton absorption decreases and also the HH, LH separation becomes small. Optimum well sizes are of the order of $\sim$100 Å for most quantum well structures.

We note that the matrix element in the excitonic absorption obeys the same polarization selection rules as the ones we discussed in Chapter 9 for band to band transitions. In Fig. 10.15, we show the TE and TM absorption spectra calculated for a 100 Å GaAs/Al$_{0.3}$Ga$_{0.7}$As quantum well. As can be seen, in the TE mode, the LH exciton strength is approximately a third of the HH exciton strength. In the TM mode on the other hand, there is no HH transition allowed in accordance with our discussions in Chapter 9. Polarization-dependent optical spectra measured in similar structures is shown in Fig. 10.16.

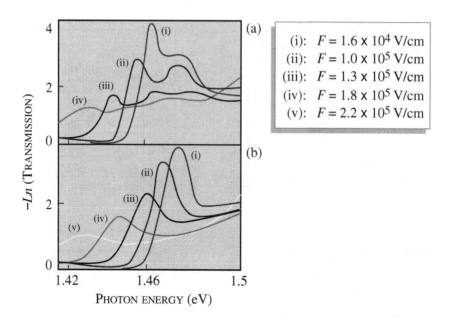

Figure 10.16: Measured polarization dependent transmittances in GaAs/AlGaAs (100 Å) multiquantum well structures when light is coming in the waveguide geometry. (a) Incident polarization parallel to the plane of the layers . (b) Incident polarization perpendicular to the plane of the layers. (After D.A.B. Miller, et al., *IEEE J. Quantum Electronics*, QE-22, 1816 (1986).)

In the context of polarization dependence of the absorption spectra, it is very interesting to see the effect of strain on the spectra. We noted in Chapter 3 that epitaxial strain can lift the degeneracy between the HH and LH states. In fact, a compressive in-plane strain pushes the LH below the HH state while a tensile strain does the opposite. Now the quantum confinement also pushes the LH state below the HH state. Thus, in principle a properly chosen tensile strain can cause a *merger* of the HH and LH states at the zone center in a quantum well.

EXAMPLE 10.5 A 100 Å GaAs/Al$_{0.3}$Ga$_{0.5}$As MQW structure has the HH exciton energy peak at 1.51 eV. A transverse bias of 80 kV/cm is applied to the MQW. Calculate the change in the transmitted beam intensity (there is no substrate absorption) if the total width of the wells is 1.0 μm and the exciton linewidth is $\sigma = 2.5$ meV. The photon energy is $\hbar\omega = 1.49$ eV.

The transmitted light is

$$I = I_o \exp(-\alpha d)$$

At zero bias, we have (see Example 10.2)

$$\alpha(V = 0) = \frac{2.9 \times 10^4}{2.5} \exp\left(\frac{-(1.49 - 1.51)^2}{1.44(2.5 \times 10^{-3})^2}\right)$$

$$\sim 0$$

At a bias of 80 kV/cm, the exciton peak shifts by ~20 meV, as can be seen from Fig. 10.17. The absorption coefficient is

$$\alpha(F = 80 \text{ kV/cm}) = \frac{2.9 \times 10^4}{2.5} \exp\left(\frac{-(1.49 - 1.49)^2}{1.44(2.5 \times 10^{-3})^2}\right)$$

$$= 1.2 \times 10^4 \text{ cm}^{-1}$$

The ratio of the transmitted intensity is

$$\frac{I(F = 80 \text{ kV/cm})}{I(F = 0)} = 0.3$$

Thus, an ON/OFF ratio is 3.3:1 for this modulator.

EXAMPLE 10.6 A small tensile strain is used to cause the HH and LH excitonic states to merge in a 100 Å GaAs/Al$_{0.3}$Ga$_{0.7}$As well. The quantum well is grown along the (001) direction. Calculate the absorption coefficient for light polarized along the z-direction and along the x (or y) direction. The exciton linewidth is $\sigma = 1.0$ meV.

When the HH and LH states are coincident, the total coupling of the z-polarized light gives, for the momentum matrix element, (see Chapter 9)

$$\frac{2}{3} P_{cv}^2$$

and for the x-polarized light,

$$\frac{1}{2} P_{cv}^2 + \frac{1}{6} P_{cv}^2 = \frac{2}{3} P_{cv}^2$$

In Example 10.2, we calculated the excitonic transition strength using $a_p = \frac{1}{2}$ for x-polarized light, coupling only to the HH state. The absorption strength thus increases to 3.9 $\times 10^4$ cm^{-1} for the peak value.

10.7.3 Optical Effects in Polar Heterostructures

We have noted that in the "square" quantum well the first order Stark effect is zero due to the inversion symmetry of the well. It is possible to design quantum wells that are not "square" by choosing a variable composition. Also in polar heterostructures there is a a strong built-in electric field which can remove the inversion symmetry. In Chapter 1 we have discussed semiconductors in which there is a net polarization due to a shift between the cation and anion sublattices. Heterostructures made from such polar materials can yield very interesting band profiles with built-in fields due to the differences in polarizations in two materials. A very important heterostructure which has large polarization (and built-in electric fields) is based on InN, GaN, and AlN.

In Chapter 1 (see Eqns. 1.25 and 1.26) we have described the electric fields produced at the interfaces of AlGaN on GaN and InGaN on GaN heterostructures grown along the commonly used (0001) direction. As a result of the large built-in field (which can be ~ 10^6 V/cm) the electron and hole wavefunctions on a quantum well are pushed apart in a quantum well. This is shown in Fig. 10.17 for a In$_{0.18}$Ga$_{0.82}$N/GaN quantum well. It is important to note that such quantum wells are used for emission of blue (and green) light. As a result this is a very important material system for display

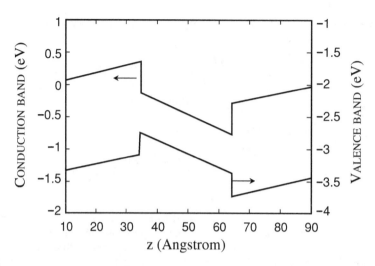

Figure 10.17: Band profile for 30 Å $In_{0.18}Ga_{0.82}N$/GaN quantum well grown lattice-matched to GaN in the (0001) orientation.

technology and high resolution optical memory (due to the short wavelength of blue light).

Due to the built-in electric field in the well, electron and hole wavefunctions are pushed towards opposite sides making the electron-hole overlap very small. The overlap depends on the well width and the InGaN composition (which controls the built-in electric field). Since the electron-hole overlap is small, the absorption coefficient and gain is suppressed in these quantum wells. On the other hand the radiative lifetime becomes correspondingly large. The small overlap also results in a very small exciton biding energy. The design of optoelectronic devices based on the InGaN quantum well must take into consideration the built-in electric field.

As electrons and holes are injected into the well, the built-in electric field is shielded due to the injected charge, and the quantum well becomes more "square" as shown in Fig. 10.18b. As a result of this the electron-hole overlap increases with injection and the radiative lifetime decreases. Note that even in "normal" square quantum wells the radiative time decreases with injection due to higher occupation factors.

As a result of the large built-in electric field in the nitride heterostructures there is a strong Stark effect in the optical transitions. This effect is linear, since at zero applied bias the quantum well does not have inversion symmetry.

10.8   EXCITON QUENCHING

In the previous section we discussed an approach used to modulate the excitonic transitions in quantum wells. There are several other approaches which can also modulate the exciton spectra and have been used to design various devices. Two important techniques are:

1. Quenching of the exciton by free carriers.

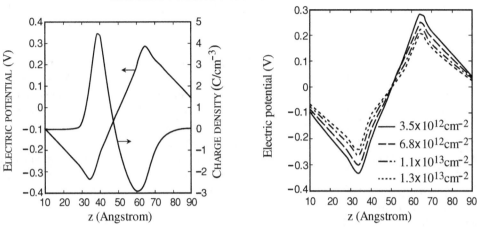

Figure 10.18: (a) Electric potential profile and charge density for 30 Å undoped $In_{0.18}Ga_{0.82}N/GaN$ quantum well at carrier density of 3.5×10^{12} cm^{-2} and $T = 300$ K. (b) Variation of electric potential at different injection levels.

2. Quenching of the exciton by creation of a high density of excitons by high optical intensity.

In the first approach, a high density of free electrons (or holes) is introduced into the quantum well. The free carriers screen the Coulombic interaction between the electron and hole, weakening the exciton binding energy and reducing the exciton oscillator strength. The physics behind the second approach is quite complex and a number of important phenomenon including bandgap renormalization, exciton phase space filling, and screening effects participate to cause the modulation of the exciton resonance. Since we are not fully equipped to treat these phenomenon which require many body theory we will only summarize the current state of knowledge.

The screening of the exciton is relatively simple to understand and has been used to design high speed modulators. In the simple theory of the effect of free carriers on Coulombic interaction, the Coulombic interaction is modified and is given by the screened Coulombic potential (see Chapter 5, Section 5.1)

$$V(\boldsymbol{r}) = \frac{e^2}{4\pi\epsilon r} e^{-\lambda r} \tag{10.71}$$

where for a nondegenerate electron gas with doping density n,

$$\lambda^2 = \frac{ne^2}{\epsilon k_B T}$$

and $N(E_F)$ is the density of states at the Fermi energy E_F. In a quantum well, one has to carry out a proper 2D treatment of the dielectric response, but the overall effect on the excitons can be understood on similar physical grounds.

As the background carrier density increases, the electron-hole attractive potential decreases due to screening and the exciton binding energy decreases. Another manifestation of this is that the exciton radius increases and the absorption coefficient decreases.

In Fig. 10.19, we show how the exciton binding energy changes as a function of carrier density for a 100 Å GaAs/Al$_{0.3}$Ga$_{0.7}$As quantum well structure. Also shown in Fig. 10.19 is the exciton radius as a function of carrier density for the same structure. We can see that the exciton essentially "disappears" once the background carrier density approaches 2–3×10^{11}cm^{-2}. This is a fairly low density and can be easily injected at high speeds into a quantum well. Since the injected charge removes the exciton peak, the absorption coefficient can be modulated rapidly by this process. We note that in addition to the effect noted above, the injected carriers also renormalize the bandgap by shrinking the gap somewhat. At the carrier concentrations of $\sim 10^{11}$cm^{-2}, the screening effects are, however, more dominant. This phenomenon has been used in creating electronic devices which can modulate an optical signal by charge injection.

Another area of interest in exciton physics is the area of nonlinear effects. At low optical intensity, the optical constants of the material are essentially independent of the optical intensity. However, as the optical intensity is increased, excess carriers are generated and these carriers affect the exciton properties. When the optical intensity generates electron-hole pairs, the optical absorption is suppressed by phase space filling. In the case of the free electron-hole pairs, this effect is simply due to the band filling effects manifested in the product term $f^e(E) \cdot f^h(E)$ which we discussed when discussing the gain in a laser.

A number of devices based on optical non-linearities have been proposed and demonstrated. These devices have been used to demonstrate various logic operations. Additionally, non-linear effects are used in characterization studies as discussed in Appendix B.

10.9   TECHNOLOGY ISSUES

The field of optics in general, and semiconductor optoelectronics in particular has bred numerous important technologies in fields ranging from medicine to entertainment to information processing. As noted in "Technology Issues" section in Chapter 9, semiconductor optoelectronics has been exploited in communications, data storage, printing, and display. Exciton based devices have been demonstrated to be able to carry out functions, like modulation, logic applications, programmable transparencies, beam steering, etc. However, apart from high speed modulation of lasers for optical communications, few devices have found commercial applications. This may be partly due to stiff competition from electronic devices which are less expensive (usually Si based), more reliable, and use well-established fabrication technologies. In fact, dreams of optical multipurpose computers based on optoelectronics is not likely to be fulfilled soon, if at all. However, by coupling the best of photonics with electronics advanced new systems are emerging.

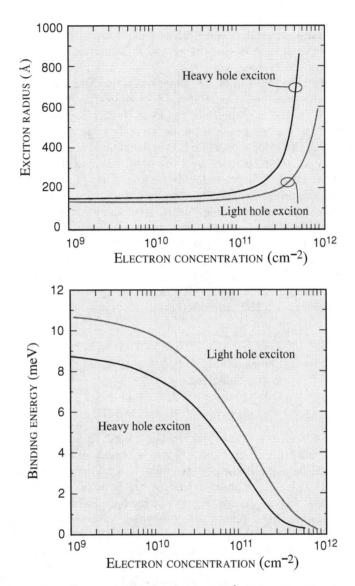

Figure 10.19: (a) Calculated exciton radii a_{ex} in a 100 Å GaAs/Al$_{0.3}$Ga$_{0.7}$As quantum well verses electron concentration n_e at $T = 300$ K. (b) Exciton binding energy as a function of carrier concentration.

10.10   PROBLEMS

10.1 Assume a simple parabolic density of states mass for the electrons and holes and calculate the exciton binding energies in GaAs, $In_{0.53}Ga_{0.47}As$ and InAs. What is the exciton Bohr radius in each case?

10.2 Assume a Gaussian exciton linewidth of 1 meV half-width at half maximum and plot the absorption coefficient due to the ground state exciton resonance in each of the three semiconductors of Problem 1.

10.3 Assume a hydrogenic form of the ground state exciton function. Using the variational method, show that a 2D exciton has four times the binding energy of the 3D exciton. Also calculate the in-plane Bohr radius of the 2D exciton.

10.4 Assume that the exciton radius a_{ex} scales roughly as inverse of the exciton binding energy. Using the in-plane dependence of the exciton-envelope function as $\exp(-\rho/a_{ex})$, use the results of Fig. 10.8 to plot the well-size dependence of the exciton-peak absorption strength of the heavy-hole ground state exciton. Use a 4 meV (HWHM) Gaussian width. In general, the linewidth may increase with decreasing well size.

10.5 Using the results of the simple perturbation theory calculate the electric field dependence of the HH exciton emission energy in a 100 Å GaAs quantum well. How does the position of the LH exciton (ground state) vary?

10.6 Design a GaAs quantum well (you may assume an infinite barrier) so that the ground state exciton resonance is at 1.5 eV. Calculate the peak absorption coefficient at this energy. If the quantum well width decreases by one monolayer, what is the absorption coefficient at 1.5 eV? Assume that:

exciton radius	$a_{ex} = 80$ Å
electron mass	$m_e^* = 0.067\ m_0$
hole mass	$m_h^* = 0.45\ m_0$
exciton halfwidth	$\sigma = 2.0$ meV
GaAs bandgap	$E_g = 1.43$ eV
1 monolayer	$d_{m\ell} = 2.86$ Å

Also assume that the exciton binding energy is 8.0 meV.

10.7 Consider a bulk GaAs electro-optic device on which an electric field is applied in the z-direction. The field is switched between 0 and 10^5 V/cm. Calculate the length of the device needed to produce a phase change of $\pi/2$ between the light waves polarized along $< 01\bar{1} >$ and $< 0\bar{1}1 >$. The wavelength of the light is 1.3 μm.

10.8 Consider a 100 Å GaAs quantum well in which the heavy-hole exciton resonance has a halfwidth of $\sigma = 3$ meV. A transverse electric field is switched between 0 and 60 kV/cm. Calculate the change in the absorption coefficient at a photon energy coincident with the peak of the exciton at zero field. If the total length of the active region is 2.0 μm, calculate the modulation depth that can be achieved (modulation depth = $I_{ph}(F = 60$ kV/cm$)/I_{ph}(F = 0)$). You may ignore the effect of the light hole exciton.

10.9 Consider a 100 Å $In_{0.53}Ga_{0.47}As$ quantum well device in which the exciton halfwidth is $\sigma = 5$ meV. An optical beam impinges on the device with an energy 10 meV below the zero field exciton resonance. The active length of the device is 3.0 μm. Calculate the minimum electric field needed to produce an optical intensity modulation of 2:1 relative to the zero field transmission intensity. Exciton radius may be taken as 120 Å in the well.

10.10 Assume that the absorption coefficient due to excitonic transitions is given adequately by the exciton radius. Calculate the change in the peak of the absorption coefficient as a function of free carriers in a 100 Å quantum well using the information provided in this chapter. The exciton halfwidth is $\sigma = 3$ meV.

10.11 In a normal incidence GaAs/Al GaAs quantum well modulator based on Stark effect, the substrate (GaAs) has to be removed. Discuss the reasons for this. Calculate the minimum composition of In in a 100 Å InGaAs/AlGaAs quantum well so that the substrate removal is not necessary.

10.12 Consider a 100 Å $In_{0.2}Ga_{0.8}As$ quantum well (grown lattice matched on a GaAs substrate) in which the ground state HH exciton at zero applied field is at 1.4 eV. Estimate the position of the ground LH exciton. Note that the strain induced splitting between HH and LH is 6 ϵ eV where ϵ is the lattice mismatch.

10.13 In problem 10.12, a transverse electric field of 10^5 V/cm is applied. Estimate the change in the refractive index for the TE and TM polarized light at $\hbar\omega = 1.3$ eV. The exciton halfwidth is $\sigma = 3$ meV and the exciton radius can be assumed to be 100 Å for both the HH and LH excitons.

10.14 A 100 Å GaAs quantum well has a heavy-hole exciton resonance at 1.5 eV with a halfwidth of 3 meV. Calculate the change in the refractive index produced by a transverse electric field of 10^5 V/cm at a photon energy of 1.4 eV. Use the Kramers Kronig relation and assume that the change in solely due to the change in the excitonic effects. Consider both HH and LH excitons and assume that $a_{ex} = 100$ Å for both cases. Consider the TE and TM polarized light separately.

10.15 In non-linear exciton "bleaching," one needs approximately 5×10^{11} cm^{-2} *e-h* pairs to cause the bleaching of the exciton. If the *e-h* recombination time is 1.0 ns, what is the optical power density needed to cause bleaching in 100 Å quantum wells? The absorption coefficient is 10^4 cm^{-1} for the impinging optical beam with energy 1.5 eV.

10.11   REFERENCES

- **General**

  - F. Bassani and G. P. Parravicini, *Electronic States and Optical Transitions in Solids*, Pergamon Press, New York (1975).

  - J. O. Dimmock, in *Semiconductors and Semimetals*, eds. R. K. Willardson and A. C. Beer, Academic Press, New York, vol. 3 (1967).

  - R. J. Elliott, in *Polarons and Excitons*, eds. C. G. Kuper and G. D. Whitfield, Plenum Press, Englewood Cliffs (1963).

  - "Theory of Excitons," by R. S. Knox, *Solid State Physics*, Academic Press, New York, vol. Suppl. 5 (1963).

  - Y. Onodera and Y. Toyozawa, *J. Phys. Soc. Japan*, **22**, 833 (1967).

  - D. C. Reynolds, C. W. Litton, E. B. Smith, and K. K. Bajaj, *Sol. St. Comm.*, **44**, 47 (1982).

  - M. D. Sturge, *Phys. Rev.*, **127**, 768 (1962).

- **Excitons in Quantum Wells**

  - G. Bastard, E. E. Mendez, L. L. Chang, and L. Esaki, *Phys. Rev. B*, **26**, 1974 (1982).

  - "Nonlinear Interactions and Excitonic Effects in Semiconductor Quantum Wells," by D. S. Chemla, in *Nonlinear Optics: Materials and Devices*, eds. C. Flytzanis and J. L. Oudar, Springer-Verlag (1986).

  - R. L. Green and K. K. Bajaj, *J. Vac. Sci. Technol.*, **B1**, 391 (1983).

  - J. P. Loehr and J. Singh, *Phys. Rev. B*, **42**, 7154 (1990).

  - R. C. Miller, D. A. Kleinman, W. A. Nordland, and A. C. Gossard, *Phys. Rev. B*, **22**, 863 (1980).

  - G. D. Sanders and Y. C. Chang, *Phys. Rev. B*, **31**, 6892 (1985).

- **Exciton Broadening Effects in Quantum Wells**

  - J. Lee, E. S. Koteles, and M. O. Vassell, *Phys. Rev. B*, **32**, 5512 (1986).

  - I. M. Lifshitz, *Adv. Phys.*, **13**, 483 (1969).

  - J. Singh and K. K. Bajaj, *Appl. Phys. Lett.*, **48**, 1077 (1986).

  - C. Weisbuch, R. Dingle, A. C. Gossard, and W. Wiegmann, *J. Vac. Sci. Technol.*, **17**, 1128 (1980).

- **Quantum Confined Stark Effect**

  - J. D. Dow and D. Redfield, *Phys. Rev. B*, **1**, 3358 (1970).

  - E. E. Mendez, G. Bastard, L. L. Chang, and L. Esaki, *Phys. Rev. B*, **26**, 7101 (1982).

  - D. A. B. Miller, D. S. Chemla, T. C. Damen, A. C. Gossard, W.Wiegmann, T. H. Wood, and C. A. Burrus, *Phys. Rev. Lett.*, **53**, 2173 (1984).

  - D. A. B. Miller, J. S. Weiner, and D. S. Chemla, *IEEE J. Quant. Electron.*, **QE-22**, 1816 (1986).

- **Exciton Quenching**

  - C. Flytzanis and J. L. Oudar, editors, *Nonlinear Optics: Materials and Devices*, Springer-Verlag, New York (1985).

  - H. Haug and S. Schmitt-Rink, *Prog. Quant. Electron.*, **9**, 3 (1984).

  - W. H. Knox, R. F. Fork, M. C. Downer, D. A. B. Miller, D. S. Chemla, and C. V. Shank, in *Ultrafast Phenomenon IV*, eds. D. H. Anston and K. B. Eisenthal, Springer-Verlag (1984).

- **Electro-Optic Effect**

  - J. Pamulapati, J. P. Loehr, J. Singh, and P. K. Bhattacharya, *J. Appl. Phys.*, **69**, 4071 (1991).

– J. S. Weiner, D. A. B. Miller, D. S. Chemla, T, C. Damen, C.A.Burrus, T. H. Wood, A. C. Gossard, and W. Wiegmann, *Appl. Phys. Lett.*, **47**, 1148 (1985).

– J. E. Zucker, T. L. Hendrickson, and C. A. Burrus, *Appl. Phys. Lett.*, **52**, 945 (1988).

Chapter

11

SEMICONDUCTORS IN MAGNETIC FIELDS

The general field of how semiconductor properties are modified in the presence of a magnetic field is a very wide one. To do justice to the field, one would need to devote several chapters to this area as we have done for electric field effects. However, it can be argued that from a technology point of view the response of electrons in semiconductors to electric fields and optical fields is more important. Magnetic field effects are primarily used for material characterization, although there is growing interest in magnetic semiconductors for device applications. In view of this fact we will provide an overview of how electrons in semiconductors respond to magnetic fields in just one chapter. In addition to a magnetic field, many important characterization techniques are carried out in the presence of an electric field or an optical field. We will therefore also discuss magneto-transport and magneto-optic properties. The general category of problems we will examine are sketched in Fig. 11.1.

In Fig. 11.1 we broadly differentiate between the "free" or Bloch states in semiconductors and the electron-hole coupled states like excitons or bound states. The magnetic field greatly alters the nature of the electronic states which then manifests itself in magneto-optic or magneto-transport phenomenon. It is important to realize that in many cases the physical phenomenon can qualitatively alter, depending upon the strength of the magnetic field. We will address the problem of electrons in the presence of a magnetic field in two steps. The first one is based on semi-classical ideas where the electron is treated as a point particle with the E vs k relation given by the bandstructure. In this case the magnetic field is assumed to be small enough that the concept of the bandstructure, effective mass, velocity, etc., is retained. Such low fields are used in experiments such as the Hall effect. The other case we will examine becomes important

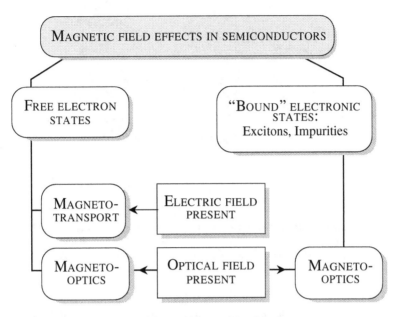

Figure 11.1: A schematic of the problems addressed in this chapter.

as the magnetic field increases. In this case the free electron states are no longer the usual planewave-like states $(u_{n\mathbf{k}}\exp(i\mathbf{k}\cdot\mathbf{r}))$, but are greatly modified. This results in new energy levels, different density of states, etc., and has important consequences on magneto-transport and magneto-optic phenomenon.

11.1 SEMICLASSICAL DYNAMICS OF ELECTRONS IN A MAGNETIC FIELD

We will first consider free electrons in semiconductors and how a magnetic field influences them. The magnetic field is assumed to be small so that the underlying picture of the electronic bandstructure remains intact and the electrons are described by the $\mathbf{k}$-vector. The equation of motion is

$$\frac{d\mathbf{k}}{dt} = \frac{\mathbf{F}_{\text{ext}}}{\hbar} \tag{11.1}$$

which becomes for a magnetic field

$$\frac{d\mathbf{k}}{dt} = \frac{e}{\hbar}\mathbf{v} \times \mathbf{B} \tag{11.2}$$

We immediately see that the change in the $\mathbf{k}$-vector is normal to $\mathbf{B}$ and also normal to the velocity. The velocity itself is normal to the constant energy surface and is given by

$$\mathbf{v_k} = \frac{1}{\hbar}\nabla_{\mathbf{k}}E(\mathbf{k}) \tag{11.3}$$

The magnetic field thus alters $\mathbf{k}$ along the intersection of the constant energy surface and the plane perpendicular to the magnetic field as shown in Fig. 11.2. The

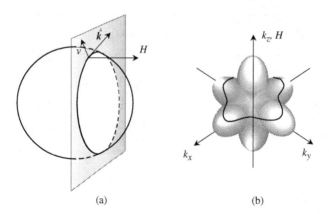

Figure 11.2: (a) A schematic showing the orbit of an electron in a magnetic field. The electron moves on a constant energy surface in a plane perpendicular to the magnetic field. (b) The orbit of an electron (or hole) on a constant energy surface that is anisotropic.

constant energy surface can, of course, be quite simple as for the conduction band of direct semiconductors or quite complex as for the valence bands. If $\boldsymbol{k}$ is the vector in this plane of intersection and $\boldsymbol{\rho}$ is the position of the electron, we have

$$\frac{d\boldsymbol{k}}{dt} = \frac{e}{\hbar}\dot{\boldsymbol{\rho}} \times \boldsymbol{B} \tag{11.4}$$

where $\boldsymbol{v} = \dot{\boldsymbol{\rho}}$ has a component, $\dot{\boldsymbol{\rho}}_\perp$, perpendicular to $\boldsymbol{B}$ and a component, $\dot{\boldsymbol{\rho}}_\parallel$, parallel to $\boldsymbol{B}$. Eqn. 11.4 shows that $\dot{\boldsymbol{k}}$ has a component, $\dot{\boldsymbol{k}}_\perp$, perpendicular to $\boldsymbol{B}$, only, while $\dot{\boldsymbol{k}}_\parallel = 0$.

The value of $\boldsymbol{\rho}_\perp(t)$ can be simply derived by taking a cross product of both sides of Eqn. 11.4 with a unit vector $\hat{B}$ parallel to the magnetic field. This gives

$$\begin{aligned}
\hat{B} \times \dot{\boldsymbol{k}} &= \frac{eB}{\hbar}\left[\dot{\boldsymbol{\rho}} - \hat{B}\left(\hat{B}\cdot\dot{\boldsymbol{\rho}}\right)\right] \\
&= \frac{eB}{\hbar}\dot{\boldsymbol{\rho}}_\perp \tag{11.5}
\end{aligned}$$

Integrating we get

$$\boldsymbol{\rho}_\perp(t) = \frac{\hbar}{eB}\boldsymbol{k}_\perp(t) + \pi/2 \text{ rotation.} \tag{11.6}$$

where the 90° rotation is included since the cross product of a unit vector with a perpendicular vector simply gives the second vector rotated by 90°. If the constant energy surface is spherical, as shown in Fig. 11.2a, then the value of $\boldsymbol{\rho}_\parallel(t)$ may be determined as well. Because no electric field is applied, the wavevector moves on a constant energy surface, where $\boldsymbol{\rho}_\parallel = \boldsymbol{v}_\parallel = 1/\hbar\hat{B}\cdot\nabla_{\boldsymbol{k}}E(\boldsymbol{k})$ is constant. Thus, by integration, $\boldsymbol{\rho}_\parallel(t) = \boldsymbol{v}_\parallel \cdot t + \boldsymbol{\rho}_\parallel(0)$. If the constant $\boldsymbol{v}_\parallel$ is nonzero, the electron moves in a helical trajectory under the application of a magnetic field.

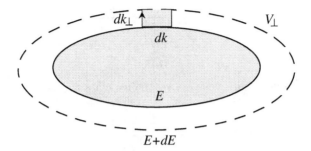

Figure 11.3: Two orbits of the electron at energies E and $E + dE$ in the magnetic field.

Let us now consider the frequency of the electron around the constant energy surface. We examine two orbits in $\boldsymbol{k}$-space with energies E and $E + dE$ as shown in Fig. 11.3. The $\boldsymbol{k}$-space separation of the orbits is

$$
\begin{aligned}
\delta k &= \frac{dE}{|\nabla_{\mathbf{k}} E|} \\
&= \frac{\delta E}{\hbar \, |\boldsymbol{v_k}|} \\
&= \frac{\delta E}{\hbar \dot{\boldsymbol{\rho}}}
\end{aligned}
\tag{11.7}
$$

The rate at which an electron moving along one of the orbits sweeps the annulus area is given by

$$
\begin{aligned}
\frac{dk}{dt} \delta k &= \frac{e}{\hbar^2} |\dot{\boldsymbol{\rho}} \times \boldsymbol{B}| \frac{\delta E}{|\dot{\boldsymbol{\rho}}|} \\
&= \frac{eB \, \delta E}{\hbar^2}
\end{aligned}
\tag{11.8}
$$

This rate is constant for constant δE and if we define by T, the time period of the orbit

$$
\begin{aligned}
\delta S &= T \times \frac{eB}{\hbar^2} \delta E \\
&= \text{area of the annulus} \\
&= \frac{dS}{dE} \cdot \delta E
\end{aligned}
\tag{11.9}
$$

where S is the area in k-space of the orbit of the electron with energy E. Thus

$$
T = \frac{\hbar^2}{eB} \frac{dS}{dE}
\tag{11.10}
$$

We can now define a cyclotron resonance frequency

$$
\omega_c = \frac{2\pi}{T}
$$

$$= \frac{2\pi e B}{\hbar^2} \frac{1}{dS/dE}$$

$$= \frac{eB}{m_c} \tag{11.11}$$

where we have introduced the cyclotron resonance mass m_c as

$$m_c = \frac{\hbar^2}{2\pi} \frac{dS}{dE} \tag{11.12}$$

The cyclotron resonance mass is a property of the entire orbit and may not be the same as the effective mass in general. For a parabolic band we have

$$E = \frac{\hbar^2 k^2}{2m^*}$$

or

$$S = \pi k^2$$
$$= \frac{2m^* \pi E}{\hbar^2}$$

and

$$\frac{dS}{dE} = \frac{2m^* \pi}{\hbar^2} \tag{11.13}$$

This gives

$$m_c = m^* \tag{11.14}$$

For a more complex band this relation will be appropriately modified. For example, the conduction band of indirect gap materials, such as Ge and Si can be represented by an ellipsoidal constant energy surface

$$E(\boldsymbol{k}) = \hbar^2 \left(\frac{k_x^2 + k_y^2}{2m_t} + \frac{k_z^2}{2m_\ell} \right) \tag{11.15}$$

where m_t is the transverse mass and m_ℓ is the longitudinal mass. The velocity components are now

$$v_x = \frac{\hbar k_x}{m_t}$$
$$v_y = \frac{\hbar k_y}{m_t}$$
$$v_z = \frac{\hbar k_z}{m_\ell} \tag{11.16}$$

If we assume that the magnetic field lies in the equatorial plane of the spheroid and is parallel to the k_x axis, we get from the equation of motion

$$\hbar \frac{dk_x}{dt} = 0$$

$$\hbar \frac{dk_y}{dt} = eBv_z$$
$$= \frac{\hbar eB}{m_\ell} k_z$$

or

$$\frac{dk_y}{dt} = \omega_\ell k_z, \text{ with } \omega_\ell = \frac{eB}{m_\ell} \quad (11.17)$$

Also,

$$\hbar \frac{dk_z}{dt} = -eBv_y$$
$$= -\hbar \frac{eB}{m_t} k_y$$

or

$$\frac{dk_z}{dt} = -\omega_t k_y, \text{ with } \omega_t = \frac{eB}{m_t} \quad (11.18)$$

If we differentiate Eqn. 11.17 with respect to time we get

$$\frac{d^2 k_y}{dt^2} = \omega_\ell \frac{dk_z}{dt} \quad (11.19)$$

Substituting for dk_z/dt from Eqn. 11.18, we get

$$\frac{d^2 k_y}{dt^2} + \omega_\ell \omega_t k_y = 0 \quad (11.20)$$

which is the equation of motion of a harmonic oscillator with frequency

$$\omega_0 = (\omega_\ell \omega_t)^{1/2}$$
$$= \frac{eB}{(m_\ell m_t)^{1/2}} \quad (11.21)$$

It can be shown also that if $\boldsymbol{B}$ is parallel to k_z, then the frequency is simply

$$\omega_0 = \omega_t$$
$$= \frac{eB}{m_t} \quad (11.22)$$

In general, if the magnetic field makes an angle θ with respect to the k_z direction we have for the cyclotron mass

$$\left(\frac{1}{m_c} \right)^2 = \frac{\cos^2 \theta}{m_t^2} + \frac{\sin^2 \theta}{m_t m_\ell}. \quad (11.23)$$

Thus, by altering the magnetic field direction, one can probe various combinations of m_ℓ and m_t. In the cyclotron resonance experiment, the cyclotron frequency can be measured directly, thus allowing measuring of the carrier masses. Results of such measurements for

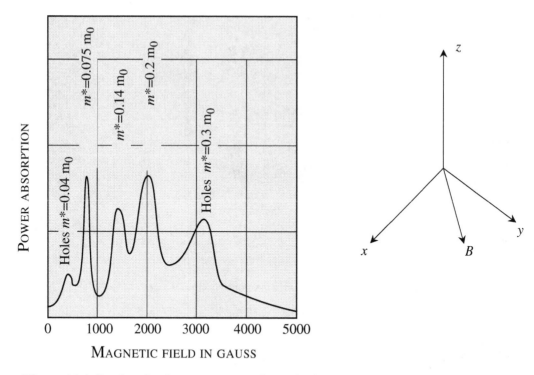

Figure 11.4: Results of cyclotron resonance absorption in germanium at 4 K, at 24 GHz. The static magnetic field is in a (110) plane at 60° from a (100) axis. Both electrons and holes are produced by illumination. The peaks correspond to the cyclotron frequency coinciding with 24 GHz. (Adapted from G. Dresselhaus, A. F. Kip, and C. Kittel, *Physical Review*, Vol. 98, 368 (1955).)

Ge electrons and holes are shown in Fig. 11.4. In Example 11.1 we describe the cyclotron resonance experiment—a very important experiment to obtain information on carrier masses. From Fig. 11.4 we see that Ge has two hole masses. The three electron masses result from different valleys and the orientation of the magnetic field with respect to these masses.

EXAMPLE 11.1 In a cyclotron resonance experiment an rf electric field is present in a direction perpendicular to the magnetic field, as shown in Fig. 11.5. When the rf frequency equals the cyclotron resonance frequency, there is resonant absorption of energy by the carriers. For absorption to occur we must also have $\omega_c \tau \geq 1$ where τ is the carrier scattering time. Consider electrons in GaAs in a cyclotron resonance experiment where a 24 GHz rf field is used. Calculate the magnetic field at which a resonance will occur. Also calculate the minimum mobility needed for the experiment to be successful.

The B-field needed is given by the relation

$$B = \frac{\omega_{rf} m^*}{e}$$

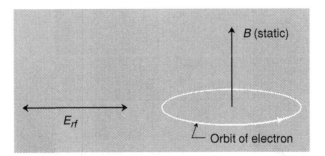

Figure 11.5: Arrangement of the magnetic field and an rf electric field during a cyclotron resonant experiment.

$$= \frac{\left(24 \times 10^{10} \times 2\pi \text{ rad/s}\right)\left(0.067 \times 9.1 \times 10^{-31} \text{ kg}\right)}{(1.6 \times 10^{-19} \text{ C})}$$
$$= 0.574 \text{ T} = 5.74 \text{ KGauss}$$

For the resonance to be observable the scattering time has to be

$$\tau \geq \frac{1}{(24 \times 10^{10} \times 2\pi \text{ rad/s})} = 6.63 \times 10^{-13} \text{ s}$$

The mobility has to be

$$\mu \geq \frac{e\tau}{m^*} = \frac{\left(1.6 \times 10^{-19} \text{ C}\right)\left(6.63 \times 10^{-13} \text{ s}\right)}{(0.067 \times 9.1 \times 10^{-31} \text{ kg})}$$
$$= 1.74 \text{ m}^2/V \cdot s = 17400 \text{ cm}^2/V \cdot s$$

The room temperature mobility of electrons in pure GaAs is $\sim 8500 \text{ cm}^2 V \cdot s$. Thus the sample has to be cooled to ~ 77 K to see the resonance.

11.1.1 Semiclassical Theory of Magnetotransport

We will now discuss the electronic system when there is a magnetic field and an electric field present. The magnetic field is again assumed to be small so that it can be treated as a small perturbation. The general formalism has already been discussed in Chapter 4, Section 4.4 and we will not repeat that analysis again in this chapter. It is important to point out that the semiclassical treatment is widely used to understand the Hall effect which is one of the most important characterization techniques for semiconductors. In Fig. 11.6 we show the configuration of the electric and magnetic fields in a typical Hall effect experiment. We showed in Chapter 4 that the Hall mobility μ_H measured in the presence of a magnetic field is related to the conductivity mobility μ by the following relation

$$\mu_H = r_H \mu \tag{11.24}$$

where r_H is the Hall factor and is given by

$$r_H = \frac{\langle\langle\tau^2\rangle\rangle}{\langle\langle\tau\rangle\rangle^2} \tag{11.25}$$

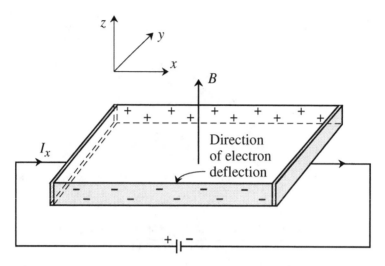

Figure 11.6: A rectangular Hall sample of an n-type semiconductor.

where

$$\langle\langle\tau\rangle\rangle = \frac{\langle E\tau\rangle}{\langle E\rangle} \tag{11.26}$$

and τ is the scattering time for various scattering processes. In particular, if it is possible to represent the energy dependence of τ by

$$\tau = \tau_0\left(\frac{E}{k_BT}\right)^s \tag{11.27}$$

then, for a Maxwellian distribution, we have

$$r_H = \frac{\Gamma(2s+5/2)\Gamma(5/2)}{[\Gamma(s+5/2)]^2} \tag{11.28}$$

where Γ is the Gamma function defined in Chapter 4. The conductivity tensor which relates the current to the applied field

$$J_i = \sigma_{ij}(\boldsymbol{B})E_j \tag{11.29}$$

is given by (for small B-fields where only terms linear in B are retained)

$$\sigma(\boldsymbol{B}) = \sigma_0 \begin{bmatrix} 1 & -\mu_H B_3 & \mu_H B_2 \\ \mu_H B_3 & 1 & -\mu_H B_1 \\ -\mu_H B_2 & \mu_H B_1 & 1 \end{bmatrix} \tag{11.30}$$

where σ_0 is the conductivity in absence of the magnetic field.

As seen in Chapter 4, when we include higher order terms in the magnetic field, we have

$$\sigma_{xx} = \sigma_{yy} = -\int \frac{d^3k}{(2\pi)^3} \frac{\partial f^0}{\partial E} \frac{e^2\tau v_x^2}{1+(\omega_c\tau)^2}$$

$$\sigma_{xy} = -\int \frac{d^3k}{(2\pi)^3} \frac{\partial f^0}{\partial E} \frac{e^3 v_x^2}{m^*} \frac{\tau^2}{1 + (\omega_c \tau)^2} \tag{11.31}$$

In the Hall experiment several quantities can be measured. These include the Hall coefficient R_H and the magneto-resistance given by

$$R_H = \frac{F_y}{J_x B}$$

$$P_H = \frac{F_x}{J_x} \tag{11.32}$$

It is easy to check that for an n-type or p-type material the Hall coefficient is simply

$$R_H = -\frac{1}{ne} \quad n\text{-type}$$

$$= \frac{1}{pe} \quad p\text{-type} \tag{11.33}$$

where n and p are the carrier densities. For materials where electron and hole densities are both significant the Hall coefficient is more complicated. From the expressions derived above it can be seen that the product of the Hall coefficient and resistivity is just the Hall mobility. The Hall mobility thus obtained is independent of carrier concentration making Hall effect a very powerful transport measurement. However, it is important to remember that Hall mobility is different from drift mobility (by the Hall factor).

In Fig. 11.7 we show results from a classic paper of Debye and Conwell on Hall measurements in n-type Ge. The samples are doped n-type at values ranging from 10^{12} cm^{-3} (sample A_1, essentially undoped) to 2×10^{17} cm^{-3} (sample E). From Fig. 11.7a we can see how the carrier concentration changes as a function of temperature. The freezeout, saturation, and intrinsic regimes are clearly seen. The Hall mobility obtained by multiplying the Hall coefficient and resistivity for each sample is shown in Fig. 11.7c. We see that for a heavily-doped sample the mobility is essentially independent of temperature, while for a lightly-doped sample the mobility increases as temperature is lowered.

An important measurement that provides insight into the scattering processes (their energy dependence) is magnetoresistance. It is important to include the $1 + (\omega_c \tau)^2$ term in the denominator of the conductivity tensor in evaluating magnetroresistance. Making the approximation

$$\frac{1}{1 + (\omega_c \tau)^2} \cong 1 - (\omega_c \tau)^2$$

it can be shown that the change in resistivity as a function of magnetic field is

$$\frac{\rho(B) - \rho_0}{\rho_0} = \frac{\Delta \rho}{\rho_0} = \frac{-\Delta \sigma}{\sigma} = \left(\frac{eB}{m^*}\right)^2 (\eta - 1)\xi \tag{11.34}$$

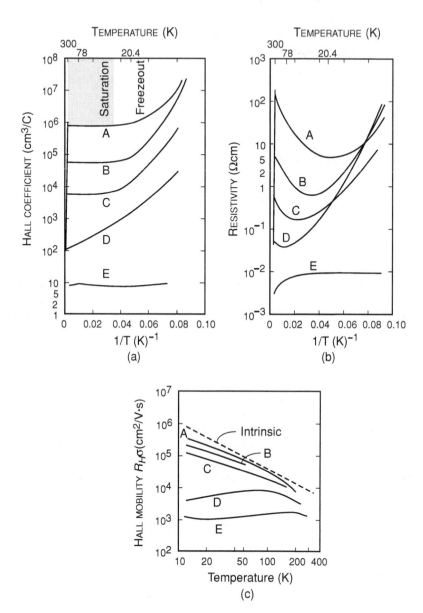

Figure 11.7: (a) Hall coefficient measurements in a series of samples with different n-type doping as a function of temperature. (b) Measurements of resistivity. (c) Hall mobility (after P. P. Debye and E. M. Conwell, *Physical Review*, Vol. 93, **693** (1954).).

where

$$\eta = \frac{\langle E\tau^3\rangle \langle E\tau\rangle}{\langle E\tau^2\rangle^2}$$

$$\xi = \frac{\langle E\tau^2\rangle^2}{\langle E\tau\rangle^2} \tag{11.35}$$

We see that if the *scattering time has no energy dependence*, $\Delta\rho$ is zero, *i.e., the resistivity of the sample has no magnetic field dependence.* In general the scattering times depend upon the carrier energy and the sample resistance increases as the magnetic field is increased.

11.2 QUANTUM MECHANICAL APPROACH TO ELECTRONS IN A MAGNETIC FIELD

We will now address the problem of electrons in a magnetic field by including the magnetic field energy in the Schrödinger equation. We will use the effective mass equation to absorb the effect of the background crystal potential. The approach is, of course, general and allows us to use the operator equivalence

$$-\frac{\hbar^2}{2m}\nabla^2 + V(r) \Rightarrow -\frac{\hbar^2}{2m^*}\nabla^2$$

provided there are no interband couplings. Here m^* is the effective mass of a particular band. The general Hamiltonian for the electrons is

$$H = \frac{1}{2m}\left(\boldsymbol{p} - e\boldsymbol{A}\right)^2 + V_c(r)$$

This equation is now written as an effective mass equation for a band with effective mass m^* (assumed isotropic and parabolic for simplicity). The resulting equation is

$$\frac{1}{2m^*}\left(\frac{\hbar}{i}\nabla - e\boldsymbol{A}\right)^2 \psi = E\psi \tag{11.36}$$

where $\psi(\boldsymbol{r})$ is now to be considered the envelope function which, when multiplied by the zone edge function of the band, gives the full wavefunction. For example, as we have discussed several times, the zone edge function of the conduction band state in the direct gap materials is an s-type state. It is important to appreciate how we have simplified Eqn. 11.36 versus how we treated the similar equation involving the electromagnetic interaction with electrons in the semiconductor. In that problem which led to the photon absorption, we did not apply the effective mass equation to the operator

$$\frac{1}{2m^*}\left(\frac{\hbar}{i}\nabla - e\boldsymbol{A}\right)^2 + V_c(\boldsymbol{r})$$

directly since the electromagnetic field does couple various bands. The effective mass approach was only applied to the operator

$$-\frac{\hbar^2}{2m}\nabla^2 + V_c(\boldsymbol{r})\left(= \frac{-\hbar^2}{2m^*}\nabla^2\right)$$

We also note that the interaction of the spin of the electron with the magnetic field is ignored at present. This interaction is $g\mu_B \boldsymbol{\sigma} \cdot \boldsymbol{B}$ where $\boldsymbol{\sigma}$ is the spin operator, $\mu_B = e\hbar/(2m)$ is the Bohr magneton and the g-factor is related to the details of the state ($g = 2$ for free electrons). This interaction is just added on to our solutions and will be discussed later.

We write the vector potential in the gauge

$$\boldsymbol{A} = (0, Bx, 0)$$

which gives a magnetic field in the z-direction

$$B = B\hat{z}$$

The equation to be solved is

$$\frac{\partial^2 \psi}{\partial x^2} + \left(\frac{\partial}{\partial y} - \frac{ieBx}{\hbar}\right)^2 \psi + \frac{\partial^2 \psi}{\partial z^2} + \frac{2m^*}{\hbar^2}E\psi = 0 \qquad (11.37)$$

where all energies are to be measured from the bandedges. Since our Hamiltonian does not involve y or z explicitly, the wavefunction can be written as

$$\psi(x, y, z) = \exp\left\{i\left(\beta y + k_z z\right)\right\} u(x) \qquad (11.38)$$

Denoting by

$$E' = E - \frac{\hbar^2}{2m^*}k_z^2 \qquad (11.39)$$

we get the equation for $u(x)$ as

$$\frac{\partial^2 u}{\partial x^2} + \left\{\frac{2m^*}{\hbar^2}E' - \left(\beta - \frac{eB}{\hbar}x\right)^2\right\} u = 0 \qquad (11.40)$$

We note that as in classical approach, the motion of the electron along the magnetic field is unaffected. The motion in the x-y plane is given by the Harmonic oscillator like equation

$$-\frac{\hbar^2}{2m^*}\frac{\partial^2 u(x)}{\partial x^2} + \frac{1}{2}m^*\left(\frac{eB}{m^*}x - \frac{\hbar\beta}{m^*}\right)^2 u(x) = E' u(x) \qquad (11.41)$$

This is the 1-dimensional Harmonic oscillator equation with a frequency eB/m^* and centered around the point

$$x_0 = \frac{1}{\omega_c}\frac{\hbar\beta}{m^*} \qquad (11.42)$$

The solution of this problem is

$$E' = \left(n + \frac{1}{2}\right)\hbar\omega_c \qquad (11.43)$$

and the total energy is

$$E = \left(n + \frac{1}{2}\right)\hbar\omega_c + \frac{\hbar^2}{2m^*}k_z^2 \qquad (11.44)$$

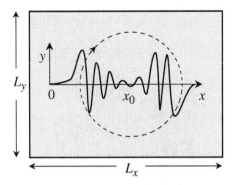

Figure 11.8: Solution of the Schrödinger equation in a magnetic field. The electron function along, say, x-axis is that of a harmonic oscillator centered around a point x_0.

The electron energy is quantized in the x-y plane and has the additional translational energy along the magnetic field. The schematic form of the wavefunction is shown in Fig. 11.8.

Since the contributions to the energy from the x-y plane motion is so drastically affected, it is important to understand what happens to the density of states of the system. Consider a box of sides L_x, L_y, L_z. From the form of the wavefunction given by Eqn. 11.38, it is clear that both k_z and β are quantized in units of $2\pi/L_z$ and $2\pi/L_y$ respectively. However, the energy of the electron has no β dependence so that all allowed β values give rise to the same energy state (for constant k_z value). However, the values of β is not infinite. As noted above, the center of the wavefunction x_0 must, of course, remain inside the dimensions of one system, i.e.,

$$0 \leq x_0 \leq L_x \tag{11.45}$$

Thus, the number of allowed values for β are

$$p = \frac{\beta_{\text{max}}}{(2\pi/L_y)} \tag{11.46}$$

where from Eqns. 11.42 and 11.45

$$\beta_{\text{max}} = \frac{\omega_c L_x m^*}{\hbar} \tag{11.47}$$

Thus, the degeneracy of a level is

$$p = \frac{m^* \omega_c L_x L_y}{2\pi\hbar} \tag{11.48}$$

One way to physically represent the effect of magnetic field is to use the schematic view in Fig. 11.9 Focusing only on the k_x-k_y plane, in absence of the magnetic field, the k_x, k_y points are good quantum numbers and the various points in Fig. 11.9a represent the allowed states. In presence of the magnetic field, various k_x, k_y points condense into

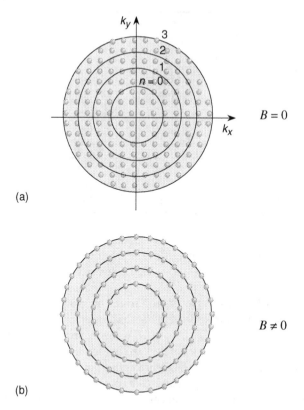

Figure 11.9: Quantization scheme for free electrons: (a) without magnetic field; and (b) in a magnetic field.

points on circles which represent constant energy surfaces with energies $\hbar\omega_c/2$, $3\hbar\omega_c/2$, etc. This rearrangement of states does not, of course, alter the total number of states in a macroscopic volume. This can be understood by examining the number of states in the presence of the magnetic field per unit area. This number is

$$
\begin{aligned}
\frac{\partial N}{\partial E} &= \frac{p}{\hbar\omega_c} \\
&= \frac{L_x L_y}{2\pi} \frac{m^* \omega_c}{\hbar^2 \omega_c} \\
&= L_x L_y \frac{m}{2\pi\hbar^2}
\end{aligned}
\tag{11.49}
$$

This is the same as the 2-dimensional density of states in the x-y plane. Thus, on the macroscopic energy scale the density of states is unaffected.

The levels that are produced for a given value of the integer n are called Landau levels. In Fig. 11.10a we show how the bandstructure of the semiconductor gets modified due to the magnetic field. We have Landau levels and only k_z is a good quantum number. The density of states of the 3-dimensional system is essentially given by the one-dimensional density of states derived in Chapter 2, weighted by the degeneracy factor p. Since k_z is still a good quantum number, the E vs. k_z given the bandstructure and the density of states. The various Landau levels are shown in Fig. 11.10.

In the 1-dimensional k_z space the density of states is (for a particular Landau level with energy E_n)

$$N_{1D}(E) = \frac{\sqrt{m^*}(E - E_n)^{-1/2}}{\sqrt{2}\hbar} \tag{11.50}$$

with the total density given by multiplying this by p and running the contribution from all Landau levels with starting energies less than E,

$$N(E) = \frac{1}{4\pi^2}\left(\frac{2m^*}{\hbar^2}\right)^{3/2} \hbar\omega_c \sum_n \left[E - \left(n + \frac{1}{2}\right)\hbar\omega_c\right]^{-1/2} \tag{11.51}$$

The density of states is shown in Fig. 11.10b where we see singularities arising from the quantization of the states. This change of the density of states due to the formation of Landau levels has important effects on the physical properties of the system. As the magnetic field is altered, the separation of the Landau level changes and the Fermi level (for a fixed carrier concentration) passes through these sharp structures in the density of states.

The treatment discussed above for 3-dimensional systems can be easily extended to the 2-dimensional system where the effects of the magnetic field become even more interesting. If we consider the magnetic field along the z-axis or the growth (confining) axis, then k_z is not a good quantum number so that not only are the x-y energies quantized, but so are the z energies. This leads to the remarkable result that the density of states becomes a series of δ-functions as shown in Fig. 11.11. The discrete energy levels are now given by

$$E_{n\ell} = E_n + \left(\ell + \frac{1}{2}\right)\frac{\hbar e B_\perp}{m_\parallel} + \frac{eB_\parallel^2}{2m_\parallel c^2}\left[(\bar{z})^2 - (\overline{z^2})\right] \tag{11.52}$$

Here the last term represents the diamagnetic effect of the parallel (in-plane) magnetic field which arises from first order perturbation theory using the perturbation Hamiltonian given by Eqn. 11.37.

$$B' = \frac{e^2 A^2}{2mc^2} \tag{11.53}$$

In Eqn. 11.52, the E_n represent the subband level energies arising from the quantization of the energy levels due to the confining potentials. Most experiments in 2D systems are carried out with $B_\parallel = 0$.

The presence of δ-function like density of states which is broadened in real systems has some very profound effects on physical phenomenon in 2-dimensional systems. The most well-known and Nobel prize winning effect is the quantum Hall effect which will be discussed later.

EXAMPLE 11.2 Calculate the magnetic field needed to cause the splitting in Landau levels in GaAs to become $k_B T$ at i) T = 4 K and ii) 300 K. The magnetic field at 4 K is given by

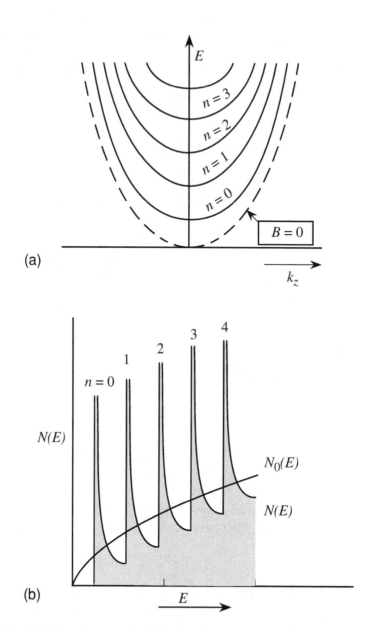

Figure 11.10: (a) Effect of magnetic field on the bandstructure of a semiconductor. Only k_z is a good quantum number. The magnetic field produces quantization in the x-y plane leading to Landau levels, The dashed curve is for zero field. (b) Density of states in a 3D system in the presence of magnetic field. The density of states develops singularities due to quantization in the plane perpendicular to the magnetic field. Also shown is the zero field density of states for comparison.

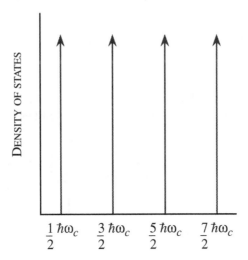

Figure 11.11: Singular density of states in an ideal 2D system in the presence of a magnetic field.

$(k_B T = 0.00033 \text{ eV})$

$$B = \frac{\Delta E m^*}{\hbar e} = \frac{\left(0.00033 \times 1.6 \times 10^{-19} \text{ J}\right)\left(0.067 \times 9.1 \times 10^{-31} \text{ kg}\right)}{\left(1.05 \times 10^{-34} \text{ J} \cdot \text{s}\right)\left(1.6 \times 10^{-19} \text{ C}\right)}$$
$$= 0.19 \text{ T}$$

At 300 K the field is 14.25 T. It is difficult to get magnetic field values much above 10 T in most laboratories.

11.3 AHARONOV-BOHM EFFECT

In Chapter 8 we have considered quantum interference devices in which an electrical signal (gate voltage) controls the interference effect. It is possible to control interference by a magnetic field as well.

Let us consider the Schrödinger equation for an electron in presence of an electromagnetic potential described by the vector potential $\boldsymbol{A}$ and scalar potential ϕ

$$\frac{1}{2m}\left(-i\hbar\nabla - e\boldsymbol{A}\right)^2 \psi + V\psi = E\psi \tag{11.54}$$

where $V = e\phi$. We assume that $\boldsymbol{A}$ and ϕ are time independent. In a region where the magnetic field is zero we can write the solution of the problem in the form

$$\psi(x) = \psi^0(x)\exp\left[\frac{ie}{\hbar}\int^{S(x)}\boldsymbol{A}(x')\cdot d\boldsymbol{s}'\right] \tag{11.55}$$

where $\psi^0(x)$ satisfies the Schrödinger equation with the same value of ϕ but with $\boldsymbol{A}(x) = 0$. The line integral in Eqn. 11.55 can be along any path as long as the end point $S(x)$

is the point x and $(\nabla \times \boldsymbol{A})$ is zero along the integral. Notice that this is essentially equivalent to making the change

$$\boldsymbol{k} \to \boldsymbol{k} - \frac{e}{\hbar} \boldsymbol{A}$$

in the usual free electron wavefunction $\exp(i\boldsymbol{k} \cdot \boldsymbol{r})$

To prove that Eqn. 11.55 satisfies the Schrödinger equation we use this solution and evaluate

$$(-i\hbar\nabla - e\boldsymbol{A})\,\psi \;=\; \exp\left[\frac{ie}{\hbar}\int^{S(x)}\boldsymbol{A}(x') \cdot d\boldsymbol{s}'\right]\left[(-i\hbar\nabla - e\boldsymbol{A}(x))\,\psi^0\right.$$

$$\left.+\;\psi^0(-i\hbar)\left(\frac{ie}{\hbar}\boldsymbol{A}(x)\right)\right]$$

$$=\; \exp\left[\frac{ie}{\hbar}\int^{S(x)}\boldsymbol{A}(x') \cdot d\boldsymbol{s}'\right]\left(-i\hbar\nabla\psi^0\right)$$

Similarly

$$(-i\hbar\nabla - e\boldsymbol{A}(x))^2\,\psi = \exp\left[\frac{ie}{\hbar}\int^{S(x)}\boldsymbol{A}(x') \cdot d\boldsymbol{s}'\right]\left(-\hbar^2\nabla^2\psi^0\right)$$

Thus, $\psi(x)$ satisfies the Schrödinger equation with $\boldsymbol{A} \neq 0$, if $\psi^0(x)$ satisfies the Schrödinger equation with $\boldsymbol{A} = 0$, but with same $V(x)$.

Let us now consider the problem described by Fig. 11.12. Here a beam of coherent electrons is separated into two parts and made to recombine at an interference region. This is the double slit experiment discussed in Chapter 8, Section 8.4, except now we have a region of magnetic field enclosed by the electron paths as shown. The wavefunction of the electrons at the point where the two beams interfere is given by (we assume that phase coherence is maintained)

$$\psi(x) \;=\; \psi_1^0 \exp\left[\frac{ie}{\hbar}\int_{\text{path }1}^{S(x)}\boldsymbol{A}(x') \cdot d\boldsymbol{s}'\right]$$

$$+\; \psi_2^0 \exp\left[\frac{ie}{\hbar}\int_{\text{path }2}^{S(x)}\boldsymbol{A}(x') \cdot d\boldsymbol{s}'\right] \tag{11.56}$$

The intensity or the electron density is given by

$$I(x) = \{\psi_1(x) + \psi_2(x)\}\{\psi_1(x) + \psi_2(x)\}^* \tag{11.57}$$

If we assume that $\psi_1^0 = \psi_2^0$, i.e. the initial electron beam has been divided equally along the two paths, the intensity produced after interference is

$$I(x) \;\propto\; \cos\left[\frac{e}{\hbar}\oint\boldsymbol{A} \cdot d\boldsymbol{s}\right]$$

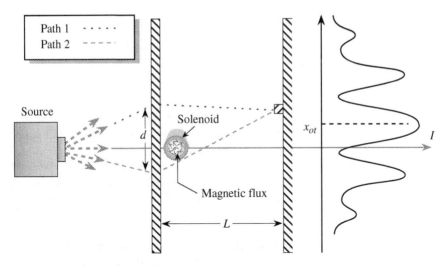

Figure 11.12: A magnetic field can influence the motion of electrons even though it exists only in regions where there is an arbitrarily small probability of finding the electrons. The interference pattern of the electrons can be shifted by altering the magnetic field.

$$= \cos\left[\frac{e}{\hbar}\int_{\text{area}} \boldsymbol{B}\cdot\boldsymbol{n}\,da\right]$$

$$= \cos\frac{e\phi}{\hbar} \tag{11.58}$$

where we have converted the line integral over the path enclosed by the electrons to a surface integral and used $\boldsymbol{B} = \nabla \times \boldsymbol{A}$. The quantity ϕ is the magnetic flux enclosed by the two electron paths. It is interesting to note that even though the electrons never pass through the $\boldsymbol{B} \neq 0$ region, they are still influenced by the magnetic field. From Eqn. 11.58 it is clear that if the magnetic field is changed, the electron density will undergo modulation. This phenomenon has been observed in semiconductor structures as well as metallic structures.

EXAMPLE 11.3 As with any quantum interference effect a key attraction of the Aharonov-Bohm effect is the possibility of its use in switching devices where a change in magnetic field can change the state of a device from 0 to 1. Consider a device of area 20 μm $\times$20 μm through which a magnetic field is varied. Calculate the field variation needed to switch the devices.

The flux "quanta" needed to switch from a constructive to destructive electron interference is

$$\Delta\Phi = \frac{\pi h}{e}$$

The magnetic field change needed for our device is

$$\Delta B = \frac{\pi h}{eA}$$

$$= \frac{\pi\left(1.05 \times 10^{-34}\text{ J s}\right)}{\left(1.6 \times 10^{-19}\text{ C}\right)\left(20 \times 10^{-6}\text{ m}\right)^2}$$

$$= 5.15 \times 10^{-6} \text{ T}$$

This is an extremely small magnetic field making a device a very low power device. However, it has not been easy to build practical circuits based on the Aharonov-Bohm effect so far.

11.3.1 Quantum Hall Effect

In Section 11.1.1 we have discussed the semiclassical theory of magnetotransport and Hall effect. At low temperatures in high quality 2-dimensional systems Hall effect takes on a rather unexpected form known as the quantum Hall effect. A two dimensional electron (or hole) gas can be created in the channel of a MOSFET or other heterostructure transistors as shown in Fig. 11.13a. While the traditional Hall effect measurements are done at very small magnetic fields, quantum Hall effect is observed under fairly high magnetic fields. At such high fields Landau levels are formed with splittings that are quite large.

In Chapter 4 we have discussed the conductivity tensor for an electronic system in the presence of an electric field and a perpendicular field. We refer to Fig. 11.13b for the geometry of the Hall effect. The general conductivity tensor is given by the relation (we ignore the averaging details of the scattering time τ).

$$\begin{pmatrix} J_x \\ J_y \\ J_z \end{pmatrix} = \frac{\sigma_0}{1 + (\omega_c \tau)^2} \begin{pmatrix} 1 & -\omega_c \tau & 0 \\ \omega_c \tau & 1 & 0 \\ 0 & 0 & 1 + (\omega_c \tau)^2 \end{pmatrix} \begin{pmatrix} F_x \\ F_y \\ F_z \end{pmatrix} \tag{11.59}$$

where the cyclotron resonance frequency is eB/m^*. It is interesting to note that in the high magnetic field limit ($\omega_c \tau \gg 1$), one has

$$\sigma_{xx} \Rightarrow 0 \tag{11.60}$$

The conductivity of a 2D system in the presence of a magnetic field can be written by the surface conductivity tensor σ with components

$$\sigma_{xx} = \frac{\sigma_0}{1 + (\omega_c \tau)^2}$$
$$\sigma_{xy} = \frac{-\sigma_0 \, \omega_c \tau}{1 + (\omega_c \tau)^2} \tag{11.61}$$

where σ_0 is the conductivity in the absence of the magnetic field and is given by

$$\sigma_0 = \frac{ne^2 \tau}{m^*}$$

where n is the density per unit area of the electrons.

The quantum Hall effect involves carrying out the usual Hall effect at very low temperatures (where the scattering time τ is long). In the limit of $\omega_c \tau \gg 1$, the conductivity components become

$$\sigma_{xx} = 0$$
$$\sigma_{xy} = \frac{-ne}{B} \tag{11.62}$$

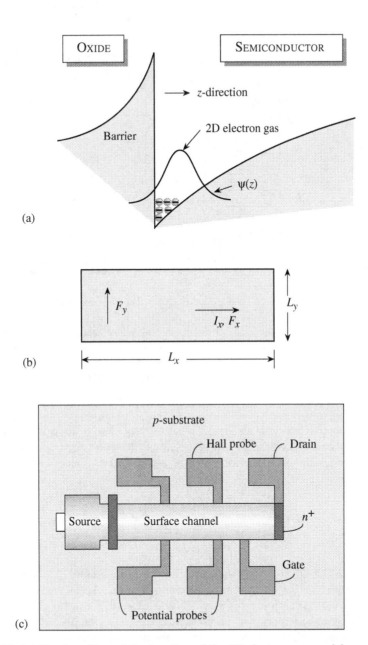

Figure 11.13: (a) Band profile and the location of the 2D electron gas used for quantum Hall effect experiments. (b) The applied electric field and drift current orientation. (c) A schematic of the positions of the various contacts used for quantum Hall effect.

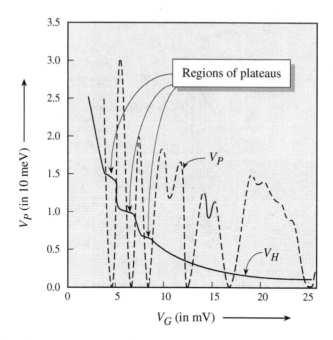

Figure 11.14: Results from the original quantum Hall effect carried out by K. von Klitzing, G. Dorda, and M. Pepper. A magnetic field of 18 T is perpendicular tot he sample and measurements area done at 1.5 K. A constant current of 1 μA is forced to flow between the source and the drain. (After K. von Klitzing, G. Dorda, and M. Pepper, *Phys. Rev. Lett.*, **45**, 494 (1980).)

In Fig. 11.13c we show an arrangement of probes that allow one to measure the Hall voltage V_H and the voltage V_p in the channel direction. In the Hall configuration the current in the y-direction is zero. The x-direction current density can then be shown to have the value

$$J_x = \left(\sigma_{xx} + \frac{\sigma_{xy}^2}{\sigma_{yy}} \right) F_x$$

$$= \sigma(eff) F_x \tag{11.63}$$

where $\sigma(eff)$ is the effective conductance. In the limit where B is very large $\sigma_{xx} = \sigma_{yy} = 0$ and the effective conductance becomes infinite. According to this result the voltage drop V_p across the probes of Fig. 11.13c will go to zero at high magnetic fields.

In their classic experiment K. von Klitzing, G. Dorda, and M. Pepper observed Hall effect in a MOSFET which was quite unexpected. The simple equations given above provided a hint at an explanation, but they were quite inadequate. In Fig. 11.14 we see several interesting features: i) the Hall voltage and the probe potential V_p do not have a monotonic behavior as the gate bias is altered (to change the sheet density), but have rich features; ii) the Hall voltage has plateaus at certain values of gate bias (i.e., sheet carrier densities); iii) the voltage across the probes becomes zero over the range where the Hall voltage shows plateaus. The most interesting observation is that the Hall resistivity is found to have the value

$$\rho_H = \frac{V_H}{I_x} = \frac{h}{\nu e^2} = \frac{25813}{\nu}\Omega \tag{11.64}$$

where ν is an integer. The value of the resistivity is so precise that quantum Hall effect has become a standard for the determination of the fine structure constant.

In contrast to what is observed experimentally, if we were to simply use the semiclassical Hall effect equations. We have from the semiclassical theory

$$
\begin{aligned}
J_x &= J_x L_y = (\sigma_{xx}F_x + \sigma_{yy}F_y)\,L_y \\
&= \frac{ne}{B}F_y L_y = \frac{ne}{B}V_p
\end{aligned}
$$

and

$$\rho_H = \frac{B}{ne} \tag{11.65}$$

Clearly the semiclassical approach does not explain the observations shown in Fig. 11.14.

Let us first consider a simplistic explanation of this phenomenon. Under strong magnetic fields we consider the description of the electronic system in terms of the Landau levels which are discrete in the 2D system as shown in Fig. 11.15a.

As the gate voltage is changed to fill the Landau levels, there will be a point where one Landau level is completely full and the next one is completely empty. In this case there can be no scattering of electrons. The condition for this to occur is from our previous discussions

$$n = \nu\frac{eB}{h} \tag{11.66}$$

where ν is an integer. The Hall resistance at this point is

$$
\begin{aligned}
\rho_H &= \frac{B}{ne} \\
&= \frac{h}{\nu e^2}
\end{aligned} \tag{11.67}
$$

However a small change in gate bias i.e., the addition of a few electrons, will alter this condition. Thus, on this basis one does not expect plateaus in the resistance.

A partial explanation of why the Hall conductivity has a plateau appears if we examine the realistic density of states in a 2-dimensional system that has some defect or disorder. As discussed in Chapter 8 and shown schematically in Fig. 11.15b the density of states is not made up of δ-functions, but is broadened. Due to the disorder present in the system, one expects extended and localized states as discussed in Chapter 8, Section 8.5. Thus, there is no reason that the Landau level will be filled (i.e., the extended states part that conducts) at precise values of the electron concentrations. This dilemma was addressed by Laughlin (1981). His arguments were based on arguments of flux quantization discussed in the section on Aharonov-Bohm effect. Laughlin divided the electronic states into extended and localized states where the localized states are

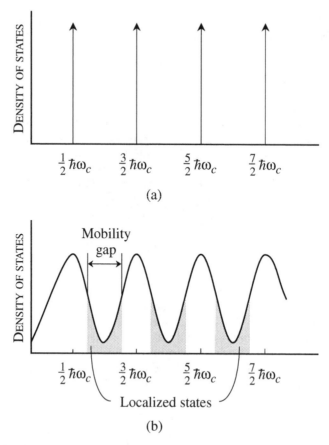

Figure 11.15: A schematic of the density of states in a 2-dimensional electron system in the presence of a high magnetic field. Case (a) is for an ideal system, while case (b) is for a real system with some disorder.

unaffected by the magnetic field. The extended states enclose the magnetic flux and can be affected by it. However, if the magnetic flux is altered by a flux quantum $\delta\phi = ch/e$, the extended states are identical to those before the flux quantum was added. Using these arguments, Laughlin was able to show that plateaus should appear in the Hall resistance with values precisely $h/\nu e^2$ regardless of the disorder provided, of course, we still had $\omega_c\tau \gg 1$.

The quantum Hall effect has been found to be even more complex with the observation of fractional quantum Hall effect. In experiments carried out at very high magnetic fields and low temperatures, one observes the Hall resistance quantized in units of $3h/e^2$ when the first Landau level is 1/3, 2/3, 2/5, 3/5, 4/5, 2/7 filled.

In these cases, the simple one electron picture of the Landau levels has broken down and many-body description of the electrons is needed. This is an ongoing area of research in condensed matter theory.

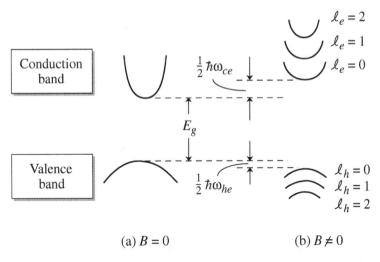

$$(a)\ B = 0 \qquad\qquad (b)\ B \neq 0$$

Figure 11.16: (a) Bands in a semiconductor in the absence of a magnetic field. (b) Bandstructure in the presence of a magnetic field.

11.4 MAGNETO-OPTICS IN LANDAU LEVELS

We have seen that in cyclotron resonance experiments an rf electrical field causes transitions when the energy is $\hbar\omega_c$. These energies are usually in the range of meV. Another important transition can occur between Landau levels in the valence band and levels in the conduction band. This effect, called the interband magnetooptic effect is very useful in providing information on the electron-hole system properties.

In Fig. 11.16 we show schematically the effect of a magnetic field on the energy bands of a semiconductor. The electron and hole levels are split into Landau levels, as has been discussed, and the energies are given by

$$
\begin{aligned}
E_e &= E_c + \hbar\omega_{ce}\left(\ell_e + \frac{1}{2}\right) + \frac{\hbar^2 k_z^2}{2m_e^*} \\
E_h &= E_v - \hbar\omega_{ch}\left(\ell_h + \frac{1}{2}\right) + \frac{\hbar^2 k_z^2}{2m_h^*}
\end{aligned}
\qquad (11.68)
$$

Here ω_{ce} and ω_{ch} represent the electron and hole cyclotron resonance frequencies, respectively.

Interband magnetooptic transitions involve optical transitions across the band between the same quantum number Landau level (i.e., $\ell_e = \ell_h$). The energy for the transitions are

$$
E_{mo} = E_g + \frac{\hbar^2 k_z^2}{2\mu} + \left(\ell + \frac{1}{2}\right)\hbar\omega_{ceh}
\qquad (11.69)
$$

where μ is the reduced mass and $\omega_{ceh} = eB/\mu$.

The density of states in the bands in the presence of a magnetic field have sharp resonances in 3D and 2D systems as shown in Figs. 11.10 and 11.11. As a result the absorption spectra shows sharp features which allows one to obtain information on the electron-hole system.

The perturbation due to the electromagnetic radiation is

$$\frac{e}{mc} \left(\boldsymbol{p} - e\boldsymbol{A}_{\text{ext}} \right) \cdot \boldsymbol{A}_{\text{em}} \tag{11.70}$$

where $\boldsymbol{A}_{\text{ext}}$ is due to the applied magnetic field and $\boldsymbol{A}_{\text{em}}$ is due to the electromagnetic field. Using the approach similar to the one used in Chapter 9 for band-to-band transitions in the presence of an electromagnetic field, we get

$$\alpha(\hbar\omega, \boldsymbol{B}) = \frac{\pi e^2 \hbar}{m_0^2 c n_r \epsilon_0} \frac{1}{\hbar\omega} \sum_{\text{all states}} |\boldsymbol{a} \cdot \boldsymbol{P}_{\text{cv}}(\boldsymbol{B})|^2 \, \delta \left(E_c(\boldsymbol{B}) - E_v(\boldsymbol{B}) - \hbar\omega \right) \tag{11.71}$$

where

$$\boldsymbol{a} \cdot \boldsymbol{P}_{\text{cv}}(\boldsymbol{B}) = \int \psi_c^*(\boldsymbol{r}, \boldsymbol{B}) \, \boldsymbol{a} \cdot \left(\boldsymbol{p} - e\boldsymbol{A}_{\text{ext}} \right) \psi_v(\boldsymbol{r}, \boldsymbol{B}) \, d^3r \tag{11.72}$$

and ψ_c, ψ_v, E_c, E_v represent the band wavefunctions and energies in presence of the magnetic field. To first order the effect of the magnetic field on the matrix element can be easily seen by neglecting $\boldsymbol{A}_{\text{ext}}$ in Eqn. 11.72 and using the effective mass equation for the representation of the electronic states and the density of states. In the usual approach of the effective mass theory, we may write

$$\psi(\boldsymbol{r}, \boldsymbol{B}) = F(\boldsymbol{r}) \psi_n(\boldsymbol{k}_c, \boldsymbol{r}) \tag{11.73}$$

where $F(\boldsymbol{r})$ is the $\psi(x, y, z)$ given in Eqn. 11.38, where we had suppressed the bandedge central cell function $\psi_n(\boldsymbol{k}_c, \boldsymbol{r})$. The matrix elements are then in k-space

$$
\begin{aligned}
\boldsymbol{a} \cdot \boldsymbol{P}_{\text{cv}}(\boldsymbol{B}) &= \int \left[\sum_{\boldsymbol{k}} F_c(\boldsymbol{k}) \psi_c(\boldsymbol{k}, \boldsymbol{r}) \right]^* \boldsymbol{a} \cdot \boldsymbol{p} \left[\sum_{\boldsymbol{k}'} F_v(\boldsymbol{k}') \psi_v(\boldsymbol{k}', \boldsymbol{r}) \right] d^3r \\
&= \sum_{\boldsymbol{k}, \boldsymbol{k}'} F_c^*(\boldsymbol{k}) F_v(\boldsymbol{k}') \, \boldsymbol{a} \cdot \boldsymbol{p}_{\text{cv}}(\boldsymbol{k}) \, \delta(\boldsymbol{k} - \boldsymbol{k}')
\end{aligned} \tag{11.74}
$$

where $\boldsymbol{k}$ and $\boldsymbol{k}'$ are nearly equal because of the near zero photon momentum. We also assume, as we did for the band-to-band transitions in Chapter 9, that $\boldsymbol{P}_{\text{cv}}(\boldsymbol{k})$ has no dependence on $\boldsymbol{k}$ and obtain the matrix elements, in terms of the real space wavefunction

$$\boldsymbol{a} \cdot \boldsymbol{P}_{\text{cv}}(\boldsymbol{B}) = \boldsymbol{a} \cdot \boldsymbol{M}_{\text{cv}} \int F_{cn'\boldsymbol{k}'}^*(\boldsymbol{r}) \, F_{vn\boldsymbol{k}}(\boldsymbol{r}) \, d^3r \tag{11.75}$$

We will assume that the bands are isotropic, in which case the envelope functions for the harmonic oscillator problem are essentially orthonormal giving the selection rule

$$
\begin{aligned}
\Delta n &= 0 \\
\Delta \boldsymbol{k} &= 0
\end{aligned} \tag{11.76}
$$

The sum over the δ-function in Eqn. 11.74 gives us the joint density of states of the electron-hole system. This joint density of states is simply

$$N_{\mathrm{cv}}(E, \boldsymbol{B}) = \frac{2eB}{\hbar^2}(2\mu)^{1/2} \sum_n \left[E - E_g - \hbar\omega_{ceh}\left(n + \frac{1}{2}\right) \right]^{-1/2} \tag{11.77}$$

where μ is the reduced mass and $\omega_{ceh} = eB/\mu$. The absorption coefficient now becomes

$$\alpha(\omega, \boldsymbol{B}) = \frac{\pi e^2 \hbar}{m_0^2 c n_r \epsilon_0} \frac{1}{\hbar\omega} \frac{2eB}{h^2}(2\mu)^{1/2} |\boldsymbol{a} \cdot \boldsymbol{p}_{\mathrm{cv}}|^2 \sum_n \left[\hbar\omega - E_g - \hbar\omega_{ceh}\left(n + \frac{1}{2}\right) \right]^{-1/2} \tag{11.78}$$

As with the density of states, the absorption profile has sharp structures. The spacings between the singularities gives the reduced mass of the electron-hole system. The results can be extended to the 2-dimensional case by following the method for the 2-D density of states. These density of states are now δ-function, as shown in Fig. 11.11, which will be broadened by an appropriate linewidth.

11.5 EXCITONS IN MAGNETIC FIELD

We will now address the problem of excitons in a magnetic field. The studies of excitons in semiconductors have been used to obtain accurate values of basic band parameters such as effective mass and g-values and to obtain information on symmetry of the states.

We have seen how electrons described by "free" Bloch states are affected by a magnetic field. What happens if the electrons are in a "bound" state? The exciton represents the electron-hole bound state with an envelope function spread over ~ 100 Å. As we have seen in Chapter 10, the exciton problems can only be addressed analytically under very simple approximations (3-dimensional, uncoupled valence bands). In the presence of a magnetic field the problem becomes difficult to solve, especially if the field is very high and perturbative techniques are invalid. Under high magnetic field conditions where magnetic field energy approaches or exceeds the exciton binding energy, variation methods are useful. We will discuss the case of uncoupled band exciton (i.e., light hole and heavy hole excitons are uncoupled) using a perturbation approach.

$$\psi_{\mathrm{ex}}(\boldsymbol{K}_{\mathrm{ex}}, \boldsymbol{r}_e, \boldsymbol{r}_h) = \sum_{\mathbf{q}} F(\boldsymbol{q}, \boldsymbol{K}_{\mathrm{ex}})\, \psi_v(\boldsymbol{q}, \boldsymbol{r}_h)\, \psi_c(-\boldsymbol{q} + \boldsymbol{K}_{\mathrm{ex}}, \boldsymbol{r}_e) \tag{11.79}$$

The equation for the Fourier transform of $F(\boldsymbol{q}, \boldsymbol{K}_{\mathrm{ex}})$ is

$$\left\{ \frac{1}{2m_e}(\boldsymbol{p}_e - e\boldsymbol{A})^2 + \frac{1}{2m_h}(\boldsymbol{p}_h - e\boldsymbol{A})^2 - \frac{e^2}{4\pi\epsilon|\boldsymbol{r}_e - \boldsymbol{r}_h|} \right\} F(\boldsymbol{r}_e, \boldsymbol{r}_h)$$
$$= EF(\boldsymbol{r}_e, \boldsymbol{r}_h) \tag{11.80}$$

where $\boldsymbol{A}$ is the vector potential due to the magnetic field. We reduce the problem to the center of mass system by the transformation

$$\boldsymbol{R} = \frac{m_e \boldsymbol{r}_e + m_h \boldsymbol{r}_h}{m_e + m_h}$$
$$\boldsymbol{r} = \boldsymbol{r}_e - \boldsymbol{r}_h$$

and performing the transformation

$$F(\boldsymbol{r}_e, \boldsymbol{r}_h) = \exp\left(i\left[\mathbf{K}_{\mathrm{ex}} - \frac{e}{\hbar}\mathbf{A}\right] \cdot \mathbf{R}\right) \Phi(\boldsymbol{r})$$

The relative motion equation is then defined by

$$\left\{\frac{p^2}{2\mu} + e\left(\frac{1}{m_h} - \frac{1}{m_e}\right)\boldsymbol{A} \cdot \boldsymbol{p} + \frac{e^2}{2\mu}\boldsymbol{A} \cdot \boldsymbol{A}\right.$$
$$\left. - \frac{e^2}{4\pi\epsilon r} - \frac{2e\hbar}{M}\boldsymbol{K}_{\mathrm{ex}} \cdot \boldsymbol{A} + \frac{\hbar^2 K_{\mathrm{ex}}^2}{2M}\right\}\Phi(\boldsymbol{r}) = E\Phi(\boldsymbol{r}) \qquad (11.81)$$

Here μ is the electron-hole reduced mass and M is the total effective mass of the system. Using the Lorentz gauge and writing

$$\boldsymbol{A} = \frac{1}{2}\boldsymbol{B} \times \boldsymbol{r}$$

the various magnetic field dependent terms of the Eqn. 11.81 take on the following values:

- The second term in parentheses can be written as

$$\frac{e}{2}\left(\frac{1}{m_h} - \frac{1}{m_e}\right)\boldsymbol{B} \cdot \boldsymbol{L}$$

  which is the Zeeman term.

- The third term in parentheses is the diamagnetic operator,

$$\frac{e^2}{8\mu}|\boldsymbol{B} \times \boldsymbol{r}|^2$$

- The fifth term depends upon the motion of the exciton and is given by

$$-\frac{e\hbar}{M}(\boldsymbol{K}_{\mathrm{ex}} \times \boldsymbol{B} \cdot \boldsymbol{r})$$

  and can be neglected due to its small value.

For optical transitions we are only interested in the $m = 0$ excitonic states (the s-states) since, as discussed in Chapter 10, these are the only allowed transitions. The equation then becomes

$$\left\{\frac{p^2}{2\mu} - \frac{e^2}{4\pi\epsilon r} + \frac{e^2 B^2}{8\mu}(x^2 + y^2)\right\}\phi(\boldsymbol{r}) = E\phi(\boldsymbol{r}) \qquad (11.82)$$

In order to define the relative strengths of the Coulombic and magnetic terms one defines a parameter

$$\gamma = \frac{\hbar\omega_c}{2R_{\mathrm{ex}}} \qquad (11.83)$$

which is the ratio of the cyclotron energy and twice the exciton binding energy in the absence of the magnetic field, R_{cx}. In terms of γ the equation for the exciton is

$$\left\{ -\nabla^2 - \frac{2}{\gamma} + \frac{\gamma^2}{4}(x^2 + y^2) \right\} \phi(\boldsymbol{r}) = E\phi(\boldsymbol{r}) \qquad (11.84)$$

where we have used $R_{ex} = \mu e^4 / \left[2\hbar^2 (4\pi\epsilon)^2 \right]$ as the units of energy and the Bohr radius $a_{ex} = 4\pi\epsilon\hbar^2/\mu e^2$ as the unit of length. We note that the Zeeman term coming from the electron and hole spins are not considered and can be simply added on to the solutions later. This contribution is

$$g_e\mu_0\boldsymbol{S}_e \cdot \boldsymbol{B} + g_h\mu_0\boldsymbol{S}_h \cdot \boldsymbol{B} \qquad (11.85)$$

For small magnetic fields, one can treat the diamagnetic term as simply a perturbation and evaluate the effect on the excitonic states, assuming that there is no mixing of the exciton states. One finds that first order perturbation theory gives

$$\begin{aligned} E_{ex}(1s) &= R_{ex}\left[-1 + \frac{1}{2}\gamma^2 \right] \\ E_{ex}(2s) &= R_{ex}\left[-\frac{1}{4} + 7\gamma^2 \right] \end{aligned} \qquad (11.86)$$

Magnetic field dependence of excitonic transitions is primarily done to shed light on the nature of the excitonic state. From the shift in the excitonic resonances information on the nature of the exciton wavefunction can be obtained.

11.6 MAGNETIC SEMICONDUCTORS AND SPINTRONICS

In traditional semiconductor devices, the density of spin up and spin down electrons is the same unless a magnetic field is applied to select a particular state. The contacts used to inject electrons also usually have no spin selectivity. However, it is possible to grow magnetic semiconductors and to use ferromagnetic contacts to inject electrons with spin selectivity. Notable examples of magnetic semiconductors are InGaAsMn, CdMnTe, Zn-MnSe, and HgMnTe. These semiconductors, known as diluted magnetic semiconductors, and their heterostructures with other semiconductors can now be fabricated and they offer a unique opportunity for the combined studies of semiconductor physics and magnetism. There is a strong exchange interaction between the magnetic moments of the magnetic ions and the spins of the band electrons giving rise to large Zeeman splittings. In a heterostructure environment, the bandstructure can be tailored so that both the electronic and magnetic properties can be tailored. The magnetic semiconductors are fabricated by the usual epitaxial techniques like MBE or MOCVD and Mn is introduced as an extra ingredient. The Mn composition is usually $\leq 20\%$.

In recent years there has been a growing interest in a field known as "spintronics" (after spin and electronics). In conventional electronics, electron density is modulated to create devices for digital and analog applications. In spintronics one modulates the spin of electrons. As in quantum interference devices discussed in Chapter 8, such a possibility promises very low power, high density devices. An important point to note in spin dependent devices is that scattering mechanisms discussed in Chapters 5 and

6 cause only very weak spin scattering. Thus an electron can maintain its spin value for several microns (or even 100 microns at low temperature). However, this does not mean that spin based transistors can function at high temperatures or for long channel lengths. Non-spin altering scattering processes are still important in spintronic devices.

In our discussions on magnetic field effect on free carrier properties we have largely ignored the electron spin. Of course, the spin orbit interaction is responsible for the light-hole, heavy-hole bands and the various selection rules in optical transitions. The main reason we have not worried about electron spin is that usually density of spin-up and spin-down electrons is the same and the spin splitting in the presence of a magnetic field is small. However, it is possible to prepare a semiconductor sample in a state where electrons in the conduction band have a much higher density of spin-down electrons. This can be done by using optical injection or electronic injection. Before discussing the details of important issues in spintronics we will remind the reader of some basic spin-magnetic field properties. Electrons (or other charged particles) interact with a magnetic field via a magnetic moment which is written as

$$\mu_s = -g\mu_B \mathbf{S} = \gamma\hbar\mathbf{S} \tag{11.87}$$

where $\mathbf{S}$ is the spin of the particle; g is known as the g-factor and characterizes the particle. The constant μ_B is known as the Bohr magneton and has a value

$$\mu_B = \frac{e\hbar}{2m} \tag{11.88}$$

The constant γ is called the gyromagnetic or magnetogyric ratio. The magnetic moment associated with the spin then gives the usual term in the hamiltonian:

$$H_{\text{spin}} = -\mu_s \cdot \mathbf{B} \tag{11.89}$$

11.6.1 Spin Selection: Optical Injection

To understand how electrons with preferential spin can be injected into a semiconductor, let us review how a magnetic field splits the conduction and valence band levels accounting for spin. Including the spin and orbital angular momentum of the electronic state, the perturbation due to a magnetic field is

$$\Delta E = \frac{\mu_B}{\hbar}\langle \mathbf{L} + 2\mathbf{S}\rangle \cdot \mathbf{B} \tag{11.90}$$

where the $\langle\rangle$ brackets represent the expectation value of the orbital angular momentum ($\mathbf{L}$) and spin ($\mathbf{S}$). It can be shown using quantum mechanics of angular momentum states that (m is the total angular momentum along the B-field direction)

$$\Delta E = \mu_B g m B$$

where g is the Landau factor given by

$$g = 1 + \frac{j(j+1) + S(S+1) - \ell(\ell+1)}{2j(j+1)} \tag{11.91}$$

$$\Delta E = \frac{\mu B}{h} \langle L+2S \rangle \cdot B = \mu_B \; gmB$$

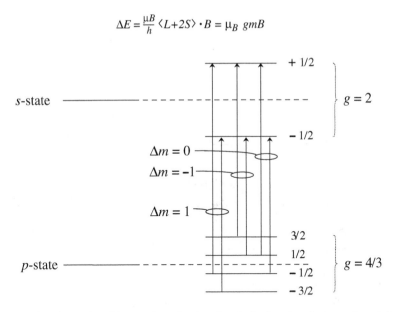

Figure 11.17: A schematic of how s (conduction bandedge) and p (valence bandedge) states split as a result of an applied magnetic field.

It can be seen that for s-states (conduction bandedge states) $g = 2$ and for $j = 3/2, \ell = 1$ states (p-states near valence bandedge) $g = 4/3$.

In Fig. 11.17 we show the splitting of conduction and valence bandedge states in a material like GaAs. Note that the splitting of the s-states is larger than the splitting in the p-states. From Fig. 11.17 we can see how optical transitions using polarized light can be used to preferentially create electrons with spin-down state. A right circular polarized light beam causes a $\Delta m = 1$ transition allowing electrons transferred to the conduction band to have $-1/2$ spin. Such spin selection has been demonstrated in a number of experiments. The selectivity can be enhanced by using quantum wells or strained quantum wells where the light-hole heavy-hole splitting can be altered.

11.6.2 Spin Selection: Electrical Injection and Spin Transistor

It is possible to inject electrons or holes in a spin selected state using ferromagnetic contacts. In Fig. 11.18a we show a spin-transistor in which spin selected electrons are injected from an Fe contact acting as a source. The magnetized iron contact injects electrons with spin selected by the magnetization field and maintain this spin state as they travel throughout the device. The device operation will be discussed later in this section.

In Fig. 11.18b we show how hole injection from a ferromagnetic semiconductor (GaMnAs) can be used to inject spin selected holes into a quantum well of InGaAs. These holes then combine with electrons injected from a "normal" n-type contact to emit polarized light.

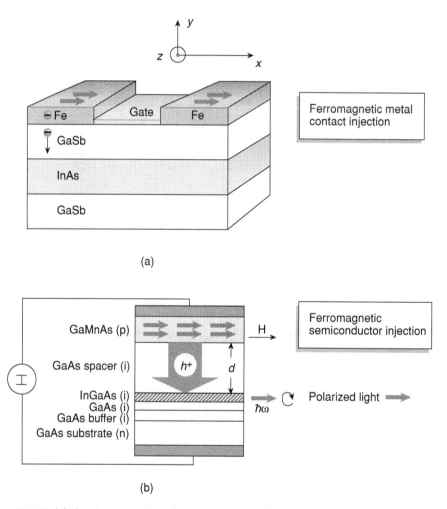

(a)

(b)

Figure 11.18: (a) A schematic of a spin transistor in which electrons with a selected spin are injected into a 2-dimensional channel. (b) Use of a ferromagnetic semiconductor injector to select holes with prechosen spin for a light emission device.

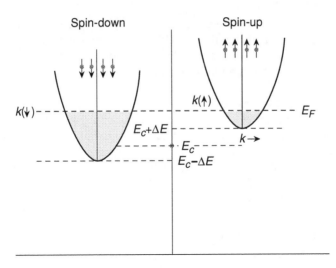

Figure 11.19: A schematic of the band profile of spin-up and spin-down electrons. At the Fermi energy the k-vector for spin-up and spin-down electrons are different.

The spin transistor exploits quantum interference effects with two nuances: i) spin select electrons can be injected into a transistor channel; ii) spin splitting of spin-up and spin-down states causes the two spin state electrons to have a different k-vector which can be controlled by a gate bias to create interference effects.

Using the geometry shown in Fig. 11.18a, electrons are injected into the 2-dimensional channel with a spin polarized along the $+x$ direction. These electrons may be written in terms of the spin-up (positive z-polarized) and spin-down (negative z-polarized) states

$$|x\rangle \to \frac{1}{\sqrt{2}} (|\uparrow\rangle + |\downarrow\rangle) \tag{11.92}$$

Now consider the possibility where the energy of the spin-up and spin-down electrons is different as shown in Fig. 11.19. The splitting in the spin-up and spin-down states can occur due to external magnetic fields or internal spin-orbit effects combined with lack of inversion symmetry. These effects are strongest in narrow bandgap semiconductors where the conduction band states are influenced by the p-type valence band states.

The position of the Fermi level is the same for the spin-up and spin-down states as shown in Fig. 11.19. We have

$$\begin{aligned} E_F &= E_c - \Delta E + \frac{\hbar^2 k^2(\downarrow)}{2m^*} \\ &= E_c + \Delta E + \frac{\hbar^2 k^2(\uparrow)}{2m^*} \end{aligned} \tag{11.93}$$

As the electrons move down the channel the phase difference between spin-up and spin-down electrons changes according to the usual wave propagation equation

$$\Delta\theta = [k(\uparrow) - k(\downarrow)] L \tag{11.94}$$

where L is the channel length. The drain contact acts as a spin filter and only accepts electron states with spin in the x-direction. Thus the current flows if $\Delta\theta = 2n\pi$. Otherwise the current value is lower. Thus the spin transistor essentially behaves as an electrooptic modulator where the phase is controlled by the gate voltage which controls E_F.

In considering spin based devices for new electronic or optoelectronic devices it has to be remembered that while electrons (holes) can maintain their spin states over long distances, this does not directly translate into superior quantum interference devices. If devices depend upon phase coherence, non-spin scattering events like defect or phonon scattering can still degrade the device performance. However, a number of novel devices are being considered where the electron spin state plays a dominant role and these look very promising.

11.7 TECHNOLOGY ISSUES

As noted several times in this chapter, magnetic effects in semiconductors have been extensively used to obtain information on bandstructure, scattering processes, carrier type, and concentration, etc. Quantum Hall effect has been instrumental in obtaining new understanding of many-body physics. However, although in theory, a number of logic and memory devices could be based on magnetic effects, in practice, these devices have not made it into the commercial world so far. Magnetic memories have, of course, been part of information technology. However, these are not based on semiconductor technology. The incorporation of ferromagnetic materials into semiconductor technology has given hope that magnetic semiconductor devices will indeed pose a challenge to traditional semiconductor devices. However, there are still many challenges—the biggest one being traditional devices keep getting better!

An area that has received a great deal of attention as a possible source of new technology is the area of spintronics (see, for example, "The Quest for the Spin Transistor," Glenn Zorpette, *IEEE Spectrum*, December 2001). Advances in spin selective injection of electrons, demonstration of long spin lifetimes and the marriage of ferromagnetic and semiconductor materials have made the spin transistor a possibility. Whether it leads to a viable technology is still not clear. However, it certainly is an exciting field!

11.8 PROBLEMS

Section 11.1

11.1 In a cyclotron resonance experiment done on electrons in Ge, it is found that when the magnetic field orientation is varied the maximum and minimum cyclotron resonance masses are 0.08 m_0 and 0.36 m_0. Calculate the transverse and longitudinal masses for electrons.

11.2 Consider the data shown in Fig. 11.4. Verify that the carrier masses given in the figure are correct.

11.3 The Hall mobility of a GaAs sample at 77 K is found to be 50000 cm^2/V·s. Assume that the mobility is dominated by acoustic phonon scattering. What is the drift mobility?

11.4 Using the data plotted in Fig. 11.7 calculate the temperature dependence of the carrier concentration and drift mobility in samples A and D. Assume that in sample A the dominant scattering is acoustic phonon scattering and in sample D it is ionized impurity scattering.

11.5 Use the data plotted in Fig. 11.7 to calculate the doping density and donor ionization energy (use data for sample B).

Section 11.2

11.6 Plot the density of states of electrons in the conduction band of GaAs (in units of eV^{-1} cm^{-3}) at applied magnetic fields of: (i) 0; (ii) 5.0 T; and (iii) 10 T. The effective mass of electrons is 0.067 m_0.

11.7 Consider a two-dimensional electron gas in a 100 Å GaAs/AlGaAs quantum well structure. Assume an infinite potential model with $m_e^* = 0.067$ m_0 to calculate the electronic properties. Plot the density of states of the system (up to 0.5 eV from the bottom of the well) at (i) $B = 0$ and (ii) $B = 10$ T.

11.8 Consider a 100 Å GaAs/AlGaAs quantum well where the electron effective mass is 0.067 m_0. There are 10^{12} electron/cm^2 in the well. At low temperatures, calculate the magnetic field at which the Fermi level just enters the fourth Landau level of the ground state sub-band.

11.9 Consider a GaAs conduction band quantum well in which the electron effective mass is 0.067 m_0. At what magnetic field is the separation between the first and second Landau levels equal to $k_B T$ at 300 K?

11.10 The conduction band of Si is described by six equivalent valleys along (100), ($\bar{1}$00), (010), (0$\bar{1}$0), etc., directions. The effective mass in these valleys is given by the longitudinal (along the k-direction of the valley) and transverse (perpendicular to the direction in which the valley occurs) effective masses. In silicon these masses are $m_\ell^* = 0.98$ $m_0, m_t^* = 0.19$ m_0. Calculate the Landau level splitting of the lowest sub-band formed in a Si MOS structure.

11.11 In order to observe Landau levels and related effects, the separation of the levels should be comparable to (or larger than) the thermal energy $k_B T$. Calculate the magnetic field needed to satisfy this condition in GaAs ($m^* = 0.067$ m_0) at 4 K, 77 K and 300 K. Repeat the calculations for the valence band ($m^* = 0.4$ m_0).

11.12 Consider a GaAs quantum well in which there are 10^{12} electrons cm^{-2}. Plot the position of the Fermi level and the Landau levels as a function of magnetic field at 4 K as the field changes from 0 to 10 T. Only consider the Landau levels arising from the ground state subband in the well.

Section 11.3

11.13 Discuss the role of scattering of electrons in the Aharonov-Bohm effect.

11.14 Consider a digital device based on the Aharonov-Bohm effect. The area enclosed by the two arms of the device is 5 μm × 5 μm. Estimate the minimum switching energy needed to switch the device from ON to OFF if the volume over which the B-field is to be altered is 10^{-9} cm^3.

11.15 In crystalline semiconductors the resistance of a sample increases as the magnetic field increases. Why does this occur? In amorphous materials, the resistance of a sample decreases (i.e., conductivity improves) as magnetic field increases. Provide some reasons for this.

Section 11.4

11.16 Biaxial strain is known to alter the effective masses of holes (with little effect on electron mass) for quantum well structures, as discussed in Chapter 3. Use the calculations given in Fig. 3.18 for change in hole mass as a function of strain.

Calculate the positions of magneto-optic levels that would be observed in a 100 Å $In_xGa_{1-x}As/Al_{0.3}Ga_{0.7}$ As quantum well as x goes from 0 to 0.2 at a field of 10 T. Only consider the E1-HH1 transitions. Assume any reasonable parameters for the InGaAs well which is pseudomorphically on a GaAs substrate.

11.9 REFERENCES

- **Semiclassical Approach**

    – Kittel, C., *Introduction to Solid State Physics* (John Wiley and Sons, New York, 1986).

    – Kittel, C., *Quantum Theory of Solids* (John Wiley and Sons, New York, 1987).

    – Lundstrom, M., *Fundamentals of Carrier Transport* (Modular Series on Solid State Devices, edited by G. W. Neudeck and R. F. Pierret, Addison-Wesley, Reading, 1990), vol. X.

    – Ziman, J. M., *Principles of the Theory of Solids* (Cambridge, 1972).

- **Quantum Mechanical Approach**

    – Feynman, R. P., R. B. Leighton and M. Sands, *The Feynman Lectures on Physics* (Addison-Wesley, Reading, 1965), vol. 3.

    – Kittel, C., *Quantum Theory of Solids* (John Wiley and Sons, New York, 1987).

    – Ziman, J. M., *Principles of the Theory of Solids* (Cambridge, 1972).

- **The Quantum Hall Effect**

    – Kittel, C., *Introduction to Solid State Physics* (John Wiley and Sons, New York, 1986).

    – Laughlin, R. B., Phys. Rev. B, **23**, 5632 (1981).

    – Stormer, H. L. and D. C. Tsui, Science, **220**, 1241 (1983).

    – Von Klitzing, K., G. Dorda and M. Pepper, Phys. Rev. Lett., **45**, 494 (1980).

- **Magneto Optics Studies**

    – Bassani, F. and G. P. Parravicini, *Electronic States and Optical Transitions in Solids* (Pergamon Press, New York, 1975).

- Jones, E. D., R. M. Biefeld, J. F. Klem and S. K. Lyo, *GaAs and Related Compounds 1989* (Inst. Phys. Conf. Ser. 106, Institute of Physics, Bristol, 1989), p. 435.

- Ziman, J. M., *Principles of the Theory of Solids* (Cambridge, 1972).

• **Excitons and Impurities in a Magnetic Field**

- Baldereschi, A. and F. Bassani, *Proc. X Int. Conf. Phys. Semiconductors* (Cambridge, Massachusetts, 1970), p. 191.

- "Excitons and Impurities in Magnetic Fields," by Baldereschi, A., in *Theoretical Aspects and New Developments in Magneto-Optics* (edited by J. T. Devreese, Plenum Press, New York, 1980). Also, see other articles in this book.

- Stillman, G. E., D. M. Larsen and C. M. Wolfe, Phys. Rev. Lett., **27**, 989 (1971).

- "Magneto-optical Studies of Impurities," by Stradling, R. A., in *Theoretical Aspects and New Developments in Magneto-Optics* (edited by J. T. Devreese, Plenum Press, New York, 1980).

• **Magnetic Semiconductors**

- "Optical and Magnetic Properties of Diluted Magnetic Semiconductor Heterostructures," by Chang, L. L., D. D. Awschalom, M. R. Freeman and L. Vina, in *Condensed Systems of Low-Dimensionality* (edited by J. L. Beeby, Plenum Press, 1991). See other references in this article.

- Gunsher, R. L., L. A. Kolodziejski, N. Otsuka and S. Datta, Surf. Sci., **174**, 522 (1986).

- Kolodziejski, L. A., T. C. Bonsett, R. L. Gunshor, S. Datta, R. B. Bylsma, W. M. Becker and N. Otsuka, Appl. Phys. Lett., **45**, 440 (1989).

• **Spintronics**

- S. Datta and B. Das, *Applied Physics Letters*, vol. 56, **665** (1990).

- G. Prinz *Physics Today*, **48** (4), 58 (1995).

- M. Johnson, *IEEE Spectrum*, **31** (5), 47 (1995).

- D. D. Awschalom and J. Kikkawa, *Physics Today*, **52** (6), 33 (1999).

- H. Ohno, *J. Magn. Magn. Materials*, **200** 110 (1999).

- J. M. Kikkawa and D. D. Awshcalom, *Nature*, **397**, 139 (1999).

- D. Hagele *et al.*, *Applied Phys. Lett.*, **73**, 1580 (1998).

Appendix

A

STRAIN IN
SEMICONDUCTORS

Incorporation of strain in heterostructures has become an accepted technique to modify bandstructure. A number of important electronic and optoelectronic devices exploit strain in their design. In Chapter 1 and Chapter 3 we have seen how strain epitaxy occurs and how bandstructure is modified by strain. In this Appendix we will establish the basic equations for stress-strain relations and strain energy in a semiconductor. We will address the cubic structure, although similar treatment can be established for other lattices.

A.1 ELASTIC STRAIN

In this section we will establish some basic expressions for strain in crystalline materials. We will confine ourselves to small values of strain. In order to define the strain in a system we imagine that we have a set of ortho-normal vectors $\hat{x}, \hat{y}, \hat{z}$ in the unstrained system. Under the influence of a uniform deformation these axes are distorted to $\boldsymbol{x}', \boldsymbol{y}', \boldsymbol{z}'$, as shown in Fig. A.1. The new axes can be related to the old one by

$$
\begin{aligned}
\boldsymbol{x}' &= \left(1 + \epsilon_{xx}\right)\hat{x} + \epsilon_{xy}\hat{y} + \epsilon_{xz}\hat{z} \\
\boldsymbol{y}' &= \epsilon_{yx}\hat{x}\left(1 + \epsilon_{yy}\right)\hat{y} + \epsilon_{yz}\hat{z} \\
\boldsymbol{z}' &= \epsilon_{zx}\hat{x} + \epsilon_{zy}\hat{y} + \left(1 + \epsilon_{zz}\right)\hat{z}
\end{aligned}
\tag{A.1}
$$

The coefficients $\epsilon_{\alpha\beta}$ define the deformation in the system. The new axes are not orthogonal in general. Let us consider the effect of the deformation on a point $\boldsymbol{r}$ which in the unstrained case is given by

$$
\boldsymbol{r} = \boldsymbol{x}\hat{x} + y\hat{y} + z\hat{z}
\tag{A.2}
$$

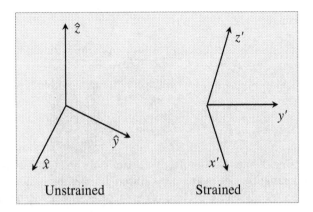

Figure A.1: Effect of deformation on a coordinate system x, y, z.

After the distortion the new vector is given by

$$\boldsymbol{r}' = x\boldsymbol{x}' + y\boldsymbol{y}' = z\boldsymbol{z}' \tag{A.3}$$

Note that by definition the coefficients x, y, z of the vector are unchanged. The displacement of the deformation is then

$$
\begin{aligned}
\boldsymbol{R} &= \boldsymbol{r}' - \boldsymbol{r} \\
&= x\left(\boldsymbol{x}' - \hat{x}\right) + y\left(\boldsymbol{y}' - \hat{y}\right) + z\left(\boldsymbol{z}' - \hat{z}\right)
\end{aligned} \tag{A.4}
$$

or

$$
\begin{aligned}
\boldsymbol{R} &= \left(x\epsilon_{xx} + y\epsilon_{yx} + z\epsilon_{zx}\right)\hat{x} \\
&+ \left(x\epsilon_{xy} + y\epsilon_{yy} + z\epsilon_{zy}\right)\hat{y} \\
&+ \left(x\epsilon_{xz} + y\epsilon_{yz} + z\epsilon_{zz}\right)\hat{z}
\end{aligned} \tag{A.5}
$$

or by defining quantities, u, v, w

$$\boldsymbol{R}(\boldsymbol{r}) = u(\boldsymbol{r})\hat{x} + v(\boldsymbol{r})\hat{y} + w(\boldsymbol{r})\hat{z} \tag{A.6}$$

For a general non-uniform distortion one must take the origin of $\boldsymbol{r}$ close to the point and define a position-dependent strain. In the small strain limit:

$$x\epsilon_{xx} = x\frac{\partial u}{\partial x} \; ; \; y\epsilon_{yy} = y\frac{\partial v}{\partial y} \; ; \; z\epsilon_{zz} = z\frac{\partial w}{\partial z} \; , \; \text{etc.} \tag{A.7}$$

Rather than using the ϵ_{xx} to describe the distortion we will use the strain components which are defined as:

$$e_{xx} = \epsilon_{xx} = \frac{\partial u}{\partial x} \; ; \; e_{yy} = \epsilon_{yy} = \frac{\partial v}{\partial y} \; ; \; e_{zz} = \epsilon_{zz} = \frac{\partial w}{\partial z} \tag{A.8}$$

The off-diagonal terms are:

$$e_{xy} = x' \cdot y' \approx \epsilon_{yx} + \epsilon_{xy} = \frac{\partial u}{\partial y} + \frac{\partial v}{\partial x} \tag{A.9}$$

$$e_{yz} = y' \cdot z' \approx \epsilon_{zy} + \epsilon_{yz} = \frac{\partial v}{\partial z} + \frac{\partial w}{\partial y} \tag{A.10}$$

$$e_{zx} = z' \cdot x' \approx \epsilon_{zx} + \epsilon_{xz} = \frac{\partial u}{\partial z} + \frac{\partial w}{\partial x} \tag{A.11}$$

The off-diagonal terms define the angular distortions of the strain. It is useful to define the net fractional change in the volume produced by the distortion. This quantity is called dilation. The initially cubic volume of unity after distortion has a volume:

$$\begin{aligned}
V' &= x' \cdot (y' \times z') \\
&= \begin{vmatrix} 1 + \epsilon_{xx} & \epsilon_{xy} & \epsilon_{xz} \\ \epsilon_{yx} & 1 + \epsilon_{yy} & \epsilon_{yz} \\ \epsilon_{zx} & \epsilon_{zy} & 1 + \epsilon_{zz} \end{vmatrix} \\
&\cong 1 + e_{xx} + e_{yy} + e_{zz}
\end{aligned} \tag{A.12}$$

The dilation is then:

$$\delta = \frac{V' - V}{V} \approx e_{xx} + e_{yy} + e_{zz} \tag{A.13}$$

In order to produce a distortion in the crystalline unit cell, it is important to define the stress components responsible for the distortion. The force acting on a unit area in the solid is called stress and we can define the nine stress components X_x, X_y, X_z, Y_x, Y_y, Y_z, Z_x, Z_y, Z_z. Here the capital letter indicates the direction of the force and the subscript is the direction of the normal to the plane on which the stress is acting, as shown in Fig. A.2. The number of stress components reduces to six if we impose the condition on a cubic system that there is no torque in the system (i.e., the stress does not produce angular acceleration). Then: $X_y = Y_x$: $Y_z = Z_y$; $Z_x = X_z$. We then have the six independent components: X_x, Y_y, Z_z; Y_z, Z_x, X_y.

A.2   ELASTIC CONSTANTS

Elastic constants are defined through Hooke's Law which states that for small distortion the strain is proportional to the stress.

$$\begin{aligned}
\epsilon_{xx} &= s_{11}X_x + s_{12}Y_y + s_{13}Z_z + s_{14}Y_z + s_{15}Z_x + s_{16}X_y \; ; \\
\epsilon_{yy} &= s_{21}X_x + s_{22}Y_y + s_{23}Z_z + s_{24}Y_z + s_{25}Z_x + s_{26}X_y \; ; \\
\epsilon_{zz} &= s_{31}X_x + s_{32}Y_y + s_{33}Z_z + s_{34}Y_z + s_{35}Z_x + s_{36}X_y \; ; \\
\epsilon_{yz} &= s_{41}X_x + s_{42}Y_y + s_{43}Z_z + s_{44}Y_z + s_{45}Z_x + s_{46}X_y \; ; \\
\epsilon_{zx} &= s_{51}X_x + s_{52}Y_y + s_{53}Z_z + s_{54}Y_z + s_{55}Z_x + s_{56}X_y \; ; \\
\epsilon_{xy} &= s_{61}X_x + s_{62}Y_y + s_{63}Z_z + s_{64}Y_z + s_{65}Z_x + s_{66}X_y
\end{aligned} \tag{A.14}$$

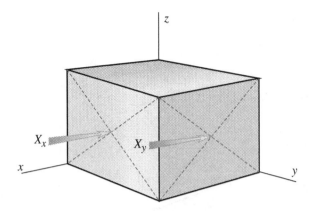

Figure A.2: Stress component X_x is a force applied in the x-direction to a unit area of a plane whose normal lies in the x-direction; X_y is applied in the x-direction to a unit area of a plane whose normal lies in the y-direction.

Conversely, the stress components are linear functions of the strain components:

$$
\begin{aligned}
X_x &= c_{11}e_{xx} + c_{12}e_{yy} + c_{13}e_{zz} + c_{14}e_{yz} + c_{15}e_{zx} + c_{16}e_{xy} \ ; \\
Y_y &= c_{21}e_{xx} + c_{22}e_{yy} + c_{23}e_{zz} + c_{24}e_{yz} + c_{25}e_{zx} + c_{26}e_{xy} \ ; \\
Z_z &= c_{31}e_{xx} + c_{32}e_{yy} + c_{33}e_{zz} + c_{34}e_{yz} + c_{35}e_{zx} + c_{36}e_{xy} \ ; \\
Y_z &= c_{41}e_{xx} + c_{42}e_{yy} + c_{43}e_{zz} + c_{44}e_{yz} + c_{45}e_{zx} + c_{46}e_{xy} \ ; \\
Z_x &= c_{51}e_{xx} + c_{52}e_{yy} + c_{53}e_{zz} + c_{54}e_{yz} + c_{55}e_{zx} + c_{56}e_{xy} \ ; \\
X_y &= c_{61}e_{xx} + c_{62}e_{yy} + c_{63}e_{zz} + c_{64}e_{yz} + c_{65}e_{zx} + c_{66}e_{xy}
\end{aligned}
\tag{A.15}
$$

The 36 elastic constants can be reduced in real structures by invoking various symmetry arguments. Let us examine the elastic energy of the system. The energy is a quadratic function of the strain, and can then be written as

$$
U = \frac{1}{2} \sum_{\lambda=1}^{6} \sum_{\mu=1}^{6} \tilde{c}_{\lambda\mu} \, e_\lambda \, e_\mu
\tag{A.16}
$$

where $1 = xx$; $2 = yy$; $3 = zz$; $4 = yz$; $5 = zx$; $6 = xy$. The $\tilde{c}$s are related to the cs defined above.

The stress component is found by taking the derivative of U with respect to the associated strain component.

$$
\begin{aligned}
X_x &= \frac{\partial U}{\partial e_{xx}} \\
&\equiv \frac{\partial U}{\partial e_1} \\
&= \tilde{c}_{11}e_1 + \frac{1}{2} \sum_{\beta=2}^{6} \left(\tilde{c}_{1\beta} + \tilde{c}_{\beta 1} \right) e_\beta
\end{aligned}
\tag{A.17}
$$

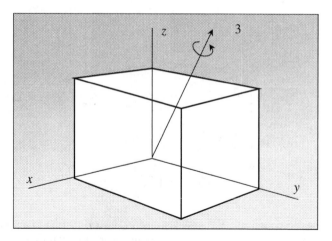

Figure A.3: Rotation by $2\pi/3$ about the axis marked 3 changes $x \to y; y \to z; \ z \to x$.

Note that we will always get the combination $(\tilde{c}_{\alpha\beta} + \tilde{c}_{\beta\alpha})/2$ in the stress-strain relations. It follows that the elastic stiffness constants are symmetrical.

$$
\begin{aligned}
c_{\alpha\beta} &= \frac{1}{2}\left(\tilde{c}_{\alpha\beta} + \tilde{c}_{\beta\alpha}\right) \\
&= c_{\beta\alpha}
\end{aligned}
\tag{A.18}
$$

Thus the 36 constants reduce to 21.

The number of force constants is further reduced if one examines the cubic symmetries. We will show that for cubic systems, in the elastic energy U, the only terms that occur are:

$$
\begin{aligned}
U &= \frac{1}{2}c_{11}\left(e_{xx}^2 + e_{yy}^2 + e_{zz}^2\right) + c_{44}\left(e_{yz}^2 + e_{zx}^2 + e_{xy}^2\right) \\
&+ \ c_{12}\left(e_{yy}e_{zz} + e_{zz}e_{xx} + e_{xx}e_{yy}\right)
\end{aligned}
\tag{A.19}
$$

Other terms of the form

$$
\left(e_{xx}e_{xy} + \ldots\right), \left(e_{yz}e_{zx} + \ldots\right), \left(e_{xx}e_{yz} + \ldots\right), \text{ etc.}
$$

do not occur.

If one examines a cubic system, the structure has a 4-fold symmetry. Focusing on the [111] and equivalent directions, if we rotate by $2\pi/3$ we get the following transformations: (see Fig. A.3)

$$
\begin{aligned}
x &\to y \to z \to x \\
-x &\to z \to -y \to -x \\
x &\to z \to -y \to x \\
-x &\to y \to z \to -x
\end{aligned}
\tag{A.20}
$$

The terms in the energy density must be invariant under these transformations. Note that

$$e_{xx}e_{xy} \rightarrow -e_{xx}e_{x(-y)} \tag{A.21}$$

and therefore any term which is odd in any of the indices will vanish.

$$\frac{\partial U}{\partial e_{xx}} = X_x$$
$$= c_{11}e_{xx} + c_{12}(e_{yy} + e_{zz}) \tag{A.22}$$

comparing the various terms:

$$c_{12} = c_{13}; \quad c_{14} = c_{15} = c_{16} = 0 \tag{A.23}$$

In Chapter 1 several useful references have been given which the reader can examine to learn more about the elastic properties of materials.

Appendix

B

EXPERIMENTAL TECHNIQUES

In this Appendix we will review several important experimental techniques that allow us to determine structural, electronic, and optical properties of semiconductor structures. The intent of this review is not to present details on these experimental techniques, but to present the reader with an overview of what their capabilities are and what the difficulties are.

B.1 HIGH RESOLUTION X-RAY DIFFRACTION

Diffraction experiments are essential tools to determine the structural quality of crystalline materials. These techniques allow one to measure the lattice parameters of crystals and also provide information on structural imperfection. The basis for all diffraction experiments is the Bragg law which in its simplest form is

$$2d \sin \theta = n\lambda \qquad \text{(B.1)}$$

where d is the spacing between identical planes, λ is the wavelength and θ is the diffraction angle. With cleverly designed experiments and sophisticated data analysis this simple equation can form the basis of very detailed structural information. In Figure B.1 we show a schematic of some important pieces of structural information that are obtained from diffraction experiments. Of course, the diffraction experiments also reveal information on lattice constants for bulk semiconductors.

- **Lattice Mismatch:** Growth of epilayers on substrates that have a lattice constant different from the epilayers constant is becoming increasingly important. As discussed in this chapter if the film is relaxed the epilayer will produce a shift $\Delta \theta$ in the diffraction pattern with respect to the substrate.

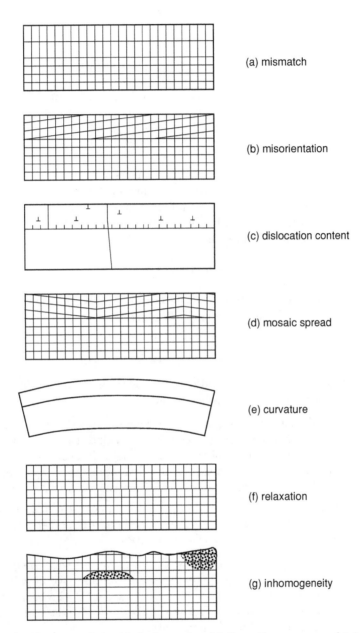

Figure B.1: A schematic of the kind of structural information one can obtain from X-ray diffraction studies.

If (a_f and a_s are the film and substrate lattice constants)

$$m = \frac{a_f - a_s}{a_s} = \frac{\Delta a}{a_s}$$

we have

$$\frac{\Delta \theta}{\theta} = \frac{\Delta a}{a_s} \tag{B.2}$$

If the film is coherently strained the inplane lattice constant of the film fits the substrate, while the out-of-plane constant is different. If m^* is this misfit we have

$$m^* = \frac{\Delta \theta}{\theta} = \frac{\Delta a_\parallel}{a_s} \tag{B.3}$$

- **Tilt:** In some film growth it is possible that the overlayer grows at a tilt with respect to the substrate. This produces a shift of diffraction peaks with respect to the peaks of the substrate. This shift $\Delta \theta$ has nothing to do with misfit effects. How does one distinguish a tilt splitting from a misfit splitting? The solution is fairly simple—one first rotates the sample by 180° in its plane. The average value

$$\Delta \theta = \left(\frac{\Delta \theta_0 + \Delta \theta_{180}}{2} \right) \tag{B.4}$$

provides the splitting due to lattice effects (misfit). To find the tilt angle a measurement is taken with a 90° rotation. We then have

$$\begin{aligned} \Delta \theta_o &= \beta \cos \omega \\ \Delta \theta_{90} &= \beta \cos(90 + \omega) = -\beta \sin \omega \end{aligned} \tag{B.5}$$

where β is the true tilt angle and ω is the splitting when the tilt angle is zero. From these measurements β can be obtained.

- **Dislocation Content:** It is common to have dislocations in strained epitaxy. These are commonly present in two regions—at the interface between the film and the substrate and in the bulk of the film. Interface dislocations give measurable shifts in the positions of the peaks in the diffraction measurements. It is then possible to estimate the dislocation density from the net strain at the interface.

  Dislocations in the bulk of the film lead to broadening of the diffraction peaks. This broadening is due to the finite extent of crystalline regions between dislocations and can be used to quantify dislocation densities.

- **Mosaic and Curvature:** A film may have a curvature which may extend over small regions (forming a mosaic) or over the entire film. As a result of the curvature the diffraction angle changes as one goes from one end of the beam to the other. This produces a shift $\Delta \theta$ which is related to the beam diameter. If D is the beam diameter and R is the radius of the curvature

$$\Delta \theta = \frac{D}{R} \tag{B.6}$$

Other structural features listed in Fig. B.1 can also be quantified by X-ray diffraction studies.

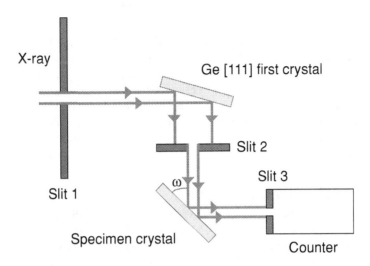

Figure B.2: A schematic of an X-ray double-crystal diffractometer. A high-quality crystal is used to select an extremely coherent beam.

B.1.1 Double Crystal Diffraction

Double crystal diffraction is an extremely powerful measurement technique which is often used to study very detailed aspects of crystalline materials. Due to its sensitivity, it is often used to study heterostructures where it can detect small strain values, non-epitaxial regions, and different thicknesses making up the heterostructure.

A typical setup for this technique is shown in Fig. B.2. An X-ray beam first impinges on a high quality crystal from which it diffracts. The diffracted beam then impinges on the sample. When the Bragg angles for the two crystals are identical, very narrow diffraction peaks are observed. To obtain the rocking curves, which provide information on the long-range structural order of the crystal, the specimen is rotated by a small angle ω, and the differential beam intensity is recorded. High quality crystals have peaks with widths of a few second of an arc. This technique is particularly useful for epitaxial layers grown on a thick substrate. In general, for such cases, two diffraction peaks are observed, one from the substrate and one form the epilayer. The difference $\Delta\omega$ in the crystal setting angle for Bragg reflection of the substrate and layer has two components, $\Delta\theta$ and $\Delta\phi$. The difference in the lattice plane spacing $\Delta d/d$ for corresponding lattice planes of layer and substrate causes a difference in Bragg angle $\Delta\theta$. The second component of $\Delta\omega$ is the difference $\Delta\phi$ in the inclination with the surface of corresponding lattice planes of layer and substrate.

B.2 DRIFT MOBILITY AND HALL MOBILITY

We have seen from our discussion of carrier transport that when an electric field is applied mobile carrier gas has a net drift velocity which is (at low fields) proportional to the applied field. The drift mobility is a very important parameter playing a key role in electronic devices. It also provides a window on various scattering processes in the material.

In addition to drift mobility we have discussed Hall mobility which is related to the drift mobility by the Hall factor. Hall mobility is the outcome of an experiment known as Hall effect. In addition to Hall mobility, Hall effect provides information on mobile carrier density and type of charge (electrons or holes). As a result Hall effect is the most common experiment carried out on semiconductors to determine transport related parameters. However, it is important to emphasize that Hall mobility is different from drift mobility. Depending upon which scattering mechanisms dominate, the difference can be quite large.

B.2.1   Haynes-Shockley Experiment

The Haynes-Shockley experiment is used to obtain the velocity-field relation for minority carriers in semiconductors. In the experiment a pulse of mobile charge s injected optically or electrically and moves under a uniform electric field. In Fig. B.3 we show a schematic of an experimental set-up for the Haynes-Shockley experiment. A constant electric field is maintained in the semiconductor samples by applying a bias. A pulse generator (as shown) or an optical signal injects a pulse of (minority) charge. The pulse drifts under the influence of the electric field to the other and of the sample where it is detected by an oscilloscope connected to the collector. The time delay of the pulse, t_d, and the width of the pulse is measured. The drift velocity is simply given by

$$v_d = \frac{L}{t_d} \tag{B.7}$$

where L is the distance between the point of injections and the detection.

As the injected pulse propagates, it spreads into a Gaussian pulse. The pulse width, as shown in Fig. B.3, provides information on the diffusion coefficient mobile charge. The diffusion coefficient is related to Δt by (after a detailed analysis)

$$D = \frac{(\Delta t L)^2}{11 t_d^3} \tag{B.8}$$

where Δt is the width of the pulse measured at points where the signal has fallen to $1/e$ times the peak value. The expression for the diffusion coefficients is a good approximation if there is negligible recombination of the minority charge during the pulse propagation.

A more sophisticated approach that has been used to obtain velocity-field curves in semiconductors is sketched in Fig. B.4a. The experimental set-up consists of an electron gun which is capable of sending out an electron beam with energy of ~ 10 kV. The beam impinges on a semiconductor sample and generates a thin sheet of electron-hole pairs. The incident electron beam is incident on the sample for 10^{-8} s and the frequency of the pulse is $\sim 1 \mu s$. This conditions are established by appropriate sinusoidal signals on the deflection plates as shown.

A typical sample could be a reverse biased Schottky barrier diode as shown in Fig. B.4. There are several important issues to consider in sample preparation:

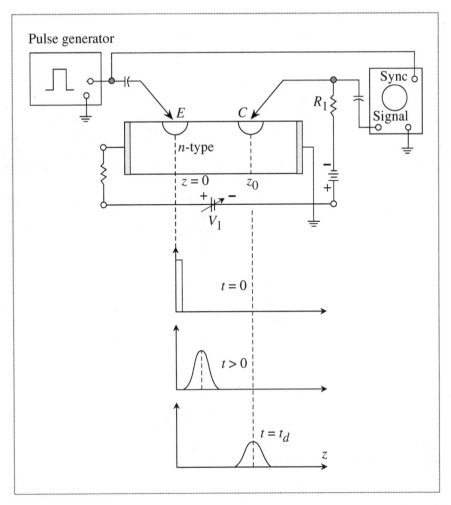

Figure B.3: A schematic of the Haynes-Shockley measurement set-up. The pulse width Δt and the propagation time t_d provide information on diffusion constant and mobility.

- The metal Schottky barrier may need to be deposited on the high resistivity semiconductor with a thin (~ 200 Å) insulators such as SiO_x. The insulator will prevent any injection of carriers from the metal into the semiconductor. For the same reason the Schottky diode is reverse biased.

- The sample is chosen to be intrinsic or highly resistive to ensure that the electric field in the material is as uniform as possible. This allows the electron (or hole) pulse to travel in a uniform field which can be altered by changing the bias on the Schottky barrier.

We have seen from our discussion of electron transport in direct gap semiconductors that there is a region of velocity-field curve where the differential resistance is negative.

When the sample is biased at field where the differential resistance is negative a charge fluctuation can grow in time and oscillations ca develop in the circuit. Having a high resistivity material helps avoid these oscillations and the resulting non-uniform field distributions.

The high energy electron beam generates an electron beam that propagates through the sample. the traveling pulse of electrons induces a current in the oscilloscope. In Fig. B.4b we show a typical pulse form that is detected. The pulse appears at time 0 and lasts for a time $t_{tr} = L/v$ where L is the thickness of the semiconductor and v is the drift velocity. The pulse that is detected is not an abrupt square pulse, but has a finite rise and fall time related to the capacitance and impedance of the Schottky barrier structure.

As noted earlier, it is essential that the electric field in the sample remain constant. A high resistivity sample helps achieve this goal. The use of short pulses also helps in maintaining a uniform field. The pulses also allow one to correlate the output signal to the input signal.

B.2.2   Hall Effect for Carrier Density and Hall Mobility

As we have seen above, it is possible to obtain drift mobility in semiconductor samples by injecting a pulse of charge and observing its progress. One may wonder why one doesn't just measure the current flowing in a semiconductor and use the relation

$$J_x = ne\mu F_x \qquad (B.9)$$

If the field F_x can be measured, one can measure the mobility directly. A key difficulty (in addition to obtaining an accurate value of the electric field) is the difficulty in finding the carrier density. The Hall effect allows one to find the carrier density as well as the Hall mobility in a single experimental set-up. Of course, one has to remember that Hall mobility and drift mobility are different and are related by the Hall factor

$$\mu_H = r_H \mu \qquad (B.10)$$

The Hall effect has been discussed in some detail in Chapter 11, so we will refer the reader to that discussion.

B.3   PHOTOLUMINESCENCE (PL) AND EXCITATION PHOTO-LUMINESCENCE (PLE)

Photoluminescence is one of the most widely used and useful optical techniques to characterize semiconductors and their heterostructures. In the PL experiment a light source with photon energies larger than the bandgap is used to create electron-hole pairs. As shown in Fig. B.5a the electrons and holes thermalize by emitting phonons and reach lower lying energies. Electrons and holes recombine, producing photons which are then detected. Electrons and holes can recombine through a number of processes: i) In band-to-band recombination electrons and holes recombine directly from the conduction and valence bands; ii) Recombination after formation of "free" excitons; and iii) Excitons

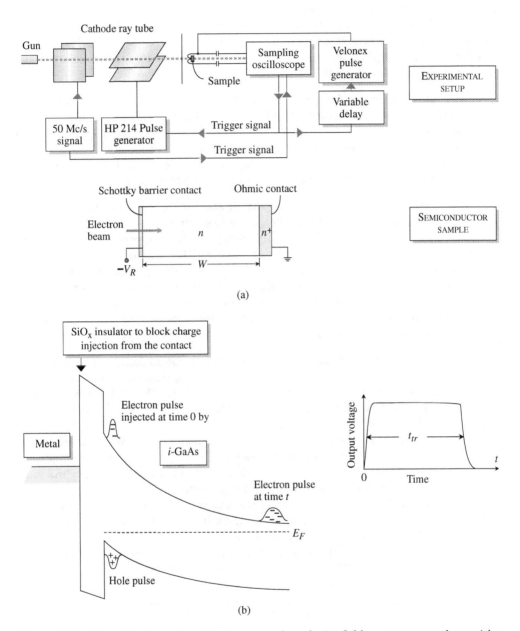

Figure B.4: (a) A detailed schematic of a set-up for velocity-field measurement along with a detail of the semiconductor sample. (b) Band profile showing electron and hole injection. Also shown is a typical output.

can be trapped by impurities such as donors and acceptors creating transitions from donor-bound-excitons or acceptor-bound-excitons.

The emission spectrum is observed via a suitable detector. Various features of the emission spectrum provide important information on structural and other physical parameters of the sample being studied. PL experiments are usually done while varying one or more of these conditions:

- Input identity: At low intensity the carriers thermalize to the lower lying states and there is essentially no emission from the higher lying states. Impurity bound excitons are quite prominent at low intensities. As the input intensity is increased defect related transitions saturate (since there is a fixed number of defect sites) and free excitons and higher level transitions (e.g., light hole excitons on higher level subband transitions in quantum wells) can be observed.

- Temperature dependence: An important variable for PL studies is the sample temperature. At low temperature phonon assisted scattering processes are suppressed allowing the transitions to be relatively sharp. At low temperatures the width of various excitonic transitions are determined by inhomogeneous broadening, i.e., local structural fluctuations. The band to band transitions emission is essentially proportional to the electron occupation and the tail thus provides information on electron temperature (usually hole masses are larger and electron properties determine the lineshape). As the sample temperature is raised transitions get broader due to the addition of phonon related homogeneous broadening.

Figs. B.5b and B.5c show a spectrum in bulk GaAs and in a sample with multiquantum wells. As can be seen from Fig. B.5c each quantum well produces its own transition. The position of the transition correlates with the well width. As noted in Fig. B.5a the width and structure in the exciton peaks provide information on the structural quality of the heterointerfaces.

In Fig. B.6 we show a schematic of a typical PL apparatus. While in principle a white light source can be used to excite a PL spectrum, in practice a laser with photon energy above the effective bandgap of the sample is used. A high quality monochromator is needed to resolve the various structure in the emission.

Excitation photoluminescence or PLE is a variation of PL which provides essentially the same information as absorption spectrum In this approach the input photon energy is varied over a range (making sure the source provides the same photon particle current at all energies). The entire light output (i.e., photons of all energies) is now collected. The plot of incident energy versus total output signal provides a curve that is essentially the absorption spectrum of the sample. The reason is that the number of electron-hole pairs generation rate at a photon energy $\hbar\omega$ is simply given by

$$G = \alpha(\hbar\omega)\frac{P_{op}(\hbar\omega)}{\hbar\omega} \tag{B.11}$$

The e-h pairs thermalize and generate the optical signal which is proportional to $\alpha(\hbar\omega)$.

The brief discussion presented here on PL and PLE shows that the PL signal is dominated by lower energy transitions while the PLE is proportional to the absorption

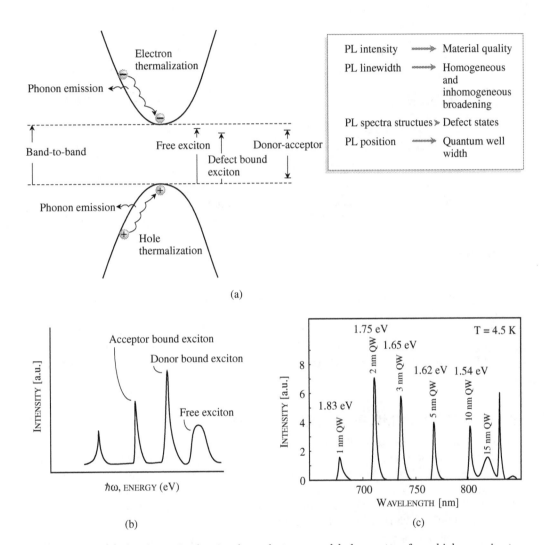

Figure B.5: (a) A schematic showing how electrons and holes scatter from high energies to thermalize. Various energies for excitonic and other features are shown. (b) A typical photoluminescence spectrum. (c) Photoluminescence from a multiquantum well sample.

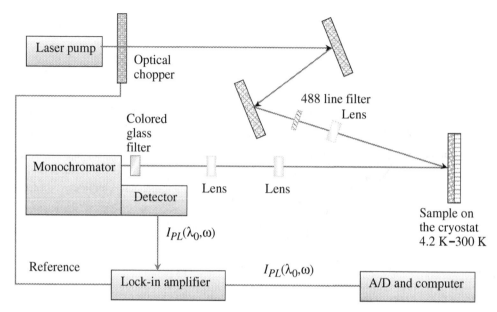

Figure B.6: Experimental setup for a typical photoluminescence measurement.

spectrum. In particular in samples with defects or disorder the defect related transitions dominate the PL emission spectrum even though due to the weal oscillators strength these transitions will not register on the PLE or the absorption spectrum. As shown in Fig. B.7 there is a shift between the PL peak and the first (free exciton) peak of the PLE signal. This shift is called the Stoke's shift and provides useful information on sample quality.

B.4   OPTICAL PUMP PROBE EXPERIMENTS

Pump probe techniques provide very useful information on carrier dynamics over a wide range of electronic energies. As the name implies a pump beam is used to disturb (or pump) a system. With a controlled delay time the system is then probed. In general, the pump beam wavelength, λ_{pump}, and the probe beam wavelength, λ_{probe}, can be different. In many experiments the pump beam is monochromatic while the probe beam is a white light source (i.e., it probes all wavelengths).

The key to a pump probe experiment is a short pulse laser source. Important parameters of the source are:

- **Pulse duration:** This determines the time resolution of the experiment. Since time duration of several scattering processes can be as small as 0.1 ps, it is necessary to pulse with durations of 100 fs or less. In Fig. B.8a and B.8b we show the autocorrelation traces of a high quality pulse in both time and wavelength domain. A narrow time domain pulse will result in a broader wavelength domain pulse.

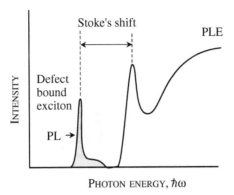

Figure B.7: A schematic of photoluminescence and excitation photoluminescence results. The Stoke shift provides input into the quality of the sample.

- **Pulse energy spectrum:** The pulse energy determines which electronic levels are examined in the experiment. In Fig. B.8b we show a typical spectrum for pump beam. Ideally one should have as narrow a linewidth (ΔE) as possible in the pump beam. However, one has the uncertainty relation $\Delta E \Delta t \sim \hbar$ which implies that if the value of ΔE is too narrow, Δt (pulse duration) will be large.

- **Tunability:** A tunable source would allow one to alter the initial conditions of the experiment and perturb the electronic system at different points in energy.

- **Repetition rate:** The repetition rate determines how rapidly the data can be collected. The repetition rate should not be so fast that the system is still disturbed before the next pulse comes in. Also if the repetition rate is too slow the total signal detected over time will be small and the sensitivity of the measurement will suffer.

In Fig. B.9 we show a typical experimental setup for a pump-probe measurement. The pump beam generates electrons and holes thus populating certain electronic states. As a result of this population the absorption coefficient (probed by the probe beam) is suppressed. We have

$$\alpha(\hbar\omega, t) = \alpha_0(\hbar\omega) \left[1 - f_e \left(E^e, t \right) - f_h \left(E^h, t \right) \right] \tag{B.12}$$

The probe beam allows one to measure the differential transmission with and without the pump pulse. The differential transmission is

$$\frac{\Delta T}{T_0} = \frac{T - T_0}{T_0} \tag{B.13}$$

where T_0 is the transmission in the absence of the pump pulse, while T is the transmission with the pump pulse. We have

$$
\begin{aligned}
T_0(\hbar\omega) &= \exp\left\{ -\alpha_0(\hbar\omega)L \right\} \\
T(\hbar\omega) &= \exp\left\{ -\alpha(\hbar\omega, t)L \right\}
\end{aligned}
\tag{B.14}
$$

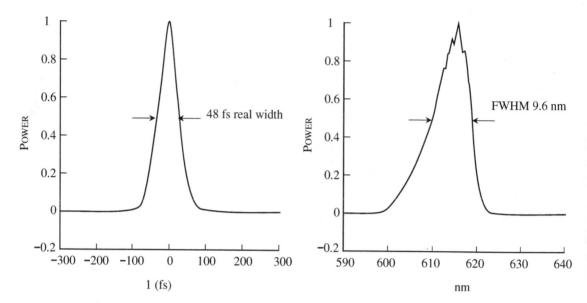

Figure B.8: (a) Autocorrelation trace from a dye laser. The real pulse is 48 fs assuming sech^2 pulse shape. (b) Spectrum of the pulse in wavelength domain. The product $\tau\Delta\nu = 0.37$, slightly larger than the transform limit.

This gives

$$\frac{\Delta T}{T} = \exp\left\{\alpha_0 L(f_e - f_h)\right\} - 1 \tag{B.15}$$

In the experiments an equal number of electrons and holes are produced. However, since hole density of states are much larger than electron density of states we can make the assumption that $f_h \ll f_e$. Moreover, the pump signal is usually weak and $\alpha_0 L f_e \ll 1$. With these assumptions

$$\frac{\Delta T(\hbar\omega, t)}{T} \sim \alpha_0 L f_e(E^e, t) \tag{B.16}$$

Thus the probe allows us to see the evolution of f_e as a function of time.

In Fig. B. 10 we show typical experimental results for experiments done on GaAs. In this experiment electrons are generated at energies well above the bandedge by the short pump pulse as shown. These "hot" carriers scatter and lose their initial energy through phonon emission processes. As the carriers trickle down, the differential transmission signal increases near bandedge energies. The various scattering processes shown in Fig. B.10 can be probed by altering the pump beam energy. Thus electrons can be placed in the Γ-valley or the higher L-valley (or X-valley) and the scattering times can be studied.

In Fig. B.11 we show results for a quantum well system (a 100 Å GaAs/ Al$_{0.3}$Ga$_{0.7}$As structure). In Fig. B.11a we show the absorption spectra in the absence of the pump beam. The results show the heavy hole and light hole peaks as well as the band to band absorption. In Fig. B.11b we show the pump-probe results. Carriers are

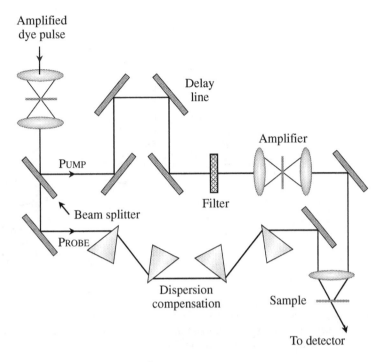

Figure B.9: Schematic of a pump-probe set-up.

injected 60 meV above the bandedge and their occupation is studied as a function of time.

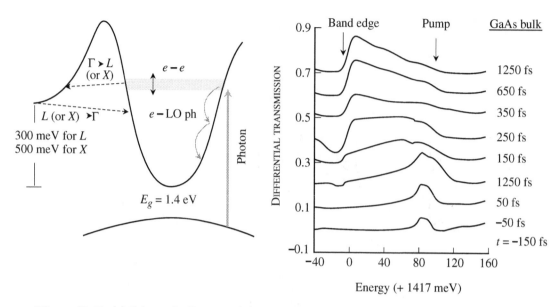

Figure B.10: (a) Schematic diagram of the various scattering processes in GaAs. The valence band is simplified. (b) Differential transmission spectra for carrier relaxation studies in bulk GaAs. The pump wavelength is indicated as the arrow.

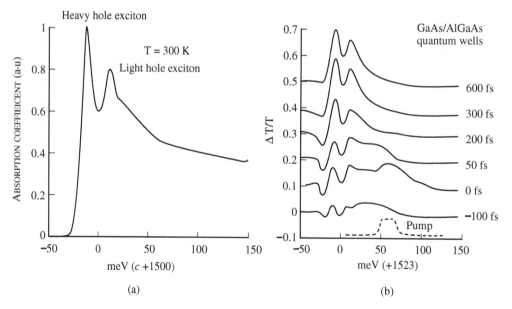

Figure B.11: (a) Absorption spectrum of the GaAs/AlGaAs quantum well sample. (b) Low temperature differential transmission spectra with the quantum well sample. The pump spectrum is shown at the bottom of the figure. Hot carriers with 60 meV excess energy are initially generated.

Appendix

C

QUANTUM MECHANICS: USEFUL CONCEPTS

In this text we have used a number of very important quantum mechanics concepts to understand the electronic and optoelectronic properties of semiconductor structures. While this appendix cannot substitute for a text on quantum mechanics, the reader may find it useful to review concepts he/she has learned. We will review the following problems which are quite useful in semiconductor physics: i) Density of states in bulk and lower dimensional systems; ii) time independent perturbation theory; iii) time dependent perturbation theory and Fermi golden rule; and iv) numerical solution for electronic bound states in an arbitrary shaped quantum well.

C.1 DENSITY OF STATES

We have seen that essentially all properties of semiconductors are related to the density of states. Using the effective mass picture the Schrödinger equation for electrons can be written as a "free" electron problem with a background potential V_0,

$$\frac{-\hbar^2}{2m^*} \left(\frac{\partial^2}{\partial x^2} + \frac{\partial^2}{\partial y^2} + \frac{\partial^2}{\partial z^2} \right) \psi(r) = (E - V_0)\psi(r) \tag{C.1}$$

A general solution of this equation is

$$\psi(r) = \frac{1}{\sqrt{V}} e^{\pm ik \cdot r} \tag{C.2}$$

and the corresponding energy is

$$E = \frac{\hbar^2 k^2}{2m} + V_0 \tag{C.3}$$

where the factor $\frac{1}{\sqrt{V}}$ in the wavefunction occurs because we wish to have one particle per volume V or

$$\int_V d^3r \mid \psi(r) \mid^2 = 1 \tag{C.4}$$

We assume that the volume V is a cube of side L.

To obtain macroscopic properties independent of the chosen volume V, two kinds of boundary conditions are imposed on the wavefunction. In the first one the wavefunction is considered to go to zero at the boundaries of the volume, as shown in Fig. C.1a. In this case, the wave solutions are standing waves of the form $\sin(k_x x)$ or $\cos(k_x x)$, etc., and k-values are restricted to positive values:

$$k_x = \frac{\pi}{L}, \frac{2\pi}{L}, \frac{3\pi}{L} \cdots \tag{C.5}$$

Periodic boundary conditions are shown in Fig. C.2b. Even though we focus our attention on a finite volume V, the wave can be considered to spread in all space as we conceive the entire space was made up of identical cubes of sides L. Then

$$
\begin{aligned}
\psi(x, y, z + L) &= \psi(x, y, z) \\
\psi(x, y + L, z) &= \psi(x, y, z) \\
\psi(x + L, y, z) &= \psi(x, y, z)
\end{aligned}
\tag{C.6}
$$

Because of the boundary conditions the allowed values of k are (n are integers— positive and negative)

$$k_x = \frac{2\pi n_x}{L}; \quad k_y = \frac{2\pi n_y}{L}; \quad k_z = \frac{2\pi n_z}{L} \tag{C.7}$$

If L is large, the spacing between the allowed k-values is very small. Also it is important to note that the results one obtains for properties of the particles in a *large volume are independent of whether we use the stationary or periodic boundary conditions*. It is useful to discuss the *volume in k-space that each electronic state occupies*. As can be seen from Fig. C.2, this volume is (in three dimensions)

$$\left(\frac{2\pi}{L}\right)^3 = \frac{8\pi^3}{V} \tag{C.8}$$

If Ω is a volume of k-space, the number of electronic states in this volume is

$$\frac{\Omega V}{8\pi^3} \tag{C.9}$$

The reader can verify that stationary and periodic boundary conditions lead to the same density of states value as long as the volume is large.

Density of States for a Three-Dimensional System

The concept of density of states is extremely powerful, and important physical properties in materials such as optical absorption, transport, etc., are intimately dependent upon

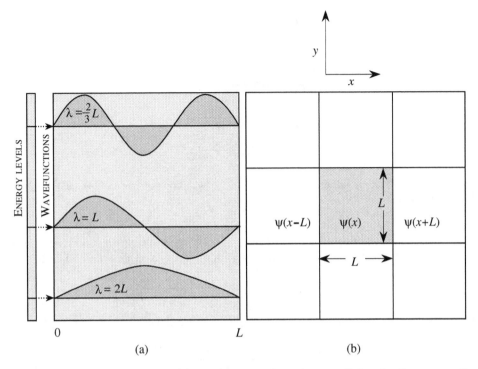

Figure C.1: A schematic showing (a) the stationary boundary conditions leading to standing waves and (b) the periodic boundary conditions leading to exponential solutions with the electron probability equal in all regions of space.

this concept. Density of states is the number of available electronic states *per unit volume per unit energy* around an energy E. If we denote the density of states by $N(E)$, the number of states in a unit volume in an energy interval dE around an energy E is $N(E)dE$. To calculate the density of states, we need to know the dimensionality of the system and the energy versus k relation that the particles obey. We will choose the particle of interest to be the electron, since in most applied problems we are dealing with electrons. Of course, the results derived can be applied to other particles as well. For the free electron case we have the parabolic relation

$$E = \frac{\hbar^2 k^2}{2m^*} + V_0$$

The energies E and $E + dE$ are represented by surfaces of spheres with radii k and $k + dk$, as shown in Fig. C.3. In a three-dimensional system, the k-space volume between vector k and $k + dk$ is (see Fig. C.3a) $4\pi k^2 dk$. We have shown in Eqn. C.9 that the k-space volume per electron state is $\left(\frac{2\pi}{L}\right)^3$. Therefore, the number of electron states in the region between k and $k + dk$ is

$$\frac{4\pi k^2 dk}{8\pi^3}V = \frac{k^2 dk}{2\pi^2}V$$

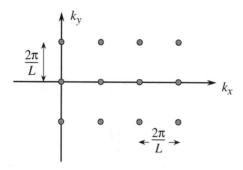

Figure C.2: k-Space volume of each electronic state. The separation between the various allowed components of the k-vector is $\frac{2\pi}{L}$.

Denoting the energy and energy interval corresponding to k and dk as E and dE, we see that the number of electron states between E and $E + dE$ per unit volume is

$$N(E)\,dE = \frac{k^2 dk}{2\pi^2}$$

Using the E versus k relation for the free electron, we have

$$k^2 dk = \frac{\sqrt{2}m^{*3/2}(E - V_0)^{1/2}dE}{\hbar^3}$$

and

$$N(E)\,dE = \frac{m^{*3/2}(E - V_0)^{1/2}dE}{\sqrt{2}\pi^2\hbar^3} \tag{C.10}$$

The electron can have a spin state $\hbar/2$ or $-\hbar/2$. Accounting for spin, the density of states obtained is simply multiplied by 2

$$N(E) = \frac{\sqrt{2}m^{*3/2}(E - V_0)^{1/2}}{\pi^2\hbar^3} \tag{C.11}$$

Density of States in Sub-Three-Dimensional Systems

In quantum wells electrons are free to move in a 2-dimensional space. The two-dimensional density of states is defined as the number of available electronic states *per unit area per unit energy* around an energy E. Similar arguments as used in the derivation show that the density of states for a parabolic band (for energies greater than V_0) is (see Fig C.3b)

$$N(E) = \frac{m^*}{\pi\hbar^2} \tag{C.12}$$

The factor of 2 resulting from spin has been included in this expression.

Finally, we can consider a one-dimensional system often called a "quantum wire." The one-dimensional density of states is defined as the number of available electronic states *per unit length per unit energy* around an energy E. In a 1D system or a

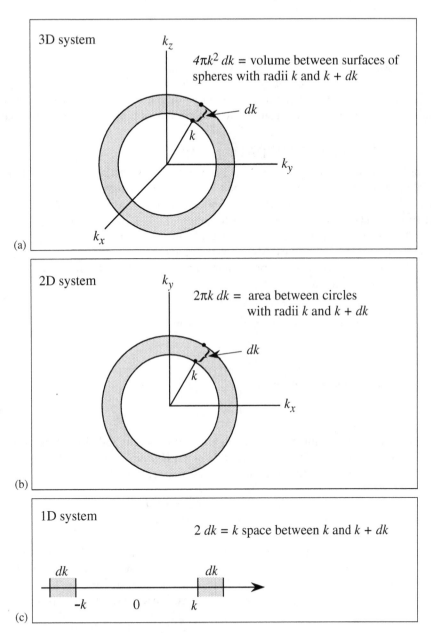

Figure C.3: Geometry used to calculate density of states in three, two, and one dimensions. By finding the k-space volume in an energy interval between E and $E + dE$, one can find the number of allowed states.

"quantum wire" the density of states is (including spin) (see Fig. 3.3c)

$$N(E) = \frac{\sqrt{2}m^{*1/2}}{\pi\hbar}(E - V_0)^{-1/2} \tag{C.13}$$

Notice that as the dimensionality of the system changes, the energy dependence of the density of states also changes. As shown in Fig. C.4, for a three-dimensional system we have $(E - V_0)^{1/2}$ dependence, for a two-dimensional system we have no energy dependence, and for a one-dimensional system we have $(E - V_0)^{-1/2}$ dependence.

C.2   STATIONARY PERTURBATION THEORY

Few problems in quantum mechanics can be solved exactly and it is extremely important to develop an understanding of important practical problems which cannot be solved exactly. The perturbation (or approximation) methods are, therefore, of critical importance. In this section we will summarize some key results from stationary perturbation theory. Details of the derivation can be found in most quantum mechanics books (e.g., L. I. Schiff, *Quantum Mechanics,* McGraw-Hill, New York, 1968).

A general overview of the perturbation approach is shown in Fig. C.5. Consider a Hamiltonian of the form

$$H = H_0 + H' \tag{C.14}$$

where the solutions of H_0 are known and are given by

$$H_0 u_k = E_k u_k \tag{C.15}$$

We seek the solution of the problem

$$H\psi = E\psi \tag{C.16}$$

We are interested in solving for ψ and the eigenvalues of the full Hamiltonian, in terms of the eigenfunctions and eigenvalues of the known Hamiltonian H_0. The eigenfunction ψ and the eigenvalue W are written in terms of a parameter λ

$$
\begin{aligned}
\psi &= \psi_0 + \lambda\psi_1 + \lambda^2\psi_2 + \lambda^3\psi_3 + \cdots \\
W &= W_0 + \lambda W_1 + \lambda^2 W_2 + \lambda^3 W_3 + \cdots
\end{aligned} \tag{C.17}
$$

where we replace H' by $\lambda H'$ in Eqn. C.14 and let λ go to zero. Then $\psi_0, \psi_1, \ldots$ and $W_0, W_1, \ldots$ represent the various orders of corrections, due to the perturbation H'. In general, the zero order state ψ_0 could be either nondegenerate or degenerate. The perturbation approach is different for the two cases. We shall discuss the results for both cases.

C.2.1   Nondegenerate Case

The first order corrections to a state $\psi_0 = u_m$ (or $|m\rangle$ in the Dirac notation) are

$$W_1 = \langle m|H'|m\rangle \tag{C.18}$$

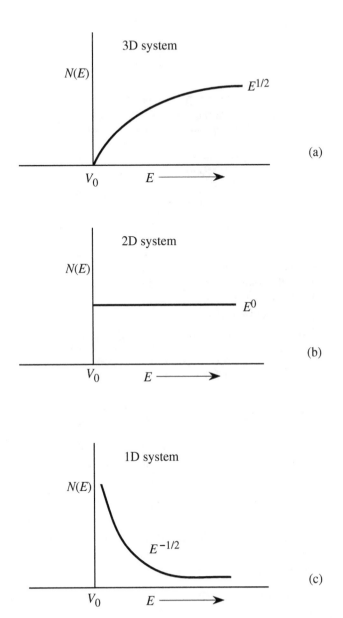

Figure C.4: Variation in the energy dependence of the density of states in (a) three-dimensional, (b) two-dimensional, and (c) one-dimensional systems. The energy dependence of the density of states is determined by the dimensionality of the system.

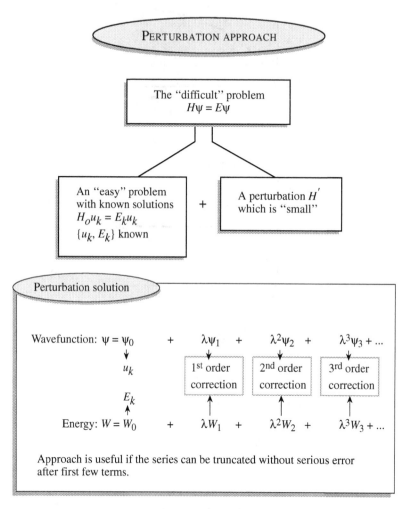

Figure C.5: A schematic of the genral approach used in perturbation theory.

The correction to the wavefunction is

$$\psi_1 = \sum_n a_n^{(1)} |n\rangle \tag{C.19}$$

where

$$eqnE.8a_k^{(1)} = \frac{\langle k|H'|m\rangle}{E_m - E_k} \text{ for } k \neq m \tag{C.20}$$

In second order we have

$$W_2 = {\sum_n}' \frac{|\langle n|H'|n\rangle|^2}{E_m - E_n} \tag{C.21}$$

The prime denotes $m \neq n$. (H' is assumed to be hermetian.) The second order correction to the wavefunction is given by

$$\psi_2 = {\sum_n}' a_n^{(2)} u_n,$$
$$a_m^{(2)} = 0 \tag{C.22}$$

where

$$a_k^{(2)} = {\sum_n}' \frac{\langle k|H'|n\rangle\langle n|H'|m\rangle}{(E_m - E_k)(E_m - E_n)} - \frac{\langle k|H'|m\rangle\langle m|H'|m\rangle}{(E_m - E_k)^2} \tag{C.23}$$

To second order we then have for the unperturbed level $|m\rangle$

$$
\begin{aligned}
W &= E_m + \langle m|H'|m\rangle + {\sum_n}' \frac{|\langle m|H'|n\rangle|^2}{E_m - E_n} \\
\psi &= u_m + {\sum_k}' u_k \left[\frac{\langle k|H'|m\rangle}{E_m - E_k} \left(1 - \frac{\langle m|H'|m\rangle}{E_m - E_k}\right) \right. \\
&\quad + \left. {\sum_n}' \frac{\langle k|H'|n\rangle\langle n|H'|m\rangle}{(E_m - E_k)(E_m - E_n)} \right]
\end{aligned}
\tag{C.24}
$$

The wavefunction is *not* normalized now and one must renormalize it.

C.2.2 Degenerate Case

In the above derivation we have assumed that the initial state $\psi_0 = u_m$ is nondegenerate. Let us now assume that u_m and u_ℓ are degenerate and have the same unperturbed energy. Then the coefficient

$$a_k^{(1)} = \frac{\langle k|H'|m\rangle}{E_m - E_k} \tag{C.25}$$

causes difficulty unless $\langle \ell|H'|m\rangle = 0$.

Let us consider the case where $\langle \ell|H'|m\rangle \neq 0$, so that our previous results are not valid. We assume that H' breaks the degeneracy at some order of perturbation. The two nondegenerate states then approach a linear combination of u_m and u_ℓ as the value of $\lambda \to 0$. Let us write

$$
\begin{aligned}
\psi_0 &= a_m u_m + a_\ell u_\ell \\
W_0 &= \boldsymbol{E_m} \\
&= \boldsymbol{E_\ell}
\end{aligned}
\tag{C.26}
$$

The first order correction is given by

$$
\begin{aligned}
W_1 &= \frac{1}{2}\left(\langle m|H'|m\rangle + \langle \ell|H'|\ell\rangle\right) \\
&\pm \frac{1}{2}\left[\left(\langle m|H'|m\rangle - \langle \ell|H'|\ell\rangle\right)^2 + 4|\langle m|H'|\ell\rangle|^2\right]^{1/2}
\end{aligned}
\tag{C.27}
$$

The two values of W_1 are equal if

$$
\langle m|H'|m\rangle = \langle \ell|H'|\ell\rangle
$$

and

$$
\langle m|H'|\ell\rangle = 0
\tag{C.28}
$$

If these are not satisfied, the degeneracy is lifted and the values of a_m and a_ℓ can be obtained.

If the values of W_1 obtained from Eqn. C.27 are the same, one must go to the second order to see if degeneracy is lifted. The second order correction is given by the eigenvalue equations

$$
\begin{aligned}
\left(\sideset{}{'}\sum_n \frac{|\langle m|H'|n\rangle|^2}{\boldsymbol{E_m} - \boldsymbol{E_n}} - W_2\right) a_m + \sideset{}{'}\sum_n \frac{\langle m|H'|n\rangle\langle n|H'|\ell\rangle}{\boldsymbol{E_m} - \boldsymbol{E_n}} a_\ell &= 0 \\
\sideset{}{'}\sum_n \frac{\langle \ell|H'|n\rangle\langle n|H'|m\rangle}{\boldsymbol{E_m} - \boldsymbol{E_n}} a_m + \left(\sideset{}{'}\sum_n \frac{|\langle \ell|H'|n\rangle|^2}{\boldsymbol{E_m} - \boldsymbol{E_n}} - W_2\right) a_\ell &= 0
\end{aligned}
\tag{C.29}
$$

The secular equation allows us to then determine W_2. Unless

$$
\sideset{}{'}\sum_n \frac{|\langle m|H'|n\rangle|^2}{\boldsymbol{E_m} - \boldsymbol{E_n}} = \sideset{}{'}\sum_n \frac{|\langle \ell|H'|n\rangle|^2}{\boldsymbol{E_m} - \boldsymbol{E_n}}
\tag{C.30}
$$

and

$$
\sideset{}{'}\sum_n \frac{\langle m|H'|n\rangle\langle n|H'|\ell\rangle}{\boldsymbol{E_m} - \boldsymbol{E_n}} = 0
\tag{C.31}
$$

the degeneracy is lifted in second order.

C.3 TIME DEPENDENT PERTURBATION THEORY AND FERMI GOLDEN RULE

We have extensively used the time dependent perturbation theory to address the problem of scattering of electrons. The general Hamiltonian of interest is once again

$$H = H_0 + H'$$

where

$$H_0 u_k = \boldsymbol{E}_k u_k$$

and $\boldsymbol{E}_k$, u_k are known. The effect of H' is to cause transitions between the states u_k. The time dependent Schrödinger equation is

$$i\hbar \frac{\partial \psi}{\partial t} = H\psi \tag{C.32}$$

The approximation will involve expressing ψ as an expansion of the eigenfunctions $u_n \exp(-i\boldsymbol{E}_n t/\hbar)$ of the unperturbed time dependent functions

$$\psi = \sum_n a_n(t) u_n e^{-i E_n t/\hbar} \tag{C.33}$$

We write

$$\omega_{kn} = \frac{\boldsymbol{E}_k - \boldsymbol{E}_n}{\hbar} \tag{C.34}$$

We assume that initially the system is in a single, well-defined state given by

$$
\begin{aligned}
a_k^{(0)} &= \langle k | m \rangle \\
&= \delta_{km}
\end{aligned}
\tag{C.35}
$$

To the first order we have at time t

$$a_k^{(1)}(t) = \frac{1}{i\hbar} \int_{-\infty}^{t} \langle k | H'(t') | m \rangle \, e^{i\omega_{km}t'} \, dt' \tag{C.36}$$

We choose the constant of integration to be zero since $a_k^{(1)}$ is zero at time $t \to -\infty$, when the perturbation is not present.

Consider the case where the perturbation is harmonic, except that it is turned on at $t = 0$ and turned off at $t = t_0$. Let us assume that the time dependence is given by

$$\langle k | H(t') | m \rangle = 2\langle k | H'(0) | m \rangle \sin \omega t' \tag{C.37}$$

Carrying out the integration until time $t \geq t_0$ in Eqn. C.36, we get

$$a_k^{(1)}(t \geq t_0) = -\frac{\langle k | H'(0) | m \rangle}{i\hbar} \left(\frac{\exp[(\omega_{km} + \omega)t_0] - 1}{\omega_{km} + \omega} - \frac{\exp[(\omega_{km} - \omega)t_0] - 1}{\omega_{km} - \omega} \right) \tag{C.38}$$

The structure of this equation says that the amplitude is appreciable, only if the denominator of one or the other term is practically zero. The first term is important

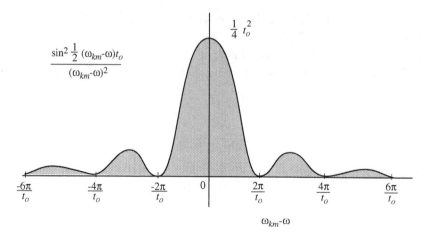

Figure C.6: The ordinate is proportional to the probability of finding the system in a state k after the perturbation has been applied to time t_0.

if $\omega_{km} \approx -\omega$, or $E_k \approx E_m - \hbar\omega$. The second term is important if $\omega_{km} \approx \omega$ or $E_k \approx E_m + \hbar\omega$.

Thus, the first order effect of a harmonic perturbation is to transfer, or to receive from the system, the quanta of energy $\hbar\omega$. If we focus on a system where $|m\rangle$ is a discrete state, $|k\rangle$ is one of the continuous states, and $E_k > E_m$, so that only the second term is important, the first order probability to find the system in the state k after the perturbation is removed is

$$\left| a_k^{(1)}(t \geq t_0) \right|^2 = 4 |\langle k|H'(0)|m\rangle|^2 \frac{\sin^2\left[\frac{1}{2}(\omega_{km} - \omega)t_0\right]}{\hbar^2(\omega_{km} - \omega)^2} \tag{C.39}$$

The probability function has an oscillating behavior as shown in Fig.C.6. The probability is maximum when $\omega_{km} = \omega$, and the peak is proportional to t_0^2. However, the uncertainty in frequency $\Delta\omega = \omega_{km} - \omega$, is non-zero for the finite time t_0. This uncertainty is in accordance with the Heisenberg uncertainty principle

$$\Delta\omega \, \Delta t = \Delta\omega \, t_0 \sim 1 \tag{C.40}$$

If there is a spread in the allowed values of $(\omega_{km} - \omega)$, which may occur either because the initial and/or final states of the electron are continuous, or because the perturbation has a spread of frequencies ω, it is possible to define a scattering rate per unit time. The total rate per unit time for scattering into any final state, is given by

$$W_m = \frac{1}{t_0} \sum_{\text{final states}} \left| a_k^{(1)}(t \geq t_0) \right|^2 \tag{C.41}$$

If t_0 is large, the sum over the final states only includes the final states where $\omega_{km} - \hbar\omega = 0$, i.e., energy is conserved in the process.

If we assume that $|\langle k|H'|m\rangle|^2$ does not vary over the (infinitesimally) small spread in final states, we can write

$$x = \frac{1}{2}(\omega_{km} - \omega)t_0 \tag{C.42}$$

and use the integral

$$\int_{-\infty}^{\infty} x^{-2}\sin^2 x\, dx = \pi$$

to get

$$W_m = \frac{2\pi}{\hbar} \sum_{\text{final states}} \delta\left(\hbar\omega_{km} - \hbar\omega\right) |\langle k|H'|m\rangle|^2 \tag{C.43}$$

This is the Fermi golden rule, which is used widely in our text. An identical expression occurs for scattering rate from fixed defects when Born approximation is used for scattering.

In a given scattering problem, one has to pay particular attention to the conditions under which the golden rule has been derived. The more important condition is that the time of interaction of the pertubation be essentially infinite. If the interaction time is finite, the δ-function of the Golden Rule changes over to the broadened function of Fig.C.6. Thus, the energy conservation is not strictly satisfied and the final state of the electron is not well-defined.

C.4 BOUND STATE PROBLEM: MATRIX TECHNIQUES

Matrix solving techniques are widely used in quantum mechanics. Such Eigenvalue solvers are generally available at most computer centers. These libraries provide subroutines which can be called to solve the matrix for both eigenvalues and eigenfunctions.

The equation to be solved is

$$H\Psi = E\Psi \tag{C.44}$$

where H is an $n \times n$ matrix, E is an eigenvalue which can have n values, and Ψ and is an n-dimensional vector. In general, the eigenfunction Ψ can be expanded in terms of an orthonormal basis set $\{\psi_n\}$

$$\Psi = \sum_n a_n \psi_n \tag{C.45}$$

A wavefunction Ψ_i is known when all the a_n's are known for that function. Substituting Eqn. C.45 in Eqn. C.44, multiplying successively by $\psi_1^*, \psi_2^*, \cdots, \psi_n^*$ and integrating, we get a set of equations

$$\begin{bmatrix} H_{11} - E & H_{12} & H_{13} & \cdots & H_{1n} \\ H_{21} & (H_{22} - E) & H_{23} & \cdots & H_{2n} \\ & & \vdots & & \\ & & \vdots & & \\ H_{n1} & N_{n2} & H_{n3} & \cdots & (H_{nn} - E) \end{bmatrix} \begin{bmatrix} a_1\psi_1 \\ a_2\psi_2 \\ \vdots \\ \vdots \\ a_n\psi_n \end{bmatrix} = 0 \tag{C.46}$$

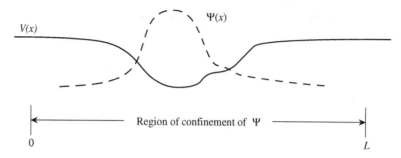

Figure C.7: A potential profile in which a wavefunction is confined.

where

$$H_{mn} = \int \psi_m^* \, H \, \psi_n \, d^3r \qquad (C.47)$$

This is the standard form of the eigenvalue problem.

The general Schrödinger equation (the approach can be used for the wave equation also)

$$-\frac{\hbar^2}{2m}\nabla^2\Psi + V\Psi = \boldsymbol{E}\Psi \qquad (C.48)$$

can be expressed in terms of a matrix equation. This approach is very useful if we are looking for bound or quasi-bound states in a spatially varying potential V. Let us consider the scalar Schrödinger equation (or the single band equation often used to describe electrons in the conduction band). Let us assume that the eigenfunction we are looking for is confined in a region L as shown in Fig.C.7. We divide this region into equidistant ℓ mesh points x_i, each separated in real space by a distance Δx. The wavefunction we are looking for is now of the form

$$\Psi = \sum_n a_n\psi_n \qquad (C.49)$$

where ψ_n are simply functions at the mesh point which are normalized within the interval centered at x_n and are zero outside that interval.

We can also write the differential equation as a general difference equation

$$-\frac{\hbar^2}{2m}\left[\frac{\Psi(x_i-1) - 2\Psi(x_i) + \Psi(x_i+1)}{\Delta x^2} + V\Psi\right] = \boldsymbol{E}\Psi \qquad (C.50)$$

Once again, substituting for the general wavefunction Eqn. C.26 and taking an outer product with $\psi_1, \psi_2, \cdots, \psi_\ell$, we get a set of ℓ equations (remember that we are assuming $a_0 = a_{\ell+1} = 0$, i.e., the wavefunction is localized in the space L) which can be written in the matrix form as

$$
\begin{bmatrix}
A(x_1) & B & 0 & \cdots & 0 & 0 \\
B & A(x_2) & B & \cdots & 0 & 0 \\
0 & B & A(x_3) & \cdots & 0 & 0 \\
& & \vdots & & & \\
& & \vdots & & & \\
0 & 0 & 0 & \cdots & B & A(x_\ell)
\end{bmatrix}
\begin{bmatrix}
a_1\,\psi_1 \\
a_2\,\psi_2 \\
a_3\,\psi_3 \\
\vdots \\
\vdots \\
a_\ell\,\psi_\ell
\end{bmatrix}
= 0
\qquad (C.51)
$$

with

$$
A(x_i) = \frac{\hbar^2}{m\Delta x^2} + V(x_i) - \boldsymbol{E}
$$

$$
B = -\frac{\hbar^2}{2m\Delta x^2}
$$

This $\ell \times \ell$ set of equations can again be solved by calling an appropriate subroutine from a computer library to get the eigenvalues $\boldsymbol{E}_n$ and wavefunctions ψ_n. In general we will get ℓ eigenvalues and eigenfunctions. The lowest lying state is the ground state, while the others are excited states.

Appendix

D

IMPORTANT PROPERTIES OF SEMICONDUCTORS

The data and plots shown in this Appendix are extracted from a number of sources. A list of useful sources is given below. Note that impact ionization coefficient and Auger coefficients of many materials are not known exactly.

- S. Adachi, *J. Appl. Phys.*, 58, R1 (1985).

- H.C. Casey, Jr. and M.B. Panish, *Heterostructure Lasers*, Part A, "Fundamental Principles;" Part B, "Materials and Operating Characteristics," Academic Press, N.Y. (1978).

- Landolt-Bornstein, *Numerical Data and Functional Relationship in Science and Technology*, Vol. 22, Eds. O. Madelung, M. Schulz, and H. Weiss, Springer-Verlog, N.Y. (1987). Other volumes in this series are also very useful.

- S.M. Sze, *Physics of Semiconductor Devices*, Wiley, N.Y. (1981). This is an excellent source of a variety of useful information on semiconductors.

- "World Wide Web;" A huge collection of data can be found on the Web. Several professors and industrial scientists have placed very useful information on their websites.

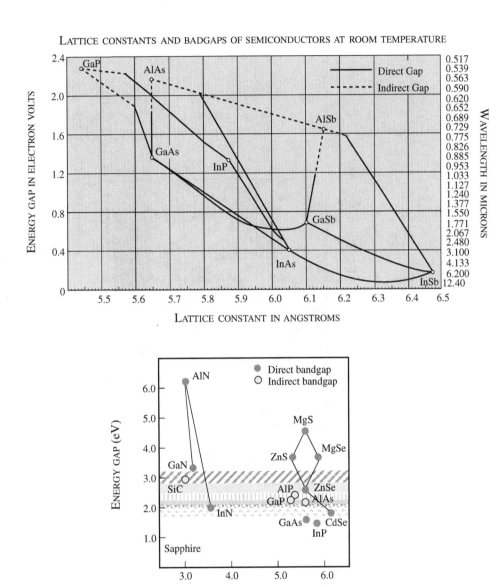

Figure D.1: Lattice constants and bandgaps of semiconductors at room temperature.

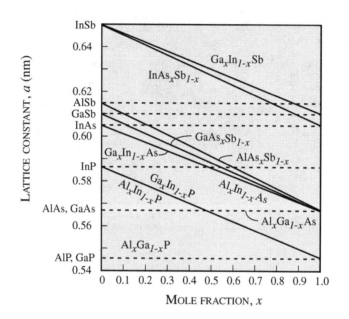

Figure D.2: Lattice constants of several alloy systems.

Semi-conductor	Type of Energy Gap	Experimental Energy Gap E_g (eV)		Temperature Dependence of Energy Gap (eV)
		0 K	300 K	
AlAs	Indirect	2.239	2.163	$2.239 - 6.0 \times 10^{-4}T^2/(T + 408)$
GaP	Indirect	2.338	2.261	$2.338 - 5.771 \times 10^{-4}T^2/(T + 372)$
GaAs	Direct	1.519	1.424	$1.519 - 5.405 \times 10^{-4}T^2/(T + 204)$
GaSb	Direct	0.810	0.726	$0.810 - 3.78 \times 10^{-4}T^2/(T + 94)$
InP	Direct	1.421	1.351	$1.421 - 3.63 \times 10^{-4}T^2/(T + 162)$
InAs	Direct	0.420	0.360	$0.420 - 2.50 \times 10^{-4}T^2/(T + 75)$
InSb	Direct	0.236	0.172	$0.236 - 2.99 \times 10^{-4}T^2/(T + 140)$
Si	Indirect	1.17	1.11	$1.17 - 4.37 \times 10^{-4}T^2/(T + 636)$
Ge	Indirect	0.66	0.74	$0.74 - 4.77 \times 10^{-4}T^2/(T + 235)$

Table D.1: Energy gaps of some semiconductors along with their temperature dependence.

Material	Electron Mass (m_0)	Hole Mass (m_0)
AlAs	0.1	
AlSb	0.12	$m_{dos} = 0.98$
GaN	0.19	$m_{dos} = 0.60$
GaP	0.82	$m_{dos} = 0.60$
GaAs	0.067	$m_{lh} = 0.082$ $m_{hh} = 0.45$
GaSb	0.042	$m_{dos} = 0.40$
Ge	$m_l = 1.64$ $m_t = 0.082$	$m_{lh} = 0.044$ $m_{hh} = 0.28$
InP	0.073	$m_{dos} = 0.64$
InAs	0.027	$m_{dos} = 0.4$
InSb	0.13	$m_{dos} = 0.4$
Si	$m_l = 0.98$ $m_t = 0.19$	$m_{lh} = 0.16$ $m_{hh} = 0.49$

Table D.2: Electron and hole masses for several semiconductors. Some uncertainty remains in the value of hole masses for many semiconductors.

Compound	Direct Energy Gap E_g (eV)
$Al_xIn_{1-x}P$	$1.351 + 2.23x$
$Al_xGa_{1-x}As$	$1.424 + 1.247x$
$Al_xIn_{1-x}As$	$0.360 + 2.012x + 0.698x^2$
$Al_xGa_{1-x}Sb$	$0.726 + 1.129x + 0.368x^2$
$Al_xIn_{1-x}Sb$	$0.172 + 1.621x + 0.43x^2$
$Ga_xIn_{1-x}P$	$1.351 + 0.643x + 0.786x^2$
$Ga_xIn_{1-x}As$	$0.36 + 1.064x$
$Ga_xIn_{1-x}Sb$	$0.172 + 0.139x + 0.415x^2$
GaP_xAs_{1-x}	$1.424 + 1.150x + 0.176x^2$
$GaAs_xSb_{1-x}$	$0.726 + 0.502x + 1.2x^2$
InP_xAs_{1-x}	$0.360 + 0.891x + 0.101x^2$
$InAs_xSb_{1-x}$	$0.18 + 0.41x + 0.58x^2$

Table D.3: Compositional dependence of the energy gaps of the binary III-V ternary alloys at 300 K. (After Casey and Panish (1978).)

SEMICONDUCTOR	BANDGAP (eV) 300K	MOBILITY AT 300 K (cm^2/Vs) ELECTRONS	HOLES
C	5.47	800	1200
Ge	0.66	3900	1900
Si	1.12	1500	450
α-SiC	2.996	400	50
GaSb	0.72	5000	850
GaAs	1.42	8500	400
GaP	2.26	110	75
InSb	0.17	8000	1250
InAs	0.36	33000	460
InP	1.35	4600	150
CdTe	1.56	1050	100
PbTe	0.31	6000	4000

Table D.4: Bandgaps, electron, and hole mobilities of some semiconductors.

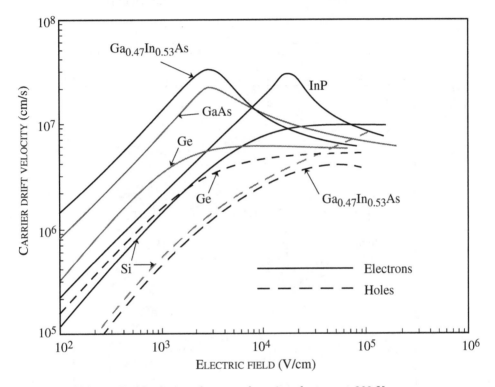

Figure D.3: Velocity-Field relations for several semiconductors at 300 K.

BREAKDOWN ELECTRIC FIELDS IN SEMICONDUCTORS

MATERIAL	BANDGAP (eV)	BREAKDOWN ELECTRIC FIELD(V/cm)
GaAs	1.43	4×10^5
Ge	0.664	10^5
InP	1.34	
Si	1.1	3×10^5
$In_{0.53}Ga_{0.47}As$	0.8	2×10^5
C	5.5	10^7
SiC	2.9	$2\text{-}3 \times 10^6$
SiO_2	9	10^7
Si_3N_4	5	10^7

Table D.5: Breakdown electric fields in some semiconductors.

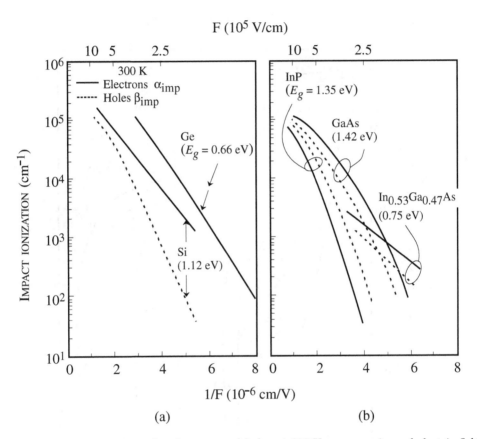

Figure D.4: Ionization rates for electrons and holes at 300 K versus reciprocal electric field for Ge, Si, GaAs, In$_{0.53}$Ga$_{0.47}$ and InP. (Si, Ge results are after S.M. Sze, *Physics of Semiconductor Devices*, John Wiley and Sons (1981); InP, GaAs, InGaAs results are after G. Stillman, *Properties of Lattice Matched and Strained Indium Gallium Arsenide*, ed. P. Bhattacharya, INSPEC, London (1993).

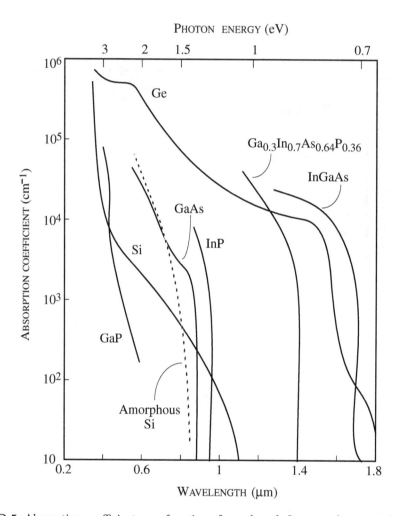

Figure D.5: Absorption coefficient as a function of wavelength for several semiconductors.

MATERIAL	b (eV)	d (eV)	dE_g/dp (10^{-12} eV cm^2/dyne)	Ξu (eV)
Si	−1.50 300 K	−3.40 300 K	−1.41 300 K	9.2 295 K
Ge	−2.20 300 K	−4.40 300 K	5.00 300 K	15.9 297 K
AlSb	−1.35 77 K	−4.30 77 K	−3.50 77 K	6.2 300 K
GaP	−1.30 80 K	−4.00 80 K	−1.11 300 K	6.2 80 K
GaAs	−2.00 300 K	−6.00 300 K	11.70 300 K	
GaSb	−3.30 77 K	−8.35 77 K	14.00 300 K	
InP	−1.55 77 K	−4.40 77 K	4.70 300 K	
InAs			10.00 300 K	
InSb	−2.05 80 K	−5.80 80 K	−16.00 300 K	

Table D.6: Strain parameters of some semiconductors. The temperature is specified.

Material	Auger Coefficient $(\text{cm}^6\text{s}^{-1})$	Comments
$\text{In}_{0.53}\text{Ga}_{0.47}\text{As}$	$\sim 10^{-28}$ at 300 K	For a compilation of results in the literature, see J. Shah in "Indium Gallium Arsenide," ed. P. Bhattacharya, *INSPEC*, London (1992).
GaInAsP	$\sim 6 \times 10^{-28}$ at $E_g = 0.8$ eV at 300 K $\sim 1.2 \times 10^{-27}$ at $E_g = 0.7$ eV at 300 K	Based on work of A. Sugimura, *Quantum Electronics*, QE-18, 352 (1982).
GaInAsSb	$\sim 6 \times 10^{-27}$ at $E_g = 0.4$ eV at 300 K	

Table D.7: Auger coefficients of some semiconductors. Considerable uncertainty still exists in the Auger coefficients. The values given are only rough estimates.

Index

CPSIA information can be obtained
at www.ICGtesting.com
Printed in the USA
LVHW100357181220
674455LV00009B/133

9 780521 035743